DYNAMIC ASPECTS OF BIOCHEMISTRY

BY

ERNEST BALDWIN, Sc.D., F.I.Biol.

Professor of Biochemistry at University College in the
University of London, formerly Fellow of
St John's College, Cambridge
Fellow of the New York Academy of Sciences

FIFTH EDITION

D0068953

CAMBRIDGE
AT THE UNIVERSITY PRESS
1967

Published by the Syndics of the Cambridge University Press
Bentley House, 200 Euston Road, London, N.W. 1
American Branch: 32 East 57th Street, New York, N.Y. 10022

Library of Congress Catalogue Card Number: 67–26065

First Edition 1947   Reprinted 1948, 1949
Second Edition 1952   Reprinted 1953
Third Edition 1957   Reprinted 1959, 1960
Fourth Edition 1963
Fifth Edition 1967

Translations

Russian 1948
Italian 1951, 1966
Spanish 1952   Reprinted 1957
Japanese 1954   New edition in preparation
German 1957
Serbo-Croat 1960   New edition in preparation
Polish 1960, 1967

Awarded the Cortina-Ulisse Prize 1952

Printed in Great Britain
at the University Printing House, Cambridge
(Brooke Crutchley, University Printer)

To

HOPPY

(Sir Frederick Gowland Hopkins, O.M., F.R.S.)

WITH RESPECT, ADMIRATION
AND AFFECTION

'The difference between a piece of stone and an atom is that an atom is highly organised, whereas the stone is not. The atom is a pattern, and the molecule is a pattern, and the crystal is a pattern; but the stone, although it is made up of these patterns, is just a mere confusion. It's only when life appears that you begin to get organisation on a larger scale. Life takes the atoms and molecules and crystals; but instead, of making a mess of them like the stone, it combines them into new and more elaborate patterns of its own.'

ALDOUS HUXLEY, *Time must have a Stop*

CONTENTS

PART I. ENZYMES

Chapter 1. The general behaviour and properties of enzymes

Introduction 3; Nomenclature and classification of enzymes 8; Specificity 8; The chemical nature of enzymes 13; Summary 20

Chapter 2. The nature of the catalytic process

The union of the enzyme with its substrate 21; Influence of concentrations of the enzyme and its substrate 22; Competitive inhibition 28; Activation of the substrate 30; Activators and coenzymes 32; Prosthetic groups 37; Quantitative characterization of enzymes 38; Summary 41

Chapter 3. Biological energetics

The concept of free energy 42; Breakdown and synthesis in biological systems 45; Reversibility of biological reactions 48; Energetics of synthetic reactions 50; Properties and functions of adenosine triphosphate 52; The biological energy cycle 56; The storage of high-energy phosphate 58; Summary 59

Chapter 4. Hydrolases

General introduction 60; Peptide hydrolases (peptidases) 61; Glycoside hydrolases (carbohydrases) 70; Ester hydrolases 78; Other hydrolases 82

Chapter 5. Transferases

General introduction 84; Group transfer by hydrolases 85; Transphosphorylation: kinases 88; Transglycosylation 90; Transpeptidation 97; Transamination 99; Transamidination 102; Transcarbamoylation 103; Transthiolation 103; Transacetylation 103; Transketolation 105; Transaldolation 106; One-carbon transferases 108; Transmethylation 110

Chapter 6. Lyases and isomerases

General introduction 113; Lyases 113; Biotin-dependent carboxylyases 118; Isomerases 120

Chapter 7. Oxidoreductases: oxidases

Chapter 8. Oxidoreductases: dehydrogenase systems

PART II. METABOLISM

Chapter 9. Methods employed in the investigation of intermediary metabolism

Chapter 10. Food, digestion and absorption

Chapter 11. General metabolism of amino-acids

Chapter 12. Special metabolism of the amino-acids

Chapter 13. Excretory metabolism of proteins and amino-acids

Chapter 14. Formation and functions of some nucleotides

Chapter 15. Metabolism of nucleic acids

Chapter 16. Carbohydrate production in the green plant

Chapter 17. Anaerobic metabolism of carbohydrates: alcoholic fermentation

Chapter 18. Anaerobic metabolism of carbohydrates in muscle

Chapter 19. Aerobic metabolism of carbohydrates

Chapter 20. The citric acid cycle

Chapter 21. The metabolism of fats

PREFACE TO THE FIFTH EDITION

This new fifth edition is destined to appear almost exactly twenty years after this book was first published. To keep pace with even a small part of biochemistry over these twenty years has been an arduous undertaking even though I have had throughout the benefit of the advice and the loyal support of many colleagues, now too numerous to list. These twenty years have indeed been years of unparalleled growth and development. The first edition of this book appeared two years before even the first International Congress of Biochemistry was held, at that time under the auspices of the International Union of Chemistry. But within a year or two a new and independent International Union of Biochemistry had already been established and International Congresses have been held at regular intervals ever since in many parts of the world, from Moscow to New York.

Another index of the rate of growth is to be seen in the still increasing membership of the Biochemical Society in this country, from 1,223 in 1947 it has reached 4,259 at the time of writing. The efforts of that same Society have led in turn to the formation of a flourishing Federation of European Biochemical Societies, as well as a number of more domestic and highly specialized research groups. This is the case of only one of the numerous biochemical societies the world over.

Many changes have necessarily been made in successive past editions of this book but inevitably some apparently outdated material has survived numerous revisions. The preparation of this new edition, wholly re-set and in its new format, has provided an opportunity for detailed scrutiny and extensive revision, but I have been surprised by the extent to which much of the old basic matter has remained substantially unchanged. Evidently we built better than we knew, even twenty years ago.

However, much material of mainly historical interest has now disappeared and reappraisals of old ideas have become necessary. For example, the chapter dealing with the metabolism of fats called for so much revision that it has been practically re-written, especially in relation to the biosynthesis of lipids and fatty acids. Another chapter (Special Aspects of Nitrogen Metabolism), in which nobody apart from the author himself seems to have been very much interested, has accordingly been removed but its salient features are transplanted to other parts of the text. A judicious rearrangement of some portions of the first part of the book has provided opportunity for more discussion of biotin-dependent and tetrahydrofolate-dependent and other enzymes and related topics, expecially one-carbon metabolism, all of which received only scant attention in earlier editions. The chapters on nucleotides and nucleic acids have been substantially reorganised and necessarily expanded to give an account of protein synthesis. The chapter on the citrate cycle has also been substantially changed to take account of newer outlooks in that area. Apart from these somewhat broad lines, a very large number of small but nevertheless material changes have been made throughout the text to bring it up to date. I hope that in its new form the book will be as free from factual and verbal errors as a book can reasonably hope to be.

One problem has given me much concern, namely the nomenclature of enzymes. As yet few of the text-books that abound today have given total respect to the Report of the International Union on this subject. Nor in fact have I. Many recommended names are too lengthy and complex to give much mental or spiritual comfort to the student, especially in the early stages of his biochemical career. Indeed, I personally feel that, to use these names in such a book as this would be positively sadistic. For the most part therefore I have again used approved trivial nomenclature and given acceptable synonyms in parentheses on many occasions. I remain unrepentant in this matter, in the hope that the day will come when something simpler will replace the very cumbrous nomenclature which has come upon us. The enthusiastic purist who feels a possibly masochistic compulsion to learn the currently official names can, after all, buy a copy of the International Union's Report for a very modest sum.

After twenty years of intimate acquaintance with this book I cannot fail to be reminded of Mark Twain's advice to those about to get married: *Don't*. I would be inclined to give the same advice to anyone who might be tempted to write such a book as this. If the product turns out to be a failure the only result is depression and misery, but these soon pass; if however it achieves some success, it imposes a constantly recurring load on its probably already overburdened author. Certainly this has been my own experience and the time that has necessarily been devoted to it has, alas, taken me much away from the bench. Which of the two activities is the more desirable may be a matter for debate. I am myself reasonably satisfied that the time that has gone into the propagation and dissemination of biochemistry by way of this book over the years, has perhaps been of more value to the subject than any original contributions to knowledge which I might have hoped to make, had I devoted the same time to research.

I offer my most grateful thanks to Drs B. R. Rabin, A. P. Mathias, K. L. Manchester, D. Gompertz, G. Offer and M. Hollaway, all of whom read larger or smaller parts of the text and have given generous help and advice. I must too record my thanks to many other friends and colleagues whose help, though smaller in amount, has nonetheless been freely and readily given. My wife has once again been most tolerant of an author in his throes and given me all manner of support, despite the many months of prepossession spent in the processes of production. I must also give sincere thanks to Margaret Anderson who compiled the index.

My thanks are due also to Dr F. H. C. Crick and Scientific American Inc. for permission to reprint Table 25 and to Dr H. R. V. Arnstein and the British Council for permission to reproduce Figs. 73 and 74. The Cambridge University Press has, as always, been helpful, skilful and ingenious. Indeed without their support and expeditious backing at every stage of the job, I doubt whether this fifth edition would ever have seen the light of day.

E. B.

London
March 1967

PREFACE TO THE FIRST EDITION

In spite of war-time difficulties and restrictions, Biochemistry has continued to expand more and more rapidly each year, in stature as well as in scope. It has been impossible for many years past for any one worker to read more than a small fraction of the new output, even when foreign journals were available. Without the invaluable aid of the *Annual Reviews of Biochemistry*, to whose authors and editors I and every other biochemist must pay high tribute, the preparation of this new book would have been impossible. Even with their help, some sections of the book will probably be out of date by the time it appears in print, and may well be out of date already in certain respects. But the last few decades have seen the establishment of a considerable body of information which, though it may change considerably in detail, will perhaps not change significantly in substance during the next few years. I venture to hope that a new edition may be called for before the present contents become wholly archaic, so that there will be opportunity to correct the many faults which have doubtless escaped notice and to bring the whole volume up to date.

The subject-matter of biological chemistry, like that of biology itself, can be roughly divided into two parts; the static, or morphological, and the dynamic, or physiological. Knowledge of the latter demands as an essential pre-requisite a knowledge of the former, and it is therefore a matter for rejoicing that many organic chemists of the present day are devoting their attention to the chemical constitution and configuration of the organic *Bausteine* that form the material basis of living cells. In some fields, notably in that of protein chemistry, the interdependence and collaboration between the organic and the biochemist are so intimate that it is impossible to say which is the organic chemist and which the biochemist. If any differentiation were necessary it could best be made, probably, in terms of their respective attitudes towards the *Bausteine*. For the organic chemist the main focus of attention is the structure and configuration of these materials while, for the biochemist, the main problems are those of the behaviour and function of these substances in organized, biological systems.

The more static aspects of these *Bausteine* are already fairly well covered by monographs and review articles, some of which I have indicated in the bibliography; the essentially dynamic aspects, on the other hand, have hitherto been but inadequately described and, in view of their wide importance and interest, I believe that a real demand exists at the present time for a book of this kind.

Elementary Biochemistry is taught in this University in two courses. The first and older of these, Chemical Physiology, forms part of the course in mammalian Physiology, and caters primarily for the needs of medical and veterinary students. For these there already exists a wealth of text-books, which the present volume neither hopes nor desires to supplant. In the second course, much more recently introduced, Biochemistry is taught as an independent scientific discipline, and without that emphasis on clinical problems with which it has usually been associated in the past, and which properly finds a place in Chemical Physiology. For students taking this second course there exists no suitable text-book, and it is primarily with their needs in mind that the present book has been written. I hope, however, that it will also help to open up new horizons to those whose interest in Biochemistry is primarily that of the organic chemist or the clinician in training. Perhaps it will serve too as an introduction to others who, wishing to take an advanced degree course here or elsewhere, find it difficult at the present time to discover suitable elementary reading. With the needs of such students as these in mind I have included a short bibliography of review articles and books, mostly of recent date and written by experts in their respective fields.*

In Biochemistry, as in any young but rapidly expanding branch of science, there are fields in which facts are scanty, evidence contradictory and speculation rife. I have tried to avoid such topics, but where this has not been possible I have attempted to give a critical account of the facts but, at the same time, to speculate a little. I cannot wholly subscribe to the doctrine that speculation is out of place in an elementary text-book, for there are many gaps in the subject, and unless these can in some way be bridged it is difficult or impossible to give a coherent account. My experience as a teacher has been that coherence is essential in an elementary exposition. Speculation plays and has always played an important part in the advancement of scientific knowledge, for no research worker gropes blindly after he knows not what; he invariably begins with certain reasonable possibilities in mind. In short, he speculates. To speculate unreasonably is worse than not to speculate at all, but providing certain tests of reasonableness and compatibility are applied beforehand, speculation is a valuable tool, and one which finds a place in every scientific workshop. The danger is that speculation is not always recognized as what it is, and I have, therefore, tried to distinguish clearly between fact and fantasy, hoping in this way to steer a middle course between unbridled imagination on the one hand, and an equally undesirable hypertrophy of the critical faculty on the other.

A word of explanation is perhaps necessary for the use in these pages of

* The bibliography was removed in the fourth edition and has not now been reinstated.

somewhat novel and certainly unorthodox methods of writing the equations of certain chemical reactions and groups or sequences of reactions. I have adopted them only after a long period of trial. They give a distinctly pictorial representation of chemical events, and many students find such a picture more easily comprehended and remembered than the more formal representations usually adopted. I trust that the reader will exercise the little patience necessary to become familiar with these 'whirligigs', for they have great advantages in cases where a long chain of successive chemical events has to be described briefly and as a whole.

The writing of this book has been largely a spare-time occupation, and there has been little enough spare time during the war years. Progress has often been slow, therefore, and the task has nearly been abandoned more than once. I owe the fact of its eventual completion to the kind encouragement given to me by my friends and colleagues. Particular thanks are due to Dr D. J. Bell and Dr E. Watchorn, who have read the whole of the manuscript, and to Prof. A. C. Chibnall, who read the proofs. I am glad also to acknowledge the help I have had from Miss V. Moyle. These, and others who have read particular sections and chapters, have all given precious advice and valuable criticisms. My task has been simplified in many ways by Prof. J. B. S. Haldane's *Enzymes* and by Dr D. E. Green's *Mechanisms of Biological Oxidations*, and particular thanks are due to Dr Malcolm Dixon, who has given me much from his great personal store of information.

I should also like to record my thanks to Dr J. C. Boursnell, who heroically undertook the preparation of the index, to Mr H. Mowl, who prepared the drawings for Fig. 30, and to members of the Cambridge Part I Biochemistry Class of 1945–6, who have allowed me to make use of some of their experimental data in the preparation of Figs. 1, 3 and 7.

To my wife, who prepared the work for publication, and to all departments of the Cambridge University Press I wish to express my humble and hearty thanks for their patience, consideration and expert workmanship.

E. B.

Cambridge
January 1946

ACKNOWLEDGEMENTS

The author's thanks are due to the following for permission to reproduce figures: Cambridge University Press for Figs. 2 and 3C; Dr H. Fraenkel-Conrat and the *Journal of Biological Chemistry* for Fig. 3B; Drs F. Schlenk and F. Lipmann and the University of Wisconsin Press for Figs. 39, 42 and 84; Messrs Longmans Green & Co., Ltd. for Figs. 4, 5, 10 and 19; and Messrs W. Heffer & Co., Ltd. for Fig. 18; Dr F. H. C. Crick and Scientific American Inc. for Table 25; and Prof. H. R. V. Arnstein and the British Council for Figs. 73 and 74.

PART I
ENZYMES

1

THE GENERAL BEHAVIOR AND
PROPERTIES OF ENZYMES

INTRODUCTION

Wherever we turn in the world of living things we find chemical changes taking place. Green plants, together with certain bacteria, are capable of fixing solar energy and synthesizing complex organic substances of high energy content from very simple starting materials, namely, water, carbon dioxide and small amounts of inorganic substances such as nitrates and phosphates. Other living organisms possess the ability to decompose these complex materials and to exploit for their own purposes the energy that is locked up within them, and it is in this way that animals, for instance, obtain the energy they expend in the discharge of their bodily functions; reproduction, growth, locomotion and so on. Now it is a significant fact that nearly all the chemical changes that go on in living tissues are changes which, left to themselves, proceed too slowly to be measurable or even, in many cases, detectable at all. How, then, does it happen that living animals can obtain energy and expend it as fast as they do? The answer is that living organisms possess numerous catalysts which speed up chemical reactions to the rates they achieve in biological systems. Whether we consider digestion, metabolism, locomotion, fermentation or putrefaction, chemical changes are going on, and these chemical changes are catalysed. It is the purpose of this book to give some account of these changes and of the many and diverse mechanisms at present known to participate in their catalysis.

A catalyst, in the classical definition of Ostwald, is 'an agent which affects the velocity of a chemical reaction without appearing in the final products of that reaction'. Examples of catalysis are familiar to every student of chemistry, and one of the most striking is that commonest of all chemical reagents, water. As is well known, hydrogen and chlorine react together with explosive violence if exposed to sunlight, yet perfectly dry hydrogen and perfectly dry chlorine fail to react together at all. Numerous familiar reactions do not proceed except in the presence of traces of water, and water is, in fact, a very important catalyst. Finely-divided metals, such as platinum, nickel and palladium, are also capable of catalysing a wide range of reactions, and Wieland, for instance, found that on the addition of colloidal palladium to aqueous solutions of various simple organic compounds, a catalytic oxidation (dehydrogenation) of the compounds ensues. Again, chemical reactions as a

3

whole proceed more rapidly at higher than at lower temperatures. But living organisms do not have at their disposal the powerful reagents, the high temperatures and the other artifices which are available to a chemist working in a laboratory, yet the synthetic facility of living cells and tissues far surpasses that of the chemist.

We must know something about simple catalysts and their mode of action before turning to the more complex catalysts and catalytic systems that we find in living tissues. There are many resemblances between catalysis as effected by more or less complex chemical reagents on the one hand and by biological systems on the other, but differences also exist. In the first place, a catalyst, of whatever kind, only affects the *rate* of the reaction which it catalyses. This fact is particularly well illustrated in the case of a reaction such as the hydrolysis of an ester, which is reversible. If we take ethyl acetate, for example, and heat it with water, the ester is slowly hydrolysed, but the reaction stops before hydrolysis is complete. On the other hand, if we start with equivalent proportions of ethyl alcohol and acetic acid and heat these together, we find that they react to form ethyl acetate, but once again the reaction stops before reaching completion. Indeed, from whichever side we start, the composition of the final reaction-mixture is always the same, the system attaining a state which can be represented by the following equilibrium:

$$C_2H_5OH + CH_3COOH \rightleftharpoons CH_3CO.OC_2H_5 + H_2O.$$

If we employ a biological catalyst such as liver esterase the final composition of the equilibrium-mixture is the same once more. These facts point to several important conclusions: first, that only the reaction velocity, and not the extent to which the reaction proceeds, is affected by the catalyst, and secondly, that in the case of a reversible reaction (and on theoretical grounds it is necessary to assume that all reactions are reversible) the catalyst influences the reaction velocity equally in both directions. The direction in which such a reaction will proceed is determined, of course, by mass-law considerations and by the availability of free energy. We must therefore suppose that a catalyst which accelerates the decomposition of a given substance must also be capable of catalysing its synthesis. But it does not by any means follow that the necessary conditions can be experimentally realized at the present time.

A second important feature of the phenomenon of catalysis is that the effect of the catalyst is normally out of all proportion to the amount used. A minute quantity of colloidal platinum is sufficient to catalyse the decomposition of an unlimited amount of hydrogen peroxide, provided that nothing happens to interfere with its catalytic properties. In practice, however, it frequently happens that catalysts are inhibited ('poisoned') by the presence of extraneous material. Thus, in the example just given, minute quantities of hydrocyanic acid, mercuric chloride or certain other substances suffice to destroy the catalytic properties of the platinum. This 'poisoning' is often a serious nuisance in commercial processes, but in many cases the catalytic activity can be restored relatively easily. In biological systems, too, we find that a comparatively small concentration of the catalytic material is all that is necessary, and that the catalysts are easily inhibited, sometimes reversibly and sometimes irreversibly, in a variety of ways which we shall discuss in later sections.

According to Ostwald's definition, the amount and chemical composition of a catalyst is the same at the end of its period of activity as it was at the beginning, though it is frequently found that its physical properties have been changed. Here is what at first sight appears to be a fundamental difference between catalysts such as colloidal metals and the catalytic agents we find in living tissues, but the difference is more apparent than real. Biological catalysts commonly lose much of their activity as the reactions which they catalyse proceed, but in such cases it usually appears that the catalyst has undergone inhibition by the products of its own activity, or else that its physical state has been modified in such a way that its catalytic properties have been destroyed.

Another apparent difference between the two types of catalysts is that whereas in the ordinary way a catalyst such as platinum black does not initiate a reaction but only accelerates one which already proceeds, albeit very slowly, in its absence, biological systems do in certain cases give the appearance of initiating new processes. For example, living yeast cells catalyse an almost quantitative conversion of glucose into ethyl alcohol and carbon dioxide according to the well-known equation:

$$C_6H_{12}O_6 = 2C_2H_5OH + 2CO_2.$$

By contrast, certain bacteria, e.g. *Streptococcus faecalis*, catalyse the conversion of glucose into lactate:

$$C_6H_{12}O_6 = 2CH_3CH(OH)COOH.$$

Other organisms again yield yet other products. Now glucose itself does not show any propensity to decompose spontaneously into either alcohol or lactate. Nevertheless, there is no real theoretical difficulty here. As is well known, it is the rule rather than the exception in organic chemistry that side-reactions take place, indicating that organic substances tend to decompose or react in more ways than one. Let us suppose, therefore, that glucose can decompose into a series of different products, A, B, C, D and so on, each product arising by its own chain of intermediate reactions. Under ordinary conditions the conversion of glucose into A, B, C, etc., proceeds only at imperceptible speed, but, under the influence of the catalysts of yeast, one of the possible modes of breakdown is selectively accelerated to such an extent that it is followed almost quantitatively. The catalysts of *S. faecalis*, by contrast, selectively accelerate another and a different mode.

5

This last case serves to illustrate what is perhaps the most striking feature of biological catalysis. Whereas a catalyst such as platinum black can catalyse any of a rather wide range of reactions, it is characteristic of biological catalysts that they catalyse only one kind of reaction and even, as often as not, one particular reaction and one only. But this is a difference only in their degree of specificity, or exclusiveness, and cannot be reckoned as evidence that biological catalysts differ essentially or fundamentally from catalysts of other kinds.

Although the effects of biological catalysts have long been familiar and have been deliberately used by mankind, probably even before the dawn of history, for the production of cheese, alcoholic beverages, vinegar, and the like, it is only in comparatively recent years that we have acquired any knowledge or understanding of their mode of action. The celebrated Italian physiologist, Spallanzani, was among the first to make a deliberate study of one of these catalysts, and this he did by feeding hawks with pieces of meat enclosed in small wire boxes, which were later regurgitated and found to be empty. In this way he demonstrated that the gastric juice of hawks contains something which brings about the liquefaction of meat. But as yet the nature of the responsible agent, which we now know under the name of pepsin, could not even be guessed.

It was Louis Pasteur who laid the foundations of our present knowledge. In the course of his famous researches on fermentation he demonstrated that solutions of organic materials such as glucose are perfectly stable if carefully sterilized and stored in sealed vessels. If, however, air is allowed to gain access to the solutions, fermentation sets in, and this, Pasteur showed, is due to contamination with living yeast cells which come in with the air. So long as these micro-organisms are carefully excluded no fermentation takes place. Similarly, Pasteur showed that the souring of wine, a troublesome phenomenon which he was commissioned by the then government of France to investigate, is attributable to the presence of certain other micro-organisms. These and other observations of a like kind led Pasteur to conclude that processes such as alcoholic fermentation, the souring of wine and milk are due to, and inseparable from, the vital activities of certain particular micro-organisms, which he accordingly named 'ferments'.

Pasteur's views received a severe blow when it was discovered by the brothers Buchner that if yeast is macerated with sand and kieselguhr (a siliceous earth) and submitted to high pressures, a juice can be expressed from it which contains no living cells whatever but is nevertheless capable of fermenting sugar with the production of alcohol and carbon dioxide. The Buchners, in fact, succeeded in demonstrating what Pasteur had regarded as an impossibility, the fermentation of sugar in the complete absence of living cells. Yeast juice clearly contains the catalyst or catalysts by means of which living yeast accomplishes the alcoholic fermentation of sugar, and to describe

6

this catalytic agent the term 'enzyme' was coined, from the Greek ἐν ζύμῃ, literally 'in yeast'. When other similar catalysts were later dicovered and studied, the term enzyme was taken over as a collective title and the yeast-juice enzyme received the distinguishing name of zymase.

The discovery of zymase was a fundamental advance. It had hitherto been possible to study fermentation and kindred processes only in the presence of living cells, but living cells multiply, die off, use up some chemical substances and produce others so that, superimposed on fermentation proper, there are many other chemical changes. With the newly discovered yeast juice, how-ever, the chemistry of fermentation could be studied in isolation, quite apart from all the other chemical operations carried out by the intact organism. As we now know, 'zymase' is not a single enzyme or catalyst, but rather a com-plex system of catalysts, and similar juices can be prepared from many kinds of cells. The Buchners made their fundamental discovery as recently as 1897, and progress thereafter was rapid. Urease, the first enzyme to be obtained in the pure, crystalline state, was isolated only some 30 years later, in 1926, and since that time more than 200 different enzymes have been purified and iso-lated.

Certain important discoveries were made comparatively early in the rather meteoric history of enzymology. Thus it was found that zymase loses its activity completely if boiled, and that if it is dialysed its activity is similarly lost. After dialysis, though not after boiling, activity could be restored by adding the dialysate, i.e. the small-molecular materials removed in the process of dialysis, or by the addition of a little boiled yeast juice. These observations show that, in addition to the thermolabile, non-dialysable enzymes, yeast juice also contains thermostable, dialysable factors in the absence of which fermentation cannot go forward. Thus there arose the concept of enzymes as thermolabile substances of high molecular weight, and of a second group of catalytic materials, called coenzymes, which consist of small, thermostable molecules. Both are necessary if fermentation is to take place. Just as we now know that zymase is in reality a complex mixture of enzymes, so, too, the dialysable complement is known to contain more than one coenzyme, and we shall have a great deal to say about this particular case in a later chapter.

Even this brief review has revealed a number of the most important pro-perties which characterize enzymes. *They are colloidal materials of high mole-cular weight, are thermolabile and highly specific, and can be extracted from the cells in which they are produced.* Although this broad definition covers the vast majority of enzymes, there are exceptions; for example a few thermo-stable enzymes have been discovered in recent years.

7

NOMENCLATURE AND CLASSIFICATION OF ENZYMES

Enzymes may be classified in any of several ways. All enzymes, so far as we know, are produced inside living cells, and the majority of them do their work inside the cells which produce them, though they can usually be extracted and their activity studied independently. In simple animal organisms such as *Amoeba* the processes of digestion are preceded by the phagocytic ingestion of food particles, which then undergo intracellular digestion, but in more highly organized forms of animal life it is commonly found that digestive enzymes are secreted into the digestive cavity, so that digestion is extra-cellular. Thus we can distinguish between intracellular and extracellular enzymes. This mode of classification is sometimes useful and is likely to be considerably extended in the future as more information is gained about the precise localization of enzymes within the cell, e.g. in the nucleus, nucleolus, mitochondria, microsomes, cytoplasm and so on.

More usually, enzymes are named with reference to the reaction or reactions which they catalyse, but a practicable system of rigid nomenclature has only recently been devised. We can distinguish, for example, a large and important group of enzymes which catalyse the hydrolysis of their substrates, i.e. the substances upon which their catalytic influence is exerted, and these enzymes are accordingly termed *hydrolases*. This group includes all the extra-cellular enzymes concerned with digestion, and many intracellular enzymes besides. Once upon a time individual enzymes were named by adding *-ase* to the names of their respective substrates and many of the old names have persisted in spite of newer innovations. For example, enzymes which catalyse the hydrolysis of starch were, and still are, collectively called amylases (Latin *amylum* = starch), and different individual amylases are distinguished by reference to the sources from which they are obtained. Thus we find salivary and pancreatic amylases among the digestive enzymes of mammals. Similarly, enzymes which catalyse the hydrolysis of proteins are known as peptidases, and those which act upon fats as lipases. The group of hydrolases also includes many non-digestive enzymes, such, for instance, as urease, which catalyses the hydrolytic breakdown of urea into ammonia and carbon dioxide; and arginase, which catalyses the hydrolysis of the amino-acid arginine into ornithine and urea. We shall return of necessity to the question of classification later on (p. 60).

SPECIFICITY

One of the most striking properties of enzymes is their specificity. By this we mean that a given enzyme can catalyse only a comparatively small range of reactions, and even, in many cases, one reaction and one only. It is possible to distinguish fairly sharply between a number of different degrees and types of specificity, and to make this clear we may consider first of all what is known as optical, or better, as *stereochemical specificity*.

The majority of chemical substances formed and broken down in metabolic processes are optically active and, of the two possible stereo-isomeric forms in which such substances can exist, only one is usually found on any large scale in natural materials and processes. Of the hexose sugars, for example, we normally encounter only the D-isomers, though it is true that their enantiomorphs are occasionally found. Thus L-galactose has been isolated from various plant materials and from the molluscan polysaccharide, galactogen. None the less, it remains a fact that there is an overwhelming preponderance of D-sugars in nature.† Similarly, of the α-amino-acids only the L-members occur extensively in nature: cases of the occurrence of D-amino-acids have been reported, in certain antibiotics for example, but are relatively rare. Perhaps, therefore, it is not surprising to find, as we do, that most enzymes show a strong and usually a complete selectivity for one member of a pair of optical isomerides, and are therefore said to exhibit stereochemical specificity.

The phenomenon of stereochemical specificity is well illustrated in the case of the hydrolytic enzyme arginase. This enzyme acts upon L- but not upon D-arginine, producing L-ornithine and urea:

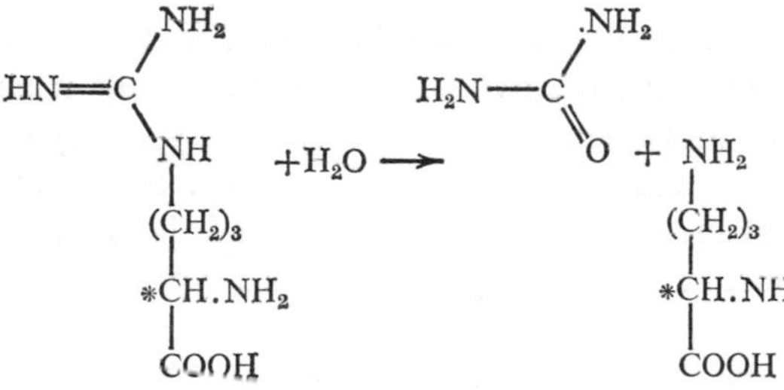

$$
\begin{array}{ccc}
\underset{\substack{\|\\ NH \\ | \\ (CH_2)_3 \\ | \\ *CH.NH_2 \\ | \\ COOH}}{HN\!=\!C\!\!\nearrow\!\!NH_2} \quad +H_2O \longrightarrow & \underset{\substack{\|\\ O \\ | \\ (CH_2)_3 \\ | \\ *CH.NH_2 \\ | \\ COOH}}{H_2N\!-\!C\!\!\nearrow\!\!NH_2} \;+\; NH_2 &
\end{array}
$$

L-*arginine*   *urea*   L-*ornithine*

Similarly, the lactate dehydrogenase of muscle can catalyse the dehydrogenation of L-(+)- but not that of D-(−)-lactate to yield pyruvate:

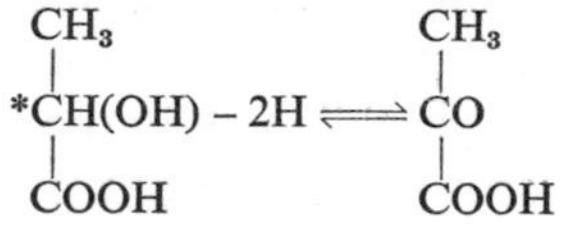

$$
\begin{array}{cc}
\underset{\substack{| \\ *CH(OH) \\ | \\ COOH}}{CH_3} - 2H \rightleftharpoons & \underset{\substack{| \\ CO \\ | \\ COOH}}{CH_3}
\end{array}
$$

L-*lactic acid*   *pyruvic acid*

The same enzyme can also work in the reverse manner, acting upon the optically inactive pyruvate (asymmetric carbon atoms are marked with asterisks) and catalysing its reduction to yield L-lactate only. D-Lactate is never formed by this enzyme. In many micro-organisms, however, we find a lactate dehydro-

† A discussion of the use and significance of the D- and L-notation can be found on pp. 236–8.

genase which is specific for the D-form of lactate, and this is true, for example, of *Bacillus delbrückii*, an organism once employed for the commercial production of lactic acid.

Stereochemical specificity of another kind is also known. The enzyme succinate dehydrogenase, for example, catalyses the oxidation of succinate to fumarate but never yields its geometrical isomer, maleate:

$$\begin{array}{ccc} \text{CH.COOH} & \text{CH}_2\text{COOH} & \text{HOOC.CH} \\ \parallel \quad \leftarrow\parallel- & \mid \quad \rightarrow & \parallel \\ \text{CH.COOH} & \text{CH}_2\text{COOH} & \text{CH.COOH} \end{array}$$

maleic acid (cis-)   *succinic acid*   *fumaric acid* (trans-)

Again, aconitase acts upon *cis*-aconitate, converting it into either citrate or *iso*-citrate, but has no action upon the *trans*- form of its substrate, by which it is actually inhibited.

Over and above the stereochemical specificity which is to be observed in the majority of enzymes, other types of specificity can be recognized. These other types differ mainly in their degree of exclusiveness. If, for the sake of simplicity, we consider only hydrolytic enzymes for the moment, the reaction catalysed by any given enzyme can be represented thus:

$$A\!-\!B + H_2O \rightleftharpoons A.OH + B.H.$$

The molecule of the substrate can be considered as consisting of three characteristic fragments, the two parts of the molecule itself, A and B, and the linkage which joins them. Three main types of specificity can be described with reference to these constitutional fragments. In the first type only the nature of the linkage is important, in the second the linkage and one-half of the molecule must be 'right', while in the third type all three fragments must be 'right'.

In the first type the precise nature of A and B is relatively unimportant, except that, if they are derived from optically active compounds they must have the appropriate stereochemical configuration, a condition which is already imposed by the stereochemical requirements of the enzyme. What is important, however, is that the linkage joining A to B shall be of the right kind, i.e. it must be an ester linkage in the case of a lipase or an esterase, a peptide link in the case of a peptidase (although some peptidases also show some esterase activity), or a glycosidic link in the case of a glycosidase. If an enzyme is specific only towards the nature of the linkage it is said to exhibit *low specificity*. This type of specificity is probably very rare, for a number of what formerly appeared to be enzymes of low specificity have proved to be mixtures, each component of which is more specific than the original complex. A close approximation to low specificity is found, however, among the lipases, which have so far largely defied attempts to fractionate them.

The second type of specificity is more exclusive, for here the enzyme can only act upon substances in which the right chemical linkage is present and in

10

which one of the two parts, *A* and *B*, is also of the right kind. As an example we may consider the case of the digestive enzyme usually called 'maltase', since it catalyses the hydrolysis of maltose (glucose-4-α-glucoside). Maltase obtained from the intestinal juices of a mammal will catalyse other reactions however; its action upon maltose is only one example of its catalytic action upon α-glucosides in general, which may be expressed in the following manner:

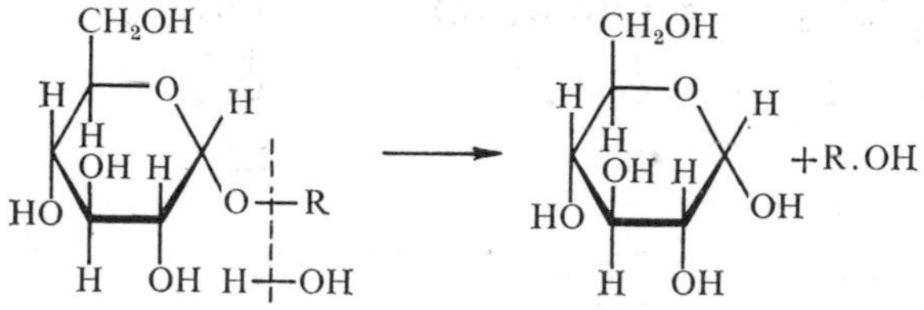

R-α-glucoside   α-glucose

The specificity requirements of this particular enzyme are as follows. An *α-glycosidic link* is required in the substrate; compounds containing a β-glycosidic linkage are not attacked. Furthermore, the α-glycosidic residue must be derived from D-*glucose*, and replacement of the glucose unit by one derived from another sugar yields a product which is resistant to this particular enzyme. Thus the nature of the linkage *and* that of one-half of the molecule must be 'right' in every detail, though the nature of the 'R' group is a matter of relative indifference. An enzyme of this kind may be said to show *group specificity*, to indicate that it can act upon any one of a group of closely related substrates, in this case a group of α-glucosides. Strictly speaking, therefore, this particular enzyme ought not to be called 'maltase', since its action is not uniquely confined to maltose; it is, in fact, an α-glucosidase. This kind of specificity is common among carbohydrases, for there also exist β-glucosidases, β-galactosidases and so on, each demanding its own particular kind of glycosidic linkage, together with a sugar residue of the appropriate type.

The third and commonest kind of specificity is the most exclusive of all. Here *both parts* of the substrate molecule must be 'right', together with the linkage, and the enzyme is therefore said to show *absolute specificity*. To take an example we may consider arginase once again. This enzyme requires for its action that its substrate shall consist of unmodified L-arginine (I). Many substances derived from and closely related to L-arginine have been prepared and submitted to the action of this enzyme, but it fails always to act. D-Arginine is not attacked nor are the following L-derivatives; α-*N*-methyl arginine (II), δ-*N*-methyl arginine (III) and agmatine (IV). Urease similarly requires that the structure of its substrate, urea, shall be intact and unsubstituted, and none of a considerable number of derived ureas that have been tested has been found to undergo hydrolysis under its influence.

11

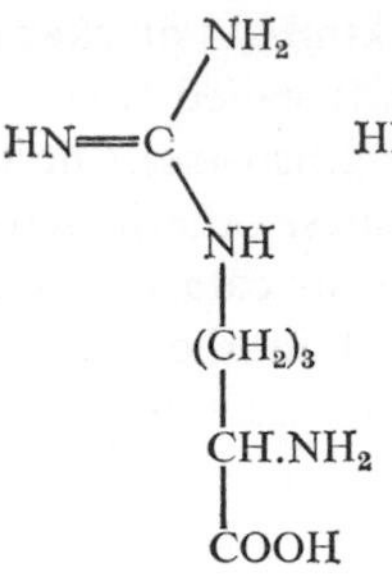

$$
\begin{array}{cccc}
\underset{\text{HN}=\text{C}}{\overset{\text{NH}_2}{\diagup}} & \underset{\text{HN}=\text{C}}{\overset{\text{NH}_2}{\diagup}} & \underset{\text{HN}=\text{C}}{\overset{\text{NH}_2}{\diagup}} & \underset{\text{HN}=\text{C}}{\overset{\text{NH}_2}{\diagup}} \\
\text{NH} & \text{NH} & \delta\ \text{N.CH}_3 & \text{NH} \\
\text{(CH}_2)_3 & \text{(CH}_2)_3 & \text{(CH}_2)_3 & \text{(CH}_2)_3 \\
\text{CH.NH}_2 & \alpha\ \text{CH.NH(CH}_3) & \text{CH.NH}_2 & \text{CH}_2\text{NH}_2 \\
\text{COOH} & \text{COOH} & \text{COOH} & \\
\text{I} & \text{II} & \text{III} & \text{IV}
\end{array}
$$

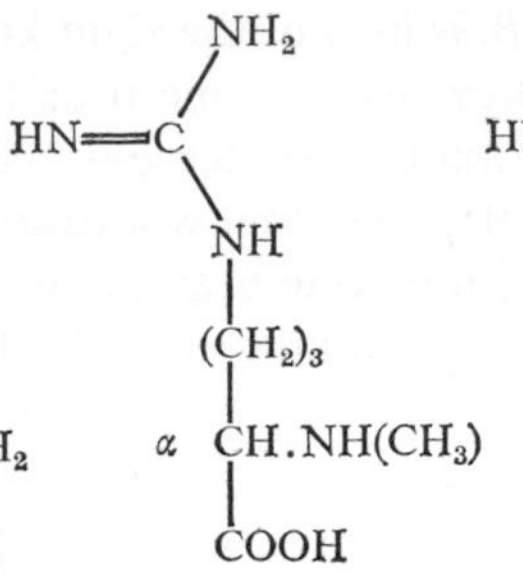

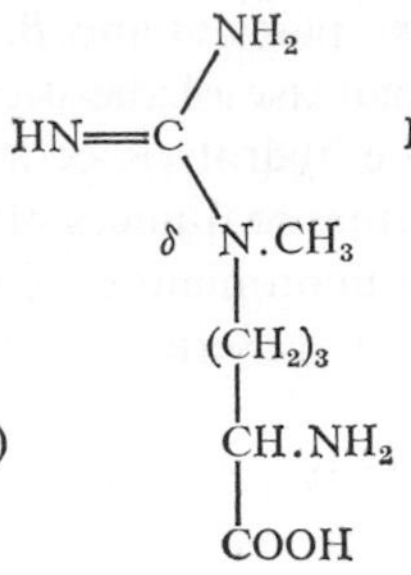

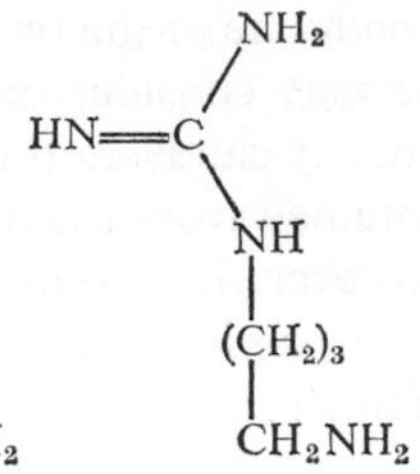

Most of the examples so far mentioned have been chosen from the group of hydrolytic enzymes, but similar phenomena are to be seen in other groups. Thus succinate dehydrogenase acts only upon succinate (V) and is without action upon the closely related malonate (VI) by which, indeed, it is strongly inhibited. Succinate dehydrogenase is therefore absolutely specific, like urease and arginase. Some oxidizing enzymes, however, are group-specific, and as an example we may take the case of the aldehyde oxidase of liver. Given suitable conditions, this enzyme can catalyse the dehydrogenation of many different aldehydes, but its action does not extend to other groups of compounds such, for instance, as the alcohols.

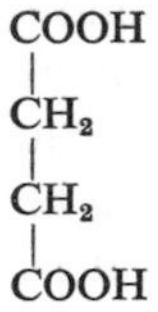

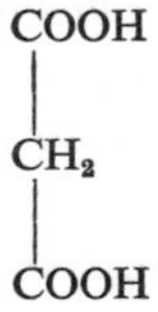

$$
\begin{array}{cc}
\text{COOH} & \text{COOH} \\
\text{CH}_2 & \\
\text{CH}_2 & \text{CH}_2 \\
\text{COOH} & \text{COOH} \\
\text{V} & \text{VI}
\end{array}
$$

In conclusion we may consider a very unusual case. Milk contains the so-called Schardinger enzyme, which catalyses the oxidation of a very large number of different aldehydes to yield the corresponding acids, and is therefore group-specific. But milk also contains a factor which catalyses the oxidation of the purine derivatives hypoxanthine and xanthine to uric acid (for formulae see p. 287), and this factor has received the name of xanthine oxidase. Many naturally occurring purines other than hypoxanthine and xanthine have been submitted to its action and found not to be attacked, so that the specificity of xanthine oxidase is very nearly absolute. The curious fact here is that the Schardinger enzyme and xanthine oxidase are demonstrably identical, so that in this case we have an enzyme which possesses two widely different spheres of specificity, one with respect to aldehydes, for which it is group-specific, and another with respect to purines, for which its specificity is virtually absolute. Indeed, it has been found in recent years that a number of enzymes can catalyse two or more totally unrelated reactions; we have in fact already mentioned the esterase activity of some peptidases.

It is clear, therefore, that the specificity of any given enzyme cannot necessarily be assigned to one or other of the types we have discussed. Low, group and absolute specificities are merely convenient standards of reference; many intermediate grades exist, just as, in the solar spectrum, we can distinguish between red, orange, yellow, green, blue and violet, although the colours themselves merge into one another and form a continuous whole.

Finally, and most important of all, there is the fact that specificity is in reality a measure of the structural specialization that an enzyme requires in its substrate, and this argues a corresponding degree of structural specialization in the enzyme itself.

THE CHEMICAL NATURE OF ENZYMES

The fact that enzymes are not dialysable suggested long ago that they must be related to substances of high molecular weight such as the polysaccharides and proteins, and even before any enzyme had been obtained in the pure state there was a considerable mass of circumstantial evidence to show that they are proteinaceous in nature.

In recent years many enzymes have been concentrated, purified, crystallized, and finally isolated, and in every case the product has proved to be a protein. Among oxidizing enzymes, there is often attached to the protein part of the molecule a non-protein fragment, known as a *prosthetic group* so that the enzyme is a conjugated protein. There is no doubt that every enzyme so far isolated is a protein of some kind. It is generally agreed today that all enzymes are proteins, although not all of the enzymes known to exist have so far been isolated.

Indirect information regarding the chemical nature of enzymes has been obtained in many different ways, and much of it from considerations of the influence upon enzymic activity of environmental factors such as temperature, pH and the presence of foreign materials of various kinds.

(i) *The measurement of enzymic activity.* The activity of an enzyme can be determined by measuring the amount of chemical change it catalyses under any given set of conditions. If we incubate an enzyme together with its substrate under constant conditions of temperature, pH and so on, we can withdraw samples of the reaction mixture from time to time and follow the course of the reaction by analysing the samples. Thus, if we choose yeast saccharase (sucrase, invertase) as our enzyme and sucrose as the substrate, we can measure the amount of chemical change at any given moment in terms of the amount of reducing sugars (glucose and fructose) that have been formed from the original non-reducing sucrose. It is necessary at the same time to carry out a 'control' experiment in which the active enzyme is replaced by a

13

previously boiled sample; in this way we can correct our experimental results for any changes due to spontaneous transformation of the substrate or to any other process that is not catalysed by the enzyme. Similar methods can be devised and used for the study of other enzymes, and the results of a typical experiment are shown in Fig. 1.

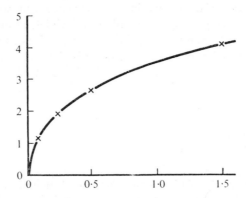

Fig. 1. A typical 'progress curve'; tryptic digestion of casein; data from a class experiment. Ordinate: increase in formol titre (ml. NaOH). Abscissa: time (hr.)

It will be observed that the reaction velocity soon begins to decrease and eventually the process stops altogether. Now while the reaction is proceeding, a number of changes are taking place in the reaction mixture. Substrate is disappearing, the products of the reaction are being formed, and the forward reaction may be opposed by the reverse process. In some cases other factors too may be at work, such, for instance, as a change of pH due to the formation or utilization of acid or alkali. All enzymes are sensitive to changes of pH in their immediate environment, all enzymes are influenced by the concentration of substrate available and by a considerable number of other factors; in fact many enzymes are actually inhibited by the products of their own activity. If, therefore, we wish to obtain a reliable measure of the activity of an enzyme under any given set of conditions, it will be necessery either to avoid these changes in the reaction mixture or else to make some suitable allowance for them.

Two main procedures are available. In the first place we can measure the *length of time required to produce a given amount of chemical change.* In this case the amount of substrate used up, the amounts of products formed, the change of pH if any, and the extent of other changes likely to interfere with the enzyme will be the same in every experiment, so that different experiments will be comparable one with another, *always provided that the enzyme is stable under the conditions selected.* In such cases we can use time as a measure of the activity of the enzyme: actually, of course the activity of the enzyme will be proportional to the reciprocal of the time since an enzyme preparation that is half as active will take twice as long to produce the same amount of chemical change.

The second method is usually preferred, and consists in measuring the *amount of chemical change taking place over a very short interval of time from the start of the reaction.* Provided that the time interval can be made short enough, the changes in the composition of the reaction mixture will be small enough to be neglected. Ideally we should measure the *instantaneous initial velocity.* To do this the usual procedure is to follow the course of the reaction during its early stages, preferably by a continuous method of some kind, e.g. by a spectrophotometric recording device. Backward extrapolation of the resulting curve to zero time then gives the instantaneous initial reaction velocity. Many excellent micro-methods are now available by the use of which we can obtain very good approximations to the instantaneous initial reaction velocity, and hence to the activity of an enzyme under any given set of experimental conditions.

(ii) *The influence of temperature.* Most chemical reactions are influenced by temperature, the reaction velocity increasing with rising and decreasing with falling temperature. Enzyme-catalysed reactions are no exception to this general rule, but, because enzymes are very susceptible to thermal inactivation, the higher the temperature becomes, the more rapidly are the catalytic properties of the enzyme destroyed. For any given set of experimental conditions, therefore, it is possible to find what is called an *optimum temperature,* i.e. a temperature at which the greatest amount of chemical change is catalysed under that particular set of conditions. At suboptimal temperatures the enzyme is relatively more stable and its activity therefore lasts longer, but the reaction which it catalyses proceeds more slowly. At temperatures above the optimum, on the other hand, the reaction takes place more rapidly but the catalyst is more rapidly inactivated at the same time.

If we work over a period of a few seconds the optimum temperature may be very high indeed, because the catalytic properties of the enzyme do not need to be long lived. If, on the other hand, we choose to work over a period of a few hours or days, a much lower optimum will be found since the activity of the enzyme must now last for a much longer period. It follows, therefore, that the time factor must be taken into account when we seek to determine the optimum temperature of any given enzyme, and that *time and temperature are interdependent variables.* The relationship between time and the optimum temperature of the digestive peptidase of an ascidian (sea squirt), *Tethyum,* is shown in Fig. 2.

With crude preparations such as were used in the earlier work on enzymes, the activity was usually of a rather low order, and it was therefore necessary to incubate the enzyme with its substrate for an hour or more in order to get a measureable amount of chemical change. Under conditions of this kind most enzymes show an optimum temperature of about 30–40° C. This has

15

led to the suggestion that, when animals became homoiothermic, they settled on a body temperature of the order of 37° C. because their enzymes would 'work better' in that neighbourhood than in any other. But consider the case of *Tethyum*. Over a period of 2 hr. the optimum temperature of this digestive enzyme is of the order of 50° C., which is well above the thermal death-point of this species. *Tethyum* normally lives at temperatures in the neighbourhood of 15° C., and the digestion of its food takes about 50–60 hr. under natural conditions. If the optimum temperature is determined for a period of 55 hr. the value found is about 20° C., so that there is, after all, a nice adjustment of

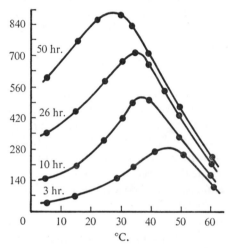

Fig. 2. Influence of temperature on digestive peptidase of *Tethyum*. Ordinate: mg. amino-acid nitrogen per litre. (Substrate, gelatin: after Berrill, 1929.)

the enzyme to the biological requirements of the animal. This seems fairly generally to be the case, for 'there is evidence that the time taken for the passage of food through the gut at any normal temperature corresponds to the period which is optimal for enzymatic action at that temperature' (Yonge).

The thermal inactivation of enzymes is interesting from the physico-chemical viewpoint as well as from that of the biological behaviour of enzymes, for it yields important clues to the chemical nature of enzymes themselves. For most chemical reactions we find a temperature coefficient, represented by Q_{10}, of approximately 2; i.e. the rate of the reaction is approximately doubled for a rise in temperature of 10° C. But if we determine the rate of thermal inactivation of enzymes in the neighbourhood of 70–80° C. we find Q_{10} values of the order of several hundreds. Temperature coefficients of this order are known for only a few reactions, including the thermal inactivation of enzymes and the thermal denaturation of proteins. There is here, therefore, a striking indication that enzymes may well be of protein nature, and that the process of thermal inactivation is analogous to, if it does not actually consist of, denaturation.

(iii) *The influence of* pH. The catalytic powers of an enzyme are, as a rule, exercised only over a somewhat restricted range of pH. Within this range the activity passes through a maximum at some particular pH, known as the *optimum* pH, and then falls off again. Fig. 3 illustrates the activity/pH relationships of several enzymes.

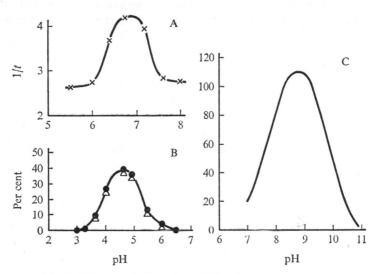

Fig. 3. Influence of pH on the activity of some enzymes.
A, *Salivary amylase* (substrate starch + NaCl). Ordinate: reciprocal of time taken to reach the achromic point. Results of a class experiment. B, *Papain-cysteine* (synthesis of benzyloxy-carbonyl-glycylanilide). Ordinate: yields as % of theoretical maximum, estimated by isolation (●) and titration (△). After Bergmann & Fraenkel-Conrat (1937). C, D-*Amino-acid oxidase* (substrate DL-alanine). Ordinate: oxygen uptake in 10 min. (μl.). Results of Krebs (1935), after Green.

Generally speaking, and other things being equal, the optimum pH is characteristic of a given enzyme, though under certain special conditions and in certain groups of enzymes the optimum pH may vary. This is true of some proteolytic enzymes, for example; pepsin has an optimum pH that varies between 1·5 and 2·5 or thereabouts, different optima being found with different protein substrates. A given carbohydrase, on the other hand, has always the same optimum pH, even when acting upon different substrates. Even so, the optimum pH may sometimes be modified by environmental factors such as the concentrations of the reactants and the ionic strength of the medium.

In its general form the pH/activity curve of a typical enzyme closely resembles that obtained by plotting the degree of ionization of a simple ampholyte such as glycine against pH. It will be recalled that most of the physical properties of solutions of ampholytes such as proteins and amino-acids—such properties as solubility, osmotic pressure, conductivity, viscosity and so on—pass through either a maximum or a minimum at some particular pH, the co-called isoelectric pH. These changes are attributable to changes in the ionic condition of the ampholytes themselves. Being multipolar, any given protein can exist in a number of different ionic forms, and one of these, the isoelectric form, possesses a number of special and peculiar properties. Here is further evidence that an enzyme may be regarded as a protein, and that of all the ionic forms in which it can therefore exist, only one particular ionic species possesses catalytic properties, this being the species which preponderates at the optimum pH.

Generally speaking, enzymes are most stable in the neighbourhood of the optimum pH, so that the observed optimum does not vary with time. The optimum pH of an enzyme is therefore a somewhat more characteristic feature than its optimum temperature. But if, as is sometimes the case, the enzyme is one which is very unstable at or near its pH optimum, the value observed will, of course, vary with the duration of the experiment. Accurate determinations of the optimum pH can only be made in such cases by working over very short intervals of time. In the case of arginase, for example, the optimum pH is about 7–8 for a period of an hour or so, but the true optimum lies at about 10, a pH at which arginase is very unstable indeed. The case of arginase, like that of pepsin, is an unusual one: most enzymes have their pH optimum not very far from neutrality, most commonly between pH 5 and 7.

(iv) *The influence of protein precipitants.* A further indication of the proteinaceous nature of enzymes is that extremes of acidity and alkalinity, which lead to the irreversible denaturation of proteins, lead also to the inactivation of the majority of enzymes. Moreover, these are irreversible changes, unlike those which are observed in the immediate vicinity of the optimum pH and which are for the most part reversible.

Enzymes are inhibited by many different groups of chemical reagents, as well as by such physical factors as high temperatures, violent mechanical agitation, ultra-violet radiation and so on, all of which lead to the denaturation of proteins. Many protein precipitants also lead to the inactivation of enzymes. Special attention may be drawn to the effects of two groups of enzyme inhibitors which form insoluble salts with proteins, viz. the salts of heavy metals on the one hand and, on the other, the so-called 'alkaloidal' or 'acid' reagents. The former precipitate proteins by virtue of the heavy, positively charged ions to which they give rise in solution, and the 'acid' reagents, which include such substances as trichloroacetic, tannic and phosphotungstic acids, act by virtue of their heavy, negatively charged ions. That all these agents are also powerful inhibitors of enzymic activity again suggests that enzymes are proteins.

18

More precise indications to the same effect are to be had by studying the effects of small concentrations of inhibitors of this kind. If it is indeed true that enzymes are proteins, we should expect to find that, in common with the proteins, they will be positively charged in acid solutions and therefore susceptible to the action of the negatively charged ions of phosphotungstic acid, for example. In alkaline solutions, on the other hand, they would be expected to be negatively charged and susceptible therefore to the action of positively charged ions, for example Ag^+. This problem has been carefully investigated in a few cases, and the results obtained with yeast saccharase are

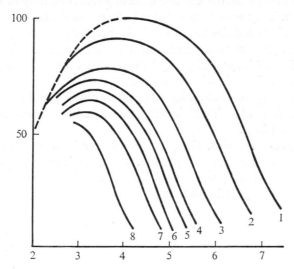

Fig. 4. Influence of small increasing concentrations of silver ions on activity of yeast saccharase. Ordinate: initial velocity of hydrolysis of sucrose. Abscissa: pH. (After Haldane, from Myrbäck, 1926.)

Curve	Conc. Ag^+	Curve	Conc. Ag^+
1	0	5	4×10^{-6}M
2	5×10^{-7}M	6	10^{-5}M
3	10^{-6}M	7	2×10^{-5}M
4	2×10^{-6}M	8	10^{-4}M

shown in Fig. 4. It will be seen that the effects of gradually increasing concentrations of silver ions are most marked on the alkaline side of the optimum pH. Phosphotungstic acid produces similar effects on the acid side. These results show that the behaviour of yeast saccharase with respect to these inhibitors is consistent with the view that this enzyme is a protein.

In practice, however, *the concentration of silver ions required to produce complete inhibition of yeast saccharase is much smaller than that needed actually to precipitate the protein*, and this suggests that the effect of Ag^+ is not a general one upon *all* the negatively charged centres of the protein molecule, but a localized and very specific one upon particular regions which are responsible for the catalytic properties of the presumptive saccharase protein. There thus emerges the notion that enzymic activity is not a property of the protein molecule as a whole, but rather that it is associated with certain special 'active' groups or centres.

SUMMARY

1. Enzymes are complex, organic catalysts of high molecular weight, produced by living cells but capable of acting independently of the cells that produce them. They are characteristically thermolabile and highly specific.

2. Several kinds and degrees of catalytic specificity can be recognized. The majority of enzymes show stereochemical (optical or geometrical) specificity, but, in addition, their specificity may be low or very high with reference to the chemical constitution of their substrates.

3. Enzymes are profoundly affected by many physical and chemical factors, and determinations of their activity must therefore be made under very closely controlled conditions.

4. Enzymes are proteins. Every enzyme so far isolated has proved to be either a simple or a conjugated protein. The behaviour of enzymes towards heat, changes of pH and protein precipitants is consistent with the supposition that they consist of protein material.

2

THE NATURE OF THE CATALYTIC PROCESS

THE UNION OF THE ENZYME WITH ITS SUBSTRATE

It is difficult to imagine how a catalyst of any kind can influence the rate of a chemical reaction unless it actually participates in that reaction. Most authorities agree that catalysts do in some manner combine with the substance or substances upon which their catalytic influence is exerted, but there has in the past been much difference of opinion as to whether the union is of a 'physical' or adsorptive kind, or whether it is to be regarded as 'chemical'. But it is difficult to maintain that there is any fundamental difference between these types of unions: rather must they be regarded as two extremes of one and the same phenomenon. Calcium carbonate is a good example of what we should call a chemical compound, formed by the chemical union of carbon dioxide and calcium oxide. Yet at high enough temperatures the product dissociates freely, as though, by raising the temperature, we have converted a chemical into an adsorptive union.

While it is true that adsorption is often relatively unspecific, there is evidence in plenty that it can be very specific indeed. Thus we find that a positively charged material such as magnesium oxide will adsorb negatively charged dyes like eosin from aqueous solution, but fails to take up a positively charged dye such as methylene blue. Similarly, a protein will take up negatively charged dyes in solutions acid to its isoelectric pH, in which it is positively charged, while on the alkaline side it takes up positively but not negatively charged dyes. At or very near the isoelectric pH it will usually take up a little of both, since, being multipolar, it carries an equal number of positive and negative charges at one and the same pH.

Clearly, therefore, several factors have to be taken into account when we are considering adsorption. *The nature of the surface* at which adsorption takes place is certainly of importance. Carotenoid pigments, for example, are adsorbed at a magnesium oxide/petrol ether interface but not at a magnesium oxide/alcohol interface. Charcoal can be used to adsorb coloured impurities of many kinds from aqueous solution, but is relatively useless in chloroform, and so on. The second important factor is, of course, *the chemical nature of the material being adsorbed*. It is not difficult to understand that a given surface may be so specialized, whether by virtue of its charge or for some other reason, as to be capable of taking up, i.e. reacting with, substances of one particular kind. Nor, if we allow for its possible topographical specialization,

21

is it difficult to imagine that a particular surface may be capable of reacting with one particular substance and one only.

There is nothing inherently improbable in the idea that an enzyme actually unites with its substrate, and it is difficult, indeed, to imagine how the facts of enzyme specificity could otherwise be accounted for. Studies of the kinetics of enzyme-catalysed reactions have made it clear that the assumption of a union is in fact warranted.

There is direct evidence for the formation of an enzyme-substrate complex between peroxidase and hydrogen peroxide. If peroxidase is added to its substrate in the presence of a suitable hydrogen donor such as pyrogallol, a vigorous reaction ensues, in which the pyrogallol is oxidized and the hydrogen peroxide reduced. In the absence of any hydrogen donor, however, the hydrogen peroxide does not undergo reduction. Now peroxidase is an iron-porphyrin derivative and as such has a strong absorption spectrum, displaying four bands at 645, 583, 548 and 498 mμ respectively. If hydrogen peroxide is added to a strong solution of the enzyme there is a sharp change in colour and the spectrum changes completely. Only two bands at 561 and 530·5 mμ respectively can now be seen. This can only mean that some kind of reaction has taken place between the enzyme and its substrate. Moreover, the amount of hydrogen peroxide required just to convert the whole of the enzyme into the new compound is equivalent to exactly one molecule of hydrogen peroxide for each atom of peroxidase-iron.

Somewhat similar observations have been made with catalase. If hydrogen peroxide is added to catalase a violent reaction takes place, the substrate being converted into water and molecular oxygen. If, however, the enzyme is first inhibited with sodium azide there is only a slow reaction when hydrogen peroxide is added. The catalase-azide complex has a strong absorption spectrum with bands at 624, 544 and 506·5 mμ, slowly changing on addition of hydrogen peroxide to a spectrum with two bands at 588 and 547 mμ. The original spectrum reappears when all the hydrogen peroxide has been decomposed, but the addition of more substrate then restores the two-banded spectrum. These observations again show that the (inhibited) enzyme reacts in some way with its substrate.

INFLUENCE OF CONCENTRATIONS OF THE ENZYME AND ITS SUBSTRATE

The rate of any enzyme-catalysed process depends, other things being equal, upon the concentrations of the enzyme and of its substrate, and an examination of the effects of these and other factors is very important for any understanding of enzymic catalysis. In the vast majority of cases we find that, with a fixed quantity of enzyme, the initial reaction velocity increases with increasing substrate concentration until a limiting value is reached. Fig. 5

shows the results of a typical experiment carried out along these lines. The magnitude of the limiting velocity finally attained depends upon the concentration of the enzyme used and is, in fact, proportional to that concentration. These observations can be accounted for in terms of the theory first brought forward by Michaelis and later developed and expanded by Briggs & Haldane.

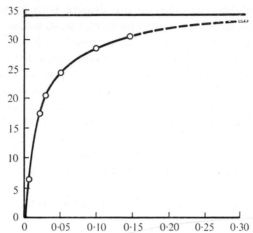

Fig. 5. Influence of substrate concentration on activity of yeast saccharase. Ordinate: initial velocity of hydrolysis. Abscissa: molar concentration of sucrose. (After Haldane, from Kuhn's data, 1923.)

For the purposes of the argument it is assumed that the enzyme and its substrate react together in some way to form an enzyme-substrate complex, which then breaks down to yield the reaction products. If we choose a case such as the hydrolysis of sucrose by saccharase, these assumptions can be expressed in the following equations:

(i) $E + S \rightleftharpoons ES$,

(ii) $ES + H_2O \rightarrow E + P + Q$.

The enzyme is represented here by E, the substrate by S and the intermediate enzyme/substrate complex by ES, while P and Q are the products of the process.

If we represent the *total* enzyme concentration by $[E_0]$ it follows that, since a certain amount $[ES]$ is bound up in the form of ES, the concentration of *free* enzyme will be equal to $[E_0] - [ES]$.

The reaction velocity, which we will call v, is the rate at which the products are formed, and this will clearly be proportional to the concentration of the complex ES. We are now in a position to apply the principles of the mass law to our equations and in this way to make predictions which, if they prove to

23

be in accordance with experimental observations, will provide evidence of the soundness or otherwise of the assumptions epitomized in equations (i) and (ii).

It is necessary, before going further, to realize clearly that the concentrations represented by $[E_0]$, $[ES]$ and $[S]$ must, if we are to apply the mass law, be expressed in *molar* and not in percentage concentrations. This fact is doubly important here because we are dealing with enzymes, which are colloidal materials, having very great molecular weights. Let us for a moment consider the enzyme urease, which is a protein with the comparatively modest molecular weight of about 480,000, and compare it with its substrate, urea, with a molecular weight of 60. If we were to prepare 1 *per cent* solutions of the pure enzyme and of its substrate, the *molar* concentration of the substrate solution would be no less than 8000 times that of the solution of enzyme. This point is of considerable theoretical importance, as we shall see, but it serves also to emphasize the relatively enormous activity of enzymes. They occur in living cells and tissues in amounts so small that their molar concentrations are minute, yet it is upon their catalytic activity that the life of the cells depends.

Returning now to our theory we see that the following statements can be made:

For equation (i) rate of forward reaction $= [S]([E_0]-[ES])k_1$

rate of reverse reaction $= [ES]k_2$

where k_1 and k_2 are the velocity constants of the forward and backward reactions respectively.

For equation (ii) rate of reaction $= [ES]k_3$

and k_3 is the velocity constant of the breakdown of the enzyme-substrate complex. Now as long as the rate of reaction (ii) is constant the value of $[ES]$ will remain constant and hence

$$[S]([E_0]-[ES])k_1 = [ES](k_2+k_3),$$

therefore
$$\frac{[S]([E_0]-[ES])}{[ES]} = \frac{k_2+k_3}{k_1} = K_m. \tag{A}$$

Here K_m, the ratio of two constants, is itself a constant and is called the Michaelis constant. Its particular significance will be considered later.†

We are very seldom in a position to evaluate either $[E_0]$ or $[ES]$, since even if we have a perfectly pure enzyme at our disposal its molecular weight may

---

† This treatment embodies the assumption that $[ES]$ will be constant throughout the course of the reaction and this 'ain't necessarily so'. However, provided that the reaction velocities actually measured are the *initial* reaction velocities (cf. p. 15) this oversimplification of treatment is of no serious consequence.

be unknown. These terms, $[E_0]$ and $[ES]$, must therefore be eliminated from our equations, and this can be done through the following considerations.

The reaction velocity, v, for the decomposition of ES (equation (ii)) will be proportional to $[ES]$ and also to the concentration of water, but since the concentration of water in the system does not change appreciably we can write

$$v = k[ES], \tag{B}$$

where k is a constant. By combining this with equation (A) we could eliminate $[ES]$, but the term $[E_0]$ would still remain, and this, like $[ES]$ itself, we are usually unable to evaluate. But let us consider a special case in which there is a large excess of substrate. This case is not a fictional invention since, on account of the great disparity of molecular weight between E and S, $[S]$ will usually be much greater than $[E_0]$. In the presence of a large excess of substrate, therefore, $[S]$ will be very much greater than $[E_0]$ so that virtually all the enzyme will be converted into ES, when $[ES] = [E_0]$. Now we have just seen (equation (B)) that the reaction velocity is proportional to $[ES]$, and in the presence of a large excess of substrate $[ES]$ will attain a maximum value of $[E_0]$. Consequently, in the presence of a large excess of the substrate, the reaction velocity will attain a similarly maximum or limiting value which may be called V. We thus have, for this special case, a third equation:

$$V = k[E_0]. \tag{C}$$

Dividing (B) by (C) we get

$$\frac{v}{V} = \frac{[ES]}{[E_0]}. \tag{D}$$

It is now possible to get rid of the unwanted terms from equation (A). We can rewrite (A) as follows:

$$[S][E_0] - [S][ES] = K_m[ES]$$

therefore

$$[S][E_0] = [ES]([S] + K_m)$$

and dividing by $[E_0]$,

$$[S] = \frac{[ES]}{[E_0]}([S] + K_m).$$

Substituting for $[ES]$ and $[E_0]$ from equation (D) we get

$$[S] = \frac{v}{V}([S] + K_m)$$

and hence the Michaelis or Briggs & Haldane equation:

$$v = \frac{V[S]}{[S] + K_m}. \tag{E}$$

This equation allows us to predict the manner in which the reaction velocity should be influenced by substrate concentration. It is, in fact, the equation of a rectangular hyperbola with the following properties (see Fig. 6):

25

(a) the limiting velocity V, is the asymptotic value to which the reaction velocity tends as the concentration of the substrate is increased, and

(b) the Michaelis constant (K_m) corresponds to that substrate concentration at which half the limiting velocity is developed. This fact is readily understood if we substitute $V/2$ for v in equation (E) itself:

$$\frac{V}{2} = \frac{V[S]}{[S]+K_m},$$

therefore
$$\frac{1}{2} = \frac{[S]}{[S]+K_m}$$

and
$$2[S] = [S]+K_m$$

so that
$$K_m = [S],$$

when the reaction velocity is one-half the limiting velocity.

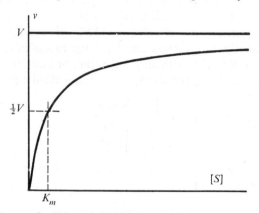

Fig. 6. Theoretical curve for Briggs & Haldane's equation,

$$v = \frac{V[S]}{[S]+K_m},$$

where v = initial reaction velocity, $[S]$ = concentration of substrate. Ordinate: initial reaction velocity. Abscissa: concentration of substrate.

If the rectangular hyperbola of Fig. 6 is now compared with the curve of Fig. 5, which portrays the results of experimental observations, there can be no doubt that the theory is in excellent agreement with the experimental results. Moreover, we have seen (equation (C)) that, according to this theory, the limiting velocity attained in the presence of an excess of substrate should be proportional to the concentration of enzyme, and this also is in agreement with the results of experimental enquiry (see Fig. 7).

Atypical results are not infrequently obtained in experiments designed to test the validity of predictions based on this theory, but these are usually due to interference by some factor or other, e.g. impurity of the enzyme or inhibition of the enzyme by the products of its own activity. When due allowance is made for this interference the corrected results are usually found to agree well with theoretical prediction.

The agreement between theoretical requirements and experimental observation goes far towards justifying the assumption upon which the theory was originally based, namely, that the enzyme actually combines with its substrate to form an unstable and correspondingly reactive complex. It may, of course, be argued that the same theoretical equations might be derived equally on the basis of different assumptions, but there is other and more direct evidence for the formation of enzyme-substrate compounds, to some of which we have already referred (p. 22).

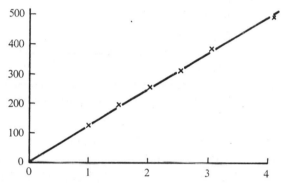

Fig. 7. Influence of enzyme concentration; yeast saccharase. Data from a class experiment. Ordinate: mg. invert sugar formed. Abscissa: ml. saccharase solution.

The Michaelis constant deserves a little further consideration. Let us suppose that we have *an enzyme of low or of group specificity,* and let us consider its activity towards *two different substrates, a and b.* If the relationships between reaction velocity and substrate concentration are experimentally determined for both substrates we get a pair of hyperbolic curves like those of Fig. 8. For each of the two substrates there is a K_m value, and in the figure it will be observed that for substrate b the value (K'_m) is greater than for substrate a (K_m). This means that in order to get the same velocity out of a given concentration of this particular enzyme, b must be taken at a higher concentration than must a. This must mean that the enzyme has a smaller affinity for b than for a: in other words a high K_m is indicative of a low enzyme-substrate affinity, and vice versa. Thus, not only does our theory furnish us with evidence that the formation of an enzyme-substrate compound is an essential part of the catalytic process: it provides at the same time a means whereby the affinity of an enzyme for its substrate can be numerically evaluated.

As we go on we shall see that the behaviour of enzymes is best explained on the supposition that they react with their respective substrates to form

27

reactive complexes. In addition to the evidence afforded by the applicability of the Michaelis–Briggs–Haldane theory there is a considerable mass of indirect evidence from other sources, as well as the direct evidence to which we have already referred (pp. 13–20).

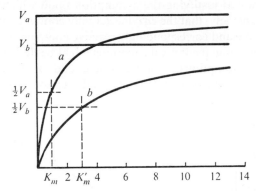

Fig. 8. Action of a group-specific enzyme upon two different substrates; for explanation see text. Ordinate: reaction velocity. Abscissa: concentration of substrate in arbitrary units.

COMPETITIVE INHIBITION

A great deal of information regarding the nature of enzymes and their mode of action has been gained by considering the influence of a variety of inhibitors upon them. Studies of this kind did much in the early days of enzymology to support the view that enzymes are made up essentially of protein material and this even before any single individual enzyme had actually been isolated.

Many enzymes are inhibited by the products of their own activity, and many more by substances which are structurally related to their substrates. In many such cases the inhibition is of what is known as the *competitive* type. A well-known and very important case is found in the competitive inhibition of succinate dehydrogenase by malonate.

If we take succinate together with succinate dehydrogenase we have a system in which, under suitable conditions, it is easy enough to measure the rate of oxidation of the succinate. If now malonate is added, the rate of oxidation promptly diminishes, but increases again if more succinate is added. Malonic acid is a dicarboxylic acid, the structure of which is closely related to that of succinic acid itself:

$$
\begin{array}{cc}
\begin{array}{c}
\text{COOH} \\
| \\
\text{CH}_2 \\
| \\
\text{CH}_2 \\
| \\
\text{COOH}
\end{array}
&
\begin{array}{c}
\text{COOH} \\
| \\
\text{CH}_2 \\
| \\
\text{COOH}
\end{array}
\\
\textit{succinic acid} & \textit{malonic acid}
\end{array}
$$

28

Because of this structural similarity malonate, the inhibitor, is able to combine with the enzyme, just as does the substrate, succinate. But whereas the enzyme-succinate complex breaks down to yield the reaction products, the enzyme-malonate complex contributes nothing to the reaction velocity. A part of the enzyme is bound in the form of enzyme-inhibitor complex, and so is not available for the catalysis of succinate oxidation, and the reaction velocity accordingly diminishes when the inhibitor is added.

This system may be more precisely described in the following manner. We have the following reactions to consider:

$$E + S \rightleftharpoons ES \rightarrow E + F + G,$$
$$E + M \rightleftharpoons EM.$$

The rate of oxidation of the succinate is determined by $[ES]$, and this can be increased *either* by increasing $[S]$ *or* by decreasing $[M]$, which indicates that these two substances 'compete' for possession of the enzyme. The observed effects depend upon the *ratio* of the competing substances, not upon the *absolute* amount of either. If the two compounds reacted at different sites on the enzyme molecule there is no reason why they should not both be accommodated at the same time, each independently of the other, allosteric effects apart. The fact that they do however compete shows that both unite with the enzyme molecule in precisely the same region.

Many other cases of the same kind are known. That competitive inhibition exists at all is an indication that the substrate does not unite at arbitrary groups on the enzyme molecule, but *only at certain particular sites, and not elsewhere* (cf. p. 20). Thus these considerations not only confirm the view that a union is set up between an enzyme and its substrate when the two are brought together: they go further, by showing that *the union is very specific not only in nature but also in locality.*

It is interesting to notice in passing that the bacteriostatic activity of sulphanilamide depends upon a competition between the drug itself and *p*-aminobenzoic acid. The latter is an important growth factor for many bacteria, and the structural resemblances between the drug and the growth factor are great enough to result in competition between the two:

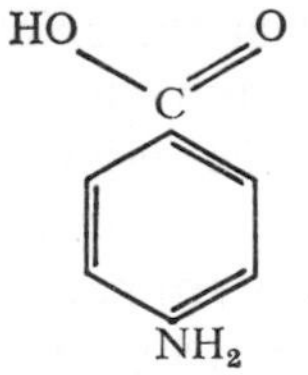

p-*aminobenzoic acid* 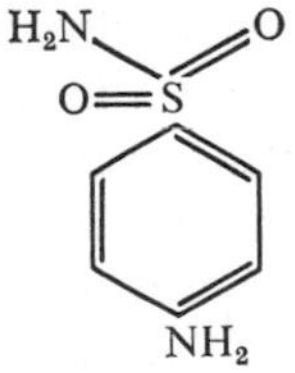   *sulphanilamide*

If *p*-aminobenzoic acid wins, the bacteria multiply: if sulphanilamide wins, multiplication is checked, and the rate of multiplication depends upon the ratio between the concentrations of the two substances.

ACTIVATION OF THE SUBSTRATE

The combination of an enzyme with its substrate seems to be the fundamental and essential step in the catalytic process, for it is as a result of this union, apparently, that the substrate molecule becomes more chemically reactive than it was in the free, uncombined state, and is more easily split, oxidized, reduced or whatever the case may be. We refer to this increase in chemical reactivity by saying that the enzyme has 'activated' its substrate, or that the substrate has undergone 'activation'. We do not know precisely what intra-molecular changes underlie activation, but we do know of other cases in which something rather similar takes place.

Let us consider the behaviour of haemoglobin. This compound consists of a protein, globin, into which is built *haem*, a complex, tetra-pyrrollic ring-structure containing an atom of ferrous iron. The special and peculiar property of haemoglobin is its ability to react reversibly with oxygen, taking it up when the partial pressure is high and giving it off again at low partial pressures. The important point for our present argument is that haem, by itself, does not possess this property. It acquires this property when, and only when, it is combined with the appropriate protein, globin. Free haem is very insoluble in water, and reacts spontaneously with oxygen to undergo oxidation to the ferric compound, haematin. Globin, by combining with haem, confers upon it a large measure of solubility in water, together with the new-found property of reacting reversibly with oxygen but without any change in the valency of the iron.

In addition to haemoglobin a number of other haem and haematin com-pounds with very special and peculiar properties are known, such, for example, as cytochrome, catalase and peroxidase. In each of these the haem or haematin system is present, but in none do we find the ability to combine reversibly with oxygen. The haem, presumably, is all right, but the protein is wrong. Thus the haem of haemoglobin possesses certain latent properties which only become apparent when the haem nucleus is combined with the right kind of protein. While few, probably, would venture to assert that globin 'activates' haem in the same sense that an enzyme activates its sub-strate, this case does show how the properties of a given substance can be profoundly and very specifically modified when the substance concerned enters into combination with the 'right' kind of protein.

One way of explaining activation starts from the very reasonable supposi-tion that there is an energy peak which any given substrate molecule must reach and pass over before it can enter the downward path of its appropriate reaction. This can be illustrated by the full line in Fig. 9, in which the free energy of activation is represented by ΔF^*. Substrate molecules must possess or acquire this amount of energy before they can enter the downward slope leading to the reaction products, in the course of which free energy equal to $(\Delta F^* + \Delta F)$ is set free.

Now in any given population of substrate molecules the free energy—chemical, kinetic, thermal or otherwise—will normally be spread over a fairly wide spectrum. A few of these molecules might even be 'energetic' enough to reach and surmount the peak in the activation curve, in which case the reaction would proceed, however slowly, even without the enzyme. But it is certain that the combination of enzyme with substrate enables *all* the substrate molecules to surmount—or circumvent—the activation peak. It is unlikely that the enzyme can produce the requisite free energy from nowhere; what is far more probable is that the enzyme, when it combines with its

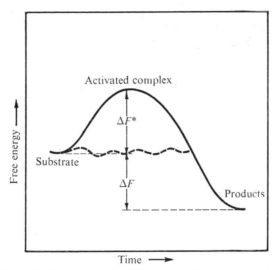

Fig. 9. Schematic formulation of an enzyme-catalysed reaction (adapted from Eyring). For explanation see text.

substrate, opens up new, alternative reaction pathways (broken line in Fig. 9) along which no high activation peaks are encountered, so that, even though one or more minor peaks may exist along the way, the substrate molecules are all or nearly all 'energetic' enough to surmount them. From the standpoint of thermodynamics it matters little or not at all what pathway is traversed between the substrate and the reaction products, so that the idea of alternative pathways has the merit of being at least reasonable.

We know a certain amount about the phenomenon of activation but too little to make a simple picture of the certainly complex changes which underlie it, but it is an interesting and highly significant fact that the hydrolytic processes catalysed by enzymes can frequently be imitated by means of dilute acids, alkalis, or both, and sometimes merely by boiling water.

Many enzymes possess built-in acidic and basic groups so arranged that their reaction with the specifically bound substrate is greatly facilitated.

31

Whether we treat a dilute solution of sucrose, for example, with hot, dilute mineral acid or with saccharase prepared from yeast, we get precisely the same products, viz. glucose and fructose in equimolecular proportions. Saccharase, therefore, does not induce any new kind of reactivity in its substrate, but only facilitates a tendency to react that is already inherent in the sucrose molecule itself.

ACTIVATORS AND COENZYMES

An enzyme must be considered as having at least two reacting centres, both of which are involved in its specific catalytic activity. There must be a *catalytic site* which promotes the ensuing reaction and a *substrate-binding site*, at which the union of the substrate with the enzyme protein takes place.

Activation of the substrate is an indispensable part of the chemical process catalysed by any enzyme, but it can take place without necessarily being followed by the hydrolysis, oxidation or other chemical modification of the substrate. Thus if we add peroxidase to a solution of hydrogen peroxide, the two unite to form an addition compound, as is shown by the resulting change in the absorption spectrum (p. 22). In the absence of other materials, there the matter ends. But, if some substance capable of being oxidized is also added, AH_2 say, there begins a rapid transference of H atoms to the activated hydrogen peroxide so that AH_2 is oxidized and the peroxide reduced, thus:

$$A\Big\langle\begin{matrix}H\\H\end{matrix} \quad + \quad \begin{matrix}O\!-\!H\\|\\O\!-\!H\end{matrix} \quad = A \ + \ 2H\!-\!O\!-\!H.$$

This example suffices to show that, while activation is an essential part of the process of enzymic catalysis, activation only makes it *possible* for the reaction to take place: whether or not the reaction actually occurs often depends upon the presence of other materials, over and above the substrate and its activating protein, the enzyme. It is, in fact, true that in the vast majority of enzyme-catalysed reactions, substances other than the substrate and its activating enzyme-protein must also be present before any chemical change can be brought about. In processes of hydrolysis, for instance, water molecules form an indispensable part of the reacting system. The enzyme must therefore be considered as only a part, albeit the most important part from the biological point of view, of the whole reacting system. Similarly, in oxidation and reduction reactions, the majority of which are accomplished by the transference of pairs of hydrogen atoms from the substance being oxidized (the 'hydrogen donor') to the substance being reduced (the 'hydrogen acceptor'), we find that both substances must be present, together with the appropriate enzyme.

It has been known for many years that a considerable number of enzymes are unable to exert their catalytic influence except in the presence of certain appropriate materials which have become known as 'coenzymes' or 'activators'. It will be remembered that zymase, for instance, loses its activity if dialysed, and that this loss of activity is attributable to the removal from the juice of certain small, thermostable ions or molecules in the absence of which fermentation cannot proceed. Recent work has shown that even such seemingly innocent substances as the ions of potassium, calcium, magnesium, chloride, phosphate and the like play indispensable parts in many enzyme-catalysed processes. In such cases it is clear that the inability of the enzyme to act in the usual way might be due to one or other of two causes. Either (a) the enzyme cannot activate its substrate because some accessory part of the enzyme itself has been removed, or (b) by contrast, the enzyme is capable of activating its substrate but no reaction takes place because some substance with which the substrate ordinarily reacts has been removed. It is possible to distinguish more or less sharply therefore between two groups of accessory substances, those which are *parts of the activating system* on the one hand, and on the other, those which are *a part of the reaction system* but play no part in activation. Although it is difficult to justify any distinction in many cases, the tendency is to refer to accessory substances which are in effect a part of the activating system and are required before the enzyme can activate its substrate, as 'activators'. The term 'coenzyme', on the other hand, tends to be reserved for substances which play some part in the reaction catalysed by the enzyme, but not in the activation of the substrate. It must, however, be remembered that the *activation of an enzyme* by the appropriate 'activator' is quite distinct from the *activation of the substrate* that takes place as a result of its union with the specific activating enzyme-protein.

It is sometimes found that enzymes are secreted in a form in which they have no catalytic activity whatever, i.e. in the form of enzyme-precursors, or 'pro-enzymes'. The classical case to consider here is that of trypsinogen. The juice secreted by the pancreas of vertebrates contains a pro-enzyme, trypsinogen, which is devoid of action upon proteins, but when the pancreatic juice enters the small intestine, trypsinogen is converted into the active proteolytic enzyme, trypsin. The change is attributed to the presence of an enzyme present in the intestinal juice to which the name of enteropeptidase has been given. The change from trypsinogen to trypsin is due to the removal from the pro-enzyme of a portion of the molecule which acts, so to speak, as a 'mask' covering the reactive centres of trypsin itself. This 'mask' is a polypeptide, and its removal is due to peptidase action on the part of enteropeptidase. This kind of activation, which for want of a better term we may refer to as '*unmasking*', is known to occur in several peptidases, and will be dealt with at greater length when we come to consider them in more detail.

A second kind of activation, and one which is commonly encountered, might similarly be referred to as '*de-inhibition*'. Many enzymes are readily inhibited by mild oxidizing agents, and it frequently happens in the course of attempts to isolate them that much activity is lost in the process as the result of oxidation by atmospheric oxygen, catalysed as a rule by traces of heavy metals present in the material or derived from mincing machines or other metallic devices used in the preparation. In such cases the activity can very often be recovered by adding reducing agents such as cysteine or reduced glutathione. It was first shown by Hopkins that in the case of succinate dehydrogenase, though the same is not true of the lactate enzyme, any treatment tending to oxidize the —SH groups of the enzyme-protein so that —S—S— cross-linkages are formed between adjacent molecules of the enzyme, results in the loss of dehydrogenase activity. These —S—S— linkages can be reduced again by means of —SH compounds, e.g. reduced glutathione, and the dehydrogenase activity of the enzyme returns therewith.

Activation by de-inhibition is a common process and may often be accomplished merely by removing inhibitory material. Cytochrome oxidase, an enzyme of central importance in respiratory metabolism, is powerfully inhibited by carbon monoxide in the dark, and a very striking case of activation by de-inhibition can be demonstrated in this case by exposing the preparation to strong light, which causes the carbon monoxide/oxidase compound to dissociate. Many other examples might be quoted.

Activation by unmasking or de-inhibition is due, apparently, to the removal of material that inhibits by blocking the active sites on the enzyme. In other cases, however, it is less clear how the activator functions. Many enzymes concerned with phosphorylation, for example, require the presence of magnesium ions, but we do not know for certain what part these play. It is believed that the magnesium plays some part in the binding of the substrate. If this is the case, it follows that the enzyme-protein alone is unable to activate its substrate and that the magnesium must be regarded as a part of the activating machine. That this is so seems very probable, since one such enzyme, enolase (phosphopyruvate hydratase), has been crystallized in the form of a magnesium-containing protein.

Many different metallic ions are now known to function as activators of one enzyme or another and the relationship between the enzyme and the metallic ion is often very specific. As is well known, minute quantities of numerous 'trace elements', including Mn, Mo, Zn and Co for example, are indispensable nutritional factors for living organisms, ranging from bacteria to mammals. It is becoming increasingly clear that they play the part of specific activators for particular enzymes. In general they appear to be parts of the activating system, i.e. the enzyme-protein fails to activate the substrate unless the appropriate metallic ion is also present.

Another much-quoted case is that of salivary and similar amylases. If a preparation of salivary amylase is dialysed it loses its power to digest starch at pH 6·8, which is the optimal value under normal conditions. Activity is regained by the addition of chloride ions, but the effect is not specific, and chloride may be replaced by other univalent anions, though these are less effective (Fig. 10). Here, again, there is reason to suspect that the ionic activator is *a part of the activating system*, and that in its absence the enzyme-protein cannot activate its substrate in the normal manner.

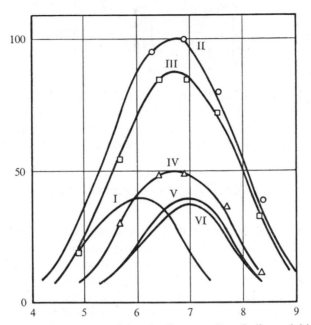

Fig. 10. Influence of anions upon activity of salivary amylase. Ordinate: initial velocity of hydrolysis. Abscissa: pH. (Substrate, soluble starch: after Myrbäck, 1926.)

Curve	Salt
I	Traces of NaCl
II	NaCl
III	NaBr
IV	KI
V	$NaNO_3$
VI	$KClO_3$

We have so far considered three main types of activators, those which act by unmasking the active groups of the enzyme, those which remove extraneous inhibitory material, and those which perhaps act because they are, in effect, a part of the enzyme. Other cases will be dealt with later when we consider individual enzymes in greater detail. We must turn now to consider accessory substances which confer activity upon inert systems because they *play a part in the chemical reaction which follows upon the successful activation of the substrate*. Substances of this kind are usually spoken of as coenzymes.

Of these the longest known, perhaps, is the co-carboxylase (diphosphothiamine, thiamine pyrophosphate) of yeast. Yeast, and the juice expressed from it by the Buchner technique, contain an enzyme known as carboxylase (pyruvate decarboxylyase) which, in the presence of *co-carboxylase* though not in its absence, catalyses the decomposition of pyruvate into acetaldehyde and carbon dioxide:

$$
\begin{array}{l}
CH_3 \\
| \\
CO \\
\cdots\cdots|\cdots\cdots \\
COO \vdots H
\end{array}
\longrightarrow
\begin{array}{l}
CH_3 \\
| \\
CHO
\end{array}
+ \quad CO_2
$$

Co-carboxylase enters into this reaction, which is called decarboxylation. It is essential also for a more complex process known as oxidative decarboxylation, a reaction that takes place on a large scale in animal tissues and in which decarboxylation is attended by a simultaneous oxidative change:

$$
R.CO.\;\overline{\vdots COO \vdots}\;H \xrightarrow{+\frac{1}{2}O_2} R.COOH + CO_2.
$$

This process, which is a good deal more complex than the 'straight' decarboxylation observed in yeast juice and involves the participation of several additional cofactors, is responsible for the production of a very large part of the carbon dioxide formed in respiration, just as the straight decarboxylation catalysed by yeast juice is the source of the carbon dioxide produced in alcoholic fermentation. In both cases we find the same substance, co-carboxylase, as an essential part of the reacting system, and in both cases decarboxylation takes place. The coenzyme must therefore play some specific part in decarboxylation reactions; it does, in fact, combine with the pyruvate in the first instance (p. 387).

We are on surer ground when we consider the coenzymes involved in many oxidative processes. The majority of biological oxidations are carried out by the transference of hydrogen atoms from the substance undergoing oxidation, the 'hydrogen donor', to another substance, the 'hydrogen acceptor'. The dehydrogenases which catalyse reactions of this kind are not only specific for the hydrogen donor but for the hydrogen acceptor as well. It follows, therefore, that in the absence of the appropriate hydrogen acceptor any given dehydrogenase, capable though it probably is of activating the hydrogen donor, cannot lead to any chemical change. This is true, for example, of the lactate dehydrogenase of muscle and of the alcohol dehydrogenase of yeast. In neither case does the substrate undergo oxidation unless the proper, i.e. specific, hydrogen acceptor is present. In these two systems the hydrogen acceptor is a substance which has recently been given the new, internationally agreed name of *nicotinamide-adenine dinucleotide* (NAD). This compound is able to take up a pair of hydrogen atoms from a suitably activated molecule

36

of, say, lactate or of alcohol, and the resulting reaction may be pictured as follows, e.g.

$$CH_3CH(OH)COOH + NAD \rightleftharpoons CH_3COCOOH + NADH_2.$$

Of all the dehydrogenases at present known, the majority require NAD as hydrogen acceptor and cannot use any other known, naturally-occurring substance in its place. Even the closely related NADP, which differs from NAD only in that it contains three instead of two phosphate residues per molecule, cannot replace NAD. Nor can NAD replace the NADP which is required as hydrogen acceptor by a smaller group of dehydrogenases, of which iso-citrate dehydrogenase may be cited as an example.

NAD and NADP are not by any means the only compounds which can act specifically as donors and acceptors of particular groupings. Indeed, numerous reactions have been discovered in which particular groups are transferred from one molecule to another, and in every case it appears that donor and acceptor substances are involved, over and above the enzyme which catalyses the transfer. These we need not discuss in any detail here since we shall refer to them frequently when we deal with intermediary metabolism.

NAD, NADP and a considerable number of other compounds discharging comparable 'carrier' functions play a vital part in metabolism. Normally they occur only in very small concentrations in living tissues—Warburg & Christian, for example, could isolate only about 20 mg. of NADP from the red blood corpuscles of some 250 l. of horse blood—but the reactions in which they participate are very rapid indeed. Since these cofactors are essential for the occurrence of these reactions, and since they are present only in small amounts, it is clear that we must regard them as true catalysts. Their catalytic influence is, in fact, no whit less important than that of the enzymes with which they collaborate.

PROSTHETIC GROUPS

A considerable number of enzymes have now been greatly concentrated and finally obtained in pure crystalline form,† and in many cases, notably among enzymes concerned with processes of oxidation and reduction, the enzyme molecule has been found to contain a non-protein moiety in addition to its protein component. Enzymes of this kind, therefore, are conjugated proteins, and the non-protein fragment is called the *prosthetic group* in each case.

The question arises whether or not there is any essential resemblance between the functional behaviour of a substrate, a coenzyme and a prosthetic group. In substances such as haemoglobin, haemocyanin and the like it has long been known that the non-protein part of the molecule is firmly attached

† It should be noticed that crystallinity is no criterion of purity where enzymes and other proteins are concerned; they commonly form mixed crystals with contaminating proteins while still far from chemically pure.

37   4-2

to the protein component, and the special name of *prosthetic group* was coined to describe it. In more recent times we have made the acquaintance of conjugated proteins of which the prosthetic groups are relatively much more loosely attached. Thus there exists in the eggs of the lobster a green chromoprotein, ovoverdin, the prosthetic group of which, astaxanthin, can be removed by heating to about 60° C., but reunites with the protein on cooling. Still more striking, perhaps, are the visual chromoproteins, rhodopsin and porphyropsin, which dissociate on exposure to light, but reunite in the dark. Many such cases are now known, and the notion that a prosthetic group is necessarily firmly attached or screwed down to its protein partner has to be abandoned.

If we import the same notions into the field of biological catalysis we find that, in the main, it is possible to distinguish between substrates and coenzymes, which are only loosely and temporarily attached to the catalytic proteins with which they co-operate, and prosthetic groups, which are relatively firmly fixed to their protein partners. The part played by the prosthetic group is precisely known in many cases. Certain oxidizing enzymes have a prosthetic group which functions as a 'built-in' hydrogen acceptor, taking over a pair of hydrogen atoms from the activated substrate and subsequently passing them on to another acceptor. In such cases the enzyme behaves as an activating protein and hydrogen acceptor rolled into one, and the functional behaviour of the prosthetic group in such a case is therefore analogous to that of the coenzyme of a typical dehydrogenase. The essential difference is that, whereas the partnership between the activating protein and the prosthetic group of an enzyme such as catalase is a relatively permanent affair, the partnership between, say, lactate dehydrogenase and lactate, or between the dehydrogenase and NAD, is only a loose and a temporary one on account of the relatively small affinity between the partners.

The difference between coenzymes, substrates and prosthetic groups is therefore one of degree rather than of kind. Whether we consider NAD as a temporary prosthetic group of lactate dehydrogenase, or haematin as a permanent or built-in coenzyme of peroxidase, matters little so long as the functional significance of the various parts of the system is clear. What does matter is that we shall realize that the old, sharp distinction that seemed to exist between enzymes and carriers, and between substrates, coenzymes and prosthetic groups cannot now be justified, a fact which brings a new unifying influence to bear on our knowledge of biological catalysis.

QUANTITATIVE CHARACTERIZATION OF ENZYMES

Certain features of enzymic catalysis already alluded to in the first chapter of this book may now be considered in a little more detail. Enzymes in general may be considered under two headings, according as their substrates do or do not ionize. As examples of the former type we may consider the peptidases, and of the latter the carbohydrases. In all these cases the activity of the e̶ zyme is profoundly affected by pH, and since the substrates of the carbo- hydrases, for example, do not ionize, the influence of pH upon these enzymes must be entirely due to its influence upon the catalytic proteins.

Knowing that enzymes are proteins, we may infer that they carry numerous ionizable groups, the ionic state of which depends upon the pH of the sur- rounding medium. Since there is some particular pH at which the enzyme is more active than at any other, we may suppose that, of all the possible ionic forms in which the enzyme-protein can exist, only one possesses catalytic properties, and that it is this form that predominates at the optimum pH. Michaelis & Davidson suggested that, since a change of pH in either direction away from the optimum leads to a diminution of catalytic activity, two kinds of ionizable groups must be involved in determining activity, one being acidic and the other basic in nature. The enzyme, like any other protein, must be considered as an ampholyte, and in view of the close resemblance that exists between the dissociation curve of a simple ampholyte such as glycine on the one hand, and the pH/activity curve of a typical enzyme on the other, Michaelis & Davidson went on to suggest that the two halves of the pH/ activity curve must correspond to the dissociation curves of the two particular groups or sets of groups, upon the ionic condition of which the catalytic activity of the protein depends. For any given enzyme, therefore, the form and position of the pH/activity curve should be constant, even if the enzyme acts upon several different substrates, always provided that the substrates them- selves do not ionize. This seems generally to be true.

If, therefore, we determine the pK values for the dissociation of the two sets of ionizable groups, which we can do by carefully plotting and inspecting the pH/activity curve, we can determine in quantitative terms two constants that are characteristic of the enzyme. Although this has been done in several cases, the method of inspection is unreliable; in any case the results depend to some extent upon the substrate, temperature, nature and concentration of buffer employed and on the ionic strength of the medium. It is considerably easier, though less precise, to determine the maximum of the resultant of the two dissociation curves, i.e. the optimum pH.

In the case of enzymes that act upon ionizable substrates, the position is complicated by the fact that changes of pH will influence the ionic conditions both of the enzyme and of its substrate. Further, if we change the substrate, the shape and position of the pH/activity curve will be expected to change too, and we do in fact find that enzymes such as pepsin and trypsin show different pH optima when acting upon different proteins. Nevertheless, if we stipulate some particular substrate in any particular case we can determine the optimum pH or the two pK values for that particular enzyme-substrate pair.

Another characteristic property of enzymes that can be measured and expressed in quantitative terms is the Michaelis constant K_m. This, it will be remembered, is that concentration of substrate at which the reaction velocity attains half its limiting value. The Michaelis constant differs from enzyme to enzyme, and varies also from substrate to substrate when the enzyme's specificity is not absolute. If we have an enzyme preparation that acts upon several glycosides, for example, we commonly find a different K_m for each substrate, but this does not tell us whether two or more different enzymes are concerned, or whether a single enzyme of group-specificity is at work. Preparations of yeast saccharase, for example, act upon both sucrose and raffinose, and the question arises whether yeast contains a specific raffinase as well as a saccharase. The K_m value for sucrose is about one-sixteenth as great as that for raffinose, but the ratio is always the same, no matter how the enzyme preparation may be prepared or purified. If two enzymes were concerned we should expect them to be present in different proportions in different preparations made by different procedures, and it therefore follows that probably only one enzyme is concerned, a conclusion which is strengthened by the fact that the pH/activity curve has the same pK values and the same optimum pH whether sucrose or raffinose is the substrate.

A fourth characteristic constant can be determined in certain cases. It will be remembered that in the presence of a large excess of substrate, the reaction velocity of an enzyme-catalysed reaction reaches a limiting value V. If the (molar) concentration of the enzyme is known and is equal to $[E]$, then

$$V = k.[E],$$

where k is the velocity constant of the reaction. We can evaluate $[E]$ in cases where the molecular weight of the enzyme is known and the enzyme is available in a chemically pure form, and the value of k is characteristic for a given enzyme acting upon a given substrate. The best we can do in other cases is to determine V in the presence of an arbitrarily defined concentration of the enzyme. If the enzyme is one that acts upon more than one substrate, say on two compounds a and b, the ratio V_a/V_b will be constant if the enzyme concentration is the same in both cases. Even this second-best determination has proved itself valuable in deciding the identity of pairs of enzymes. Consider once more the case of yeast saccharase. This enzyme attacks sucrose twice as fast as raffinose, and the ratio remains the same from one preparation to another. If two enzymes were concerned we should expect the ratio to vary from case to case, but this does not happen, thus adding still more to the evidence for the identity of the presumptive raffinase with saccharase.

To establish the identity of one enzyme with another, even in relatively crude extracts, a number of ways are thus open and those considered here by no means exhaust the possibilities. We can see whether the two behave in the same manner with respect to inhibitors and activators, and we can find out whether their specificities are similar or different. These, however, are properties that cannot be exactly defined or numerically expressed.

40

SUMMARY

Summarizing our conclusions as to the nature of the enzyme-substrate union it may be said first of all that there is every reason to believe that such a union does in fact take place. The union is a very specific one; a given enzyme can combine with and activate only a limited number of substrates and, often enough, only one substrate. There is reason to think that the reaction takes place at certain definite sites on the surface of the enzyme, rather than at any arbitrary site or sites, and it seems that the specificity of an enzyme is really a measure of the extent to which the enzyme and the substrate 'fit' at the points through which they unite.

Even when the activating protein fits the substrate well enough to allow union to take place between them, the 'fit' may be thought to be still slightly imperfect, so that either the enzyme or its substrate or both undergoes an intramolecular rearrangement of some kind which results in the change of chemical reactivity to which we refer as activation.

The essential function of an enzyme is that of activating its substrate. It may lose the power to do this for any of a large number of reasons, but in many cases the lost activity can be recovered.

Even when activation has been accomplished, the substrate does not necessarily undergo any chemical change, since substances other than the substrate and its activating protein are required in many cases. We have considered ways in which the power of activation may be restored to an enzyme that has lost it, and we have considered some cases in which accessory substances, coenzymes or prosthetic groups, enter into the reactions taking place after the substrate has been successfully activated.

Finally, we have seen that it is possible to obtain quantitative data which are characteristic of individual enzymes.

3

BIOLOGICAL ENERGETICS

THE CONCEPT OF FREE ENERGY

Before going further it is necessary to know something about the conditions which determine whether or not, and if so to what extent, any given chemical reaction will proceed, whether it be catalysed or uncatalysed. To analyse these conditions completely would require lengthy thermodynamic arguments, but for present purposes use may be made of a simple mechanical analogy.

Let us consider a perfectly smooth body standing on a perfectly smooth plane. This body has a certain amount of gravitational potential energy, but this energy is not available for the performance of work of any kind unless the plane is tilted. Suppose now that the plane is slightly inclined. The body begins to slide downwards because some of its potential energy has become available to push it down the plane. When the body slides, work can be done (e.g. if the body is attached by means of a string to some suitable machine), and the total amount of work done will be thermodynamically equivalent to the amount of gravitational potential energy lost by the body. How much work can be done by this system depends upon the system itself, for while it is theoretically possible for the body to go on sliding down an inclined plane of indefinite length until the whole of its potential energy has been converted into work, this is not a case of much practical interest. Generally speaking *the properties of natural systems are such that only a part of their total potential energy is available for the performance of work*. A larger or smaller part of the total energy is always unavailable except in theoretical cases. Hence we must distinguish between the 'free' or available energy and the total energy of the system.

Now we know as a matter of everyday practical experience that a body will never move *up* a plane so long as the system is left to itself; an upward movement can only be accomplished by supplying energy to the system from an external source. The body will always slide *down* an inclined plane if left to its own devices, provided that the frictional forces opposing the tendency to slip are not too great. This simple case exemplifies the general rule that, in any self-operating system, *free energy always tends to be lost*.

Let us now consider a heat engine supplied with an amount of heat energy equal to Q. This energy is supplied at a high temperature T_1 (measured on the absolute scale), and conducted to a lower temperature T_2. It can be shown on theoretical grounds that the amount of work done by such an engine cannot

42

exceed an amount W, where

$$W = Q\frac{T_1 - T_2}{T_1}.$$

This equation can be transformed to give

$$W = Q - Q\left(\frac{T_2}{T_1}\right).$$

Thus the maximum *available* work will always be less than the total *possible* work by an amount given by $Q(T_2/T_1)$. It follows that no work at all can be done by an engine of this kind *unless the heat can be conducted from a higher to a lower temperature* since, if $T_1 = T_2$, the second term becomes arithmetically equal to Q but opposite in sign. This point is of great importance for biochemical systems, which usually operate at a virtually constant temperature. Furthermore, an engine of this kind can only convert its *total* heat energy into work if $Q(T_2/T_1) = 0$, i.e. when $T_2 = 0$, and the 'exhaust' or condenser of the engine is maintained at the absolute zero of temperature. At all other temperatures a part of the total energy will be unavailable, and the magnitude of the unavailable energy at any temperature T_2 is determined by the product of that temperature and the factor (Q/T_1). The latter is known as the entropy of the system and is usually represented by S. Entropy measures the extent to which the total energy of a system is unavailable for the performance of work.

Similar considerations apply to systems other than heat engines and, in fact, for any self-operating system working at a temperature T we can write the following general equation:

$$F = H - T.S,$$

where F represents the free energy of the system (represented by G in the British usage), i.e. the amount of energy available for the performance of work, H is the total heat energy of the system and S is its entropy.

It is not possible to determine the absolute values of these variables apart, of course, from the temperature and this temperature, T, is virtually constant in any given biological system. We can, however, observe the changes that H and S undergo when the system passes from its original state into a new condition represented by

$$F' = H' - T.S'.$$

Subtracting from the former equation we get

$$(F - F') = (H - H') - T(S - S'),$$

or, in the usual terminology,

$$\Delta F = \Delta H - T.\Delta S.$$

i.e. the *difference* in F is equal to the *difference* in H, minus the product of the temperature and the *difference* in S.

In a chemical as in any other kind of system *there is always a tendency for free energy to decrease* when any change takes place in the system. A chemical change that is accompanied by a loss of free energy can therefore proceed without external assistance. On the other hand, a reaction involving an increase of free energy can only proceed if it is in some way coupled to a suitable source of free energy, which may itself be another chemical reaction.

Now chemical changes as a whole are accompanied by thermal changes, and are most commonly exothermic. But *the heat evolved during an exothermic process does not correspond to the loss of free energy*, the change of entropy has also to be reckoned with. In an ordinary chemical reaction, therefore, all we can measure directly is ΔH, the change in total heat energy; ΔF can only be arrived at indirectly. This is not the place to deal with procedures whereby values for ΔF can be determined or calculated, but one method of particular importance may be mentioned in passing. This depends upon accurate measurements of the equilibrium constant, K, of a reversible reaction, say

$$A + B \rightleftharpoons C + D.$$

K is defined by the equation

$$K = \frac{[C][D]}{[A][B]}.$$

In such a case, it can be shown that

$$\Delta F^0 = -RT \log_e K,$$

where R is the gas constant and T the absolute temperature. The term ΔF^0 is the *standard* free energy change, a quantity which need not concern us at this stage of our argument.

Sometimes the change of entropy is small, so that ΔF is approximately equal to ΔH, but it is equally possible to have a reaction in which the change of entropy is so large that the reaction is actually endothermic but results nevertheless in a loss of free energy. We see then, that the total heat change in an exothermic or endothermic reaction gives no indication as to whether or not that reaction can take place spontaneously or requires to be assisted by other processes going on in the environment; the only reliable guide is a knowledge of ΔF.

If ΔF is zero the system is in chemical equilibrium: if it is positive (i.e. free energy goes into the system) the reaction cannot take place except with external aid: but if it is negative (i.e. free energy is lost from the system) the reaction can proceed of its own accord. Processes which are attended by a loss of free energy (i.e. $\Delta F < 0$) are said to be *exergonic*, while those which involve an increase of free energy (i.e. $\Delta F > 0$) are described as *endergonic* changes.

A chemical reaction is thermodynamically possible, then, if it is exergonic, i.e. is attended by a loss of free energy. Whether or not it actually takes place however, depends upon other factors. A body will not slide down a rough plane if the frictional forces are greater than the forces exerted by its free gravitational potential energy. Similarly, while a chemical reaction *can* take place if it entails a loss of free energy, it will not actually *do* so if the 'frictional forces' tending to oppose it are too large. In other words, *a chemical reaction requires for its accomplishment that the molecules shall be in a reactive state.* This requirement is taken care of in biological systems by the enzymes they contain. For example, we may keep a neutral, aqueous solution of sucrose almost indefinitely without appreciable hydrolysis, for although hydrolysis is thermodynamically possible, it does not actually take place because the molecules are not sufficiently reactive. If we add a small amount of saccharase or, alternatively, a little dilute mineral acid, the sugar is hydrolysed. By activating the molecules these catalysts overcome the 'frictional forces' opposing hydrolysis; these 'frictional forces' correspond to the activation energy, already discussed (p. 31).

It follows from all this that no catalyst can initiate or accelerate a reaction that is not already possible on energetic grounds: all that it can do is to influence the velocity at which a thermodynamically possible reaction actually takes place.

BREAKDOWN AND SYNTHESIS IN BIOLOGICAL SYSTEMS

The numerous and diverse chemical processes which underlie the activities of a living organism collectively constitute its *metabolism*. For purposes of discussion the overall metabolism of a cell or tissue can be considered as consisting of two parts, *katabolism*, which involves the chemical degradation of complex materials into simpler products, and *anabolism*, which involves the elaboration of complex products from simpler starting materials. Katabolism and anabolism are usually defined in terms of chemical complexity, but this is a somewhat unsatisfactory variable because we have at present no quantitative means of measuring or expressing it. But, in a general kind of way, increases in chemical complexity are associated with increases of free energy, and for our present purposes we may conveniently define as *katabolic* those processes which involve a loss of free energy and are therefore exergonic, and as *anabolic* those in which free energy is gained and which are therefore endergonic.

Now, animal and plant tissues as a whole are known to contain numerous intracellular enzymes which can be extracted and shown to catalyse the katabolic breakdown of proteins and other high-molecular materials into simpler units. After death these enzymes do in fact lead to the digestion of much of the tissue substance, a process known as autolysis, and this is why game and certain kinds of meat are allowed to 'hang' before being cooked. This same process of autolysis is the first stage in the decomposition of dead organisms; bacteria and the biblical worms come into the picture considerably later.

45

During life, however, anabolic (i.e. synthetic) processes are numerous and of very great importance, especially during growth and in tissue repair after injury. They include the synthesis of catalytic and tissue proteins from amino-acids; polysaccharides from simpler sugars; oils, fats and waxes from their constituent alcohols and fatty acids, to make no mention of the elaboration of other more or less highly specialized products such as hormones, pigments and so forth. But, in addition, we have good reason to believe that the tissue constituents themselves are not permanent, static structures, but rather that they are constantly in a state of breakdown balanced by synthesis. This is true even of hard structures such as the bones and the teeth. It seems, there-fore, that we must envisage the possibility that intracellular enzymes are concerned both with the breakdown *and* with the synthesis of proteins, fats, carbohydrates and other cell constituents.

On theoretical grounds we must believe that an enzyme which can catalyse the hydrolysis of its substrate, for example, must also be capable of catalysing the condensation of the products, and in a number of instances such a reversi-bility of action can readily be demonstrated at least *in vitro*, e.g. among lipases and in most dehydrogenases. In many other cases, however, attempts to demonstrate synthesis by an enzyme known to catalyse the corresponding degradation have yielded only negative results. It must be remembered, how-ever, that in order that a given chemical reaction shall take place it is not enough for the molecules concerned to be in a reactive state: *the energy con-ditions also must be favourable*. It is possible, therefore, that the failure of many early experiments was due only to inability to reproduce under experi-mental conditions the proper energetic circumstances, and does not neces-sarily mean that the enzymes concerned are inherently incapable of working in both directions. For these reasons we must assume that in the cell, if not in the laboratory, the right energetic conditions are attainable, and that, in the cell, enzymes are present that can catalyse both synthesis and breakdown. But the synthetic enzymes would be expected to require the maintenance of normal physiological conditions inside the cell because, if these break down, the necessary provisions of energy might not be forthcoming so that only the phase of degradation would be observed.

The function of the intracellular enzymes might be regarded, then, as that of maintaining equilibrium between the complex materials of the cell sub-stance and their simpler constituents, so that a balance is struck between synthesis and degradation. But it does not follow that synthesis and break-down follow the same pathways or even use similar enzymes. Indeed, most biochemical evidence goes to show that, as a rule, and especially when considerable changes of free energy are involved, *synthesis and degradation follow quite different pathways and are catalysed by different enzymes*. We shall come across many cases that bear out the truth of this general theorem. We must not think of the equilibrium thus set up and maintained as in any sense a static affair but as an essentially dynamic system in which breakdown and resynthesis are proceeding simultaneously at high but equal velocity. We should do well to recall Hopkins's celebrated aphorism, that *life is a dynamic equilibrium in a polyphasic system.*

In living cells, then, katabolism and anabolism proceed together. This raises several problems of the greatest energetic importance. First, what is the biological source of the free energy required for the essentially endergonic processes of anabolism; secondly, how is that free energy harnessed or 'captured' by the cell and, thirdly, how is it brought to bear upon the processes wherein it is consumed? And it should be borne in mind that free energy is expended not only in anabolic reactions, but in the performance of muscular and all other kinds of biological work.

For many years these problems seemed so complex and so wholly baffling that many biologists were content to believe them unanswerable; that living organisms must in some way contrive to operate in defiance of the laws of thermodynamics. Although the most spectacular progress in the field of biological energetics has been made in comparatively recent years, the most fundamental experiments were undoubtedly those performed just before the end of the last century by Rubner, in Germany, and speedily confirmed in America by Atwater, Rosa & Benedict. These workers constructed special calorimeters capable of containing dogs, and later even men, in conditions of comparative comfort for periods of days or even weeks. The apparatus, which will not be described here since there are admirable descriptions in many physiological and biochemical text-books, was constructed in such a way that the following measurements could be accurately and simultaneously carried out:

(*a*) Total energy output, measured as heat.

(*b*) Total oxygen consumption.

(*c*) Total carbon dioxide production.

(*d*) Total nitrogen excretion.

Numerous corrections had, of course, to be applied, e.g. for heat brought into the apparatus in hot food and for heat removed in the bodily excreta. And there were many engineering difficulties.

From (*b*), (*c*) and (*d*) it is possible to calculate the weights of carbohydrate, fat and protein oxidized in the course of an experiment and hence to arrive at the total yield of energy attributable to their metabolism, the calorific values of these metabolic substrates having previously been determined with a high degree of precision by combustion in the bomb calorimeter.

Of such fundamental importance is the outcome of these experiments that the reader is asked to examine with more than usual care the data of Table 1, which is compiled from the results of Rubner's experiments on two dogs. These were later followed up by experiments on human subjects in the Atwater–Rosa–Benedict calorimeter. Whether the subjects were resting, active

47

Table 1. *Calculated and observed heat outputs of dogs: Rubner's experiments.*

Diet	Dog	Duration of exp. (days)	Total heat output (k.cal.) Calculated	Determined	Differences (%)
None	A	5	1296·3	1305·2	+0·69
	B	2	1091·2	1056·6	−3·15
Fat	A	5	1510·1	1495·3	−0·97
Meat and fat	A	8	2492·4	2488·0	−0·17
	A	12	3985·4	3958·4	−0·68
Meat	A	6	2249·8	2276·9	+1·20
	B	7	4780·8	4769·3	−0·24
				Mean difference	−0·47

or even taking fairly strenuous exercise, e.g. by riding a stationary bicycle or sawing wood, the observed and calculated heat outputs agreed within limits even narrower than in Rubner's experiments. These results provide irrefutable proof that a living organism as exalted even as man himself cannot create energy out of nothingness, and no convincing evidence to the contrary has at any time been forthcoming. *Living organisms, like machines, conform to the law of the conservation of energy, and must pay for all their activities in the currency of katabolism.*

With these results before us it is clear that there can be but one answer to the most fundamental of our problems. *Energy used in anabolism must be drawn from katabolic processes: no other source is available.*† This means that there must exist some kind of coupling between katabolism and anabolism; some kind of mechanism that allows the transference of free energy from one to the other. There then remains the problem of the manner in which the free energy to which the cell gains access in the course of its katabolic operations can be captured and utilized for anabolic and other purposes. In order to deal with these problems we must necessarily enquire into the mechanisms which underlie the synthetic reactions that make up anabolism.

REVERSIBILITY OF BIOLOGICAL REACTIONS

All reactions are reversible in theory, but because they are reversible in theory it does not necessarily follow that they are reversible in practice, at any rate under biological conditions. The explosion of a hydrogen bomb might be reversible in theory, but reversal of the explosive process is something that is not very likely to be accomplished, at any rate in the foreseeable future. Much

† This statement applies, of course, to animals. Green plants and photosynthetic bacteria can use solar energy for anabolism.

the same is true of many enzyme-catalysed reactions under biological conditions; and, of course, it is with biological reactions that we are particularly concerned here. Now enzymes, in common with catalysts of other kinds, influence only the velocity and not the direction of the reactions they catalyse, and there are, as has been pointed out already, many enzyme-catalysed processes that go equally freely in either direction. But many others do not: often it happens that the products of a given reaction are removed as fast as they are formed so that, although the reaction is reversible in reality, it may nevertheless proceed in only one direction under biological conditions.

No chemical reaction is totally irreversible; on the contrary, every reaction tends towards an equilibrium, e.g.

$$A \rightleftharpoons B+C.$$

Equilibrium is rapidly reached in the presence of a suitable enzyme or other catalyst, and the composition of the equilibrated mixture is determined by the molar concentrations of the reactants and by the overall change of free energy. However, if the intrinsic free energy of A is very large indeed compared with that of the products, decomposition of A will predominate over its synthesis and, if the degree of predominance is very large indeed, as it often is in practice, the process will be virtually unidirectional and may, for all practical purposes, be regarded as irreversible. In considering the conditions required to reverse the process and so bring about the synthesis of A, we have in fact to consider the conditions required to alter the final equilibrium in such a way that synthesis becomes the predominant process.

Let us consider now a simple biological process of a virtually irreversible kind that can be represented as follows:

$$A \rightarrow B+C; \quad \Delta F = -\epsilon \text{ cal.}$$

Since the change of free energy associated with the process is less than zero ($\Delta F < 0$) the reaction is one that is thermodynamically possible and will therefore proceed spontaneously in the presence of an appropriate catalyst. When it takes place, ϵ cal. of free energy will become available for each g.mol. of substrate transformed and, if some suitable device were present to trap it, this free energy could be converted into work of some kind. But in the absence of such a contrivance this free energy is merely dissipated and the output of heat can be calculated or measured experimentally.

Since ϵ cal. of free energy are liberated when the reaction proceeds in the forward direction, it follows that it can be made to go backwards if, *and only if*, an amount of free energy equal to or greater than ϵ cal. is supplied in some way. It is useless merely to supply heat to the system because, as we have seen, heat cannot do work unless it is conducted from a higher to a lower temperature, and biological processes usually take place at virtually constant temperature. Thus, although our reaction is theoretically reversible and although

49

the enzyme concerned may be *able* to catalyse the forward and reverse processes alike, the backward reaction can only be realized if the requisite amount of free energy can be provided; and provided, moreover, *in a biological manner and under biological conditions.*

In the laboratory a chemist faced with the task of performing an energy-consuming synthesis will as a rule make use of high temperatures, strong acids, strong alkalis and similar 'powerful' reagents. But these are resources that are denied to biological systems. Living cells, by contrast, operate in the neighbourhood of pH 7 and at temperatures ranging roughly from zero to 40° on the centigrade scale, and even under these very 'mild' conditions their synthetic ability and dexterity far surpass those of the organic chemist.

ENERGETICS OF SYNTHETIC REACTIONS

Consider again the reaction

$$A \rightarrow B + C; \quad \Delta F = -\epsilon \text{ cal.}$$

One way of achieving the reverse process would be in some way to incorporate into one of the reactants, B and C, an amount of free energy equal to or greater than ϵ cal. per g.mol. This, indeed, is the method usual in biological systems, and processes whereby this is accomplished are commonly known as *priming reactions.* The usual biological procedure is that some suitable additional grouping, very commonly a phosphate group, is introduced into one of the reactants, say B, to give a product of higher free-energy content, say B—R. In the presence of a suitable catalyst a reaction leading back to the original starting material is then possible:

$$B—R + C \rightarrow A + R; \quad \Delta F < 0.$$

What happens here is that, in essence, the chemical potential of R is used to provide the driving force needed for the reverse reaction. But B—R is not chemically identical with B; and so great is the specificity of enzymes in general that an enzyme capable of activating B would almost certainly be unable to activate B—R. In other words, although the *breakdown* of A is possible in the presence of a suitable enzyme, its *synthesis follows a different chemical pathway and requires the participation of different enzymes.* Moreover, the synthesis of A involves a preliminary priming reaction whereby B or C can be raised to a high enough energy level.

A process of the kind usually represented by $X \rightleftharpoons Y$ will only be possible, then, if it entails little or no change of free energy: where there is any large change of free energy the interconversion can only be properly represented by another kind of symbolism specially designed to emphasize the fact that the forward and backward pathways are totally different.

As an example of the case in point we can take a reaction in which glucose is phosphorylated by the phosphate donor, ATP, under the influence of an enzyme known as hexokinase. The reaction is irreversible and the regeneration of free glucose requires glucose-6-phosphate phosphatase, a hydrolytic enzyme, and this reaction too is irreversible. These reactions can be represented and their irreversibilities emphasized in a diagram like that shown in Fig. 11 (a).

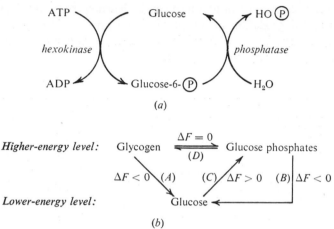

(a)

$$\Delta F = 0$$

Higher-energy level: Glycogen ⇌ Glucose phosphates
(D)

$\Delta F < 0$ (A)   (C) $\Delta F > 0$ (B) $\Delta F < 0$

Lower-energy level: Glucose

(b)

Fig. 11. (a) Synthesis and breakdown of glucose-6-phosphate. (Here ATP represents adenosine*tri*phosphate, ADP stands for adenosine*di*phosphate and ℗ for the phosphate residue, —PO_3H_2.) (b) Synthesis and breakdown of glycogen. See text for explanation.

To make these notions clearer we may take another example in which 'R' is a phosphate group.

The digestive hydrolysis of the polysaccharide glycogen is a 'downhill', i.e. an exergonic, reaction, and is irreversible. Now glycogen can also be broken down to glucose by intracellular enzymes present in the liver, and this process too is exergonic and irreversible. But certain other enzymes present in the liver can bring about the endergonic synthesis of glycogen from free glucose. The reactions involved may be diagrammatically summarized in the shorthand manner shown in Fig. 11 (b). Whether glycogen is broken down to glucose by digestive enzymes (A) or by the liver enzymes (D + B), there is a fall from a higher to a lower free-energy level, and in neither case can the pathway of breakdown be retraced. Synthesis can only proceed by a pathway involving priming reactions (C) whereby the glucose is transformed into phosphorylated sugar derivatives and thus raised to a higher free-energy level. Once this has been accomplished the phosphorylated derivatives can polymerize (D) to form glycogen with little further change of free energy, so that an equilibrium is set up. We shall return later on to a more detailed examination of this process.

Endergonic or anabolic synthesis can be accomplished, then, through the *incorporation of free energy into the reactants* by means of 'priming reactions'. Such a reaction is a necessary preliminary to the synthesis of glycogen from glucose, and similar priming reactions are now known to be involved at one or another stage in many biological synthesis, e.g. of glutamine, hippuric acid, urea and many other more complex substances, including proteins, sterols and many more.

One lesson of supreme importance can be learned from all this, that while synthesis and degradation might be catalysed on occasion by one and the same enzyme, *synthesis and breakdown commonly follow different pathways and are catalysed by different enzymes.*

PROPERTIES AND FUNCTIONS OF ADENOSINE TRIPHOSPHATE

It will be helpful now to consider certain properties of the substance known as *adenosine triphosphate*. This compound, otherwise known as ATP, is apparently a universal constituent of living stuff and plays a part of fundamental importance in biological exchanges of energy. This important compound has a very high chemical potential and provides the immediate driving force required in many synthetic and other biological operations. We shall consider its detailed structure later; for the moment it may be represented as A—ⓅⓅ—Ⓟ—Ⓟ, where A represents the nucleoside *adenosine* and Ⓟ stands for the phosphoric acid group, $-PO_3H_2$.

The terminal phosphate group of ATP can readily be removed by hydrolysis by dilute mineral acid, and the process is attended by the liberation of energy in the form of heat. The corresponding yield of free energy, allowing for changes of entropy, has been variously estimated at from about 8000 to 16,000 cal./g.mol. though these figures are open to serious doubt and should not be taken too seriously. Assuming a round figure of some 10,000 cal./g.mol. for purposes of argument the reaction may be written as follows:

$$A—Ⓟ—Ⓟ—Ⓟ + H_2O \rightarrow A—Ⓟ—Ⓟ + HO.Ⓟ; \quad \Delta F = -10,000 \text{ cal.}$$

The second phosphate group can be similarly removed:

$$A—Ⓟ—Ⓟ + H_2O \rightarrow A—Ⓟ + HO.Ⓟ; \quad \Delta F = -10,000 \text{ cal.}$$

The third phosphate group also can be removed, again by acid hydrolysis or under the influence of an appropriate enzyme, but with much more difficulty, and this time only some 2000–3000 cal. of heat are evolved, of which about 2000 cal. correspond to free energy:

$$A—Ⓟ + H_2O \rightarrow A + HO.Ⓟ; \quad \Delta F = -2000 \text{ cal.}$$

The yield of free energy in this last reaction is of the same order of magnitude as that found for the hydrolysis of most simple phosphate esters such, for example, as the glucose phosphates ($\Delta F = -3000$ to -4000 cal.) and glycerol phosphate ($\Delta F = -2200$ cal.). Again these figures are, at best, only the roughest of approximations.

Clearly however, there is some far-reaching difference between the first two phosphate groups and the third: in fact there exist in the ATP molecule two specimens of what may be called '*high energy*' or '*energy-rich*' *phosphate groups* as distinct from the third '*energy-poor*' or '*low energy*' *phosphate group*, which resembles those found in most common phosphate esters. Thus, if we use the special symbol $\sim$ ℗ to represent a high-energy phosphate group, i.e. a phosphate group of high chemical potential, the structure of ATP can be more adequately represented thus:

$$\text{A—℗} \sim \text{℗} \sim \text{℗}.$$

Table 2. *Approximate standard free energies of hydrolysis of some phosphoric acid derivatives (after Avison & Hawkins, 1951)*

Compound	ΔF^0 (cal.)	pH	°C.	Method of determination
Glycerol-1-phosphate	−2,200	8·5	38	Equilibrium measure-
Glucose-6-phosphate	−3,000	8·5	38	ments of enzymatic
Fructose-6-phosphate	−3,350	8·5	38	hydrolysis
	−3,500	8·5	38	Equilibria with gluc-
Glucose-1-phosphate	−4,750	8·5	38	ose-6-phosphate
ATP (terminal group) ($\sim$)	11,500	7·5	20	Calculation
Glyceric acid-1:3-diphosphate ($\sim$)	−16,250	6·9	25	Enzymatic equilibria
Acetyl phosphate ($\sim$)	−14,500	6·3	37	with ATP
Enol-pyruvic acid phosphate ($\sim$)	−15,900	?	20	
	−15,850	?	20	Calculation
Phosphocreatine ($\sim$)	−13,000	7·7	?	Enzymatic equilibria
Phosphoarginine ($\sim$)	−11,800	7·7	20	with ATP

We shall not trespass here into the fields of theoretical chemistry in search of reasons for these phenomena; it is sufficient for our immediate purposes to know that they exist. Similar high-energy phosphate groups are now known to be present in many other compounds, including inorganic pyrophosphates, and a list of some of these, together with approximate data for the standard free energies of their respective hydrolyses under certain closely standardized conditions, is shown in Table 2. High-energy phosphate groups are found in particular association with certain particular structures (see Table 3).

The values for ΔF^0 shown here must be regarded as only approximate as far as actual biological conditions are concerned; a high degree of accuracy is only obtainable under exceptionally favourable conditions and, as the table itself shows, different methods of estimation give different results. In any case the actual values of $-\Delta F$ under biological conditions are influenced by many factors such as molarity, degree of ionization, pH and so on, the

Table 3. *Some high-energy phosphate derivatives*

Type and identity of bond	
Esters:	

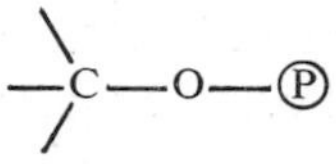

Glycerol-α-phosphate
Glucose-6-phosphate

Carbonyl phosphates:

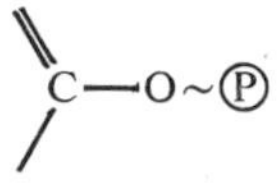

Acetyl phosphate
Enol-pyruvate phosphate
Glycerate-1:3-diphosphate

Guanidine phosphates:

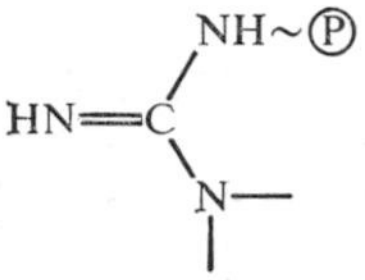

Phosphocreatine
Phosphoarginine

Polyphosphates:

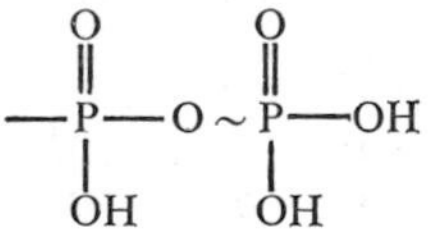

ATP, ADP

precise values of which under biological conditions are usually uncertain or unknown.

Adenosine triphosphate plays a central part in the energy transactions of many and probably all kinds of living tissues, from unicellular organisms to the highest plants and animals. Its importance lies in the ease with which its terminal phosphate group can be transferred, *together with a part or all of the free energy with which it is associated*, to other molecules. Two examples of this behaviour may be considered here.

First we have the Lohmann reaction, so named after its discoverer. This, a freely reversible reaction, is catalysed by *creatine kinase*, an enzyme that occurs in all kinds of vertebrate muscles and can be expressed as follows:

$$A—\textcircled{P}\sim\textcircled{P}\sim\textcircled{P} + C \rightleftharpoons A—\textcircled{P}\sim\textcircled{P} + C\sim\textcircled{P},$$

where C represents the guanidine derivative, creatine. When this reaction takes place there is only a small change of free energy. The free energy associated with the terminal phosphate group of the ATP is transferred, along

54

with the phosphate itself, to the creatine, and the free energy of hydrolysis of the product, phosphocreatine, is about equal to that furnished by the ATP, so that the transfer is readily reversible. A very convenient shorthand notation for reactions of this kind may be used and is frequently used throughout these pages:

ATP ⤸   ⤷ C

ADP ⤶   ⤵ CP

Double-ended arrows are used to indicate the reversibility of the reaction.

As a second example we may consider again the essentially irreversible transfer reaction that takes place between glucose and ATP in the presence of *hexokinase*. The products are glucose-6-phosphate and adenosine diphosphate (ADP). The reaction may be written in the manner shown in Fig. 12. If we consider only the conversion of ATP to ADP we see that this part of the system suffers a loss of somewhere about 10,000 cal. of free energy. Again the figures are only very rough approximations and are used to qualify rather than to quantify the process. The gain of free energy in the conversion of glucose to its phosphate amounts to only about 3000 cal., however, so that, in the complete system, a balance of some 7000 cal. of free energy is left over and spilled out. This reaction is therefore undirectional and irreversible. Although so much free energy is wasted, this is a reaction of fundamental importance in carbohydrate metabolism because glucose can neither be

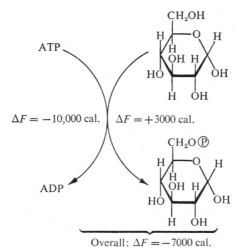

Overall: $\Delta F = -7000$ cal.

Fig. 12. The hexokinase reaction: phosphorylation of glucose by ATP. (The calorie values are only the roughest of approximations.)

55

stored nor metabolized unless it is first of all phosphorylated, and this, the *hexokinase reaction*, is the principal and perhaps the only mechanism through which this preliminary phosphorylation can be achieved; simple direct introduction of a phosphate group is energetically impossible under biological conditions since $\Delta F \geqslant 0$ for such a reaction. The hexokinase reaction is typical of priming reactions as a whole and is, in fact, the first step in the synthesis of glycogen from glucose (cf. Fig. 11).

Group transfer is a common phenomenon in biochemistry. Enzymes are known that can catalyse the transference of methyl-, amino-, amidino- and a variety of other groupings over and above the phosphate and glycoside groups that have provided our examples here, and it should be borne in mind that *whenever a group is transferred from one molecule to another, there is always a strong probability that it will carry with it a larger or smaller quota of free energy.*

THE BIOLOGICAL ENERGY CYCLE

In both the creatine kinase and the hexokinase reactions, ATP loses phosphate and energy in favour of its reaction partner and in both cases an endergonic reaction is achieved by coupling it to one that is exergonic, the chemical potential of one ~ ℗ of ATP providing the driving force necessary for the phosphorylation of one molecule of creatine or glucose as the case may be. This appears to be a fundamental type of operation in the synthesis of complex biological compounds from simpler starting materials, and, as far as we know at present, anabolic, synthetic operations can only be accomplished at the expense of a high-energy group of some kind, most usually the terminal ~ ℗ of ATP. In some cases the immediate energy source is the terminal ~ ℗ of the triphosphate of some other nucleoside, e.g. guanosine triphosphate (GTP), but this rapidly recoups itself by an enzyme-catalysed reaction with ATP so that the energy comes from ATP in the end:

(i) GTP → GDP + ℗ + energy,

(ii) GDP + ATP ⇌ (GTP + ADP).

And this is not all. ATP can be split in any of a considerable variety of ways. It may yield up its terminal phosphate group and the energy with which it is associated in a reaction of the kind we have just discussed, so that free energy is transferred into the phosphate acceptor: or, under the influence of an adenosine triphosphatase of some kind, it may undergo catalysed hydrolysis to yield ADP together with free inorganic phosphate. In the latter case, however, the free energy of the terminal phosphate group would not be conserved, either in whole or in part as it is in a transfer reaction, but is liberated. If the reaction is allowed to proceed in a test-tube or flask this free energy is dissipated mainly as heat, but in an intact muscle it appears largely in the form of the *mechanical energy* of the concomitant muscular contraction. It is probably not without significance, therefore, that the adenosine triphosphatase of muscle is identical with myosin, which is itself a constituent of the actual contractile machinery of the muscle cell and plays the part of a 'transformer' which converts chemical into mechanical energy.

In the electric organs of certain fishes, e.g. *Torpedo, Raia* spp., *Gymnotus, Malapterurus*, 'transformers' of a different kind are present, for myosin is lacking in the most highly developed of these organs, despite the fact that they arise by the morphological transformation of embryonic pre-muscular cells. The *electrical energy* dissipated when these organs discharge arises, according to present knowledge, from ATP once again. The *light energy* dissipated by bioluminescent organisms such as the fire-fly (*Photinus pyralis*) similarly has its origin in ATP.

The amounts of ATP present in most tissues are not very large, and are out of all proportion to the energy turnover of these tissues. It follows, therefore, that ATP must be used and reformed over and over again if the many and diverse activities of the tissues are to be maintained. But, as we know today, *new supplies of* ~ ⑪ *arise in the course of many katabolic processes* and, indeed, it would appear that katabolism as a whole is directed above all to the generation of the richest possible harvest of new high-energy from low-energy phosphate groups. These can be transferred to ADP left after previous breakdown of ATP molecules. It is significant, however, that *newly generated* ~ ⑪ *cannot be utilized directly* for the performance of endergonic operations in the ordinary way: *they can be put to service only through the intermediation of the ADP/ATP system.* No major alternative system has so far been brought to light.

It would not be profitable to discuss the processes that lead to the generation of new ~ ⑪ until we have studied katabolic processes in some detail, but one simple example may not be out of place here. The synthesis of ATP from ADP calls for the provision of a large amount of free energy and this can be provided, for instance, by new ~ ⑪ generated in the course of the metabolic degradation of glucose or glycogen. One such group appears when one of the numerous intermediate products, glycerate-2-phosphate, undergoes dehydration under the influence of the enzyme enolase (phosphopyruvate hydratase) to yield *enol*-pyruvate phosphate:

$$
\begin{array}{ccc}
CH_2OH & & CH_2 \\
| & & \| \\
CHO⑪ & \rightarrow & CO \sim ⑪ + H_2O \\
| & & | \\
COOH & & COOH
\end{array}
$$

The removal of the elements of water from the glycerate phosphate results in structural alterations within the molecule. These are attended by a redistribution of the intrinsic free energy of the system, leading as it were to a 'concentration' of free energy in the neighbourhood of the phosphate group; in other words, to the generation of a new high-energy phosphate group. The free energy, together with the phosphate group, is then transferred to ADP and a new molecule of ATP is produced.

THE STORAGE OF HIGH-ENERGY PHOSPHATE

High-energy phosphate can be transferred from ATP to creatine, arginine and a number of other guanidine bases, to form 'phosphagens' and, in certain tissues, especially in muscle, high-energy phosphate is stored in the form of these compounds. This stored energy cannot be used directly, but only through the ADP/ATP system. It can, however, be transferred directly to ADP through the creatine kinase reaction, and the transfer is freely reversible.

When such a tissue enters into physiological activity, ATP is broken down and provides energy in the form characteristic of the particular organ. Unless or until new high-energy phosphate is generated rapidly enough to resynthesize ATP as fast as it is used up, the tissue can draw on the high energy phosphate groups of the phosphagen reserves, transferring these to ADP and thus regenerating ATP. Later, after activity has ceased, the metabolic generators continue to run for a short time, so that ATP continues to be produced and, reacting now with the free guanidine base, resynthesizes phosphagen and thus replenishes the energy store. These notions may be diagrammatically summarized in the so-called 'energy dynamo' of Fig. 13.

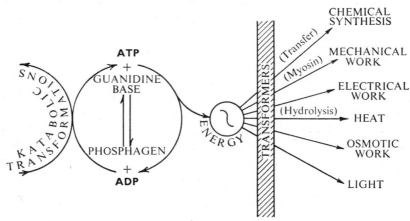

Fig. 13. The 'energy dynamo' (modified after Lipmann).

It is interesting and important to notice that the second high-energy phosphate group of ATP is not as a rule available as a direct source of energy for chemical synthesis, nor for the production of mechanical or electrical energy. It can, however, be rendered available through the activity of an enzyme, *adenylate kinase,* which specifically catalyses the transference of the terminal phosphate group from one molecule of ADP to a second, so that AMP (adenosine *mono*phosphate) and ATP are produced:

$$2A—\textcircled{P} \sim \textcircled{P} \rightleftharpoons A—\textcircled{P} + A—\textcircled{P} \sim \textcircled{P} \sim \textcircled{P}.$$

The new molecule of ATP can then be utilized in the usual manner.

SUMMARY

1. The breakdown and synthesis of biological materials may, but do not necessarily, follow a common chemical pathway. Whether the two pathways can be the same or must be different depends upon the energetic conditions that prevail.

2. Free energy can be transferred from one molecule to another and its transference is attendant upon that of a phosphate, a glycosidic or some other grouping.

3. The free energy required for anabolic processes and other energy-consuming operations arises from katabolism. Katabolic processes are so organized that they lead to the formation of new high-energy phosphate groups at the expense of the intrinsic free energy of their substrates.

4. The newly generated high-energy phosphate groups are 'captured' by transference to ADP so that new molecules of ATP are formed, and it is only through the intermediation of the ADP/ATP system that their energy can be utilized for biological purposes.

5. High-energy phosphate can be stored in certain tissues in the form of phosphagens.

6. The free energy of the terminal high-energy phosphate group of ATP can be utilized in one or another of the many energy-expending processes in which this remarkable substance participates. Among these are numerous 'priming' reactions which form essential preliminary steps in the biochemical synthesis of complex, higher-energy products from simpler starting materials of lower free-energy content.

4

HYDROLASES

GENERAL INTRODUCTION

The classification of enzymes presents a good many problems but a satisfactory though rather cumbersome system has recently been devised. Following a new internationally recommended system,† enzymes are classified under the following main headings:

1. Oxidoreductases.
2. Transferases.
3. Hydrolases.
4. Lyases.
5. Isomerases.
6. Ligases (synthetases).

Every known enzyme is allotted a number, an official title and an approved trivial name and, for the sake of brevity, these trivial names are used as a rule in this book.

We shall begin here with the hydrolases.

Hydrolases are enzymes which catalyse a hydrolytic splitting of their substrates:

$$R{:}R' + H.OH \rightleftharpoons R.OH + H.R'.$$

All digestive enzymes, whether intracellular or extracellular, fall into this class, and digestion itself may be regarded as an orderly series of hydrolytic reactions which result in the smooth, stepwise breakdown of large, complex food molecules into smaller and simpler products. Hydrolytic enzymes are capable of carrying through the processes of digestion from beginning to end without the aid or intervention of enzymes of other kinds. Apart from the digestive hydrolases there are numerous intracellular hydrolases concerned with processes other than digestion.

Certain hydrolases have been found to be capable of catalysing transfer reactions as well as hydrolysis, and this aspect of their activity will be dealt with in the next chapter.

In what follows we shall consider the properties of a number of the most widely distributed hydrolases. If the next few chapters savour somewhat of the catalogue it is because we must necessarily know a good deal about individual

† '*Enzyme Nomenclature*', *Recommendations of the International Union of Biochemistry* (1964). Elsevier: Amsterdam, London, New York. Published 1965.

enzymes before we can attempt to see how, in the living cell, tissue or organism, they are organized into the orderly catalytic systems that underlie the metabolic processes inseparable from life itself. In any case the 'catalogue' is not by any means complete for we shall deal here only with enzymes that are either very common, very important, or intriguing in one way or another.

Although reversibility signs are used in the generalized equations given here and in the next few chapters, there are many cases in which the reactions catalysed are, for energetic reasons, irreversible in practice, at any rate under biological conditions.

PEPTIDE HYDROLASES (PEPTIDASES)

As is well known, the chemical synthesis of peptides was, until about 1932, a difficult undertaking. Of the enormous variety of possible peptides, a mere handful was obtainable by synthetic chemistry. Our knowledge of the specificity of protein- and peptide-splitting enzymes was fragmentary in consequence. In more recent years, thanks at first to the ingenious methods introduced by Bergmann, peptides of many kinds hitherto unavailable have been produced and, in the meantime, a number of the enzymes themselves have been obtained in highly purified, crystalline form. It must again be emphasized at this point, and remembered hereafter, that *crystallinity is no criterion of purity where enzymes and other proteins are concerned*; they too commonly form mixed crystals while still far from chemically pure.

The older methods for the synthesis of peptides mostly involved covering the $-NH_2$ group of one amino-acid and condensing the acyl chloride of the protected product with a second amino-acid or its ester. A 'covered' dipeptide could thus be obtained from which, by removal of the covering group, the free dipeptide could theoretically be regenerated. Tri- and higher peptides could be prepared by further reactions based on the same procedure before removing the covering group. The use of the benzoyl group, introduced by Curtius, made possible the synthetic production of numerous benzoylated peptides, but attempts to remove the benzoyl group by hydrolysis resulted in a simultaneous hydrolysis of the peptide bonds, so that the yields of free peptides were negligible at best. The use of other substituent groups had little better success, and numerous other methods of synthesis have been tried.

In Bergmann's method the covering group employed is one which can be removed by reduction, a treatment which does not at the same time attack peptide links. Bergmann employed *benzyloxycarbonyl chloride*. This reagent is made by treating benzyl alcohol with phosgene in solution in toluene:

$$C_6H_5CH_2OH + COCl_2 = C_6H_5CH_2O.CO.Cl + HCl.$$

It reacts readily with the amino-group of an amino-acid to yield an *N-benzyloxycarbonyl derivative*, thus:

$$C_6H_5CH_2O.CO.Cl + H_2N.R = C_6H_5CH_2O.CO.HN.R + HCl.$$

61

Its subsequent removal can be accomplished by catalytic reduction with hydrogen in the presence of colloidal palladium, a treatment that is without action upon peptide bonds:

$$C_6H_5CH_2 \vdots O.CO \vdots HN.R + H_2 = \underset{toluene}{C_6H_5CH_3} + CO_2 + H_2N.R.$$

A newer alternative method is by treatment with HBr in glacial acetic acid. The benzyloxycarbonyl compounds are very stable, and can readily be converted into the corresponding acyl chlorides so as to facilitate condensation with a second amino-acid. There is, moreover, no racemization of the product under Bergmann's conditions.

As an example of a synthesis carried out by Bergmann's method we may take the simple case of the preparation of glycylglycine. The reactions originally used were as follows, though methodological improvements have been introduced in more recent times:

(i) $C_6H_5CH_2O.CO.Cl + H_2N.CH_2COOH = C_6H_5CH_2O.CO—HN.CH_2COOH + HCl$;
benzyloxycarbonylglycine

(ii) $C_6H_5CH_2O.CO—HN.CH_2COOH \xrightarrow{PCl_5} C_6H_5CH_2O.CO—HN.CH_2COCl$;

(iii) $C_6H_5CH_2O.CO—HN.CH_2COCl + H_2N.CH_2COO.C_2H_5$
glycine ethyl ester
$= C_6H_5CH_2O.CO—HN.CH_2CO—HN.CH_2COO.C_2H_5 + HCl$;
benzyloxycarbonylglycylglycine ethyl ester

(iv) $C_6H_5CH_2 \vdots O.CO - \vdots -HN.CH_2CO—HN.CH_2COO \vdots C_2H_5$

$\xrightarrow[\text{and reduction}]{\text{saponification}} C_6H_5CH_3 + CO_2 + H_2N.CH_2CO—HN.CH_2COOH + C_2H_5OH.$
glycylglycine

By taking suitable precautions it is possible to prepare peptides containing the dicarboxylic and dibasic amino-acids as well as mono-amino-mono-carboxylic acids by this method. With its aid peptides of many different kinds have been made available and a number of newer alternative synthetic methods have been devised.

Digestive peptidases. The proteolytic enzymes of vertebrates fall into two groups, the first of which is mainly concerned with the degradation of the large molecules of the food proteins to yield smaller fragments, the second group completing the processes initiated by the first and leading eventually to the liberation of free amino-acids. It is worthy of note that denatured proteins are more readily attacked than the native materials. The first group includes *pepsin*, which is formed from a precursor present in gastric juice, and *trypsin* and *chymotrypsin*, formed from precursors present in pancreatic juice. In the second group we have *carboxypeptidases*, contributed by the pancreatic juice, together with *aminopeptidases* and *dipeptidases*, which are present in the intestinal secretions, probably together with other carboxypeptidases.

Activation of peptide hydrolases

Four of these enzymes are actually secreted in the form of enzymatically inactive precursors which undergo activation by 'unmasking'. Pepsin is secreted by the gastric mucosa in the form of an inert pro-enzyme, *pepsinogen*, which is activated in the first instance by the hydrochloric acid of the gastric juice to yield pepsin itself. Pepsin, once formed, is capable of activating more pepsinogen, so that, once begun, activation is an autocatalytic process. Both pepsinogen and pepsin itself have been obtained in pure, crystalline form, and it has been shown that the conversion of the pro-enzyme into the active form is attended by a fall in molecular weight from some 42,000 to 38,000. A polypeptide of molecular weight 4000 or thereabouts is split off, and may be regarded as the 'masking' substance.

Trypsin and chymotrypsin are similarly secreted in the form of enzymatically inert precursors, *trypsinogen* and *chymotrypsinogen*. All four compounds have been crystallized and isolated in the pure state. Trypsinogen is activated by an enzyme called *enteropeptidase*, which is present in the intestinal secretions and acts upon trypsinogen to produce trypsin, which then activates more trypsinogen so that, as in the case of pepsinogen, once begun, activation is an autocatalytic process. In this case, however, there is a comparatively small change in molecular weight. The masking peptide has in fact been isolated and identified as valyl-(aspartyl)$_4$-lysine (mol.wt. = *ca.* 700). The carboxyl group of the lysine is engaged with the amino-group of an *iso*leucine residue in trypsin itself, and the activation process consists in the hydrolysis of the lysyl-*iso*leucyl bond.

Chymotrypsinogen differs from trypsinogen in that it is not activated by enteropeptidase. It is, however, activated by trypsin. Chymotrypsin does not activate chymotrypsinogen, and in this case, therefore, activation is not autocatalytic.

The activation of the enzyme precursors of pancreatic juice is thus started off by enteropeptidase, which activates trypsinogen with production of trypsin. The trypsin then activates more trypsinogen and chymotrypsinogen as well. The proteolytic enzymes of pancreatic juice do not therefore become active until they reach the small intestine and come into contact there with enteropeptidase, which fires off the entire activation process.

Of the other pancreatic peptidases only carboxypeptidase requires this kind of activation: the pro-enzyme is activated by trypsin, but not by enteropeptidase or chymotrypsin. Amino-peptidases and dipeptidases are not activated in this way. They lose their activity if dialysed, but activity is specifically restored by the addition of traces of specific metals such as Mn, Zn, Mg and occasionally even Co, which appear to be the natural activators for these enzymes. Some authorities consider that these metallic activators behave as loosely bound prosthetic groups for the enzymes with which they collaborate, and thus provide a means of attachment for the substrates by forming co-ordination compounds with them.

Specificity of peptide hydrolases

It was formerly held that pepsin, trypsin and chymotrypsin are able to attack proteins at more or less any point in the peptide chain. But if both pepsin and trypsin are allowed to act upon the same protein we find that both enzymes together open up more peptide linkages than either alone, and it follows that both enzymes do not act upon the same, but upon different peptide linkages. We now know a good deal about the nature of the particular bonds attacked by the various peptidases. One general fact may be emphasized at once: with certain exceptions, the peptidases as a whole act only upon normal peptide links, i.e. links formed between the α-amino- and α-carboxyl groups of L-amino-acids.

Pepsin can act only on peptide bonds of certain definite types. As Bergmann showed, it can attack a peptide link lying between an L-*dicarboxylic* and an L-*aromatic amino-acid*, given certain conditions. These are, first, that the second carboxyl group of the dicarboxylic acid residue must be free, and secondly, that there must not be a free amino-group in the immediate vicinity of the peptide linkage. Thus pepsin attacks benzyloxycarbonyl-L-glutamyl-:-L-tyrosine, glycyl-L-glutamyl-:-L-tyrosine and benzyloxycarbonyl-L-glutamyl-:-L-phenylalanine. The influence of the free γ-carboxyl group of the glutamic acid residue is neutralized if there is a free amino-group nearby, for L-glutamyl-L-tyrosine and benzyloxycarbonyl-L-glutamyl-L-tyrosine amide are resistant to pepsin. The resistance of benzyloxycarbonyl-L-glutamyl-L-tyrosine amide is not due solely to the fact that the α-carboxyl group of the tyrosine is covered, for benzyloxycarbonyl-L-glutamyl-:-L-tyrosyl-glycine is attacked, though more slowly than benzyloxycarbonyl-L-glutamyl-:-L-tyrosine. Replacement of the L-acids by their D-isomers makes the peptides resistant to pepsin, for benzyloxycarbonyl-D-glutamyl-L-tyrosine and benzyloxycarbonyl-L-glutamyl-D-phenylalanine are not attacked. These results are summarised in Table 4.

Table 4. *Action of pepsin upon synthetic peptides*

Substrate	Action of pepsin
Benzyloxycarbonyl-L-glutamyl-:-L-tyrosine	+
Glycyl-L-glutamyl-:-L-tyrosine	+
Benzyloxycarbonyl-L-glutamyl-:-L-phenylalanine	+
L-Glutamyl-L-tyrosine	−
Benzyloxycarbonyl-L-glutamyl-L-tyrosine amide	−
Benzyloxycarbonyl-L-glutamyl-:-L-tyrosyl-glycine	+
Benzyloxycarbonyl-D-glutamyl-L-tyrosine	−
Benzyloxycarbonyl-L-glutamyl-D-phenylalanine	−

Chymotrypsin resembles pepsin in attacking peptide links in which *aromatic amino-acids* are involved but, whereas pepsin attacks on the amino side of the aromatic acid, chymotrypsin acts on the carboxyl side. Thus both enzymes attack benzyloxycarbonyl-L-glutamyl-:-L-tyrosyl-:-glycine amide, but do so at different points, as follows:

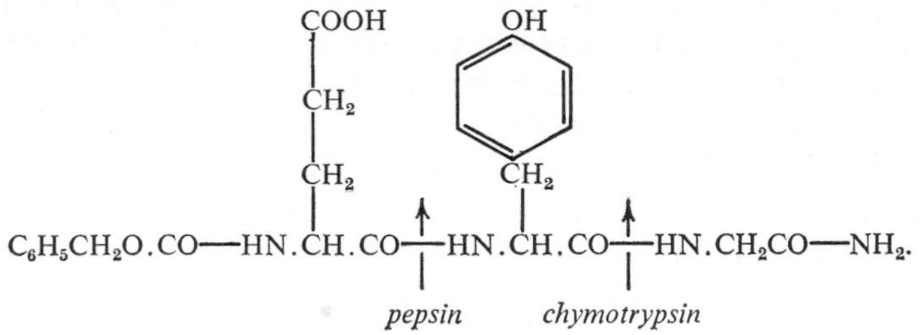

$$\underset{pepsin \qquad chymotrypsin}{C_6H_5CH_2O.CO—HN.CH.CO\uparrow HN.CH.CO\uparrow HN.CH_2CO—NH_2.}$$

Chymotrypsin also attacks benzyloxycarbonyl-L-tyrosyl-:-glycine amide and benzyloxycarbonyl-L-phenylalanyl-:-glycine amide, for example, but its action is prevented by the presence of a free carboxyl group in the immediate vicinity of the peptide link. Thus benzyloxycarbonyl-L-glutamyl-L-tyrosyl-glycine, un-like its amide, is resistant to chymotrypsin, though acted upon by pepsin. On the other hand, L-glutamyl-L-tyrosyl-:-glycine amide is attacked by chymo-trypsin and not by pepsin, since the effect of the free γ-carboxyl group of the glutamyl unit, which is required for the activity of pepsin, is neutralized by the free α-amino-group of the same amino-acid unit. These results are sum-marized in Table 5.

Table 5. *Action of pepsin and chymotrypsin upon synthetic peptides*

	Action of	
Substrate	Pepsin	Chymo-trypsin
Benzyloxycarbonyl-L-glutamyl-:-L-tyrosyl-:-glycine amide	+	+
Benzyloxycarbonyl-L-tyrosyl-:-glycine amide	−	+
Benzyloxycarbonyl-L-phenylalanyl-:-glycine amide	−	+
Benzyloxycarbonyl-L-glutamyl-:-L-tyrosyl-glycine	+	−
L-Glutamyl-L-tyrosyl-:-glycine amide	−	+

Trypsin can act at peptide linkages adjacent to either an *arginine* or a *lysine* unit and replacement of these basic amino-acid residues by others yields resistant products. The second amino-group of the dibasic amino-acid unit must be unsubstituted, for trypsin acts upon α-benzoyl-L-arginine:amide and α-benzoylglycyl-L-lysine:amide, for example, but not upon α-benzoylglycyl-(ϵ-benzyloxycarbonyl-)-L-lysine amide, in which both the amino-groups of the lysine unit are covered.

To sum up, we may say that *pepsin can act at peptide links formed between the α-carboxyl group of a dicarboxylic amino-acid and the α-amino group of an aromatic amino-acid,* but requires that the second acidic group of the dicarboxylic acid shall be free, and is inhibited if there is an amino-group nearby. *Chymotrypsin can act upon peptide bonds formed from the α-carboxyl group of an aromatic amino-acid,* but is inhibited if there is a carboxyl group in the immediate vicinity, while *trypsin can act upon peptide links formed from the α-carboxyl group either of arginine or of lysine,* but requires that the second amino-group of the dibasic amino-acid unit shall be free. This enzyme appears to be inhibited by nearby α-amino or carboxyl groups.

It must not be supposed, however, that these enzymes cannot necessarily attack peptide linkages other than those discovered by Bergmann, for in the case of pepsin it has been shown that while an aromatic amino-acid is required, the free carboxyl grouping of glutamic acid can be replaced by the sulphydryl of cysteine. Both tyrosyl-cysteine and cysteinyl-tyrosine, for example, were attacked by pepsin, though more slowly than the corresponding *N*-benzyloxycarbonyl-derivatives. Even so, the fact remains that pepsin cannot act upon any arbitrary peptide linkage, but is restricted, probably, to linkages of only a few special types. Clearly, therefore, these peptidases are very exacting indeed and, contrary to earlier opinion, able to act only upon bonds of certain types. The specificity requirements established by Bergmann are summarized in Fig. 14.

By their concerted action these peptidases divide intact protein molecules into smaller fragments, and the stage is set for the action of members of a second group of peptidases, viz. carboxypeptidases and aminopeptidases. These enzymes catalyse the splitting only of terminal peptide bonds, with consequent liberation of the terminal amino-acid units. *Carboxypeptidases* remove the terminal unit of which the carboxyl group is free, *aminopeptidases* acting at the other end of the chain, where the terminal unit has a free amino-group. Thus in L-leucyl-glycyl-L-tyrosine, for example, we have the following structural arrangement:

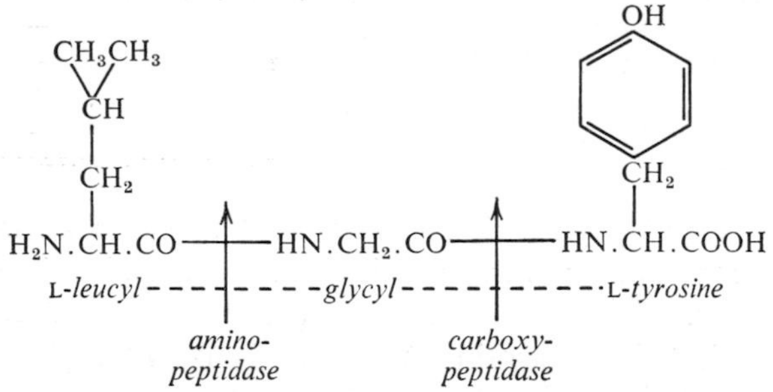

The appropriate carboxypeptidase acts upon this tripeptide to liberate tyrosine, and aminopeptidase to produce free leucine. Carboxypeptidases require a free carboxyl group for their action to take effect, aminopeptidases requiring a free amino-group; but carboxypeptidases are unable to act if there is a free amino-group nearby, while aminopeptidases are similarly affected by a free carboxyl group. Neither type of enzyme, therefore, can attack the dipeptide left after the other has attacked the original tripeptide, nor will either group attack peptides containing amino-acid residues of the D-series.

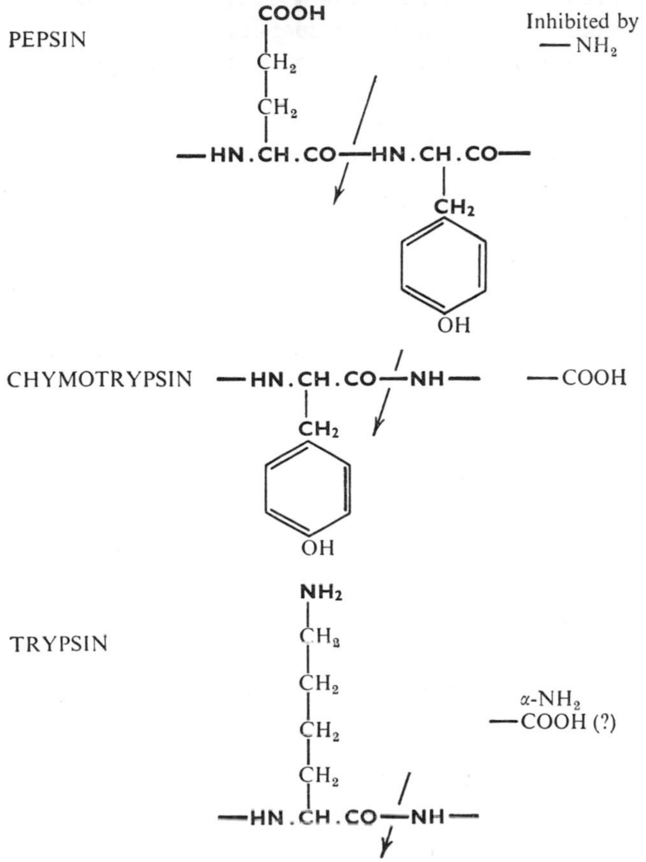

Fig. 14. Specificity requirements of endopeptidases. The essential requirements are printed in heavy type: the point of attack is indicated by an arrow in each case.

In these peptidases, then, we can recognize specificity requirements which include the presence of the right terminal group, which must be unsubstituted, and the absence from the immediate vicinity of electrically opposite groups, together with the usual stereochemical requirements. It is now certain, however, that the requirements of many peptidases are more exacting even than this, and that there exist more than one amino- and more than one carboxy-peptidase, each with specific end-group requirements.

Whereas pepsin, trypsin and chymotrypsin split large protein units into smaller fragments and produce few free amino-acid molecules in the process, the amino- and carboxypeptidases liberate free terminal amino-acid units one by one until only dipeptides remain. These, as we have seen, are not further attacked by these enzymes but are split in their turn by *dipeptidases*. Probably there are several of these enzymes collectively capable of covering the required range of specificities. For example, glycylglycine is hydrolysed by a specific dipeptidase which is rather specifically activated by Co; Mn can replace Co but is very much less effective. Again, a specific *prolidase* has been described: this attacks the bond linking the imino-group of proline or hydroxyproline to an adjacent carboxyl group and is activated by Mn. Dipeptidases are activated by metals and require that the constituent amino-acids of their substrates shall be members of the L-series, so that, of the four possible alanyl-leucines, L-L-, L-D-, D-D-, and D-L-, only one is attacked by a depeptidase, namely L-alanyl-L-leucine.

Finally, it is interesting that some peptidases have considerable esterase activity; this is not due to contamination, for esterase activity has been observed in highly purified, crystalline trypsin, chymotrypsin and carboxypeptidase.

Rennin. In the gastric juice of young mammals we find another proteolytic enzyme, rennin. Like pepsin, this enzyme is secreted in the form of an inactive precursor, *pro-rennin*, which is activated by hydrochloric acid. The optimum pH for activation is considerably higher than that for pepsin. The most characteristic feature of rennin is its milk-clotting power, and it is, in fact, the active principle of commercial preparations of 'rennet'. It catalyses the conversion of milk casein ('caseinogen') into another product, paracasein ('casein'), the calcium salt of which is insoluble so that, in the presence of the calcium of the milk, a clot or curd is formed. Rennin from the abomasum (fourth stomach) of the calf has been obtained in crystalline form, but little seems to be known about its specificity requirements. Like pepsin, rennin has proteolytic properties, but with a more alkaline pH optimum: its optimal pH when acting upon haemoglobin, for example, is 3·7 as against about 2·0 for pepsin.

Intracellular peptidases

The presence of intracellular enzymes capable of catalysing the hydrolysis of peptides of greater or less chemical complexity and molecular weight has been demonstrated in many animal and plant materials. In many cases these enzymes probably have digestive functions, especially among the lower animals, many of which produce no digestive secretions but take in particulate food by phagocytosis and digest it intracellularly. There is a considerable literature on this subject, and it may be said that enzymes resembling pepsin and trypsin, at least in their general properties, are present in some cells, but not much is known about them.

Other intracellular proteolytic enzymes include powerful plant enzymes such as papain, ficin and bromelin, obtained from the sap or latex of the paw-paw, fig and pineapple respectively. All three have many features in common. More is known about the intracellular, autolytic enzymes of animal tissues, known collectively as *kathepsin*. Kidney and spleen offer good sources of kathepsin, but it is present in many other organs and also in tumours of various kinds. Kidney and spleen kathepsins comprise at least four components, the specificity of each of which has been studied. Of these, kathepsin I is homospecific with, i.e. has the same specificity requirements as, pepsin; kathepsin II is homospecific with trypsin, while components III and IV are homospecific with aminopeptidase and carboxypeptidase respectively. The existence of these homospecificities suggests that the extracellular digestive peptidases may have had their evolutionary origin in intracellular, kathepsin-like enzymes.

The kathepsins are not activated by unmasking nor yet by heavy metals such as Co or Mn, as are the digestive peptidases. On the contrary, many intracellular peptidases are inactivated by heavy metals, but can be re-activated by the addition of cyanide, hydrogen sulphide, cysteine, glutathione and sometimes by ascorbic acid. The inhibitory action of heavy metals is probably due to their tendency to react with and block free —SH groups.

We have seen that the extracellular, digestive peptidases constitute a kit of tools whereby the proteins of the food can be completely dismantled, and, if it could be shown that the action of enzymes of this kind is reversible, we should feel more confident that the kathepsins, with which they are homospecific, constitute an outfit capable of reconstituting as well as degrading proteins. Bergmann in fact succeeded in demonstrating synthetic activity on the part of several peptidases and has shown, for example, that chymotrypsin catalyses the condensation of benzoyl-L-tyrosine with glycylanilide to yield benzoyl-L-tyrosyl-glycylanilide. The product in this case is insoluble and is precipitated, so that the hydrolytic action of the enzyme does not seriously oppose its synthetic performance. Other syntheses have been accomplished with, for example, papain-cysteine, i.e. papain activated by the addition of cysteine (an example can be seen in Fig. 3 (p. 17)), but it is doubtful in the extreme whether these enzymes contribute anything to the synthesis of proteins, at any rate in animal tissues.

Other peptidases include *carnosinase*, a dipeptidase that can be obtained from pig kidney. It acts upon the thoroughly atypical dipeptide carnosine (β-alanyl histidine) and on a few other peptides containing histidine. A similar *anserinase*, present in fish muscle, similarly hydrolyses anserine, a methylated carnosine. Of rather peculiar interest is the κ-toxin of the gas gangrene organism, *Clostridium welchii*. This toxin is a *collagenase* which, by digesting connective tissue, enables the organism to spread rapidly and extensively into the adjacent tissues. Apart from collagen this toxin is only known to attack gelatin, which is itself derived from collagen. Similar enzymic toxins are produced by other *Clostridia* also.

GLYCOSIDE HYDROLASES (CARBOHYDRASES)

Enzymes capable of catalysing the breakdown of carbohydrates are very widely distributed indeed, and occur both in digestive secretions and within the cells of animals, plants, and micro-organisms of many and perhaps all kinds. They may be considered under two main headings, the *polysaccharases*, which act upon the large molecules of polysaccharides such as starch and glycogen, and *glycosidases*, the substrates of which are smaller molecules such as di- and trisaccharides, in addition to simple glycosides of many other kinds.

Polysaccharases

Amylases

Most is known about the amylases, which act upon starch and glycogen but not upon cellulose. Plant amylases have been resolved into components known respectively as α- and β-amylases, and representatives of both types have been crystallized and highly purified. In the ordinary course of events these enzymes act jointly upon starch and glycogen to catalyse a more or less quantitative conversion into the disaccharide, maltose. They act, however, in rather a different manner and in order to appreciate the differences it is necessary to have in mind a fairly clear picture of the probable structure of starch and glycogen. In considering the modes of action of the α- and β-amylases, however, it is important to realize that the use of the prefixes α- and β- is *not* meant to imply that these enzymes act upon α- and β-glucosidic links respectively. Purification of these two amylases has shown that they are accompanied by other enzymes, which also play a part in the total digestion of starch, and their names are given in Table 6, together with the identities of the linkages upon which they are believed to act. Starch consists of a mixture of two main components, amylose, which accounts for 20–25 % of most vegetable starches, and amylopectin.

Amylose, which gives a pure blue coloration with iodine, consists mainly of 1:4-α-linked glucose units and can be attacked either by the α- or β-

Table 6. *Linkages attacked by amylolytic enzymes*

Enzyme	Linkages attacked
α-Amylase	1:4-α-
β-Amylase	1:4-α-
Dextrin-1:6-glucosidase R-enzyme	1:6-α-

amylases. Pure crystalline β-amylase, which acts at the non-reducing ends of the 1:4-α-linked glucose chains, catalyses a 70% conversion of natural amylose into maltose, but complete conversion requires the assistance of another enzyme, which is usually found in association with the amylases. This enzyme, dextrin-1:6-glucosidase, has no action upon 1:4-α-links but is known to act upon 1:6-α-links. It may therefore be concluded that amylose, while it consists mainly of 1:4-α-glucosidic units, contains a few 1:6-α-linkages as well. The whole molecule contains a large number of 1:4-α-glucosidic

Fig. 15. Diagram to illustrate structure of glycogen and amylopectin.

units, the number varying with the source from which the amylose is prepared, and probably has a long, coiled and slightly branched structure, branching taking place at α-1:6-links.

Amylopectin, on the other hand, is extensively branched and built up from inter-linked chains of α-glucosidic units, the average chain length being of the order of 20–24 1:4-α-linked glucosidic units. The terminal, reducing unit of one chain is united to a hydroxyl of a glucose unit in an adjacent chain, usually in the 6-position, for these linkages are susceptible to the action of *dextrin-1:6-glucosidase*. The structure of amylopectin is diagrammatically represented in Fig. 15. The molecular structure in the neighbourhood of a branching point, whether in amylose or in amylopectin, is as shown in Fig. 16.

Glycogen appears to contain no component analogous to amylose. It resembles amylopectin in structure except that the chains contain an average of only about twelve 1:4-α-linked glucosidic units. Whereas amylopectin gives a purplish coloration with iodine, different glycogens give different colorations; some give none at all, but the usual response to the iodine test is a reddish brown.

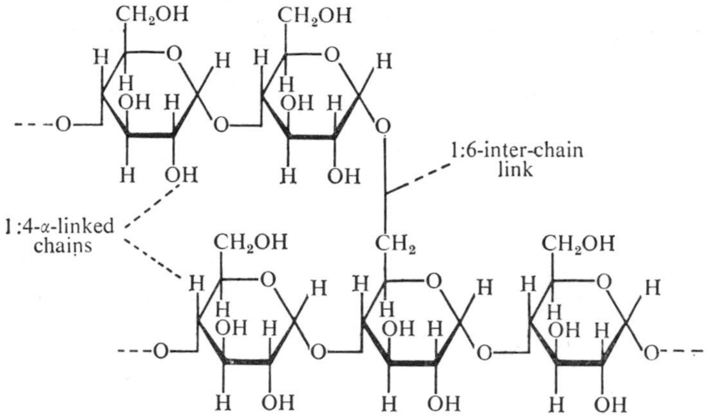

Fig. 16. Structure of a branching point in amylopectin or glycogen.

β-Amylase ('maltogenic amylase'), like the α-enzyme, appears to be absolutely specific for 1:4-α-linkages and acts freely upon amylose until a 1:6-α-link is approached, removing pairs of glucose units in the form of maltose. Similarly, it acts upon the open ends of the chains of amylopectin and glycogen, again until a branching link (1:6-α-link) is approached, and in this way about 55% of the amylopectin is converted into maltose (glucose-1:4-α-glucoside). The residual resistant product, the so-called 'limit dextrin', gives a port-wine colour with iodine. This dextrin, which is resistant to β-amylase, can be attacked by dextrin-1:6-glucosidase. It can also be further attacked by α-amylase which, by attacking the 'limit dextrin' at internal points in its structure, splits it into smaller fragments which are still of a dextrin-like nature but no longer give any colour with iodine.

α-Amylase received the alternative title of 'dextrinogenic amylase' because acting alone upon amylopectin, it yields only comparatively small dextrin-like products in the first instance, and these can then be further attacked by the β- or 'maltogenic' enzyme. But the activity of α-amylase is not much affected by the presence of linkages other than the 1:4-α-links and can 'straddle' the branching linkages of amylopectin. If allowed to act for long periods, it degrades some 80–90% of amylopectin. The main product is maltose but, since 1:6-linkages are not attacked by this enzyme, appreciable quantities of 1:6-linked sugars, especially *iso*maltose (glucose-1:6-α-glucoside) are also formed

72

at this stage, the amounts depending on the degree of branching of the parent polysaccharide.

The fission of the 1:6-branching links requires the presence of another hydrolytic enzyme, *dextrin*-1:6-*glucosidase*. This enzyme has been obtained from muscle, and closely similar enzymes have been obtained from beans and potatoes (R-enzyme). The total breakdown of starch in plants is thus a somewhat complex process in which three enzymes at least are involved, viz. the α- and β-amylases together with a dextrin-1:6-glucosidase. The chief product, maltose, is accompanied by smaller amounts of glucose.

The digestive (salivary and pancreatic) amylases of animals have not been resolved into α- and β-components. They are apparently single enzymes which act in the same sort of way as the dextrinogenic α-amylases of plants. They can carry out an extensive conversion of amylose, amylopectin and glycogen into maltose, together with small amounts of free glucose. Representatives of the group have been crystallized and highly purified and, according to one report, purified salivary amylase catalyses a large-scale (> 90 %) conversion of potato amylose into maltose and maltotriose in a ratio of 2·3:1, but stops acting when this has been achieved. Pancreatic amylase, however, is said to hydrolyse maltotriose to maltose and glucose and presumably contains other enzymic components. Thus although maltose is the main product of digestion by these two enzymes, some free glucose is also produced.

It has long been known that salivary and pancreatic amylases lose their activity if dialysed, and the case is often quoted in illustration of the importance of inorganic ions in the activation of certain enzymes. This, however, is not entirely a true bill, for while it is sure enough that dialysed salivary amylase, for example, is no longer active under conditions which were formerly optimal, it is still weakly active at more acid pH values. There is, it is also true, a large-scale loss of total activity, but the removal of chloride ions does not deprive the enzyme of its power to activate its substrate, but only appears to alter it in some way. In all probability the enzyme dissociates in a different manner in the absence of chloride ions (see p. 34) and, if this is so, it might perhaps be expected that the replacement of chloride by other ions would lead to a displacement of the optimum pH. This does in fact happen, as is shown by the curves of Fig. 10 (p. 35).

Although the enzymes that catalyse the digestive hydrolysis of starch and glycogen must, on theoretical grounds, be regarded as capable of working reversibly, it has not proved possible to synthesize polysaccharides with their aid because the synthesis is essentially an endergonic process. Indeed, as we shall see later on, the synthesis of amylose, amylopectin and glycogen proceeds by a quite different route and involves a group of quite different enzymes.

Cellulase

Although cellulose forms a very large part of the food of herbivores, remarkably few animals of any kind possess any enzyme or enzymes capable of catalysing its hydrolysis. Being built up from $1:4$-β-glucose units as it is, cellulose is resistant to the action of the usual amylases. However, cellulose-splitting enzymes have been described in the digestive secretions of a number of herbivorous gastropod snails, including terrestrial forms like *Helix* and aquatic species such as *Strombus*, *Pterocera* and *Aplysia*; cellulases appear to be present also in the digestive juices of silverfish and a few wood-eating insects. The shipworm, *Teredo*, is an interesting creature from this point of view, for its digestive gland contains cells which appear to be specialized for the phagocytic ingestion and intracellular digestion of the fine particles of wood which the animal scrapes off as it bores. But in the vast majority of cases it is nevertheless true that animals do not produce cellulases, even when they depend largely upon cellulose as a primary source of food.

The explanation of this paradox is that most cellulose-eating animals do not digest cellulose for themselves but maintain in their alimentary canals large populations of symbiotic micro-organisms, including bacteria, protozoa and yeasts, which play a very important part in their nutrition. Indeed, it has even been claimed that the snail's cellulase is not a product of the snail itself but is formed by symbiotic inhabitants of the crop and the intestine. Many intestinal micro-organisms are capable of degrading cellulose, and work on the processes of digestion in ruminants has shown that cellulose is broken down by the symbionts, with production of large amounts of lower fatty acids. Acetic and propionic acids predominate, and are accompanied by formic, butyric and valeric acids, together with large volumes of carbon dioxide, methane and hydrogen. The ruminant thus obtains fatty acids rather than sugars from the cellulose it consumes, and has to reward its micro-organisms by providing them with living space in the form of a capacious caecum, a multiple stomach, or some other commodious dilatation of the alimentary canal. Similar processes take place in many plant-eating insects and other herbivores, from cows to cockroaches.

Little is known about the enzymes whereby these symbiotic organisms break down cellulose. Presumably they must include a cellulase, and it has indeed been shown that the free-living protozoan, *Vampyrella*, secretes an extracellular cellulase which attacks the cellulose of the cell walls of the *Spirogyra* which furnishes its food. Cellulases have also been found in culture media in which such cellulolytic bacteria as *Cellulomonas* and *Clostridium thermocellum* have been grown, and the secretion of extracellular cellulases probably plays an important part in the early stages of microbial attack upon cellulose, which is so insoluble as to require some kind of extracellular comminution before it can be got into the cells for further chemical manipulation.

74

Cellulose-splitting enzymes and enzymes capable of synthesizing cellulose certainly occur in many plants and fungi, and must be of very great importance in plant economy, but we have very little information about them.

It is said that preparations containing cellulase will also act upon the animal polysaccharide, *chitin*, in which the β-linked glucose groups of cellulose are replaced by similarly linked units of *N*-acetyl glucosamine. Whether or not cellulase and chitinase are identical is uncertain.

Other polysaccharases include enzymes capable of splitting polyfructofuranosides, such as inulin and levan, and enzymes that act upon polysaccharides such as the mannans, pectins and so on.

Glycosidases

We know of many enzymes capable of splitting simple glycosides, all of which show a high order of specificity towards the glycosidic part of the substrate molecule. Of these the most common are the glucosidases and the β-fructofuranosidases (saccharases, sucrases, invertases). Some of these enzymes can catalyse transfer reactions as well as hydrolyses.

α-Glucosidases

α-Glucosidases of two types are known. The most specific of these are the '*true*' *maltases*, which act only upon a single α-glucoside, viz. maltose (glucose-4-:-α-glucoside). Enzymes of this kind are thought to occur in malt and in *Aspergillus*. The more widely distributed digestive 'maltases' of animals, and the 'maltase' of yeast, are able to act upon α-glucosides other than maltose and should therefore be called *α-glucosidases* rather than maltases. These enzymes will act upon such substances as methyl-:-α-glucoside and sucrose (β-fructofuranosido-:-α-glucoside), though they are without action upon the α-glucosidic linkages of the much larger molecules of starch and glycogen.

The order of specificity of these α-glucosidases is very high, for a completely unmodified α-glucosido-group is required in their substrates: β-glucosides, α-galactosides, α-xylosides and α-*iso*-rhamnosides are not attacked, in spite of their close structural resemblances to the α-glucosides (Fig. 17).

β-Glucosidases

Many cells and tissues have been shown to contain β-glucosidases. The classical source of such an enzyme is the 'emulsin' of bitter almonds. The latter contain amygdalin, a β-glucoside which, when attacked by the β-glucosidase component of emulsin yields mandelonitrile ($C_6H_5CH(OH).CN$), which is then attacked by another enzyme (mandelonitrilase) to give benzaldehyde and free hydrocyanic acid. Sweet almonds also contain these enzymes, but amygdalin is absent.

β-Glucosides are as common in nature, especially in plant materials, as their α-counterparts are rare, and among the more interesting of these we may mention salicin (salicyl-:-β-glucoside) and a β-glucoside of indoxyl which occurs in the indigo plant and gives rise, after hydrolysis and oxidation, to natural indigo (woad). There are many others. In addition, a number of simpler saccharides are β-glucosides, notably cellobiose (glucose-4-:-β-glucoside), which stands in the same relation to cellulose as maltose does to starch and glycogen, and gentiobiose (glucose-6-:-β-glucoside), which occurs naturally in combination with mandelonitrile in the form of amygdalin.

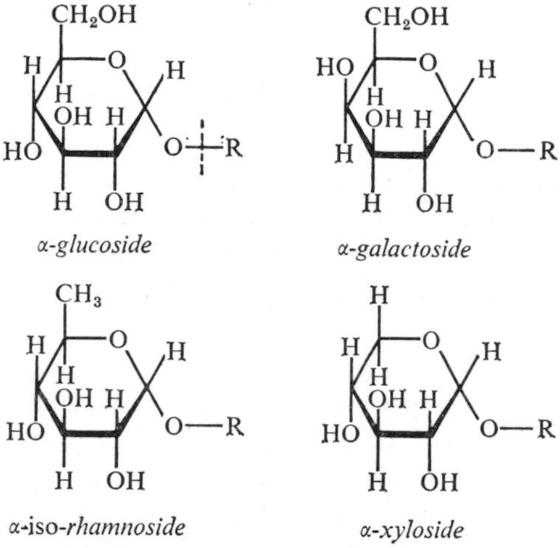

Fig. 17. Examples of glycoside structures.

β-Glucosidases obtained from different plant and animal sources have much in common. In particular, their specificity, though high, is somewhat less marked than that of the α-glucosidases. Although 'cellobiases' have been described from time to time there is no reason to think that they are comparable with the 'true' maltases of malt and *Aspergillus*, but rather that they are all group-specific β-glucosidases. The specificity requirements of the β-glucosidases do not extend as far as carbon 4, for some β-galactosides are also split. Modifications can also be made at position 6, for β-*iso*-rhamnosides and β-xylosides are split too, though less rapidly than the normal substrates.

Saccharases

Saccharases are very widely distributed indeed in the digestive secretions of animals, in plants and in many micro-organisms, though not, apparently, in the cell contents of animals. Two types can be distinguished, the *glucosaccharases* (in digestive secretions of animals and in *Aspergillus*) and the *fructosaccharases* (in yeast).

Sucrose itself is α-glucopyranosido-:-β-fructofuranoside, and the molecule may be attacked from either end. The glucosaccharases appear to be α-glucosidases, while highly purified yeast fructosaccharase is able to attack methyl-:-β-fructofuranoside as well as sucrose and is therefore a β-fructofuranosidase.

The differences in specificity between these types of saccharases are shown even more clearly by the fact that β-fructofuranosidases attack the trisaccharide raffinose (β-fructofuranosido-:-α-glucosido-6-α-galactoside), which occurs in sugar-beet and molasses, while α-glucosidases do not. Another trisaccharide, melezitose (melicitose: α-glucosido-β-fructofuranosido-6-:-α-glucoside), present in some kinds of manna, is split by α-glucosidases and not by β-fructofuranosidases.

Other glycosidases are known. They include α-galactosidases, which attack compounds such as raffinose (β-fructofuranosido-α-glucosido-6-:-α-galactoside), and β-galactosidases which act upon compounds such as lactose (glucose-4-:-β-galactoside). Brewers' yeast, for example, contains an α-galactosidase ('melibiase'), while bakers' yeast, *Aspergillus* and the digestive secretions of animals contain β-galactosidases ('lactases'). Mannosidases also are known.

All of the glycosidases mentioned so far are specific towards 'oxygen glycosides' of the type: $\qquad$ R—O—R′,

where R′ represents a glycosyl group. But there exists another group which attack 'nitrogen glycosides': $\qquad$ R—N—R′.

Here R′ represents a nitrogenous base, usually a purine or a pyrimidine. These '*N-glycosidases*' play an important part in the metabolism of nucleotides and will be mentioned again in that connection.

Finally it should be noticed once more that many glycosidases can catalyse *transfer reactions* (pp. 85–8).

Reversible action of glycosidases

A number of β-glucosides have been synthesized enzymatically with the aid of a β-glucosidase, but to achieve such syntheses it is usually necessary to employ conditions which, from the biological point of view, are grossly artificial. Thus, in the synthesis of gentiobiose, for example, concentrated solutions of the reactants are taken in order to force the reaction backwards in accordance with the principle of the mass law. β-Methyl-D-glucoside, again, has been synthesized by a β-glucosidase acting in a non-aqueous medium (dioxane). Claims have been made for the synthesis of sucrose from strong solutions of glucose and fructose, but these seem to be less well supported by the experimental evidence.

Whereas the amounts of sugars such as gentiobiose are fairly large under equilibrium conditions, the amounts of sucrose present at equilibrium are certainly small, and it seems certain that the free-energy conditions favour the hydrolysis rather than the synthesis of sucrose, just as is the case in the breakdown and synthesis of starch and glycogen from glucose. Sucrose is, nevertheless, synthesized on a large scale by many plants, but the process is not catalysed by saccharase with glucose and fructose as starting materials; initial priming reactions are necessary as we shall see (pp. 91, 329).

ESTER HYDROLASES

Under this heading we have to consider several distinct groups of enzymes. There are, first of all, the very widely distributed *lipases* and *esterases* which catalyse the hydrolysis of esters of alcohols with organic acids and, in addition, several groups which catalyse the hydrolysis of esters formed between alcohols and inorganic acids. Of these the *phosphatases* may first be mentioned, phosphatases of various kinds being almost universally distributed.

There are sharp differences of specificity with respect to the acidic component of the ester. Phosphatases are group-specific as a rule, and act only upon organic esters of phosphoric acid such, for example, as glycerol phosphate and glucose-6-phosphate, and are without action upon organic sulphates or esters of organic acids. The carboxylesterases on the other hand are enzymes of relatively low specificity. They act only upon esters of carboxylic acids, but provided that the acid is organic in nature, its precise chemical identity is a matter of relative indifference. The true lipases act preferentially upon esters of glycerol and there exist a number of other, on the whole more specific esterases, e.g. aryl- and acyl-esterases.

Lipases

Lipases occur in the gastric, pancreatic and intestinal juices of vertebrates as well as in liver and adipose tissue. They play an important part in the absorption, transport and storage of lipid material; these aspects will be discussed in a later chapter. Lipases have also been reported in the digestive secretions of many invertebrates. They also occur in plants, especially in seeds, and have been found in other organisms such as *Aspergillus* as well. Their specificity is low, and any lipase will split virtually any organic ester of glycerol. The acid component may be anything from acetic to palmitic, stearic, or even higher fatty acids, and may be saturated or unsaturated.

Pancreatic lipase acts only upon *emulsified* esters and has no action upon esters in solution. Gastric lipase and the lipase of the castor-oil bean (*Ricinus*) resemble the pancreatic enzyme in this respect, but the liver enzyme has no action on ester emulsions but acts only upon *dissolved* esters. Only short-chain esters are sufficiently soluble to be attacked by the liver lipase (esterase); long-chain esters are too insoluble for attack by this enzyme but are readily emulsified under suitable conditions and so are liable to attack by the pancreatic enzyme.

78

Although their specificity towards the structure of the substrate is so low, the lipases show marked stereochemical specificity when allowed to act upon esters containing an asymmetric carbon atom, such, for example, as the ethyl esters of the mandelic acids, $C_6H_5\overset{*}{C}H(OH)COOH$. One isomer is always more rapidly attacked than the other in cases of this kind, whether the asymmetric carbon atom is situated in the alcoholic or in the acidic component of the ester.

An interesting feature of some lipases is the readiness with which they will catalyse the synthesis as well as the hydrolysis of fats and esters though it is doubtful whether this phenomenon has any great biological significance. If a fat such as triolein is incubated under suitable conditions with say, *Ricinus* lipase, the reaction does not go to completion. Instead, an equilibrium is approached, and the composition of the final reaction mixture is the same from whichever side it is approached. Fig. 18 shows the results of experiments on the action of *Ricinus* (castor-oil bean) lipase on triolein, in one case, and upon a mixture of glycerol and oleic acid in the other.

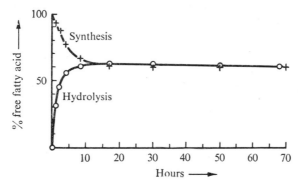

Fig. 18. Hydrolysis and synthesis of triolein by *Ricinus* lipase.
(After Parsons; data of Armstrong & Gosney.)

The three fatty acid groups of a typical triglyceride are not removed simultaneously, but come off one after another under the influence of digestive lipases. The outer ester links are preferentially attacked but that in the β-position is very resistant to hydrolysis, so that free fatty acids and monoglycerides are produced. As we shall see, this is important in connexion with the absorption of fatty food materials from the small intestine in animals.

Mention must also be made of a curious enzyme found in the venoms of the cobra and the rattlesnake. This acts upon lecithins, in which, it will be recalled, two of the alcoholic groups of glycerol are esterified by fatty acids and the third by phosphocholine. The ordinary digestive phospholipases (phospholipase B) are able to remove both of the fatty acid groups, but the so-called *phospholipase* A (lecithinase A) of these snake venoms, e.g. crotoxin from the rattlesnake, removes the terminal fatty acid only, and yields a product in which the remaining fatty acid is usually of the unsaturated type.

This compound is called *lysolecithin*, on account of its powerful haemolytic action. The undesirable consequences of being bitten by these reptiles include a large-scale lysis of the red blood cells and consequent disturbances of the respiratory functions of the blood. The venom of the viper, by contrast, causes intravenous clotting of the blood, a process which is catalysed by a proteolytic rather than a lipolytic enzyme. The toxins formed by the gas-gangrene organisms, the *Clostridia*, include lecithinases which, however, differ from the rattlesnake enzyme both in specificity and mode of attack but lead again to haemolysis. Several other phospholipases are also known including the phospholipase B of liver; this enzyme can remove both fatty acids from lecithin or the single residual fatty acid from lysolecithin. All these are important in the intermediary metabolism of lipids and will be mentioned again, along with some others, in a later chapter.

Other esterases of more specific nature include a group of *choline esterases*, enzymes of great importance in animal tissues, in which they catalyse the hydrolysis of acetylcholine, the neuro-hormone of the parasympathetic nervous system.

Phosphatases appear to be almost universally distributed. There appear to be a number of distinct types, two of which are especially common and may be termed phosphomonoesterhydrolases and phosphodiesterhydrolases, catalysing the hydrolysis of substances of the following general types:

monoesterase type   *diesterase type*

$$\text{R-O-}\!\!\mid\!\!-\overset{\displaystyle \text{O}}{\overset{\|}{\underset{\underset{\text{OH}}{|}}{\text{P}}}}-\text{O—H} \qquad \text{R-O-}\!\!\mid\!\!-\overset{\displaystyle \text{O}}{\overset{\|}{\underset{\underset{\text{OH}}{|}}{\text{P}}}}-\text{O—R'}$$

(R and R' represent alcoholic groups). The phosphomonoesterases can be further subdivided according as their pH optima lie in the region of 9 ('alkaline phosphatases'; blood, intestinal mucosa, bone) or 5 ('acid phosphatases'; prostate, semen). Some members of the alkaline phosphatase group are absolutely, and others group-specific. In addition a number of *metaphosphatases* have been discovered; these enzymes catalyse the hydrolysis of meta- to orthophosphates.

A very important example of another type of phosphoric ester hydrolase is the *adenosine triphosphatase* of muscle. This enzyme is identical with myosin, which, in combination with actin, is the contractile protein which makes up the bulk of the muscle substance. Muscle and other tissues contain other adenosine triphosphatases that are separable from myosin, and similar enzymes have been found in snake venoms, potatoes and elsewhere. Bakers' yeast contains an interesting *pyrophosphatase* which appears to be wholly specific for the hydrolysis of inorganic pyrophosphate, which arises biologically in a considerable number of reactions, but appears to be devoid of action upon adenosine triphosphate and other organic pyrophosphates. A similar enzyme has been found in potatoes.

Generally speaking, phosphatases of the monoesterase type seem not to be specific with respect to the nature of the alcoholic group, and act alike on many organic phosphates such, for instance, as glycerol-:-phosphate, glucose-6-:-monophosphate and other phosphate esters. The reactions they catalyse may be generally expressed as follows:

$$R—O\circledP + H.OH \rightarrow R—OH + HO\circledP.$$

One such enzyme plays an important part in the ossification of cartilaginous structures such as bones and teeth. Unossified cartilage and cartilaginous structures which do not undergo ossification contain no phosphatase, but the enzyme makes its appearance just at the time that ossification sets in. Acting upon organic phosphates present in the blood, the enzyme is thought to catalyse a localized liberation of phosphate ions, which unite with calcium ions, also provided by the blood, to form the insoluble calcium phosphate, $Ca_3(PO_4)_2$. This is deposited in an ordered, crystalline manner in the cartilaginous matrix. The more rapidly ossification proceeds the higher is the phosphatase activity of the tissue. An interesting point in this connexion is that the calcified structures of the cartilaginous fishes (Elasmobranchii) contain an active phosphatase, just as do those of the bony fishes (Teleostei). The differences between the mechanical properties of calcified cartilage and true bone appear to be due to differences in the manner and perhaps in the amounts in which calcium phosphate is deposited in the two cases.

In addition to being present in cells of nearly every kind, a phosphatase is present in milk. If milk is pasteurized in the correct manner the treatment accorded is just sufficient to inactivate the milk phosphatase, so that the absence of phosphatase activity may be taken as an indication that the process has been properly carried out. One method used for routine testing depends upon the hydrolysis of p-nitrophenyl-:-phosphate with formation of free p-nitrophenol, which gives a colour reaction with alkalis and so can be estimated colorimetrically.

The group of phosphatases also includes some of the specialized enzymes concerned in the breakdown and possibly with the synthesis of the nucleic acids and will be considered later.

Among the phosphatases many and perhaps all require magnesium ions as activators. Very minute concentrations of Mg^{++} are all that are required in most cases, and the mode of action of these ions is still uncertain.

In conclusion it may be mentioned that some phosphatases of plant and animal origin can catalyse transfer as well as hydrolytic reactions (pp. 85–6).

OTHER HYDROLASES

Arginase, an enzyme of which larger or smaller concentrations are present in most animal cells, catalyses the hydrolytic deamidination of arginine to yield urea and ornithine (see p. 9) and plays a central part in the mechanism whereby urea is synthesized in the mammals and other ureotelic vertebrates. It is thought to be a manganese-containing protein, and is one of the most specific hydrolases known. Arginase is very powerfully inhibited by ornithine, though not by urea, a circumstance which suggests that it must combine with the ornithine residue, especially since it is also inhibited by the closely related amino-acid, lysine. Its optimum pH lies far in the alkaline range, probably above 10, at which pH it is very unstable.

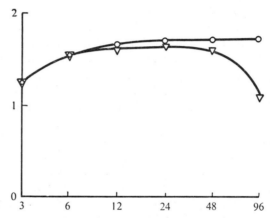

Fig. 19. Influence of urea concentration on activity of urease ($\triangledown$). The inhibitory influence of high concentrations of urea is counteracted by 0·2 % glycine ($\bigcirc$). Ordinate: initial velocity of hydrolysis. Abscissa: concentration of urea. (After Haldane, from Kato, 1923.)

Urease occurs in large concentrations in certain seeds, notably in jack- and soya-beans, from which it is usually prepared. It has been found in numerous other plant tissues and in the tissues of a few invertebrates, though it appears to be totally lacking from vertebrate organisms. Urease had the distinction of being the first enzyme to be obtained in a pure, crystalline state. It was extracted from jack-bean meal by Sumner, purified and crystallized in 1926, and shown to be a protein. It catalyses the following reaction:

$$CO(NH_2)_2 + H_2O \rightarrow CO_2 + 2NH_3.$$

Urease possesses a number of very peculiar properties. Unlike most enzymes, it is inhibited by high concentrations of its substrate, an effect which can be abolished by the addition of glycine, as is shown in Fig. 19. This phenomenon may be due to contamination of the reagents but is usually explained on the supposition that, in addition to combining with urea to form a reactive complex, *ES*, which is then hydrolysed, it goes on to combine with a second molecule of urea when the concentration of the latter is high, to form a stable complex, ES_2. Another, and possibly a unique feature is that the optimum pH of urease is proportional to the logarithm of the substrate concentration.

The specificity of urease is very high. Its action has been tested upon a large number of substituted ureas, none of which appears to be attacked, with possible though dubious exceptions in the cases of the *sym*. dimethyl- and diethyl-ureas.

Hydrolytic deaminases and deamidases other than urease are also known. *Adenase*, which catalyses the hydrolytic deamination of adenine to yield hypoxanthine, and *guanase*, which similarly converts guanine into xanthine, are present in the liver tissue of most mammals, and most probably occur elsewhere (see p. 287 for formulae of substrates). Another important enzyme concerned with the metabolism of purine derivatives is the *adenylic deaminase* of muscle. This enzyme, which is not identical with adenase, catalyses the hydrolytic deamination of adenylic acid to yield inosinic acid, in which the adenylyl group is replaced by that of hypoxanthine.

Animal tissues contain a powerful *glutaminase* and plants an homologous *asparaginase*. These enzymes catalyse the (irreversible) hydrolysis of the amides of the dicarboxylic amino-acids, e.g.

$$
\begin{array}{ccc}
CONH_2 & & COOH \\
| & & | \\
CH_2 & & CH_2 \\
| & & | \\
CH_2 + H_2O & \longrightarrow & CH_2 + NH_3 \\
| & & | \\
CH.NH_2 & & CH.NH_2 \\
| & & | \\
COOH & & COOH \\
\textit{glutamine} & & \textit{glutamic acid}
\end{array}
$$

Many other hydrolases are known and will be mentioned later in the appropriate context.

5

TRANSFERASES

GENERAL INTRODUCTION

Some of the most striking advances in biochemistry came with the discovery that there exist many enzymes capable of catalysing the transference of some group or other from one organic molecule to another. Transfer reactions of this kind are of fundamental importance in the energetics of biological systems, especially in biological synthesis. Pairs of hydrogen atoms, amino-, methyl-, phosphoryl- and glycosyl-groups are among those that may thus be catalytically transferred.

The enzymes which catalyse reactions of this kind are known as *transferases* and the reactions they catalyse can be written as follows:

$$A.G + B \rightleftharpoons A + G.B,$$

where $A.G$ stands for the donor of G, the group being transferred, and B for its acceptor. The oxidoreductases are a special group of transferring enzymes in which G corresponds to a pair of hydrogen atoms or electrons. They constitute a large and important group of functionally related enzymes concerned especially with tissue respiration, and will be considered separately from the rest.

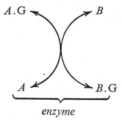

It is necessary to assume the formation of an intermediary complex of some kind in reactions of this sort because it is difficult otherwise to account for the transference of energy along with the group being transferred. At least three factors may be involved in the formation of such a complex, viz. the reactants themselves ($A.G$ and B, or A and $B.G$), and the transferring enzyme or *transferase*. This essential feature of these reactions can be expressed by writing the equations in the general form shown above on this page. At the present time we have little information about the intimate details that lie behind these reactions. Transference of a group is attended in many and perhaps most cases by the conservation of some or all of the free energy of the starting materials, and the number of ways in which this can be achieved is probably limited. In cases where free energy is quantitatively conserved it is evident that each step in the process must be reversible, but in others, e.g. in the hexokinase reaction (p. 55), some one or other of the intermediate steps must be irreversible for energetic reasons.

We may consider two possible modes of reaction which may be summarized as follows:

$$1. \quad (a) \ A.G + \text{enzyme} \rightleftharpoons A + \text{enzyme}.G,$$

$$(b) \ \text{enzyme}.G + B \rightleftharpoons \text{enzyme} + B.G.$$

This formulation assumes that the group G is first transferred to the enzyme and thence to the acceptor B, and there is evidence that this actually takes place in many cases. Any free energy associated with G is quantitatively transferred along with G itself. A second possible formulation assumes that the donor and acceptor are simultaneously accommodated on and activated by the enzyme:

$$2. \quad A.G + B \rightleftharpoons A.G...B \rightleftharpoons A...G.B \rightleftharpoons A + B.G$$
$$+ \qquad\qquad\qquad\qquad\qquad\qquad\qquad\qquad +$$
$$\text{enzyme} \qquad \text{enzyme} \qquad \text{enzyme} \qquad \text{enzyme}$$

GROUP TRANSFER BY HYDROLASES

As was pointed out in the preceding chapter, some of the enzymes there classified as hydrolases are capable of catalysing transfer as well as hydrolytic reactions. It will be well to consider some examples of this dual activity before we deal with those enzymes that catalyse transfer reactions but are devoid of hydrolytic activity.

Transphosphorylation by phosphatases

The juice of citrous fruits contains monophosphatases which appear at first sight to be typical hydrolases inasmuch as they can catalyse the hydrolysis of a variety of organic phosphate esters. But they also catalyse transfer reactions and one of the first of these to be studied was the following:

$$\text{nitrophenyl} \ \textcircled{P} + \text{methanol} \rightleftharpoons \text{nitrophenol} + \text{methyl} \ \textcircled{P}.$$

This reaction has a number of more biological counterparts, e.g.

$$\text{glucose-1-}\textcircled{P} + \text{glycerol} \rightleftharpoons \text{glucose} + \text{glycerol-}\alpha\text{-}\textcircled{P},$$
$$\text{glucose-1-}\textcircled{P} + \text{glucose-1-}\textcircled{P} \rightleftharpoons \text{glucose} + \text{glucose-1}:6\text{-di-}\textcircled{P}.$$

Reactions of this kind can also be catalysed by several other (e.g. apple) but not by all fruit phosphatases, and by phosphatases from some (e.g. prostate) but not all animal sources.

We have so far only considered the phosphatases as purely hydrolytic enzymes capable of catalysing reactions of the general type:

$$\text{R.O}\textcircled{P} + \text{H.OH} \rightleftharpoons \text{R.OH} + \text{HO.}\textcircled{P}.$$

7-2

It now appears, however, that this is only one example of an even more general case in which H.OH can be replaced by other hydroxylic compounds, such, for instance, as sugars and alcohols:

$$R.O℗ + R'.OH \rightleftharpoons R.OH + R'.O℗.$$

The overall effect of such a reaction is that the phosphate group is transferred, not to water necessarily, but to some alternative phosphate acceptor, the enzyme acting as a *phosphotransferase*.

If the phosphate donor is labelled with radioactive phosphorus and incubated with the enzyme and a suitable acceptor in the presence of inorganic phosphate, no radioactive phosphorus appears in the inorganic phosphate. This shows that the phosphate group undergoing transference does not have even a transient free existence. It seems probable, therefore, that the phosphate donor becomes attached to the enzyme E. The resulting complex might then react in one of two ways, either (*a*) with water, to undergo hydrolysis, or (*b*) with some alternative acceptor (e.g. glycerol) to give a new ester:

$$\text{(a)} \quad E—RO℗ + H.OH \rightleftharpoons E + R.OH + HO.℗,$$
$$\text{or} \quad \text{(b)} \quad E—RO℗ + R'.OH \rightleftharpoons E + R.OH + R'.O℗.$$

Now it is, as we have seen in our discussion of energetics, a universal observation that free energy always tends to be lost when a chemical reaction takes place. Since the decrease in free energy is greater in hydrolysis than in transfer, it might have been anticipated that hydrolysis rather than transfer would always result. Apparently, however, the available free energy is not lost all at once in a single reaction; rather does it seem that free energy tends when possible to be lost in consecutive steps or stages. Other examples of this tendency are known and exemplify the so-called 'law of successive reactions'.

How important these phosphatase-catalysed transfer reactions are in metabolism we do not know at the present time. It is important to realize that, biologically speaking, high concentrations of the alternative phosphate acceptor, e.g. glycerol, are often required in order to demonstrate the transfer reaction and, moreover, that transfer is usually succeeded by hydrolysis in the course of time, so that the newly formed products, e.g. glycerol phosphate, are only transient, at any rate *in vitro*. None the less, phosphate esters of glycerol, glucose and other alcoholic substances are of very general occurrence and great metabolic importance. In these phosphatase-catalysed transfers we have mechanisms for the phosphorylation of sugars, glycerol and so on whereby new low-energy phosphates can be formed from low-energy precursors without consuming the high-energy source used in, for example, the hexokinase reaction, viz. ATP. Finally, if, in fact, *in vivo* conditions are such that phosphatase-catalysed transfers can go on, it is always possible that the newly formed products might be drained off as fast as they are formed and metabolized before they can undergo hydrolysis.

Transglycosylation by glycosidases

There exist numerous enzymes capable of transferring glycosyl groups from one molecule to another; glucosyl and fructosyl groups are among those that may be thus transferred. While these transfers are catalysed for the most part by specialized enzymes known as *transglycosylases* (or glycosyltransferases), which will be discussed presently (pp. 90–6), many such transfers can be catalysed by enzymes, e.g. the β-fructofuranosidases (saccharases), usually classified as hydrolases. A great number and variety of these processes are known, some leading to the formation of di- and higher oligosaccharides and some even to that of polysaccharides.

The β-fructofuranosidases (*saccharases, invertases, sucrases*) act freely upon sucrose, the double glycoside of glucopyranose and β-fructofuranose, and yield these component sugars as hydrolysis products. At the same time, however, smaller or larger amounts of oligosaccharides are produced. These are transient products which, like the original sucrose, undergo eventual hydrolysis under *in vitro* conditions. One such reaction catalysed by the yeast enzyme may be written as follows and involves *transfructosylation*:

$$\text{sucrose} + \text{sucrose} \rightleftharpoons \text{fructosylsucrose} + \text{glucose}.$$

Glucose units are not transferred by this enzyme.

Similar enzymes obtained from moulds (e.g. *Aspergillus*) carry out similar reactions, in which sucrose itself can be replaced both as the donor and as the acceptor of fructosyl units by other sugars and sugar alcohols, and a variety of aliphatic and aromatic primary alcohols can also act as fructosyl acceptors.

Similar transfructosylations are catalysed by β-fructofuranosidases from the higher plants, including those of sugar beet, broad bean, clover, cabbage, etc., but not by extracts of plants which do not contain this enzyme. Transfructosylation seems to be the invariable rule in all these cases, but some animal enzymes, e.g. that of the honey bee, seem able to achieve transglucosylation as well.

α-*Glucosidases* (maltases) from various sources can catalyse *transglucosylation* on a large scale. Extracts from the intestines of dogs and other animals will build up a series of oligosaccharides at the expense of maltose;

$$\text{maltose} + \text{maltose} \rightleftharpoons \text{maltotriose} + \text{glucose},$$
$$\text{maltose} + \text{maltotriose} \rightleftharpoons \text{maltotetrose} + \text{glucose},$$

and so on.

Here again we have an enzyme which, in addition to its indubitable hydrolytic powers, can also act in a transglycosylating capacity and, indeed, the same is proably true of many other members of the group of glycosidases.

It seems possible that in some of these cases, as among the phosphatases, the donating sugar first 'parks' a glycosyl group on the enzyme, the product then either undergoing hydrolysis by reacting with water, or transferring the glycosyl group to some alternative acceptor. Eventually, however, the oligo-saccharides, which are only transient products, undergo hydrolysis, at any rate under *in vitro* conditions.

TRANSPHOSPHORYLATION: KINASES

A number of enzymes are known which can catalyse the transference of phosphate groups from one to the other of a pair of molecules, adenosine di- and triphosphates being employed as the carrier system. Enzymes of this kind, formerly known as 'phosphokinases', after hexokinase, the longest-known representative of the group, have now been re-named and given the shorter name of *kinases*. This title is however restricted to phosphotransferases that catalyse reactions involving the ATP/ADP system. They are very numerous and in all probability universally distributed.

It will be convenient to refer once more to the transfer of phosphate from adenosine triphosphate to creatine, or to adenosine diphosphate from phos-phocreatine under the catalytic influence of *creatine kinase*. If we represent high-energy phosphate by the usual symbol $\sim \textcircled{P}$, the structure of adenosine triphosphate may be expressed as follows:

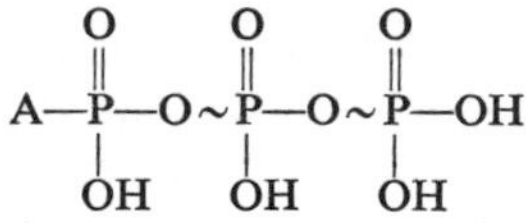

$$A{-}\overset{\overset{\displaystyle O}{\|}}{\underset{\underset{\displaystyle OH}{|}}{P}}{-}O\sim\overset{\overset{\displaystyle O}{\|}}{\underset{\underset{\displaystyle OH}{|}}{P}}{-}O\sim\overset{\overset{\displaystyle O}{\|}}{\underset{\underset{\displaystyle OH}{|}}{P}}{-}OH$$

Further, it is known that phosphocreatine also contains high-energy phos-phate, so that its structure may be written thus:

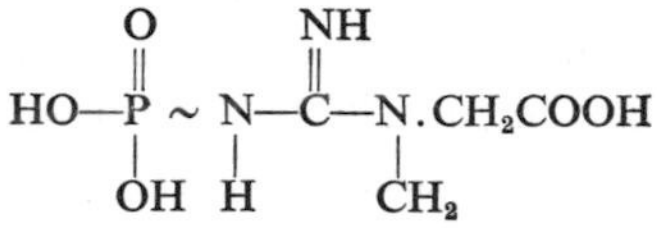

$$HO{-}\overset{\overset{\displaystyle O}{\|}}{\underset{\underset{\displaystyle OH}{|}}{P}}\sim N{-}\overset{\overset{\displaystyle NH}{\|}}{\underset{\underset{\displaystyle CH_2}{|}}{C}}{-}N.CH_2COOH$$

Accordingly, the creatine kinase reaction can be represented in the following terms:

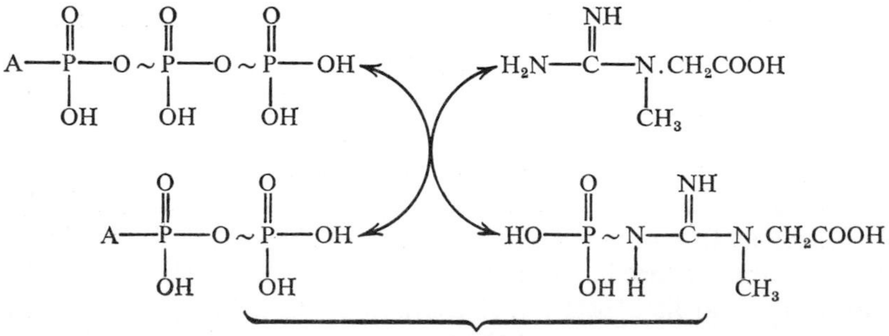

creatine kinase

We know today that both substances react together on the surface of the enzyme to give a complex; the most important feature of this reaction is that it accomplishes the transference of a high-energy phosphate group from one molecule to another.

Kinases probably occur universally among living cells and tissues. Yeast, plants and some mammalian tissues contain *hexokinases* catalysing the phosphorylation of glucose, fructose, mannose and glucosamine at the expense of ATP. Liver and muscle contain specific kinases for glucose, fructose and mannose while a specific galactokinase is also found in liver, and in yeast that has been 'trained' to ferment galactose. A kinase specific for the phosphorylation of gluconic acid has been found in some bacteria and in mammalian tissues as well. Other kinases catalyse a similar phosphorylation of fructofuranose-6- and fructofuranose-1-phosphates to give the 1:6-diphosphate and of glucose-6-phosphate to produce glucose-1:6-diphosphate. None of these reactions is reversible, however, presumably because none of the sugar phosphates contains a high-energy phosphate group. The reconversion of glucose-6-phosphate to free glucose, and that of fructofuranose-1:6-diphosphate to the 6-monophosphate, follow routes different from those of their synthesis, and are accomplished by hydrolysis catalysed by specific phosphatases.

Adenylate kinase (myokinase), a somewhat special enzyme, catalyses the

Table 7. *Some kinases*

Reaction	Enzyme
Creatine ⇌ phosphocreatine	Creatine kinase
Arginine ⇌ phosphoarginine	Arginine kinase
Glucose → glucose-6-phosphate	
Frustose → fructose-6-phosphate	Hexokinase
Mannose → mannose-6-phosphate	
Glucosamine → glucosamine-6-phosphate	
Gluconic acid → gluconic acid-6-phosphate	Gluconokinase
Fructose → fructose-1-phosphate	Fructokinase
Ribose → ribose-5-phosphate	Ribokinase
2-Deoxyribose → 2-deoxyribose-5-phosphate	D-2-Deoxyribokinase
Galactose → galactose-1-phosphate	Galactokinase
Fructose-6-phosphate → fructose-1:6-diphosphate	6-Phosphofructokinase
Fructose-1-phosphate → fructose-1:6-diphosphate	1-Phosphofructokinase
Glucose-6-phosphate → glucose-1:6-diphosphate	6-Phosphoglucokinase
Glyceric acid-3-phosphate ⇌ glyceric acid-1:3-diphosphate	Phosphoglycerate kinase
Pyruvate ⇌ phospho-*enol*-pyruvate	Pyruvate kinase
Adenosine monophosphate ⇌ adenosine diphosphate	Adenylate kinase
Adenosine → adenosine monophosphate	Adenosinekinase
Riboflavin → riboflavin phosphate	Flavokinase
NAD → NADP	NAD kinase

transference of a high-energy phosphate group from one molecule of adenosine diphosphate to a second, so that the products are adenosine monophosphate, on the one hand, and the corresponding triphosphate on the other. In this case the reaction can be represented in the following manner:

$$A - Ⓟ \sim Ⓟ + A - Ⓟ \sim Ⓟ \rightleftharpoons A - Ⓟ \sim Ⓟ \sim Ⓟ + A - Ⓟ,$$

It is a matter of definition that in all cases of kinase activity known at the present time, *adenosine triphosphate and the corresponding diphosphate are obligatory reactants*. The second reactant may be any of a considerable number of substances; the identities of some of them and the nature of the corresponding reactions are summarized in Table 7. It is quite certain that many substances other than those listed in the table can also enter into reactions involving a transfer of phosphate from ATP or to ADP, and there is evidence that the enzymes concerned are as a rule highly specific for their substrates as well as for the adenosine derivatives.

It is known that the triphosphates of nucleosides other than adenosine sometimes participate in kinase-like reactions, e.g. guanosine triphosphate (GTP) and uridine triphosphate (UTP). Again a $\sim Ⓟ$ is transferred to some appropriate acceptor leaving GDP or UTP. This is promptly re-phosphorylated at the expense of ATP by a specific member of a special group of *nucleoside diphosphate kinases* (cf. adenylate kinase):

$$G - Ⓟ \sim Ⓟ + A - Ⓟ \sim Ⓟ \sim Ⓟ \rightleftharpoons G - Ⓟ \sim Ⓟ \sim Ⓟ + A - Ⓟ \sim Ⓟ.$$

Several different nucleosides and their di- and triphosphates react similarly. To take a specific example, 6-*phosphofructokinase* can use ATP, ITP (inosine triphosphate) and UTP, all three of which are about equally active with this enzyme.

We shall have numerous occasions to refer to the other kinases listed in Table 7, for all play important parts in the metabolism of cells and tissues. But many other kinases exist. We know, for instance, that ATP can act as an energy source in the biological synthesis of NAD, NADP, flavinadenine dinucleotide, glutamine, hippuric acid, urea, glutathione and many more compounds of interest and importance. Each of these syntheses must probably involve a 'priming' reaction in which $\sim Ⓟ$ is transferred from ATP to one of the reactants, the transfer being catalysed by an enzyme which must, by definition, be a kinase of some kind.

TRANSGLYCOSYLATION

There exist numerous enzymes, the *glycosyltransferases* (transglycosylases), which are capable of catalysing the transference of glycosylic groups from one molecule to another. Some of these act also as hydrolases and have already been considered (pp. 75–8). Often, though by no means always, the starting material is α-glucose-1-phosphate and a number of the enzymes concerned exclusively with transglucosylation were formerly known as *phosphorylases*. This name has now been generally abandoned because of the confusion to which it can easily give rise.

Sucrose glucosyltransferase. Sucrose cannot be synthesized chemically or biologically by simple reversal of its hydrolysis. It has, however, been shown that certain bacteria (e.g. *Pseudomonas saccharophila*) can catalyse a synthetic production of sucrose from α-glucose-1-phosphate and free fructose. The synthetic process in this case is freely reversible.

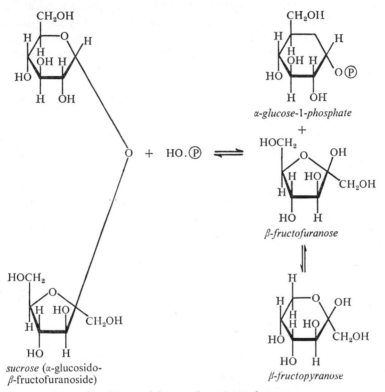

α-glucose-1-phosphate

β-fructofuranose

β-fructopyranose

sucrose (α-glucosido-β-fructofuranoside)

Fig. 20. *Breakdown* and *synthesis* of sucrose.

This is a good example of *transglucosylation* inasmuch as a glucosyl residue is transferred from one aglycone, in the form of a phosphate group, to another, which in this case happens to be glycosidic also. This second aglycone, a fructofuranosyl group, is available in aqueous solutions of fructose, which contain considerable amounts of fructofuranose in equilibrium with the more stable pyranose form.

The free energy required for the formation of the new glycosidic link is already available in that of the starting material, α-glucose-1-phosphate, and

91

is transferred along with the glucosyl group from the first aglycone to the second. Free glucose cannot serve as a starting material for this synthesis because its free-energy level is too low: it can, however, be raised to a high enough level through the hexokinase reaction, in which ATP provides the necessary free energy for the formation of glucose-6-phosphate, from which α-glucose-1-phosphate can then be formed by the action of an isomerase, phosphoglucomutase. In the end, therefore, the free energy for the synthesis of sucrose comes from ATP.

While sucrose glucosyltransferase is absolutely specific with respect to the glucosidic portion of the substrate molecule, it is only group-specific towards the aglycone. Thus the fructosidic aglycone can be replaced by phosphate or by any of a number of other substances, including certain sugars, so that new disaccharides can be synthesized by the action of this enzyme. Other aglycone units can be furnished by L-sorbose, D-xylulose and L-arabinose, for example:

$$\alpha\text{-glucosido-1-fructoside} + \text{sorbose} \rightleftharpoons \alpha\text{-glucosido-1-sorboside} + \text{fructose.}$$
(sucrose)

A number of other similar reactions have been demonstrated, so that the transglycosylic properties of sucrose glucosyltransferase cannot be held in any sort of doubt.

It should be noticed in passing that the synthesis of sucrose in plants (wheat germ) proceeds in quite another way (p. 329) and involves uridine diphosphate glucose (UDPG).

Synthesis of polysaccharides by transglycosylation. Somewhat similar transglycosylations leading to the synthesis of polysaccharides from sucrose have been observed in a variety of micro-organisms. In *Leuconostoc dextranicum*, for example, there is a transglucosylase (sucrose-6-glucosyltransferase) commonly known as *dextran sucrase*, which acts upon sucrose to yield fructose, which is metabolized by the organism, together with a *dextran* built up from $1:6\text{-}\alpha$-linked glucopyranoside units. Small amounts of sucrose are simultaneously hydrolysed. In certain spore-forming aerobes sucrose is again split into glucose and fructose, but in these cases the glucose is metabolized and the fructose units are built up into *levans* formed by a transfructosylase, *levan sucrase* (sucrose-6-fructosyltransferase), from $2:6$-linked fructofuranoside units. In both these cases the sugar units not katabolized for energy production by the organism are transferred one after another to build up these specialized polysaccharides.

Sucrose is also the starting material for the synthesis of a *glycogen*-like polysaccharide produced by another transglucosylase, *amylosucrase* (sucrose glucosyltransferase), of another micro-organism, *Neisseria perflava*: here the newly formed glucosidic linkages are mainly of the $1:4\text{-}\alpha$-type. Nor is sucrose the only disaccharide that can serve as starting material for polysaccharide synthesis by transglycosylation, especially among micro-organisms. For example, certain strains of *Escherichia coli* yield cell-free extracts containing yet another transglucosylase, *amylomaltase* (maltose-4-glucosyltransferase), which converts maltose into glucose together with a *starch-* or *glycogen*-like polysaccharide.

In all these cases the enzymes concerned show a weak hydrolytic activity as well as their transferring power; amylosucrase, for example, produces mainly fructose and its characteristic glycogen-like glucose polymer, but small amounts of glucose are set free at the same time. It may be that whereas the saccharases show a strong preference for water as their glycosyl acceptor (p. 87), amylosucrase and similar enzymes have an equally marked preference for polysaccharides as acceptors for the glycosyl units which they transfer.

Cyclical dextrans, amylose-, amylopectin- and glycogen-like polysaccharides are produced by many micro-organisms of various kinds, sometimes with, but often without, the participation of phosphorylated intermediates, but space will not allow us to discuss these interesting syntheses in detail.

Inulins and levans, the storage polysaccharides of Compositae and grasses respectively, are polyfructofuranoses. The general formulae are:

$$\text{sucrose-(fructose)}_n.$$

It appears that sucrose acts as a 'starter' and that the polysaccharide chain is assembled by the transference of fructose units from further molecules of sucrose to the 'starter' by enzymes (sucrose-6-fructosyltransferases) somewhat resembling the levan sucrase, a fructosyltransferase, already mentioned.

Synthesis of starch and glycogen

Many animal and plant tissues contain enzymes capable of hydrolysing starch and glycogen. It is certain moreover that plant and animal tissues can produce starch and glycogen from glucose and for many years it was supposed that the synthesis must involve enzymes essentially similar to the amylases already known to be concerned in the digestive breakdown of these polysaccharides. Later investigations have shown, however, that the intracellular breakdown and synthesis of starch and glycogen follow pathways that are quite different from those of digestive hydrolysis. This intracellular degradation is not, in fact, hydrolytic but transglucolytic, and yields α-glucose-1-phosphate as the first product of breakdown. The digestive amylases, it will be remembered, yield mainly maltose, a disaccharide.

Hanes and the Coris succeeded in obtaining preparations containing the transglucosylases (*α-glucan phosphorylases*) of peas, potatoes, liver, muscle and other tissues, and were able to show that they catalyse the decomposition of starch and glycogen in the presence of phosphate, as is shown in Fig. 21. Given suitable conditions this reaction is reversible. If α-glucose-1-phosphate is added in high concentrations to a suitable preparation of the enzyme, amylose-like polysaccharides can be synthesized, but these enzymes have no action upon β-glucose-1-phosphate. The crystalline α-glucan phosphorylase prepared from muscle exists in two forms, an inert form *b* which can be activated by reacting with ATP to give the active phosphorylase *a*. Pyridoxal phosphate is in some way involved in this reaction.

Work with highly purified enzymes has shown that the complete synthesis of starch and glycogen, like their digestive hydrolysis (see pp. 70–3) involves

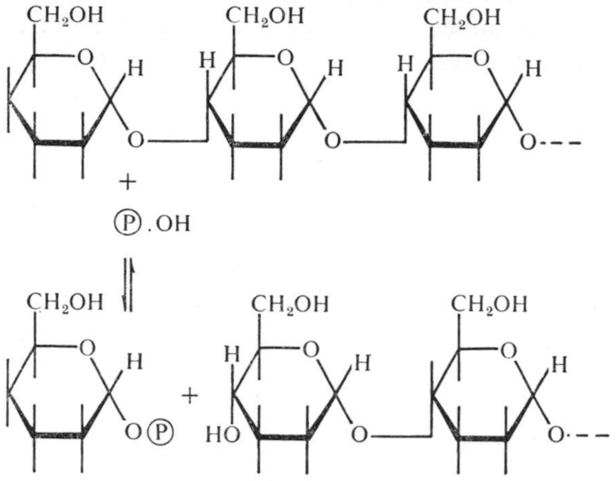

α-glucose-1-phosphate

Fig. 21. Phosphorolysis of starch by α-glucan phosphorylase.

Table 8. *Linkages formed by enzymes synthesizing amylose, amylopectin and glycogen*

Name of enzyme	Nature of linkage synthesized
α-Glucan phosphorylase (animals) ⎫ P-enzyme (plants) ⎬	1:4-α-
α-glucan-branching glycosyltransferase ('branching factor') (animals) Q-enzyme (plants) ⎫⎬⎭	1:6-α-

several different enzymes, and the names of these and the nature of the linkages synthesized by their action are summarized in Table 8.

An interesting feature of these syntheses is that the α-glucan phosphorylases do not act upon α-glucose-1-phosphate except in the presence of a suitable 'starter' such as a little added starch or glycogen: presumably therefore the enzyme needs an 'anchor' to which the new glucosidic units can be transferred from the substrate. At least 3 or better 4 condensed glucose units seem to be necessary and 1:6-linked units are as efficient as 1:4-bonded units, always provided that the number of groups is sufficiently large. Glucosyl groups are

94

then transferred from the phosphate groups of α-glucose-1-phosphate to the α-linked glucosidic units supplied by the starter, and the process is, in fact, another case of transglycosylation.

Crude preparations from peas and potatoes yield granular products closely allied to natural starch while the liver and muscle enzymes yield polysaccharides resembling natural glycogen. Purified preparations however yield products more closely resembling amylose. Crystalline muscle or potato phosphorylase yield products that give a pure blue coloration with iodine. Examination of one such product showed that it could be completely degraded to maltose by the action of crystalline β-amylase and contained about 80 glucose units, and products with up to 1000 units have been obtained. Natural amylose, however, is only 70 % degraded by crystalline β-amylase and contains 200 or more glucose units, a few of which are incorporated through 1:6-α-linkages (p. 71). But both α-glucan phosphorylase and β-amylase are wholly specific for 1:4-α-links, so that the formation of natural amylose presumably involves an additional enzyme.

Impure α-glucan phosphorylase preparations acting upon α-glucose-1-phosphate yield products which stain reddish brown or purple with iodine; evidently they do contain other factors which can catalyse the formation of branching linkages, of the usual 1:6-α-type. The presence of such *branching factors* has been demonstrated in muscle, liver, yeast and various plant materials, and a highly purified factor of this kind has been isolated from potatoes ('Q-enzyme'), which closely resembles the corresponding animal enzyme. Work with the purified branching enzyme shows that it operates without the participation of phosphate although, like α-glucan phosphorylase itself, it is a transglycosylase but possesses rather special properties.

If purified branching enzyme is allowed to act upon purified amylose, which contains mainly 1:4-α-linkages, amylopectin-like substances are formed, containing 1:6-α- as well as 1:4-α-glucosidic links. Careful studies of the products shows that two processes take place simultaneously: 1:4-linkages disappear and 1:6-linkages take their place. Apparently, therefore, the branching enzyme detaches chains of 1:4-linked glucosidic units from the main chain of amylose and transplants them into 1:6-positions. This enzyme can transfer and transplant a chain of about twenty 1:4-α-linked glucose units in one operation; as far as we know, phosphate plays no part in the process.

The reaction catalysed by the branching enzyme seems to be irreversible, for α-glucan phosphorylase plus branching enzyme cannot accomplish the complete breakdown of natural glycogen or amylopectin. α-Glucan phosphorylase, like β-amylase, ceases to act when a branching linkage is approached and it therefore seems certain that the branching enzyme cannot catalyse the breakdown of the 1:6-α-links which are formed by its own activity. Muscle, however, contains a hydrolytic *dextrin-1:6-α-glucosidase* (p. 73) which is competent to break these branching linkages, and similar 'debranching

95

factors' are present in plant materials (beans, potatoes). These enzymes can act together with either the appropriate transglucosylases or amylases to catalyse a complete degradation of glycogen or amylopectin to α-glucose-1-phosphate or maltose as the case may be.

The interconvertibility between starch or glycogen and α-glucose-1-phosphate is due to the higher free energy of the latter as compared with that of free, unphosphorylated glucose. The free energy required for the synthesis of these polysaccharides is introduced into the free glucose molecule by phosphorylation through the action of hexokinase and ATP. This yields glucose-6-phosphate, an ester, in the first instance, and from this the glucoside, α-glucose-1-phosphate, is formed by an intramolecular rearrangement catalysed by *phosphoglucomutase* (p. 122). In the end, therefore, the energy for the synthesis of these polysaccharides comes once again from ATP.

Now the equilibrium between glycogen or starch and α-glucose-1-phosphate lies very much in favour of the latter and the *in vitro* synthesis of the polysaccharides requires the presence of α-glucose-1-phosphate in high concentrations, concentrations far higher than are likely ever to be encountered in biological systems. More recent work has disclosed the participation of another system, the operation of which has the effect of shifting the equilibrium in favour of polysaccharide formation, presumably by pumping yet more free energy into the starting material. The processes of synthesis and breakdown are summarized in Fig. 22.

α-Glucose-1-phosphate reacts first with *uridine triphosphate* (UTP) to give *uridine-diphosphate-glucose* (UDPG), under the influence of a kinase-like enzyme (*glucose-1-phosphate-uridylyltransferase*)

$$\text{UTP} + \alpha\text{-glucose-1-phosphate} \rightarrow \text{UDPG} + \text{pyrophosphate.}$$

From the UDPG, glucosyl units are transferred by a specific *α-glucan-UDP-glucosyl-transferase* to the already partly formed polysaccharide chain:

$$\text{UDPG} + (\text{glucose})_n \rightarrow \text{UDP} + (\text{glucose})_{n+1}.$$

Whether or not UDPG plays any part in the breakdown as well as in the synthesis of glycogen is uncertain and unlikely but, as far as the formation of glycogen is concerned, the requisite free energy still comes from ATP in the end because the UDP left over is reconverted into UTP by reacting with ATP:

$$\text{UDP} + \text{ATP} \rightleftharpoons \text{UTP} + \text{ADP.}$$

In the ordinary way, glycogen is stored in the livers of animals and distributed to other tissues by way of the blood, in which it travels in the form of free glucose. The latter arises by the action of a specific liver phosphatase upon glucose-6-phosphate formed from α-glucose-1-phosphate, which is itself formed by the action upon glycogen of the enzymes we have been considering here. In the muscles and other tissues, where blood glucose is either metabolized or used as a source of newly-formed glycogen, it seems certain that the glucose, on entering the cells, must first of all be phosphorylated again by hexokinase and ATP to yield glucose-6-phosphate, from which α-glucose-1-phosphate can again be formed and then built up into glycogen, once more with UDPG as an intermediate reactant.

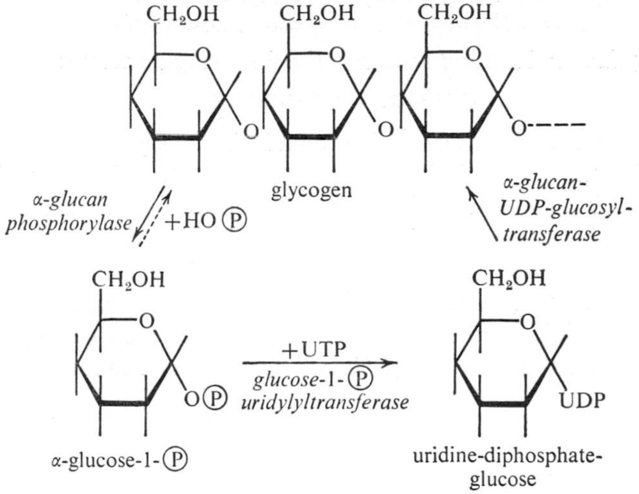

Fig. 22. *Synthesis* and *breakdown of glycogen* via α-*glucose*-1-*phosphate*. The dotted arrow refers to synthesis requiring high concentrations of α-glucose-1-phosphate.

Transribosylation is a form of transglycosylation of major importance in the metabolism of nucleosides and nucleotides. The nucleosides are β-N-glycosides of the pentose sugars, D-ribose and D-2-deoxyribose. They can be split by transribosylation in the presence of free phosphate to yield the corresponding pentose-1-phosphates, together with the parent purine, pyrimidine or other nitrogenous base. Here, therefore, we have another case in which a glycosidic unit is transferred from one aglycone, e.g. a purine or pyrimidine base, to a second, in the form of phosphate. We shall refer to this subject again (p. 292).

TRANSPEPTIDATION

Formation of peptide links

It has been suspected for some time that the synthesis of polysaccharides and disaccharides by transglycosylation might have some counterpart in the synthesis of peptides and even of proteins, though a more detailed consideration of this topic must be held over to a later chapter.

We know of several reactions in which new peptide bonds are synthesized. The synthesis of *hippuric acid* involves the formation of such a bond. Liver extracts contain an enzyme which will form hippuric acid from benzoic acid and glycine, provided that ATP is available. This time it is the adenosine monophosphate (AMP) moiety of ATP which is transferred (reaction (i)). Coenzyme A is also concerned in this process, which can probably be formulated as follows:

(i) ATP + benzoic acid → adenylbenzoate + pyrophosphate,
(ii) adenylbenzoate + Co A → benzoyl ~ Co A + AMP,
(iii) benzoyl ~ Co A + glycine → benzoylglycine + Co A.

Benzoylglycine, or hippuric acid, has the following structure:

$$C_6H_5\mathbf{CO.HN.CH_2}COOH.$$

The newly formed peptide bond is shown in heavy type.

Glutamine, the amide of glutamic acid, can readily be hydrolysed by tissue extracts, but its synthesis, which again involves formation of a new peptide bond, can only be accomplished under certain special conditions. Enzyme preparations have been obtained from bacteria and from animal tissues which, if glutamate and ammonia are available, catalyse the synthetic formation of glutamine on addition of ATP. Here again it has been suspected (though not proved) that a reactive γ-acyl-Co A derivative must be formed as an intermediary.

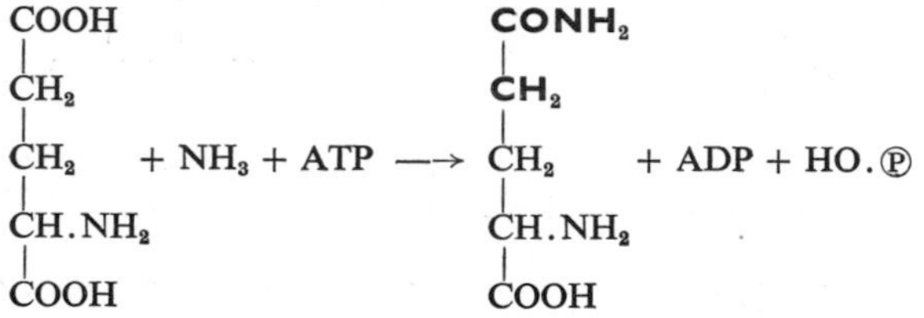

$$
\begin{array}{c}
\text{COOH} \\
|\\
\text{CH}_2 \\
|\\
\text{CH}_2 \\
|\\
\text{CH.NH}_2 \\
|\\
\text{COOH}
\end{array}
+ \text{NH}_3 + \text{ATP} \longrightarrow
\begin{array}{c}
\text{CONH}_2 \\
|\\
\text{CH}_2 \\
|\\
\text{CH}_2 \\
|\\
\text{CH.NH}_2 \\
|\\
\text{COOH}
\end{array}
+ \text{ADP} + \text{HO}.\textcircled{P}
$$

Another reaction in which peptide bonds are formed is the synthesis of ornithuric acid, a compound formed in bird kidney as a means of detoxicating benzoic acid (p. 253), and here again ATP is required as an energy source and Co A is probably implicated as well. But in none of these cases is a 'typical' peptide bond produced, i.e. a bond formed by the union of the α-carboxyl group of one amino-acid and the α-amino-group of a second. One such bond and one atypical bond are formed however in the synthesis of *glutathione* (formula on p. 240), which has been much studied and in which ATP must once more be provided. The general need for ATP in peptide-bond formation makes

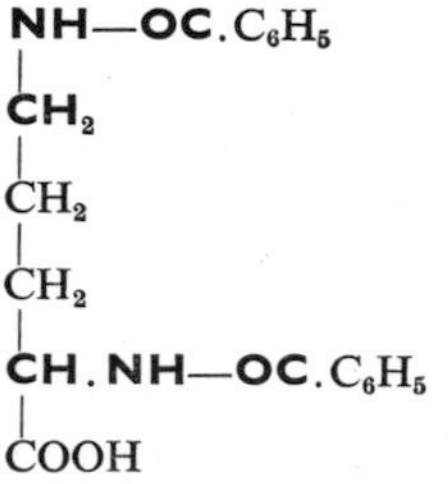

$$
\begin{array}{l}
\text{NH--OC}.C_6H_5 \\
|\\
\text{CH}_2 \\
|\\
\text{CH}_2 \\
|\\
\text{CH}_2 \\
|\\
\text{CH.NH--OC}.C_6H_5 \\
|\\
\text{COOH}
\end{array}
$$

ornithuric acid

it clear that free energy is required for the formation of bonds of this type, just as it is in the formation of glycosidic bonds.

Transfer of peptide groups. The suggestion has been made that just as certain hydrolytic phosphatases can catalyse low-energy transfer of glycosyl groups with formation of long-chain polysaccharides, so too hydrolytic peptidases might perhaps catalyse that of chains of peptide-bound amino-acid units. Intracellular peptidases are numerous and, moreover, very specific indeed, and might in fact be able to carry out highly specific transferring operations leading to the formation of correspondingly specific peptides. If we assume that transpeptidation is possible—and there is evidence that this is indeed the case—there are two ways in which the transfer might be effected. These may be expressed in the manner shown in Fig. 23:

Transfer of amino-group:

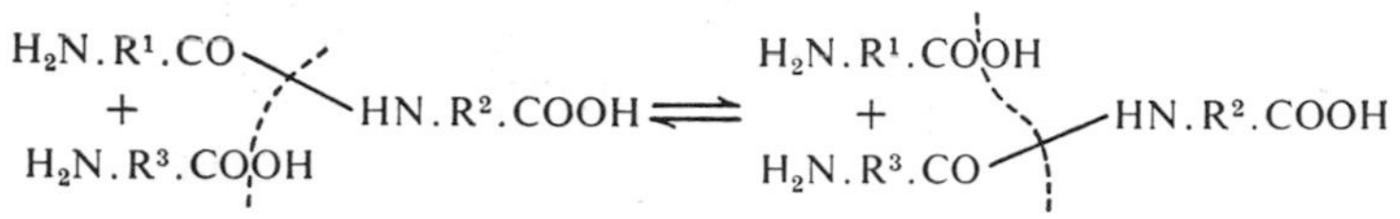

$$H_2N.R^1.CO \quad \cdots \quad HN.R^2.COOH \rightleftharpoons H_2N.R^1.COOH \quad + \quad HN.R^2.COOH$$
$$+ \qquad\qquad H_2N.R^3.CO$$
$$H_2N.R^3.COOH$$

Transfer of carboxyl group:

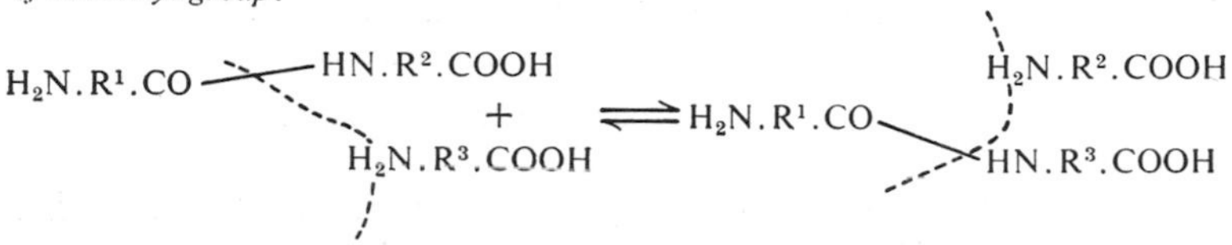

$$H_2N.R^1.CO \quad\times\quad HN.R^2.COOH \qquad\qquad H_2N.R^2.COOH$$
$$+ \qquad \rightleftharpoons H_2N.R^1.CO$$
$$H_2N.R^3.COOH \qquad\qquad HN.R^3.COOH$$

Fig. 23. Possible mechanisms of transpeptidation.

The tripeptide, glutathione, and its constituent amino-acids, glycine, cysteine and glutamic acid were incubated with enzyme preparations derived from kidney and pancreas tissue; the digests were examined chromatographically and new spots were found which indicated the formation of new peptides. New peptides were also produced by interaction between glutathione and other amino-acids, including leucine, valine and phenylalanine.

Important though transpeptidation may possibly be in the formation of simple peptides such as glutathione, it is probably safe to say that it plays no part in the biosynthesis of proteins.

TRANSAMINATION

L-*Glutamate transaminase* (aminotransferase) is an enzyme that occurs widely among micro-organisms and in plant and animal tissues. The process of transamination, which is reversible, can be expressed in the following abbreviated manner:

$$\begin{array}{cccc} R & R' & R & R' \\ | & | & | & | \\ CH.NH_2 + CO & \rightleftharpoons & CO & + CH.NH_2 \\ | & | & | & | \\ COOH & COOH & COOH & COOH \end{array}$$

Work with purified specimens of the enzyme shows that one member of the reacting pair must be either L-glutamate or α-ketoglutarate. L-Aspartate or oxaloacetate can replace these but react much more slowly, and the enzyme, L-*glutamate transaminase,* is thus highly specific towards at least one of the reactants.

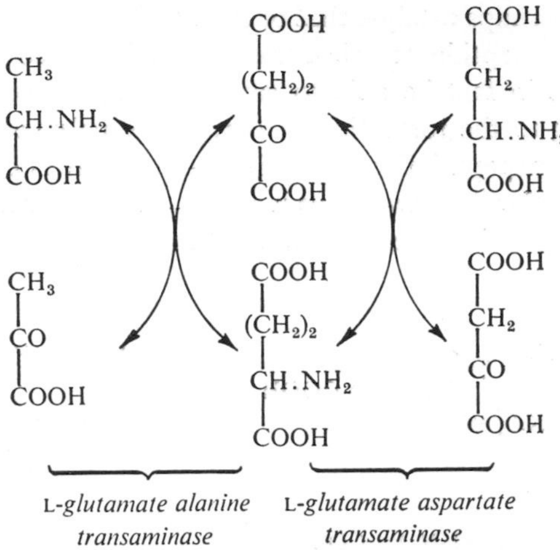

L-*glutamate alanine
transaminase*   L-*glutamate aspartate
transaminase*

Fig. 24. Mechanism of transamination.

The α-ketoglutarate-glutamate system itself can act as an amino-group-carrying system. If, for example, we take alanine and add to it a purified sample of glutamate transaminase in the presence of oxaloacetate, no direct transference of amino-groups from alanine to oxaloacetate takes place. If now a catalytic amount of glutamate or α-ketoglutarate is also added, amino-groups are transferred from the alanine to the oxaloacetate so that pyruvate and aspartate are formed. On account of the free reversibility of the system the reaction does not go to completion but an equilibrium condition is eventually attained as is shown in Fig. 24.

Glutamate transaminase has a prosthetic group which consists of a phosphorylated derivative of pyridoxal, one of the B_2 group of vitamins. Pyridoxal phosphate and the corresponding amine, pyridoxamine phosphate, have the following structures:

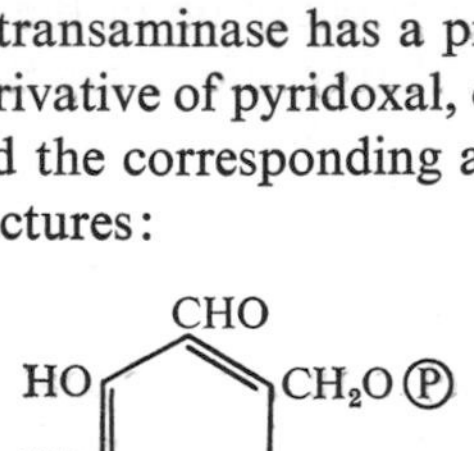

pyridoxal phosphate 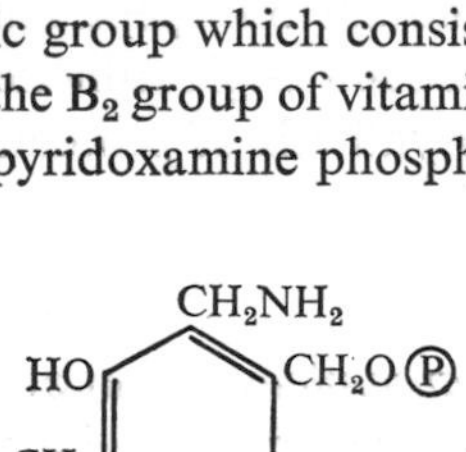   *pyridoxamine phosphate*

It seems probable that the primary reaction in transamination consists in the transference of the amino-group from an amino-acid to the prosthetic group of the enzyme, pyridoxal phosphate, in which case the first stages of the re-action will probably be as shown in Fig. 25.

Here R represents the remainder of the molecule of the phosphate of pyridoxal or pyridoxamine. By further reaction with α-ketoglutarate and following the same reactions in reverse, the pyridoxamine derivative could then revert to the aldehyde and yield glutamate.

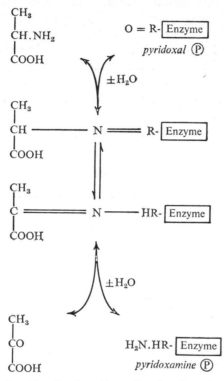

Fig. 25. Participation of pyridoxal and pyridoxamine phosphates in transamination.

Other transaminases. The transaminase system has been re-investigated with the help of freeze-dried enzyme preparations of heart, muscle, liver and kidney, dialysed and fortified with pyridoxal phosphate. In these newer experiments it has proved possible to demonstrate the transamination of twenty-two amino-acids, over and above alanine, aspartate and glutamate. It seems that the glutamate-α-ketoglutarate system plays a central part in most if not all of these processes but that a different enzyme is probably involved for each amino-acid concerned. According to some experiments α-ketoglutarate is not invariably the primary amino-group acceptor. Sometimes it is replaced by pyruvate, which is converted into alanine and the latter then reacts with α-ketoglutarate.

It would be difficult to overestimate the importance of transamination in protein and amino-acid metabolism (see p. 229).

Other experiments have shown that glutamine, as opposed to glutamate, can undergo a simultaneous deamidation and transdeamination, the amido-group being set free as ammonia while the α-amino-group is transferred to any of a large variety of α-keto-acids. In this way alanine, phenyl alanine, tyrosine and methionine have been formed from the corresponding α-keto-acids. The mechanisms and the enzymes involved appear to be quite different from the transaminases discussed in the last paragraphs. This new mechanism, together with transamination—which is reversible—supply reactions capable of synthesizing most, if not all, of the amino-acids normally involved in protein formation, always provided that the corresponding α-keto-acids are available.

Aspartate transaminase. In some plants the place of the glutamate trans-aminases is taken by similar systems in which aspartate and oxaloacetate replace glutamate and α-keto-glutarate respectively.

TRANSAMIDINATION

Enzymes exist that can catalyse the transference of the amidine group of arginine to other substances, e.g.

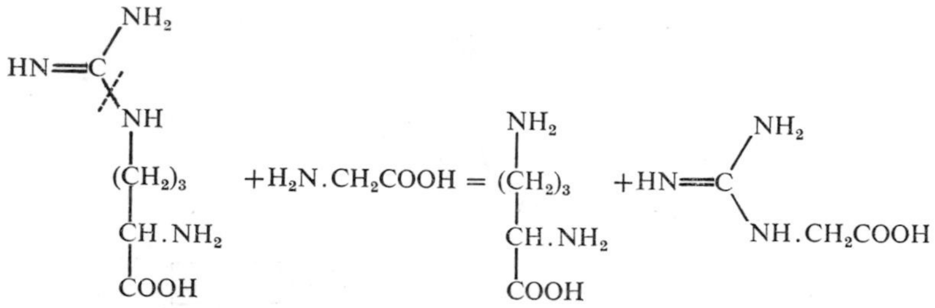

| arginine | glycine | ornithine | glycocyamine |

The product is subsequently methylated to yield *creatine* (p. 111), a well-known constituent of vertebrate muscle.

A similar reaction probably takes place between arginine and taurine yielding *taurocyamine*, at any rate in the muscles of some annelid worms.

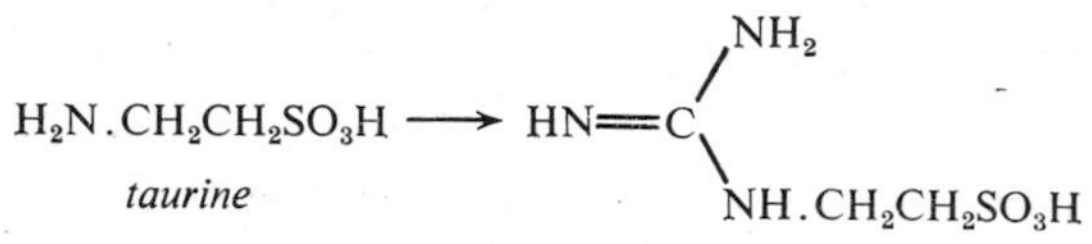

taurine

taurocyamine

Little is known at present about the enzymes involved or about the mechanisms of transfer.

102

Taurine has long been known as a constituent of the bile of some though not of all vertebrates, in which it is present in conjugation with cholic and other bile acids (p. 216). That free taurine often occurs in very notable quantities among invertebrates has been known for many years. Thus the adductor muscle of the edible mussel, *Mytilus edulis*, contains 4–5%, while the muscles of an annelid worm, *Audouinia spirabranchus*, contain 3% of taurine. It is abundant in *Abalone* also.

Asterubin, an interesting derivative of taurocyamine, is not, as its name might suggest, a pigment, but a sulphur-containing guanidine derivative which has been obtained from two species of star-fishes.

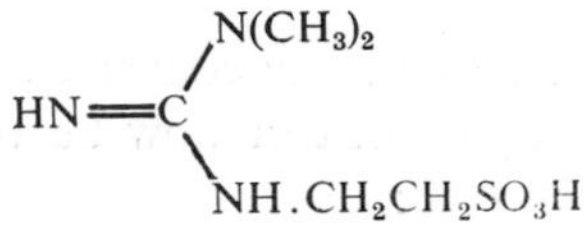

$$HN\!=\!C\begin{smallmatrix} \diagup N(CH_3)_2 \\ \diagdown NH.CH_2CH_2SO_3H \end{smallmatrix}$$

Its mode of synthesis is unknown. It is probably derived by methylation from taurocyamine, and like taurine and taurocyamine, it belongs to the class of sulphonic acids, a somewhat rare group in nature.

TRANSCARBAMOYLATION

The early stages of urea and pyrimidine synthesis include transfer reactions involving the formation and transfer of carbamoyl groups. In the presence of *N*-acetylglutamate and ATP, carbon dioxide and ammonia react together to yield carbamoyl phosphate. The carbamoyl group can then be transferred to ornithine to yield citrulline (p. 276), an important intermediate in the synthesis of urea, or to aspartate, yielding carbamoyl aspartate (p. 290); the latter is the starting material for the formation of pyrimidine bases. We shall study these processes in more detail in later chapters.

TRANSTHIOLATION

In the foregoing transfer reactions the group being transferred is most usually exchanged for a hydrogen atom. The possibility that groups other than hydrogen may be exchanged is indicated by the exchange of the —SH of homocysteine for the —OH of serine, which takes place under the influence of cystathionase, an enzyme present in rat liver (see p. 244).

TRANSACETYLATION

Acetylation is a not uncommon biological process. It is employed in the detoxication of many foreign substances such, for example, as aniline and the sulphonamides. It also plays a part in the formation of mercapturic acids

(p. 246) and is important too in the elaboration of acetyl choline from choline.

Cell-free extracts of liver contain an enzyme or enzymes which will bring about acetylation in the presence of free acetate, coenzyme A (Co A) and ATP: ATP is required to provide free energy, since acetylation is an endergonic process. The acetylating agent is acetyl-coenzyme A, which can replace free acetate plus ATP plus coenzyme A itself, and is formed by the following preliminary reactions between ATP, the coenzyme and acetate:

(i) ATP + acetate $\rightleftharpoons$ adenyl acetate + pyrophosphate,
(ii) Adenyl acetate + Co A $\rightleftharpoons$ AMP + acetyl-Co A.

Most other fatty acids can be similarly converted into their Co A derivatives.

The acetyl group of acetyl-Co A can now be transferred to choline, for example, yielding acetyl choline:

$$\overset{+}{(CH_3)_3N}.CH_2CH_2OH + CH_3CO\sim Co\ A \longrightarrow \overset{+}{(CH_3)_3N}.CH_2CH_2O.OC.CH_3 + Co\ A,$$

choline   *acetylcholine*

or to sulphanilamide:

$$CH_3CO\sim CoA + NH_2-C_6H_4-SO_2NH_2 \longrightarrow CH_3CO.HN-C_6H_4-SO_2NH_2 + CoA.$$

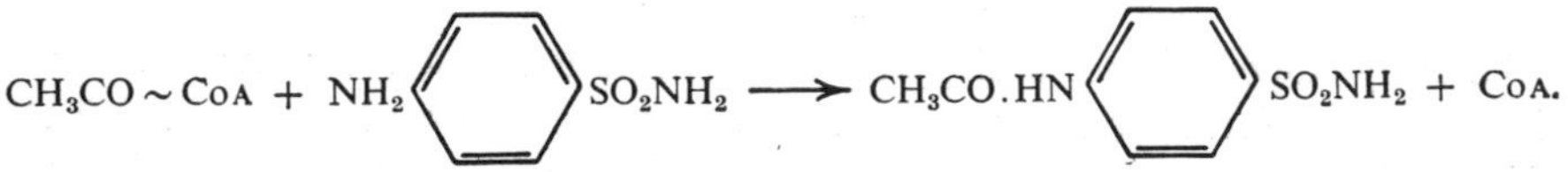

Another interesting transacetylation reaction takes place in some bacteria; here the acetyl group is transferred between Co A and inorganic phosphate:

$$CH_3CO\sim Co\ A + HO.\textcircled{P} \rightleftharpoons CH_3CO\sim O\textcircled{P} + Co\ A.$$

In addition, some bacteria contain a special *acetokinase* which forms acetyl phosphate by direct phosphorylation at the expense of ATP:

$$ATP + CH_3COOH \rightleftharpoons ADP + CH_3CO\sim O\textcircled{P}.$$

Interestingly enough the 'active' acetyl group of acetyl-coenzyme A can also react not only through its carboxyl but also through its methyl group, e.g. with *enol*-oxaloacetate to form citrate:

$$
\begin{array}{ll}
COOH & COOH \\
| & | \\
CH & CH_2 \\
\| & | \\
C(OH)COOH + H_2O \longrightarrow & C(OH)COOH + Co\ A \\
\overset{+}{CH_3} & CH_2 \\
| & | \\
CO\sim Co\ A & COOH \\
\end{array}
$$

citric acid

104

This is a basic and fundamental reaction in the metabolic breakdown of both fats and carbohydrates, while acetyl-CoA itself is a primary starting material in the synthesis of fatty acids.

TRANSKETOLATION

Liver, yeast and spinach and, doubtless, many other tissues contain enzymes known as transketolases. These enzymes catalyse reactions like that shown in Fig. 26. Thiamine diphosphate (co-carboxylase) is involved in some way, presumably in the capacity either of a prosthetic group or a coenzyme.

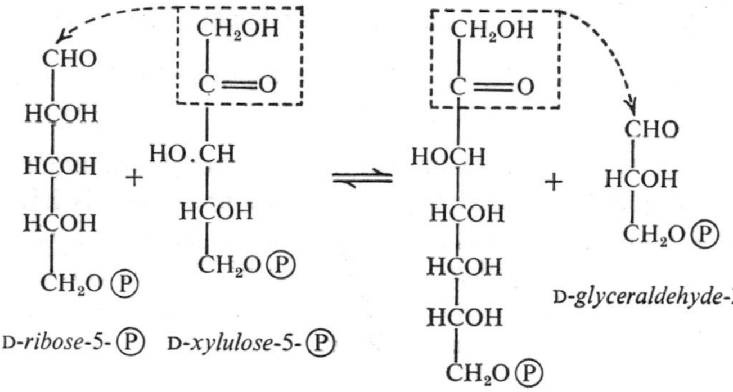

D-*ribose*-5-Ⓟ D-*xylulose*-5-Ⓟ D-*sedoheptulose*-7Ⓟ D-*glyceraldehyde*-3Ⓟ

Fig. 26. An example of transketolation.

In effect these transketolases catalyse the transfer of a *glycol aldehyde group* from a ketose to an aldose sugar, and although the mechanism of the transference is uncertain, it seems probable that the glycol aldehyde group is first transferred to the enzyme and thence to the accepting aldose:

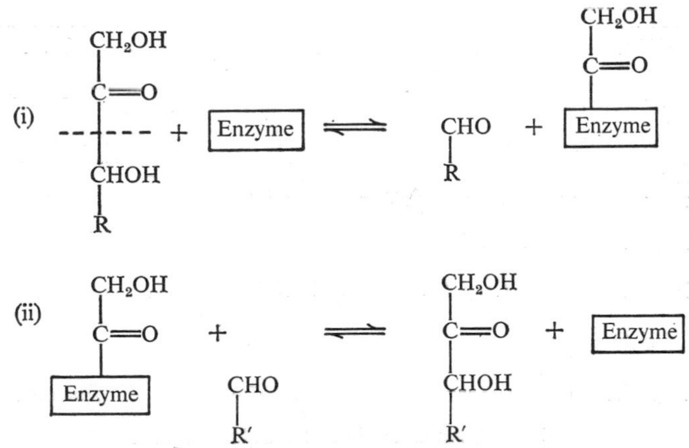

A large number of substances can participate in transketolase reactions, including:

Donors (ketoses)	*Acceptors (aldoses)*
Xylulose-5Ⓟ	D-Glyceraldehyde-3Ⓟ
L-Erythulose	L-Glyceraldehyde-3Ⓟ
D-Sedoheptulose-7Ⓟ	D-Ribose-5Ⓟ
Hydroxypyruvate	Glycolaldehyde
D-Fructose-6Ⓟ	Erythrose-4Ⓟ
	D-L-Glyceraldehyde

Evidently, therefore, the transketolases can catalyse the formation and decomposition of a very great variety of sugars and sugar phosphates. Indeed, their discovery has in effect opened a new chapter in carbohydrate biochemistry. Of particular interest are the two following reactions:

$$\text{ribose-5Ⓟ} + \text{xylulose-5Ⓟ} \rightleftharpoons \text{sedoheptulose-7Ⓟ} + \text{glyceraldehyde-3Ⓟ},$$
$$\text{fructose-6Ⓟ} + \text{glyceraldehyde-3Ⓟ} \rightleftharpoons \text{xylulose-5Ⓟ} + \text{erythrose-4Ⓟ}.$$

Ribose, xylulose, sedoheptulose and erythrose phosphates are important intermediaries in photosynthesis. Glyceraldehyde phosphate is immediately related to glyceric acid phosphate, the first product to be labelled when photosynthesis is allowed to proceed in the presence of isotopic carbon dioxide, while fructose-6 Ⓟ is an intermediary in the photosynthesis as well as in the breakdown of starch, glycogen and glucose.

TRANSALDOLATION

Transaldolases have been obtained from yeast, bacteria and plant sources. Like the transketolases they require thiamine diphosphate as a cofactor but catalyse the transference of *dihydroxyacetone groups* instead of glycol aldehyde. An example is shown in Fig. 27. Again the transfer is from ketoses to aldoses.

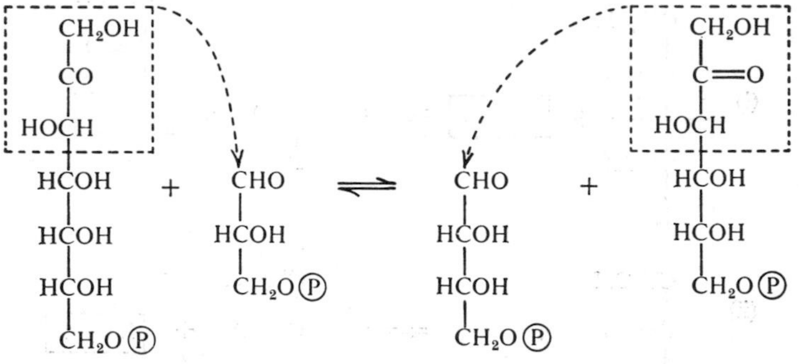

sedoheptulose-7 Ⓟ *erythrose*-4 Ⓟ *fructose*-6 Ⓟ

Fig. 27. An example of transaldolation.

Present indications are that these enzymes are probably as versatile as the transketolases and play as important a part in carbohydrate metabolism. The transfer of the dihydroxyacetone group probably takes place by way of the enzyme, forming an intermediary complex that can be roughly formulated as:

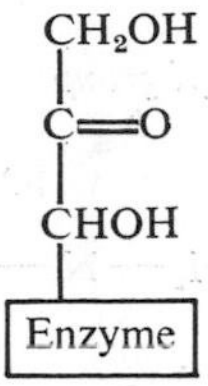

$$
\begin{array}{c}
CH_2OH \\
| \\
C{=}O \\
| \\
CHOH \\
| \\
\boxed{Enzyme}
\end{array}
$$

Before leaving these transaldolases the reader may be reminded of the fact that the *aldolases proper* (p. 119) are capable of catalysing a large variety of condensations as opposed to transfer reactions between aldehydic substances and *dihydroxyacetone phosphate*, for which latter it is highly specific. Most usually dihydroxyacetone phosphate reacts with glyceraldehyde-3-phosphate to form fructose-1:6-diphosphate in the usual way, but many other reactions are possible.

With the discovery of transketolation and transaldolation a new interest has been awakened in the long familiar aldolases. Much new information has been gained about a variety of sugars formerly considered as rather exotic and, at the same time, much has been learned about their metabolism and the interplay between them and other more familiar sugars and sugar phosphates. Nowhere, perhaps, is this interplay more important than in photosynthesis.

ONE-CARBON TRANSFERASES

One-carbon fragments of various kinds play a part of great importance in biosynthesis. For the most part they are transferred by enzymes the prosthetic groups of which consists of derivatives of *folic acid*:

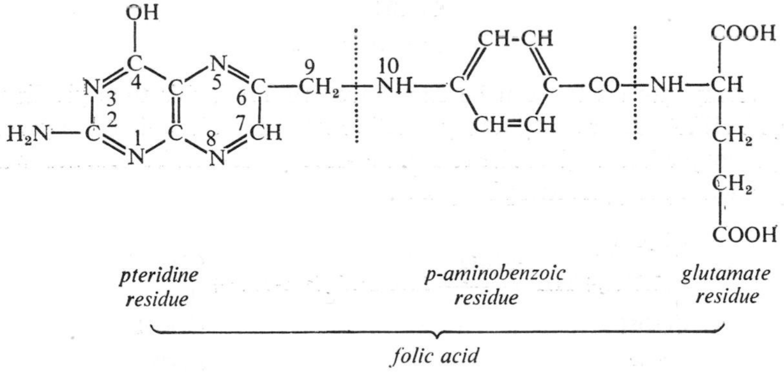

| pteridine | p-aminobenzoic | glutamate |
| residue | residue | residue |

folic acid

Folic acid exists in several forms containing one or more glutamate residues arranged in series, but all these can be reduced by a $NADPH_2$-specific dehydrogenase to 7:8-dihydrofolate and this in turn by a similar enzyme to 5:6:7:8-tetrahydrofolic acid (THF). It is in this form that folic acid discharges its transferring functions.

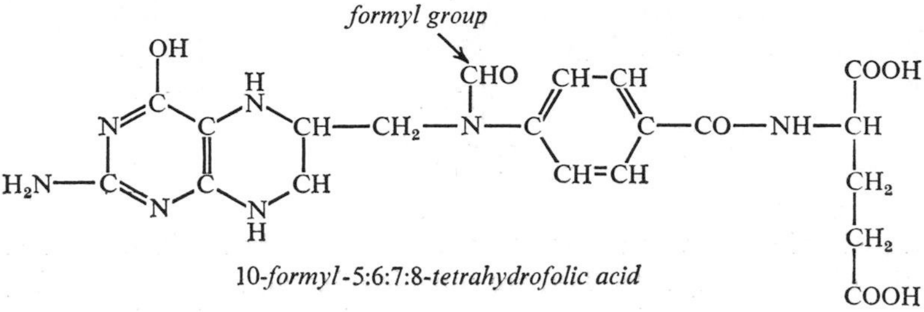

10-*formyl*-5:6:7:8-*tetrahydrofolic acid*

By a reaction with formate in an ATP-dependent system THF is converted to 10-formyl-THF, the parent of a considerable group of other derivatives. *Formate* itself arises from a number of different sources. For example, the deamination of glycine, whether by transdeamination or otherwise, gives rise to glyoxylic acid, which by decarboxylation and oxidation gives formate:

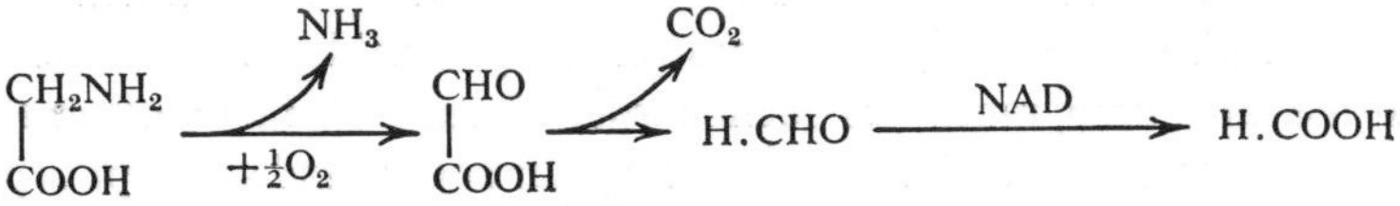

Other formyl groups are present in *formyl glutamate* for example; and

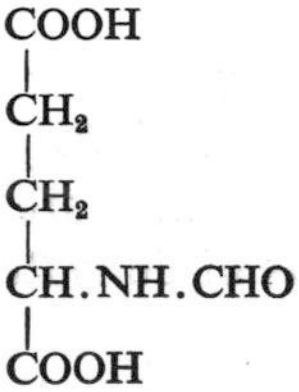

COOH
|
CH_2
|
CH_2
|
CH.NH.CHO
|
COOH

formyl glutamate

These formyl groups too can be transferred to THF, this time yielding 5-formyl-THF. Another contributory reaction takes place in which the hydroxymethyl group of serine gives rise to N^5:N^{10}-methylene-THF, a reaction which is dependent upon pyridoxal phosphate.

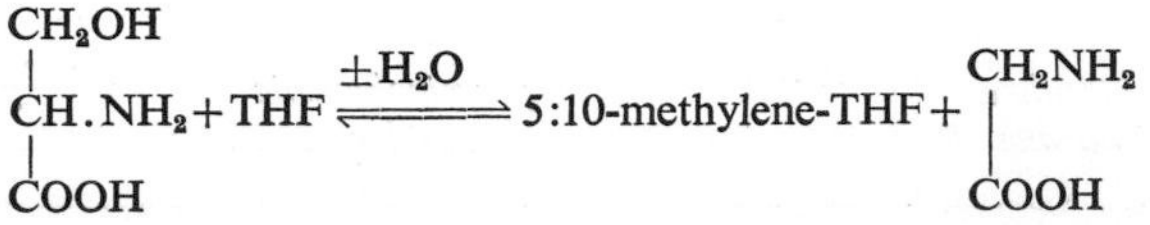

108

Beginning with formyl groups further important one-carbon units can be formed by manipulating the formyl-THF; some of these reactions are illustrated in Fig. 28.

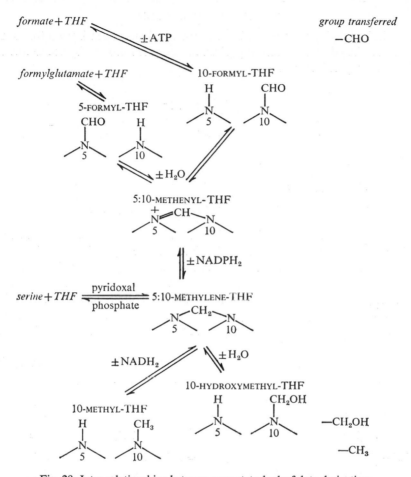

Fig. 28. Inter-relationships between some tetrahydrofolate derivatives.

Formyl groups from 10-formyl-THF enter into two steps in the synthesis of the purine ring system and into that of methylated pyrimidines (Fig. 55, p. 283) (p. 291); hydroxymethyl groups can be transferred to glycine, regenerating serine, while methyl groups are used in the production of methionine by way of S-adenosyl homocysteine (see below). These are only samples— THF-derivatives are doubtless involved in many other processes. Although many transmethylations are reversible this is by no means a general rule. Indeed, the transference of CH_3- from CH_3-THF to S-adenosyl homocysteine seems to be not only irreversible but to provide the sole source of new methyl groups in the organism, apart of course from any that may be present in the food.

TRANSMETHYLATION

Biological methylation is a well-known process. It has been known for many years that the administration of pyridine to dogs is followed by the excretion of N-methyl pyridine in the urine.

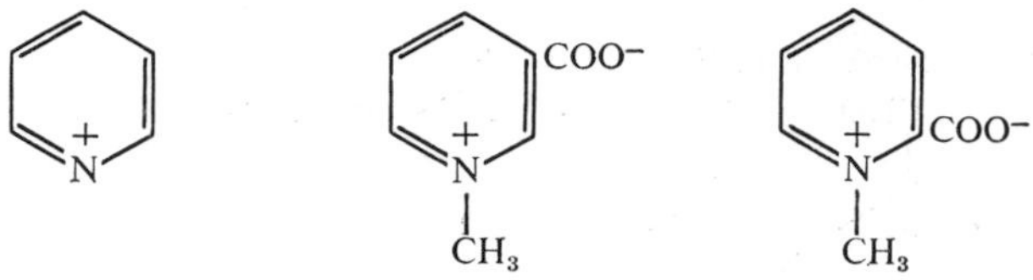

| N-methyl pyridine | trigonelline (N-methyl nicotinic acid) | homarine (N-methyl picolinic acid) |

Trigonelline also has been isolated from mammalian urine. It arises as an end-product of the metabolism of nicotinic acid. If large doses of the latter are given to mammals, a part is excreted unchanged, a part is conjugated with glycine to form nicotinuric acid, and a part undergoes N-methylation and gives trigonelline. *Homarine* is a methylated picolinic acid, isomeric with trigonelline, and may conceivably arise by biological methylation of picolinic acid, which is formed as a side product in the metabolism of tryptophan, together with nicotinic acid (p. 259).

It has been known for some time that glycocyamine undergoes biological conversion into *creatine*, and this change, too, involves methylation:

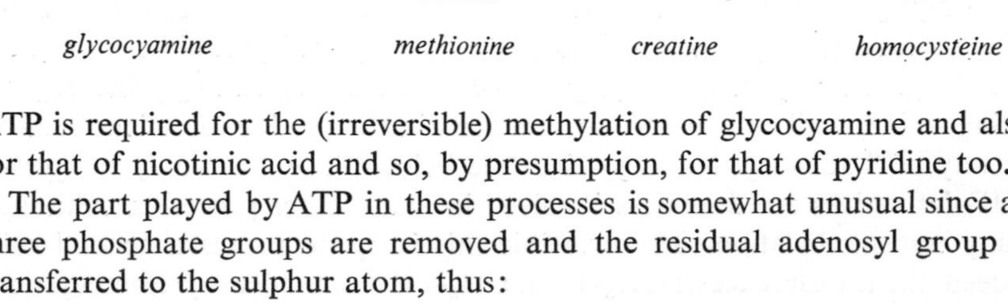

| glycocyamine | methionine | creatine | homocysteine |

ATP is required for the (irreversible) methylation of glycocyamine and also for that of nicotinic acid and so, by presumption, for that of pyridine too.

The part played by ATP in these processes is somewhat unusual since all three phosphate groups are removed and the residual adenosyl group is transferred to the sulphur atom, thus:

110

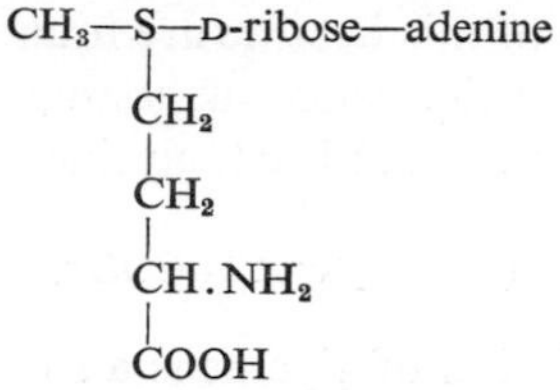

$$CH_3—S—\text{D-ribose—adenine}$$
$$\Big|$$
$$CH_2$$
$$\Big|$$
$$CH_2$$
$$\Big|$$
$$CH.NH_2$$
$$\Big|$$
$$COOH$$

S-adenosylmethionine

It is from this compound rather than from free methionine itself that the methyl group is transferred.

Many different animal tissues contain larger or smaller amounts of highly methylated nitrogenous compounds such as trimethylamine, trimethylamine oxide (see pp. 277–9) and various betaines, and the suggestion has often been made that they must arise from *choline*, $(CH_3)_3N^+.CH_2CH_2OH$, which is universally distributed as the basic constituent of phospholipids of the lecithin type. The methyl groups of choline can be transferred to S-adenosyl-homocysteine.

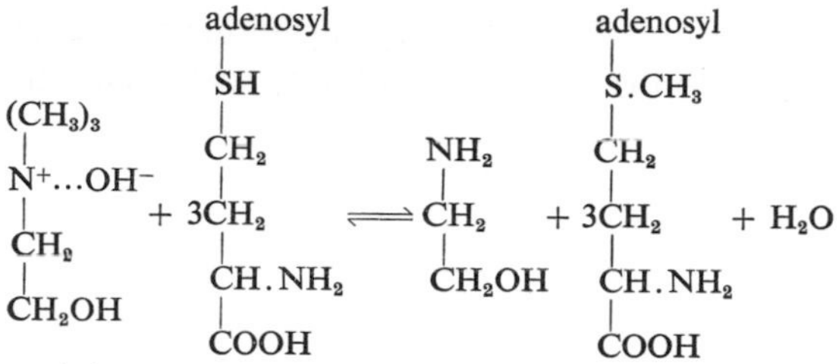

$$
\begin{array}{ccccc}
 & \text{adenosyl} & & \text{adenosyl} & \\
 & | & & | & \\
(CH_3)_3 & SH & & S.CH_3 & \\
| & | & NH_2 & | & \\
N^+...OH^- & CH_2 & | & CH_2 & \\
| & +\,3CH_2 & \rightleftharpoons CH_2 & +\,3CH_2 & +\,H_2O \\
CH_2 & | & | & | & \\
| & CH.NH_2 & CH_2OH & CH.NH_2 & \\
CH_2OH & COOH & & COOH & \\
\textit{choline} & & \textit{ethanol-} & & \\
 & & \textit{amine} & &
\end{array}
$$

The other product, *ethanolamine* (cholamine), replaces choline in phospholipids of the kephalin type and can also be formed by the decarboxylation of serine.

The discovery of a choline dehydrogenase in mammalian liver indicates that choline can be oxidized to betaine aldehyde, and the latter can then be further oxidized to *glycine betaine* itself, e.g. by the group-specific aldehyde oxidase of liver:

$$(CH_3)_3N^+.CH_2CH_2OH \xrightarrow{-2H} (CH_3)_3N^+.CH_2CHO \xrightarrow[+H_2O]{-2H} (CH_3)_3N^+.CH_2COO^-$$
$$\quad OH^- \qquad\qquad\qquad\qquad OH^-$$

choline   *betaine aldehyde*   *glycine betaine*

Betaines of various kinds are widely distributed among animals and plants alike. It is usually supposed that they are formed by the methylation of simpler substances, *glycine betaine* arising from glycine for example:

$$N_3N^+.CH_2COO^- \longrightarrow (CH_3)_3N^+.CH_2COO^-$$

glycine (zwitterion)   *glycine betaine*

111

As a rule, large amounts of this betaine are found in association with small quantities of free glycine and vice versa, suggesting that the two are metabolically interconvertible. In one case, at least, glycine has been found side by side with *sarcosine*,

$$(CH_3)H_2N^+.CH_2COO^-,$$

suggesting that the methylation of glycine is a step-wise process. The methyl groups of glycine betaine can be transferred, like those of choline, to S-adenosyl homocysteine so that transmethylation is reversible in this case.

Other interesting betaines include:

γ-butyrobetaine, $(CH_3)_3N^+.CH_2CH_2CH_2COO^-$,
crotonbetaine, $(CH_3)_3N^+.CH_2CH{=}CH.COO^-$, and
carnitine, $(CH_3)_3N^+.CH_2CH(OH)CH_2COO^-$,

which are closely related one to another, and all three are in part interconvertible, as has been shown by subcutaneous administration to dogs. It is possible that *γ-butyrobetaine* arises by methylation from γ-aminobutyrate; the latter has been found in animal materials, in which it can arise by the action of glutamate α-decarboxylyase upon glutamate (p. 251). *Carnitine* has a special part to play in the metabolism of fatty acids as we shall see (p. 426).

6

LYASES AND ISOMERASES

GENERAL INTRODUCTION

In this chapter we consider two smaller groups of enzymes. *Lyases* are enzymes that catalyse reactions of the general type

$$A.B \rightleftharpoons A+B$$

and can thus disrupt their substrates without the intervention of any other reagent. Alternatively they can synthesize AB from the separate component parts, A and B.

Isomerases, as their name suggests, catalyse reactions involving isomerization in their substrates and form a very intriguing group. They fall into two classes, *simple isomerases* and *mutases*.

LYASES

Enzymes adding or removing water (hydrolyases: hydratases)

Aconitase (aconitate hydratase) plays a part of great importance in oxidative metabolism. It might be classified either as a lyase or as an isomerase, since it catalyses the interconversion of citrate and *iso*-citrate through the intermediate stage of *cis*-aconitate. A water molecule is removed from citrate to yield aconitate. If the water molecule is then replaced the other way round, *iso*-citrate is formed:

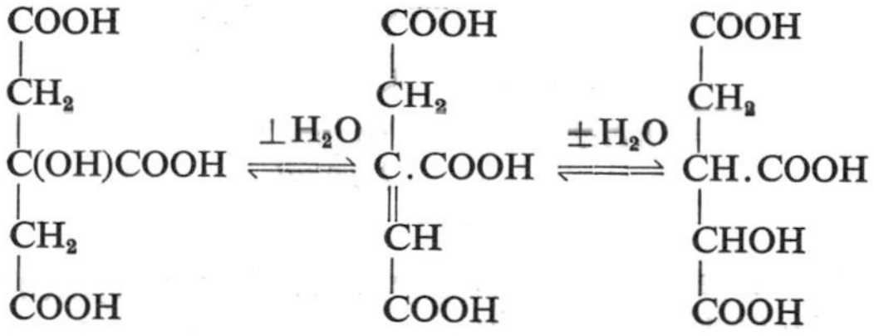

$$
\begin{array}{ccccc}
\text{COOH} & & \text{COOH} & & \text{COOH}\\
|\ \ & & |\ \ & & |\ \ \\
\text{CH}_2 & & \text{CH}_2 & & \text{CH}_2\\
|\ \ & \overset{\mp\text{H}_2\text{O}}{\rightleftharpoons} & |\ \ & \overset{\pm\text{H}_2\text{O}}{\rightleftharpoons} & |\ \ \\
\text{C(OH)COOH} & & \text{C.COOH} & & \text{CH.COOH}\\
|\ \ & & \|\ \ & & |\ \ \\
\text{CH}_2 & & \text{CH} & & \text{CHOH}\\
|\ \ & & |\ \ & & |\ \ \\
\text{COOH} & & \text{COOH} & & \text{COOH}\\
\textit{citric} & & \textit{cis-aconitic} & & \textit{iso-citric}\\
\textit{acid} & & \textit{acid} & & \textit{acid}
\end{array}
$$

Both reactions are reversible, and the enzyme is strongly inhibited by *trans*-aconitate and by fluorocitrate.

Enolase (phosphopyruvate hydratase), another enzyme of great importance in oxidative metabolism and of very wide distribution, catalyses the inter-

113

conversion of glycerate-2-phosphate and *enol*-pyruvate phosphate. The 'low-energy' phosphate of the former becomes a 'high-energy' phosphate in the latter:

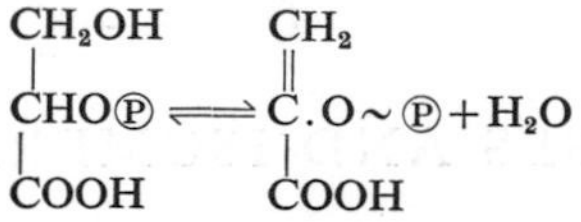

$$
\begin{array}{cc}
CH_2OH & CH_2 \\
| & \| \\
CHO\circledP & C.O\sim\circledP+H_2O \\
| & | \\
COOH & COOH
\end{array}
$$

This enzyme requires the presence of magnesium in fairly high concentrations. It has been isolated in the form of a catalytically inert mercury compound which becomes active if the mercury is removed and replaced by magnesium, manganese or zinc. The naturally occurring enzyme is believed to be the magnesium complex on account of its extreme sensitivity towards fluoride, which forms a complex, nonionized magnesium fluorophosphate in the presence of inorganic phosphate.

Fumarase (fumarate hydratase) catalyses the interconversion of fumarate and malate thus:

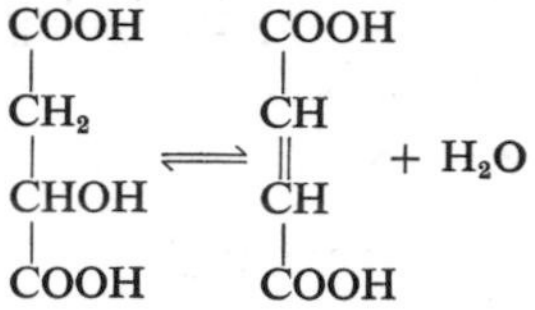

$$
\begin{array}{cc}
COOH & COOH \\
| & | \\
CH_2 & CH \\
| & \| \quad + H_2O \\
CHOH & CH \\
| & | \\
COOH & COOH
\end{array}
$$

It is very widely distributed and plays a very important part in metabolic oxidations.

Glyoxalase, an enzyme that is very widely distributed among living tissues, catalyses the conversion of methyl glyoxal to lactate:

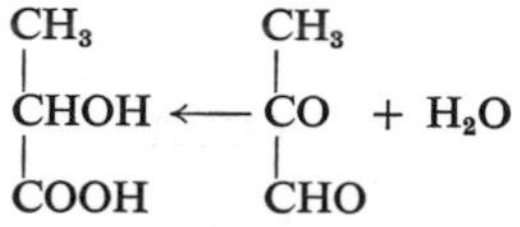

$$
\begin{array}{cc}
CH_3 & CH_3 \\
| & | \\
CHOH \longleftarrow & CO \quad + H_2O \\
| & | \\
COOH & CHO \\
\textit{lactic acid} & \textit{methyl} \\
& \textit{glyoxal}
\end{array}
$$

It requires glutathione as a specific activator. Two enzymes are involved in reality, the first of which catalyses the union of methyl glyoxal with the irreplaceable glutathione, while the second catalyses the decomposition of the resulting complex to yield lactate and regenerate glutathione. Its function is unknown.

'*Serine deaminase*' (serine dehydratase) is an enzyme that is involved in the deamination of serine. It catalyses the dehydration of serine to yield α-amino-acrylate which, by intramolecular rearrangement, yields α-iminopropionate

114

and this is spontaneously and probably simultaneously hydrolysed to give ammonia and pyruvate:

$$\begin{array}{c} CH_2OH \\ | \\ CH.NH_2 \\ | \\ COOH \end{array} \rightleftharpoons \begin{array}{c} CH_2 \\ \| \\ C.NH_2 \\ | \\ COOH \end{array} + H_2O;$$

$$\begin{array}{c} CH_2 \\ \| \\ C.NH_2 \\ | \\ COOH \end{array} \rightleftharpoons \begin{array}{c} CH_3 \\ | \\ C=NH \\ | \\ COOH \end{array} \xrightarrow{+H_2O} \begin{array}{c} CH_3 \\ | \\ CO \\ | \\ COOH \end{array} + NH_3$$

A similar but apparently not identical enzyme is involved in the deamination of *threonine*. Pyridoxal phosphate is required as a coenzyme for both of these enzymes.

Enoyl-Co A hydratase ('crotonase'), a more recently discovered lyase, catalyses reactions of the following general types:

$$R.CH=CH.CO\sim CoA + H_2O \rightleftharpoons R.CH(OH)CH_2.CO\sim CoA,$$
$$R.CH=CH.CO\sim CoA + H_2O \rightleftharpoons R.CH_2CH(OH).CO\sim CoA.$$

It plays an important part in the metabolic breakdown of fatty acids. The same enzyme can act also upon acyl CoA derivatives with a double bond in the $\beta:\gamma$ position (e.g. vinyl acetic acid).

Enzymes adding or removing carbon dioxide (carboxylases: carboxylyases)

Without doubt the most important reaction involving the fixation of carbon dioxide is that occurring in *photosynthesis*. It is discussed later (pp. 324–5).

Amino-acid decarboxylyases, each highly specific with respect to a particular amino-acid, have been found in various bacteria, many of which are able, especially in somewhat acid media, to catalyse reactions of the following general type:

$$R.CH(NH_2)\overline{[COO]}H \longrightarrow R.CH_2NH_2 + CO_2.$$

The products formed in this way include such highly toxic bodies as tyramine, histamine, putrescine and cadaverine, all of which are strongly basic and tend, therefore, to neutralize the acidity of the medium in which they are produced. The bacterial decarboxylyases require the co-operation of a 'co-decarboxylyase' which has been isolated and identified with pyridoxal phosphate. Pyridoxal, it will be remembered, is a member of the B_2 group of vitamins.

Animal tissues contain a number of enzymes comparable with the bacterial decarboxylyases; these include enzymes that are specific for the α-decarboxylation of glutamate, tyrosine, histidine and cysteic acid; all of these again require pyridoxal phosphate as a coenzyme. At present there is no evidence that any of these amino-acid decarboxylyases can act reversibly.

Carbonic anhydrase, an enzyme that plays an important part in the transport of respiratory carbon dioxide in the higher animals, may be regarded either as a hydro-lyase or, alternatively, as the prototype of the important group of carboxylyases. Carbonic anhydrase itself catalyses the splitting of carbonic acid to yield carbon dioxide and water and, in the reverse direction, the hydration of carbon dioxide to yield carbonic acid:

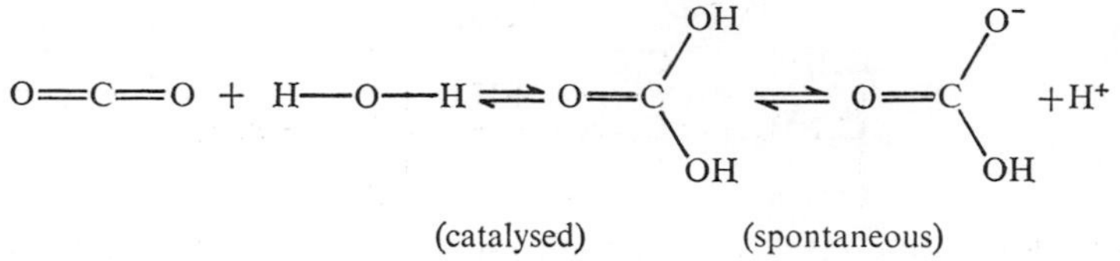

$$O{=}C{=}O + H{-}O{-}H \rightleftharpoons O{=}C\overset{OH}{\underset{OH}{\diagup\diagdown}} \rightleftharpoons O{=}C\overset{O^-}{\underset{OH}{\diagup\diagdown}} + H^+$$

(catalysed)   (spontaneous)

This enzyme has a wide distribution in animal tissues, and is especially abundant in erythrocytes, from which it has been prepared, highly purified, and shown to be a zinc-containing protein. It plays an important part in the secretion of HCl by the gastric mucosa and appears to be involved in the formation of the shells of birds' eggs. No doubt it has other important functions as well.

Carboxylase (pyruvate decarboxylyase), originally discovered in yeast, appears to occur in some other micro-organisms and also in plants. It is, however, entirely absent from animal tissues. This enzyme contains magnesium and in the presence of its coenzyme, co-carboxylase, which is identical with the diphosphate of vitamin B_1 (thiamine diphosphate, formula p. 30), catalyses a simple decarboxylation of α-keto-acids, notably that of pyruvate:

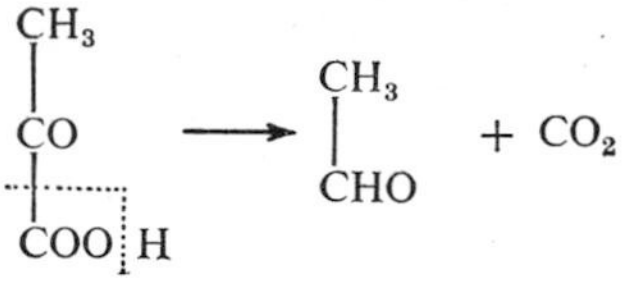

$$\begin{array}{c}CH_3\\ |\\ CO\\ |\\ COO\,|\,H\end{array} \longrightarrow \begin{array}{c}CH_3\\ |\\ CHO\end{array} + CO_2$$

There is no reason to believe that this enzyme can act reversibly but we shall discuss the mode of action of the coenzyme later on in a more convenient context.

'*Pyruvate oxidase*' contrasts sharply with carboxylyase (above), for it catalyses a complex process of decarboxylation that is essentially oxidative in nature and is for that reason known as oxidative decarboxylation. It is more fully discussed on pp. 387–9.

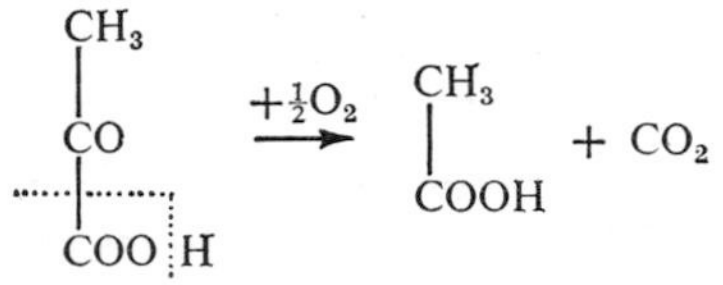

$$\begin{array}{c}CH_3\\ |\\ CO\\ |\\ COO\,|\,H\end{array} \xrightarrow{+\frac{1}{2}O_2} \begin{array}{c}CH_3\\ |\\ COOH\end{array} + CO_2$$

The so-called pyruvate oxidase is not a single enzyme but a somewhat complex catalytic system as we shall see, but it resembles pyruvate carbo-

116

xylyase in two noteworthy respects, first that it catalyses the decarboxylation of α-keto-acids but has no action upon β-keto-acids, and secondly that its activity again requires the participation of co-carboxylase (diphospho-thiamine).

Malate decarboxylyase (decarboxylating) may be mentioned at this point. This enzyme may be classed as a decarboxylyase and as a dehydrogenase, for it catalyses a simultaneous dehydrogenation and decarboxylation of malate, using NADP as its hydrogen acceptor:

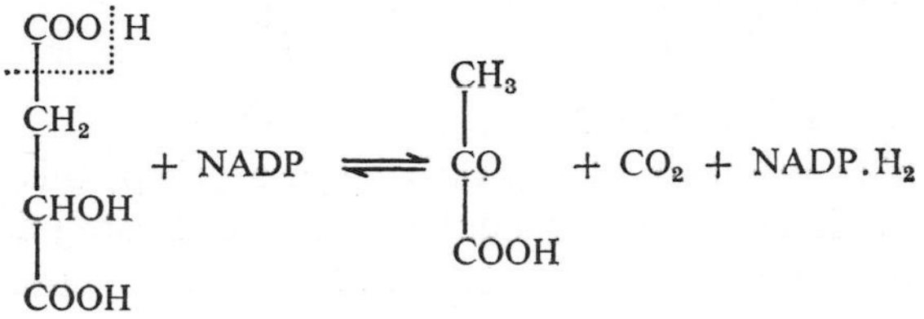

$$
\begin{array}{c}
\text{COO}\vdots\text{H} \\
\text{CH}_2 \\
| \\
\text{CHOH} \\
| \\
\text{COOH}
\end{array}
+ \text{NADP} \rightleftharpoons
\begin{array}{c}
\text{CH}_3 \\
| \\
\text{CO} \\
| \\
\text{COOH}
\end{array}
+ \text{CO}_2 + \text{NADP.H}_2
$$

In this case the reaction, an atypical oxidative decarboxylation, is reversible (cf. 'pyruvate oxidase'). This enzyme, which has been obtained from pigeon liver and highly purified, probably plays an important part in CO_2-fixation in liver tissue.

Oxaloacetate decarboxylyase is an important and widely distributed enzyme which catalyses the reversible interconversion of pyruvate and oxaloacetate by addition and removal of carbon dioxide as follows:

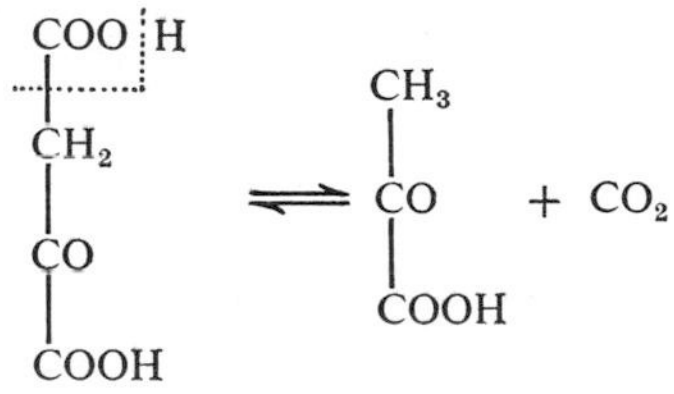

$$
\begin{array}{c}
\text{COO}\vdots\text{H} \\
\text{CH}_2 \\
| \\
\text{CO} \\
| \\
\text{COOH}
\end{array}
\rightleftharpoons
\begin{array}{c}
\text{CH}_3 \\
| \\
\text{CO} \\
| \\
\text{COOH}
\end{array}
+ \text{CO}_2
$$

The equilibrium is heavily in favour of pyruvate (cf. pyruvate carboxylyase (p. 119)). There are indications that biotin, another B_2 vitamin, is in some way implicated in the action of this enzyme (see next page).

Phosphoenol pyruvate carboxylyase catalyses the synthesis of oxaloacetate, this time from phosphoenol pyruvate and carbon dioxide, but this time guanosine triphosphate is required:

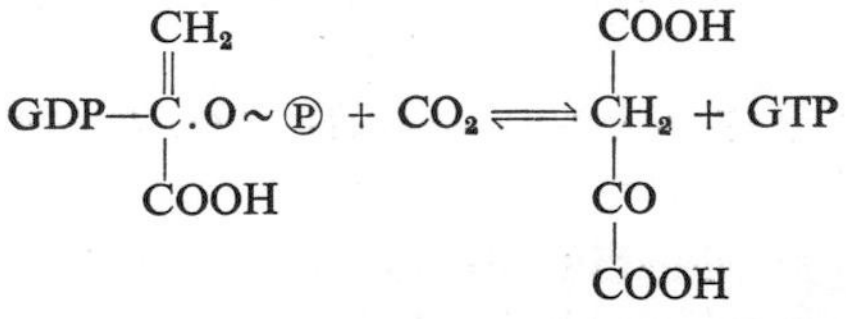

$$
\begin{array}{c}
\text{CH}_2 \\
\parallel \\
\text{GDP—C.O}\sim\circledP \\
| \\
\text{COOH}
\end{array}
+ \text{CO}_2 \rightleftharpoons
\begin{array}{c}
\text{COOH} \\
| \\
\text{CH}_2 \\
| \\
\text{CO} \\
| \\
\text{COOH}
\end{array}
+ \text{GTP}
$$

Guanosine diphosphate can be replaced by inosine diphosphate (IDP) in this reaction, and equilibrium is heavily in favour of phosphoenol pyruvate.

9-2

Oxalosuccinate decarboxylase similarly catalyses the interconversion of oxalosuccinate (α-keto-β-carboxyglutarate) and α-keto-glutarate (see also *iso*-citrate dehydrogenase, p. 166).

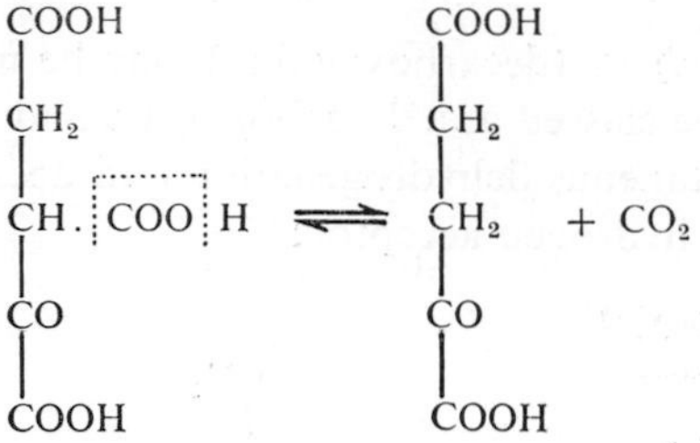

$$
\begin{array}{ccc}
\text{COOH} & & \text{COOH} \\
| & & | \\
\text{CH}_2 & & \text{CH}_2 \\
| & & | \\
\text{CH.}[\text{COO}]\text{H} & \rightleftharpoons & \text{CH}_2 \quad + \text{CO}_2 \\
| & & | \\
\text{CO} & & \text{CO} \\
| & & | \\
\text{COOH} & & \text{COOH}
\end{array}
$$

Like the other members of the carboxylase group, this enzyme has the distinction, which is shared by aldolase (below), of catalysing the formation and rupture of direct carbon-to-carbon linkages. It requires manganese ions for activation.

BIOTIN-DEPENDENT CARBOXYLASES†

A number of carboxylases require the participation of *biotin*, another vitamin of the B_2 group. It operates in conjunction with a protein (enzyme) the union lying between its carboxyl group and a lysyl residue in the protein;

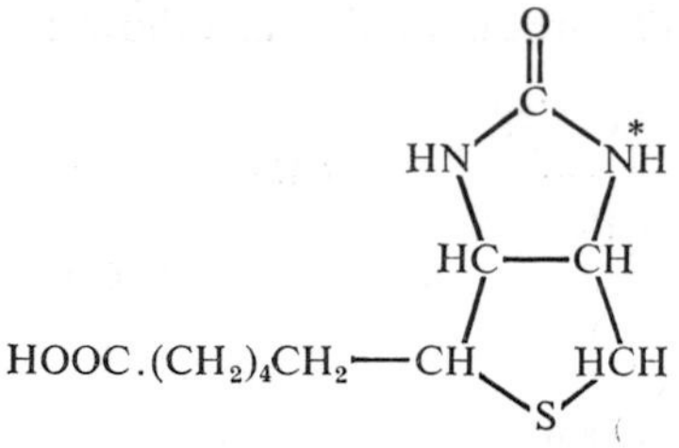

$$
\begin{array}{c}
\text{O} \\
\parallel \\
\text{C} \\
\diagup \quad \diagdown \\
\text{HN} \qquad \overset{*}{\text{NH}} \\
| \qquad\qquad | \\
\text{HC}\text{---}\text{CH} \\
\text{HOOC.(CH}_2)_4\text{CH}_2\text{---CH} \qquad \text{HCH} \\
\diagdown \quad \diagup \\
\text{S}
\end{array}
$$

biotin

ATP is required in all the reactions where biotin is concerned and is believed to react with the —NH group marked with an asterisk in the formula, giving adenyl∼biotin. The activated biotin then exchanges its adenylate residue for a molecule of CO_2 and yields 'active CO_2', in other words biotin∼CO_2. However, there is reason to think that the two reactions proceed simultaneously and in concert so that the overall reaction is:

$$
\overset{\diagdown}{\underset{\diagup}{\text{N}}}\text{H} + \text{ATP} + \text{CO}_2 \longrightarrow \overset{\diagdown}{\underset{\diagup}{\text{N}}} \sim \text{CO}_2 + \text{ADP} + \text{HO.}\textcircled{P}
$$

The activated CO_2 provided in this way takes part in a number of important synthetic reactions in which biotin, ATP and Mn^{2+} are always required.

† Officially classified as ligases (synthetases) although they act by CO_2-transfer from biotin-CO_2.

Acetyl-CoA carboxylyase catalyses the carboxylation of acetyl-CoA to produce malonyl-CoA, an important intermediate in the synthesis of fatty acids;

$$CH_3CO.CoA + CO_2 \xrightarrow{\text{biotin} \sim CO_2} \begin{array}{l} COOH \\ | \\ CH_2.CO.CoA \end{array}$$

ATP and Mn^{2+} are required in addition to biotin and CO_2.

Propionyl-CoA carboxylyase plays a part in the conversion of propionate into carbohydrate. In the first instance propionate is transformed into methylmalonyl-CoA and then follows a special isomerase reaction involving vitamin B_{12}. The product, succinyl-CoA, is a potential precursor of carbohydrate:

$$CH_3CH_2CO.CoA \xrightarrow{\text{biotin} \sim CO_2} \begin{array}{l} COOH \\ | \\ CH_3.CH.CO.CoA \end{array} \xrightarrow{B_{12}} \begin{array}{l} COOH \\ | \\ CH_2CH_2CO.CoA \end{array}$$

Pyruvate carboxylyase catalyses the fixation of CO_2 by pyruvate to form oxaloacetate:

$$\begin{array}{l} CH_3 \\ | \\ CO \\ | \\ COOH \end{array} \xrightarrow{\text{biotin} \sim CO_2} \begin{array}{l} COOH \\ | \\ CH_2 \\ | \\ CO \\ | \\ COOH \end{array}$$

This enzyme occurs in mammalian liver but not, apparently, within the extrahepatic tissues.

It is worth noticing in passing that there exists in raw egg white a substance called *avidin* which reacts strongly and specifically with biotin and consequently acts as a powerful inhibitor of biotin-catalysed reactions. In fact avidin can be and has been used as a diagnostic tool for the detection of biotin-dependent reactions.

Other lyases

Aldolase (zymohexase, cf. transaldolase, p. 106). This important enzyme was originally discovered in yeast but is now known to be very widely distributed indeed. It catalyses the splitting of one molecule of fructofuranose-1:6-diphosphate into two molecules of 'triose phosphate', i.e. one molecule each of D-glyceraldehyde-3-phosphate and dihydroxyacetone phosphate. The process is reversible:

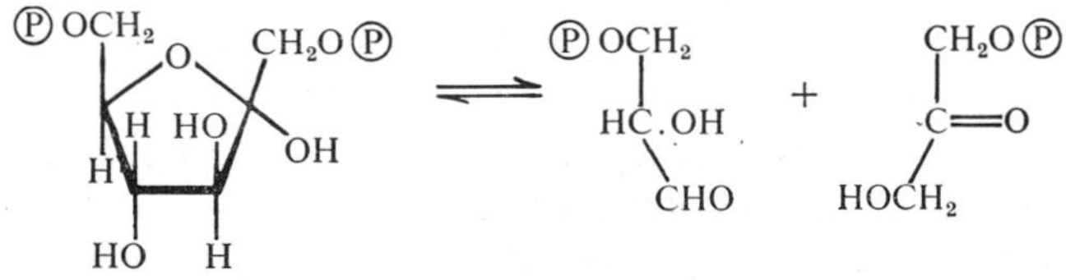

| *fructofuranose-* | D-*glyceraldehyde-* | *dihydroxyacetone* |
| 1:6-*diphosphate* | 3-*phosphate* | *phosphate* |

Aldolase is absolutely specific with respect to one of the products, namely, dihydroxyacetone phosphate, but is group-specific towards the other. Glyceraldehyde-3-phosphate can be replaced by other aldehydes, which need not necessarily be phosphorylated, so that many new compounds can be synthesized enzymatically by aldolase, including fructofuranose-1-phosphate for example.

Aspartase is an example of an enzyme that catalyses the addition and removal of ammonia:

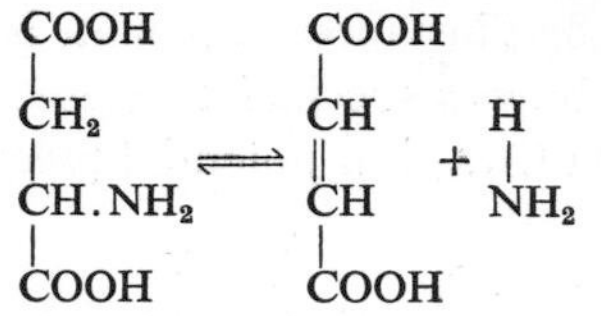

$$
\begin{array}{ccc}
\text{COOH} & \text{COOH} & \\
| & | & \\
\text{CH}_2 & \text{CH} & \text{H} \\
| & \| & + \; | \\
\text{CH.NH}_2 & \text{CH} & \text{NH}_2 \\
| & | & \\
\text{COOH} & \text{COOH} & \\
\textit{aspartic} & \textit{fumaric} & \\
\textit{acid} & \textit{acid} &
\end{array}
$$

It occurs in certain bacteria.

ISOMERASES

There remains to be considered a group of enzymes which catalyse isomerization in their substrates. There appear to be at least two types, the first, the members of which catalyse simple isomerization, are known as *isomerases*.

Triosephosphate isomerase is the longest known representative of the group, and was formerly known simply as 'isomerase'. It catalyses the interconversion of D-glyceraldehyde-3-phosphate and dihydroxyacetone phosphate:

$$
\begin{array}{cc}
\text{CH}_2\text{O}\text{\textcircled{P}} & \text{CH}_2\text{O}\text{\textcircled{P}} \\
| & | \\
\text{CHOH} \rightleftharpoons & \text{CO} \\
| & | \\
\text{CHO} & \text{CH}_2\text{OH}
\end{array}
$$

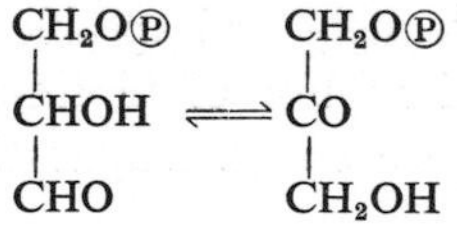

It is usually found in association with aldolase, from which it has however been separated and highly purified. Its mode of action is unknown, but conceivably the interconversion takes place by way of a hypothetical di-enol which is common to both substances:

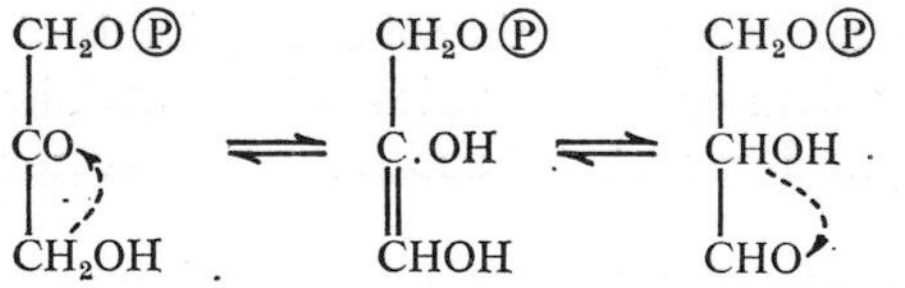

$$
\begin{array}{ccc}
\text{CH}_2\text{O}\,\text{\textcircled{P}} & \text{CH}_2\text{O}\,\text{\textcircled{P}} & \text{CH}_2\text{O}\,\text{\textcircled{P}} \\
| & | & | \\
\text{CO} \rightleftharpoons & \text{C.OH} \rightleftharpoons & \text{CHOH} \\
| & \| & | \\
\text{CH}_2\text{OH} & \text{CHOH} & \text{CHO}
\end{array}
$$

dihydroxyacetone phosphate   (di-enol)   glyceraldehyde-3-phosphate

Aconitase, which we have already mentioned (p. 113), acts by converting citrate and *iso*-citrate into a common intermediate, *cis*-aconitate, and it is possible that the isomerases in general act by converting the pairs of isomeric substances upon which they act into intermediate compounds which are common to both members of each pair.

120

Glucosephosphate isomerase (phosphohexoseisomerase) catalyses the interconversion of glucose- and fructofuranose-6-phosphates:

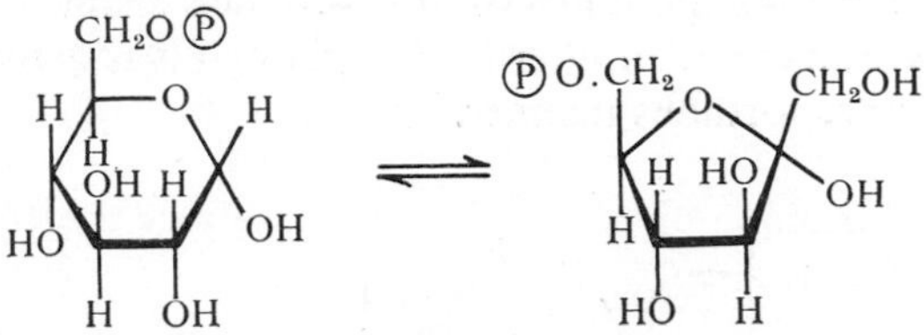
CH₂O Ⓟ
H
O
H
H
OH H
HO
OH
H OH
Ⓟ O.CH₂
O
CH₂OH
H HO
OH
H
HO H

In this case the common intermediate is presumably the open-chain form of the sugar phosphates. This reaction involves the conversion of an aldose to the corresponding ketose sugar without changing the position of the phosphate group. A *mannosephosphate isomerase* has also been described. It converts mannose-6-phosphate into fructose-6-phosphate. A *ribosephosphate isomerase*, more recently discovered, catalyses a similar reaction, viz. the interconversion of ribose-5-phosphate and ribulose-5-phosphate:

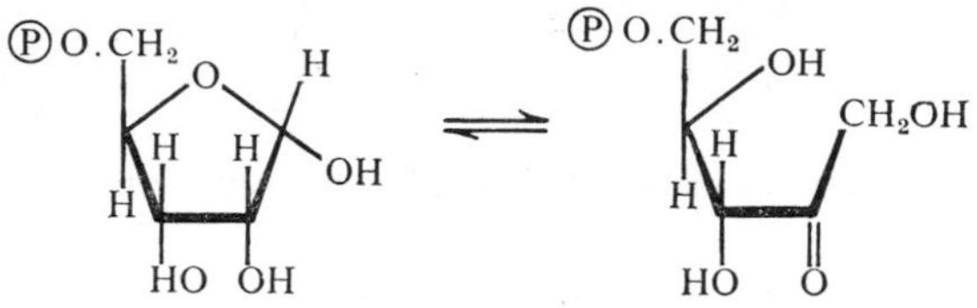
Ⓟ O.CH₂
O
H
H H
OH
H
HO OH
Ⓟ O.CH₂
OH
CH₂OH
H
H
HO O

ribose-5-phosphate *ribulose-5-phosphate*

Ribulosephosphate epimerase (phosphoketopentose epimerase) catalyses the interconversion of the 5-phosphates of ribulose and xylulose, a pair of ketopentoses, again probably by way of a common intermediate:

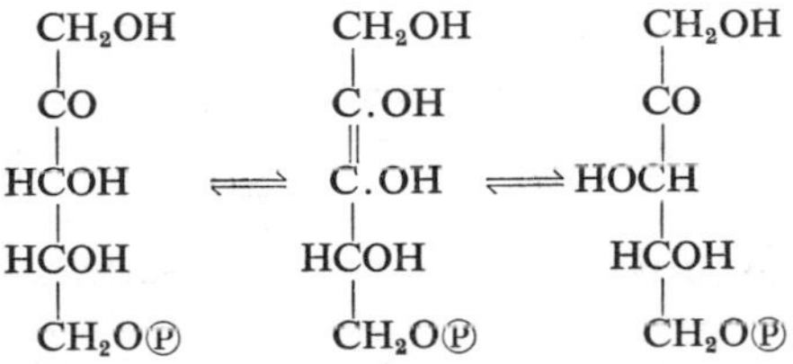
CH₂OH
CO
HCOH
HCOH
CH₂OⓅ
CH₂OH
C.OH
C.OH
HCOH
CH₂OⓅ
CH₂OH
CO
HOCH
HCOH
CH₂OⓅ

ribulose-5-Ⓟ (*di-enol*) *xylulose*-5-Ⓟ

UDP-glucose epimerase (galactose phosphate isomerase) is present in galactose-trained yeast and in mammalian liver, and is of rather special interest since it catalyses the interconversion of galactose-1-phosphate and glucose-1-phosphate, a change which corresponds to a Walden inversion at carbon 4. This enzyme requires a coenzyme which is present in yeast, mammalian liver and elsewhere. The coenzyme has been isolated and shown to be uridine diphosphate glucose (UDPG):

$$\text{galactose-1-Ⓟ} + \text{UDP–Glucose} \rightleftharpoons \text{UDP–Galactose} + \text{glucose-1-Ⓟ}.$$

This transformation also involves NAD as shown in Fig. 29. It is difficult to imagine how the inversion could be achieved without destroying the asymmetry at the carbon atom at position 4. Once again there is evidence for the existence of a common intermediate though it is arrived at this time in a somewhat more tortuous manner.

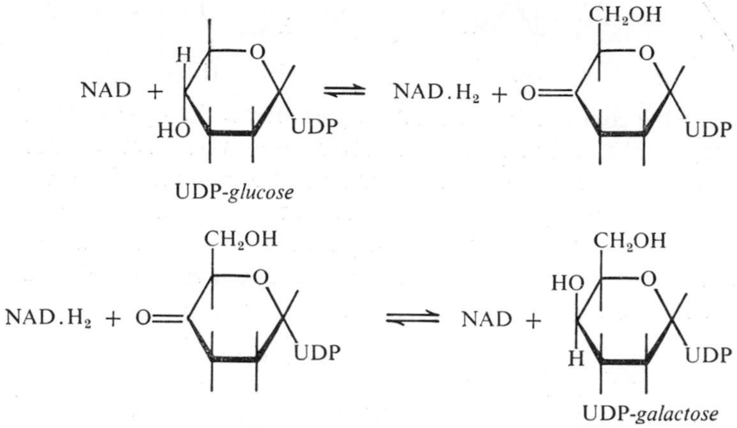

UDP-*glucose*

Fig. 29. Mechanism of action of UDP-glucose epimerase.

Phosphoglucomutase, which is found in association with glucosephosphate isomerase, is representative of isomerizing enzymes of another type, for it catalyses the interconversion of α-glucose-1-phosphate and glucose-6-phosphate:

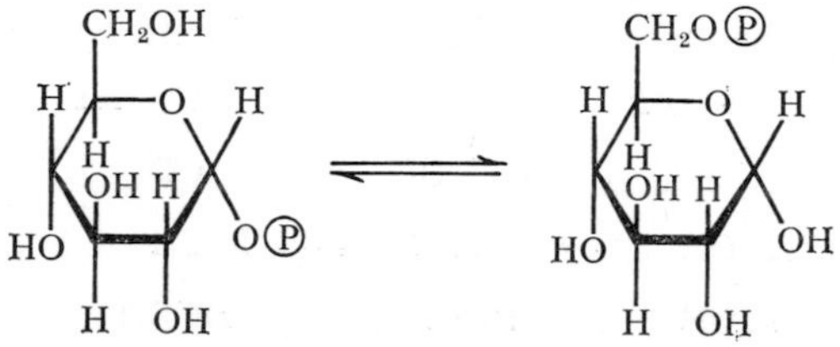

Similar enzymes react with the 1-phosphates of mannose and galactose, converting them into the corresponding 6-esters (see also the case of ribose below). In these cases the isomerization is rather more radical than in the simple isomerases since it involves a shift in the position of the phosphate group and the interconversion of an ester and a glycoside. It is usual to call isomerizing enzymes *mutases* when there is some shift, such as that indicated above, as opposed to a simple intramolecular rearrangement.

It has been shown that this enzyme requires glucose-1:6-diphosphate as a coenzyme. This substance can be produced by the action of a specific kinase upon glucose-6-phosphate in the presence of ATP, and also by a transfer reaction between two molecules of glucose-1-phosphate, in which ATP is not involved:

$$2 \text{ }\alpha\text{-glucose-1-phosphate} \xrightarrow{\text{ATP}} \text{glucose-1:6-diphosphate} + \text{glucose.}$$

It appears that each molecule of the monophosphorylated substrate can take over a second phosphate group from the doubly phosphorylated prosthetic group, which thus becomes the product of the reaction, the former substrate molecule becoming the new prosthetic group. There is thus a continuous stream of traffic across the active centres of the enzyme. Probably a common intermediate complex must be formed between the prosthetic group of the enzyme and the substrate, perhaps as follows:

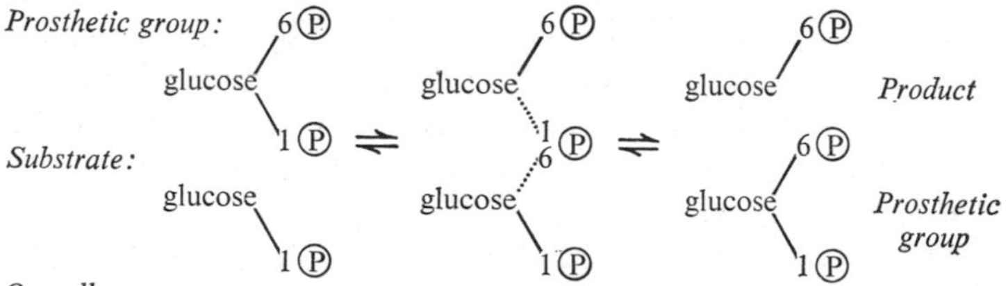

Prosthetic group:

| glucose 6-(P) | glucose 6-(P) | glucose 6-(P) | *Product* |

Substrate:

| glucose 1-(P) ⇌ | glucose 1/6-(P) ⇌ | glucose 6-(P) | |
| glucose 1-(P) | glucose 1-(P) | glucose 1-(P) | *Prosthetic group* |

Overall:

glucose-1-(P) + glucose-1:6-di-(P) ⇌ glucose-1:6-di-(P) + glucose-6-(P).

The same or a similar enzymes can also work with ribose-1:5-diphosphate, which can be produced from glucose-1:6-diphosphate and ribose-1-phosphate:

ribose-1-(P) + glucose-1:6-di-(P) ⇌ ribose-1:5-di-(P) + glucose-6-(P).

Together with this new prosthetic group the enzyme can then catalyse the interconversion of the 1- and the 5-phosphate esters of ribose:

ribose-1-(P) + ribose-1:5-di-(P) ⇌ ribose-1:5-di-(P) + ribose-5-(P).

The 1- and the 5-esters of D-2-deoxyribose are similarly interconvertible.

Phosphoglyceromutase is another enzyme of the mutase type. It catalyses the interconversion of glycerate 3-phosphate and the corresponding 2-phosphate:

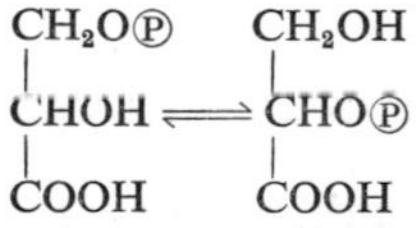

$$\begin{array}{ccc} CH_2O(P) & & CH_2OH \\ | & & | \\ CHOH & \rightleftharpoons & CHO(P) \\ | & & | \\ COOH & & COOH \end{array}$$

Here again a prosthetic group is present and consists of glycerate-2:3-diphosphate, a substance of which the biological existence but not the function has long been known. Its mode of action probably resembles that of phosphoglucomutase.

Next in order we shall deal with the oxidoreductases. These are typical transferring enzymes which deal in hydrogen atoms or electrons; they are so numerous and so closely related in function that they deserve separate chapters to themselves.

OXIDOREDUCTASES: OXIDASES

THE OXIDATION OF ORGANIC COMPOUNDS

Chemical compounds may be oxidized in a variety of different ways. Oxygen may be added to the molecule, as when hydrogen is oxidized to form water, or hydrogen may be removed, as when hydrogen sulphide is oxidized to free, elemental sulphur. Again, oxidation can be effected by the removal of electrons, as when ferrous salts containing Fe^{2+} are oxidized to the corresponding ferric compounds, in which Fe^{3+} is present. However, biological oxidation most commonly takes place by the removal of hydrogen by the process known as *dehydrogenation*.

In biological systems the hydrogen atoms are usually removed in pairs and although it is convenient to refer to each such pair as 2H, this does not tell the whole truth. Hydrogen atoms are not usually removed as such but in ionic form. In many and perhaps most cases each H^+ is accompanied by the two electrons through which it was bonded in the substrate, so that the complete assembly of $1H^+ + 2$ electrons constitutes a hydride ion;

$$H^+ + 2 \text{ electrons} = H^-.$$

The oxidation of, say, ethanol to acetaldehyde, can be written in the following manner, where the dotted arrow shows the movement of a pair of electrons to form a new double bond in the product:

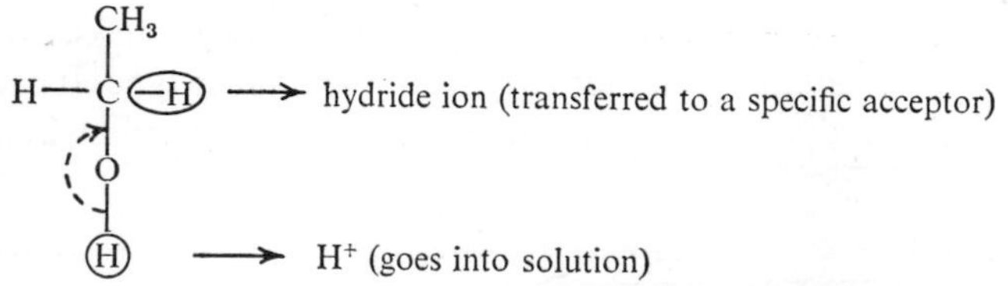

Similarly in the oxidation of acetaldehyde to acetic acid we have a preliminary and spontaneous formation of the aldehyde hydrate; then

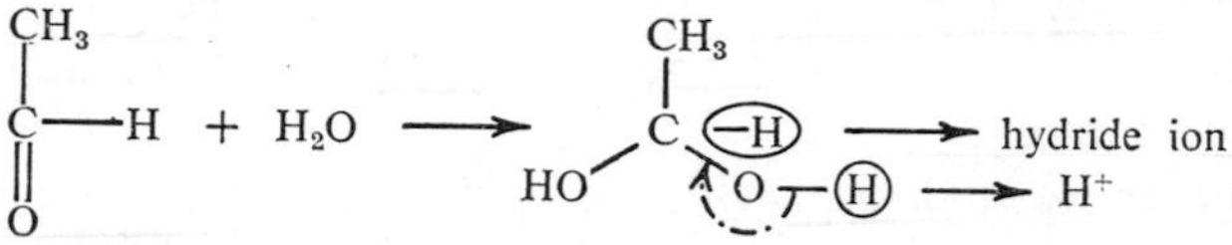

This, then, is the kind of mechanism most usually employed in biological oxidations. In the two cases just quoted, one positive hydrogen ion and one

124

negative hydride ion are removed from the substrate so that the pair, taken collectively, carry no charge, and this provides some justification for the abbreviation, 2H. Sometimes, as we shall see later on, hydrogen transfer reactions are better expressed in terms of electron transfers and in such cases the 2H notation is better expressed as

$$2H = H^- + 1 \text{ proton } (H^+) = H^+ + 2 \text{ electrons} + 1 \text{ proton } (H^+) = 2H^+ + 2 \text{ electrons.}$$

For example we might write

$$4H^+ + 4 \text{ electrons} + O_2 \longrightarrow 2H_2O$$

Similarly

$$2H^+ + 2 \text{ electrons} + O_2 \longrightarrow H_2O_2$$

or alternatively

$$1 \text{ hydride ion} + 1 \text{ hydrogen ion} + O_2 \longrightarrow H_2O_2$$
$$(H^-) \qquad\qquad (H^+)$$

It must be remembered that whenever one substance is oxidized at the expense of another, the oxidation of the first is necessarily attended by the reduction of the second. Hence it is usual to speak of an 'oxidation-reduction reaction', and enzymes that catalyse such reactions are known as '*oxido-reductases*'. If the process is a biological one it can usually be represented in general terms by the equation:

$$A.H_2 + B \rightleftharpoons A + B.H_2$$

and we can define a number of terms which are in general use in discussions of reactions of this kind. The reductant, AH_2, is known as the *hydrogen donor*, and the oxidant, B, as the *hydrogen acceptor*. A third factor is involved in biological oxidation-reduction reactions, viz, the *catalyst*.

In a classical series of experiments, Wieland showed that colloidal palladium can play the part of a combined catalyst and hydrogen acceptor in many oxidation-reduction reactions of this kind; for example,

$$CH_3CH(OH)COOH + (Pd) \longrightarrow CH_3COCOOH + (Pd).2H$$

and

$$CH_3CHO + H_2O + (Pd) \longrightarrow CH_3COOH + (Pd).2H$$

He also found that specimens of palladium that had become 'charged' with hydrogen in this way can pass on their hydrogen to certain reducible substances such as the synthetic dye, methylene blue. In the case of methylene blue itself, reduction yields a colourless substance, leuco-methylene blue, which we may write for the sake of convenience as $MB.H_2$. Thus, in the presence of an oxidizable substance, AH_2, together with methylene blue, palladium black catalyses two processes, first the dehydrogenation of AH_2 (reaction *a*), and secondly the reduction of methylene blue (reaction *b*):

(*a*) $AH_2 + (Pd) = A + (Pd).H_2$

(*b*) $(Pd).2H + MB = (Pd) + MB.H_2$

125

Hydrogen taken up from the primary hydrogen donor, AH_2, is passed on to the dye, the catalyst acting as an *intermediary carrier of hydrogen*. This significant fact becomes more apparent if we write the equations in the following unorthodox but very descriptive fashion:

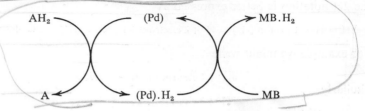

By using this method of expression we can emphasize the essentially *cyclical* manner in which the carrier catalyst acts, a small amount being alternately hydrogenated and dehydrogenated over and over again, and thus participating in a very large amount of chemical change. Alternatively, if we only knew that palladium acts catalytically but did not know how it does so, we might write the overall process as follows:

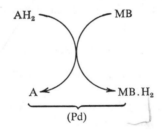

The bracket is used here to indicate that (Pd) acts as the catalyst, and the scheme should be interpreted as meaning that AH_2 and methylene blue react together under the catalytic influence of (Pd) with production of A and leuco-methylene blue.

Now living cells and extracts prepared from them by suitable methods are able to oxidize many organic compounds such, for example, as glucose, lactate and succinate. These compounds are quite stable in aqueous solution, and it follows, therefore, that the cells contain enzymes which catalyse their oxidation. Today these enzymes are studied mostly by spectrophotometric techniques or by methods using oxygen electrodes. However, the nature and distribution of these enzymes, to which the name of *dehydrogenases* has been given, were first studied extensively by Thunberg, using much simpler methods, by taking advantage of the extreme ease with which methylene blue can be reduced, and we shall do well to consider some of these early observations.

If methylene blue is added to a suspension of chopped muscle, brain, kidney, etc., or to a suspension of yeast or bacteria, the dye is rapidly reduced, thus revealing the presence of reducing systems in the cells. If such an experiment is tried in the laboratory it is usually found that the dye is decolorized in the bulk of the mixture but not at the surface. This is because leuco-methylene blue is rapidly and spontaneously re-oxidized by atmospheric oxygen. To obviate this difficulty Thunberg introduced the use of vacuum tubes of the type illustrated in Fig. 30 in which is also shown a more modern version in which the simple stopper is replaced by a curved, hollow device. Substrate can be put into the hollow stoppers of these tubes and tipped into the rest of the reaction mixture at any desired moment. Suitable tissue preparations are placed in these tubes, together with buffer, methylene blue and other

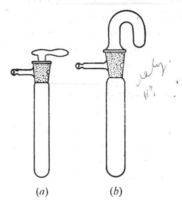

(*a*)   (*b*)

Fig. 30. Thunberg vacuum tubes. (*a*) Original type, (*b*) newer type, with hollow stopper. (Drawings by H. Mowl.)

appropriate reagents, the stoppers are greased and inserted, and the tubes are then evacuated. When exhaustion has been completed the stopper is turned and the tube is thus sealed. Alternatively, the tubes may be filled with an inert gas, such as hydrogen or nitrogen. Mixing of the contents of the tube with those of the stopper starts the reaction under investigation. In the presence of an active reducing system the methylene blue is reduced, and its reduction is, in effect, an indication of the presence of such a system.

Although it has long been superseded this method is quick, cheap and very convenient, and has the additional advantage of giving quantitative results, since the time taken to decolorize a given quantity of methylene blue under standard conditions of temperature, pH and so on gives an inverse measure of the activity of the system concerned. Other dyes such as cresyl blue, pyocyanine, and simpler organic substances such as *m*-dinitrobenzene and *o*-quinone can replace methylene blue.

With the aid of this simple but ingenious technique, Thunberg carried out many important investigations on living cells and tissues. If a sample of minced muscle tissue is placed together with buffer and methylene blue in a vacuum tube and the latter evacuated, the dye is reduced very rapidly. If now the experiment is repeated using boiled muscle the dye is no longer reduced, indicating that the catalysts concerned are thermolabile and therefore probably enzymes. If a third experiment is performed in which unboiled and unwashed muscle tissue is replaced by tissue that has been minced and well washed, the time taken for the reduction of the dye is much increased. Some part of the complete reducing system has therefore been removed by washing. If now substances such as succinate, lactate and the like are added to the mixture of washed tissue, buffer and methylene blue, a rapid reduction of the dye again takes place.

Succinate, lactate, malate, α-glycerolphosphate, glucose, aldehydes and alcohols are among the many substances that can be activated by tissue dehydrogenases from one or another source. Even molecular hydrogen can be oxidized by a bacterial enzyme. We know now that there is not one single, master dehydrogenase but a large number of different individual dehydrogenases. Some, of which the succinate enzyme is an example, act only upon one natural substrate, in this case succinate; others, such as the aldehyde oxidase (Schardinger enzyme) of milk, catalyse the oxidation of any of a wide range of substrates, in this case aldehydes, aliphatic or aromatic. The dehydrogenases thus show the phenomenon of specificity. They are thermolabile, and their activity is profoundly affected by pH. They are susceptible to the action of many enzyme inhibitors, and often to that of narcotic substances such as the higher alcohols and various substituted ureas. In short, they show all the properties characteristic of enzymes.

It was realized fairly early in the history of the dehydrogenases that some members of the group require the co-operation of coenzymes because, after exhaustive washing of the tissues, the reduction time for some substrates is very much increased. It can, however, be shortened again by adding boiled extracts of muscle, in which the necessary coenzymes are present. In other cases, however, the addition of boiled muscle-juice does not restore activity, the reason being that some dehydrogenases are themselves soluble in water and are therefore removed by vigorous washing of the tissue.

It must be emphasized once more that the use of methylene blue is not in any sense a 'natural' procedure. The dye does not occur in nature, and is used simply as a convenient hydrogen acceptor for the visual demonstration of the existence of natural reducing systems. Methylene blue merely—but very conveniently—*replaces* the natural hydrogen acceptors of the tissues, and our next inquiries must be into the nature and identity of these substances.

The first likely natural hydrogen acceptor that comes to mind is, of course, molecular oxygen, since it is at the expense of molecular oxygen that tissue oxidations are carried out in the end. But if we set up experimental systems in which a given dehydrogenase and its substrate are mixed together in the presence of oxygen instead of methylene blue, we find that while virtually all known dehydrogenase systems can reduce methylene blue, only a few actually take up oxygen. Thus, contrary to all expectation, *molecular oxygen is not the natural hydrogen acceptor for most dehydrogenase systems*, and we must look further. In the meantime, however, we can distinguish between two groups of dehydrogenases: those which can utilize molecular oxygen directly as a hydrogen acceptor and are sometimes called *aerobic dehydrogenases*, and the remainder, which operate through hydrogen acceptors other than oxygen and are known as *anaerobic dehydrogenases* (see Table 9).

Table 9 *Classification of dehydrogenating enzymes*

		H-acceptors reduced		Product
		MB	O_2	
Dehydrogenases	anaerobic	+	−	−
	aerobic	+	+	H_2O_2
Oxidases	aerobic	−	+	H_2O

Attention must also be drawn to another important group of oxidizing enzymes, the *aerobic oxidases*, which, unlike Thunberg's 'anaerobic' dehydrogenases, cannot reduce methylene blue and similar hydrogen acceptors and consequently escaped discovery for some time. It may be doubted whether their inability to reduce synthetic dyes like methylene blue constitutes adequate grounds for regarding them as essentially different from the aerobic dehydrogenases, but there are other differences too.

Another less common group comprises enzymes known as *oxygenases*; these are among the few oxidoreductases which can incorporate environmental oxygen into the products of oxidation instead of removing hydrogen from their subtrates. As a matter of convenience they will be considered here along with the aerobic dehydrogenases under the general heading of oxidases.

OXIDASES

The oxidases are distinguished from the other dehydrogenating enzymes by their ability to use molecular oxygen as a hydrogen acceptor. The aerobic oxidases are specifically confined to oxygen as a natural acceptor and cannot use other naturally occurring substances, but in the case of the aerobic dehydrogenases, molecular oxygen can be replaced for experimental purposes by methylene blue.

The oxidases may be broadly divided into two groups, the first of which catalyses the reduction of molecular oxygen to water. The second group, comprising the aerobic dehydrogenases, leads to the formation of hydrogen peroxide. The overall reactions may be generally expressed as follows:

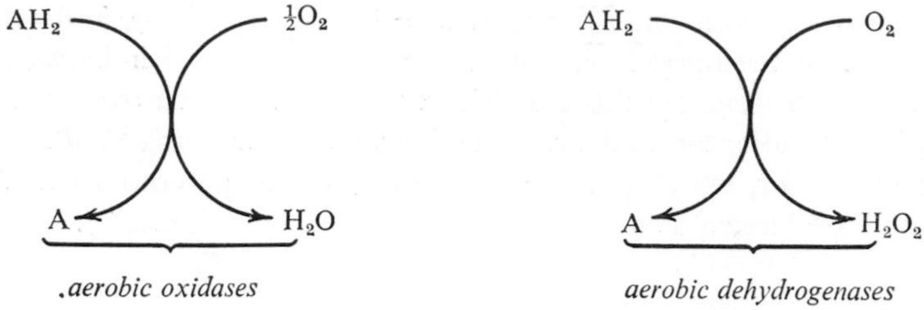

.aerobic oxidases   aerobic dehydrogenases

It seems to be characteristic of these enzymes that they are conjugated proteins, although in several cases no prosthetic group has so far been identified. In general, however, the presence of such a group has been demonstrated and its identity has been established. Among the known prosthetic materials are copper in the phenol oxidases and flavin adenine dinucleotide, often together with heavy metallic components, especially iron and molybdenum, in aerobic dehydrogenases.

Aerobic oxidases

Phenol oxidases

The phenol oxidases are a group of enzymes that catalyse the oxidation of phenolic substances. Several representatives of the group have been purified and shown to be copper-containing proteins, thus resembling oxygen-carrying pigments of the haemocyanin type. They also have a requirement for Fe^{2+}.

Monophenol oxidase, isolated from mushrooms, catalyses the oxidation of monophenols to the corresponding *o*-quinones:

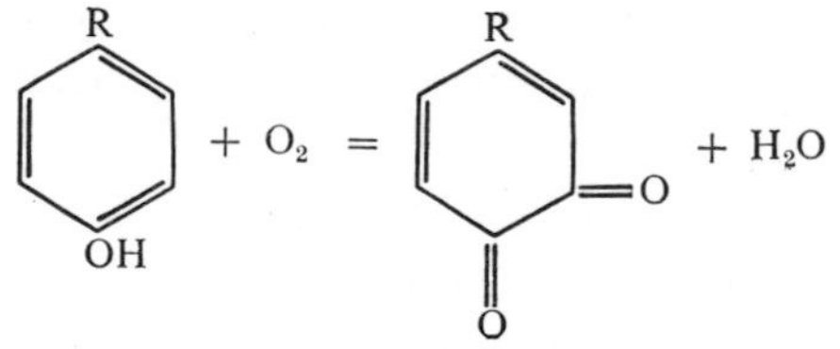

The intimate details of the process are not known. It seems unlikely that an *o*-diphenol is formed as an intermediate product, because this enzyme acts much more strongly upon mono- than upon diphenols. It uses molecular oxygen as hydrogen acceptor but cannot reduce methylene blue.

o-Diphenol oxidase (polyphenol oxidase) has been isolated from mushrooms and potatoes and this too is a copper compound. Enzymes of this kind have no immediate action upon monophenols, but act rapidly upon *o*-diphenols such as catechol to form the corresponding *o*-quinones in the first instance:

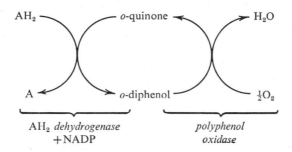

$$\tfrac{1}{2}O_2 + \text{[catechol]} = \text{[o-quinone]} + H_2O.$$

catechol   *o-quinone*

Triphenols such as pyrogallol are also attacked. Another member of this group of phenol oxidases, *laccase*, has been isolated from the latex of the lac tree, and this too is a copper protein but differs from the others in specificity in that it attacks *p*- rather than *o*-diphenols.

The primary oxidation is followed as a rule by further changes which are spontaneous, but before we consider these further reactions, attention may be drawn to one very important feature of the polyphenol oxidase system. The *o*-quinones formed can be rapidly reduced again by ascorbic acid and by certain reducing systems such, for example, as glucose-6-phosphate dehydrogenase together with its substrate and the appropriate coenzyme (NADP). If we represent the reducing substrate by AH_2 the reactions can be written in the following manner:

$$
\begin{array}{ccccc}
AH_2 & & o\text{-quinone} & & H_2O \\
 & \diagdown\diagup & & \diagdown\diagup & \\
 & \diagup\diagdown & & \diagup\diagdown & \\
A & & o\text{-diphenol} & & \tfrac{1}{2}O_2 \\
\end{array}
$$

AH₂ dehydrogenase   *polyphenol*
+NADP   *oxidase*

In this system a very small amount of the diphenol, e.g. catechol, or the corresponding quinone can be alternately reduced and oxidized many times and thus can act as an intermediate carrier of hydrogen between a theoretically unlimited amount of AH_2 on the one hand and molecular oxygen on the other. Such a system catalyses a continuous uptake of oxygen and a simultaneous oxidation of AH_2 in equivalent amount. The amount of carrier required is very small indeed, and the carrier must, in fact, be regarded as a catalyst in its own right. We have in this system a biological counterpart of the palladium systems studied by Wieland (p. 126), while Thunberg's methylene-blue systems can act in a similar way under aerobic conditions:

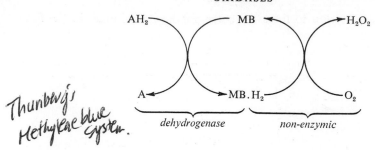

Thunberg's Methylene blue System.

dehydrogenase   non-enzymic

Systems of this kind are very interesting because they can be regarded as 'models' of the respiratory systems of living cells. The latter contain reducing substances, represented by AH_2, together with the appropriate dehydrogenases and coenzymes. These pass on hydrogen to acceptors, the identity of which we shall discuss presently, and, at the other end of the chain, molecular oxygen acts as the ultimate hydrogen acceptor and is reduced in the process.

Whether systems involving polyphenol oxidase play any important part in the respiration of plant tissues is uncertain, but the possibility has certainly not been excluded. Indeed, the respiration of spinach leaves is greatly increased by dihydroxyphenylalanine. This is a natural diphenolic constituent of the leaves and is susceptible to the action of polyphenol oxidase, which is abundant in these leaves.

Polyphenol oxidase systems are also involved in the 'browning' of potatoes, apples and many other plant materials that takes place when a cut surface is exposed to the air. It is usually found that plants containing polyphenol oxidase also contain traces of *o*-diphenols, especially of catechol. Catechol, for example, is oxidized to the corresponding *o*-quinone through the agency of the oxidase (reaction (i)) and the remaining stages take place spontaneously:

(i)   + $\tfrac{1}{2}O_2$ ⟶ + H_2O

     catalysed

catechol

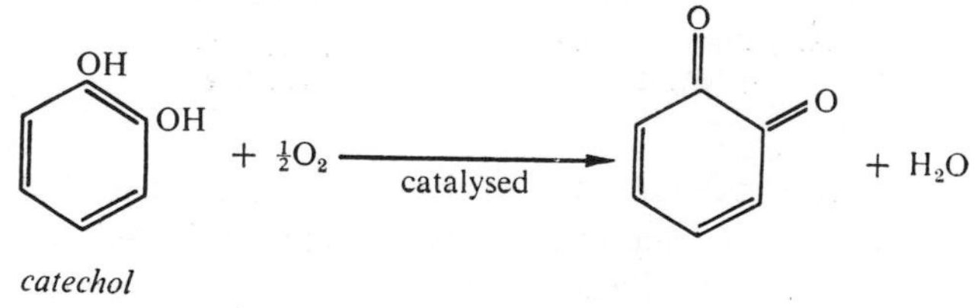

(ii)   + H_2O ⟶

     spontaneous

132

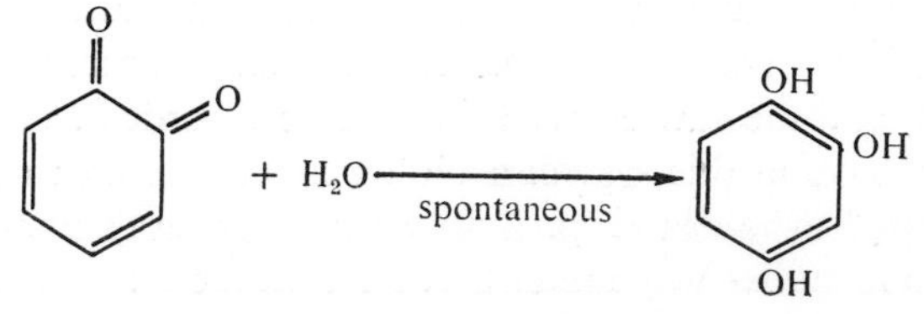

(iii)

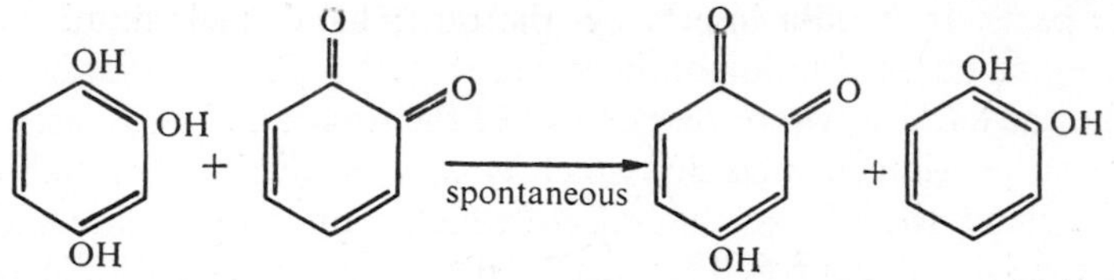

spontaneous

Finally, the hydroxyquinone undergoes polymerization to yield complex, dark-coloured, melanic products, the constitution of which is still uncertain. *o*-Quinone formed in reaction (i) is consumed in reactions (ii) and (iii), so that melanic products can continue to be formed only so long as *o*-quinone is supplied by the action of polyphenol oxidase in reaction (i).

Tyrosinase, which closely resembles and may even be identical with *o*-diphenol (polyphenol) oxidase, occurs widely in plants and animals alike. It is this enzyme that is responsible for the production of most of the dark brown and black melanins of animals. Tyrosine itself is probably the starting material

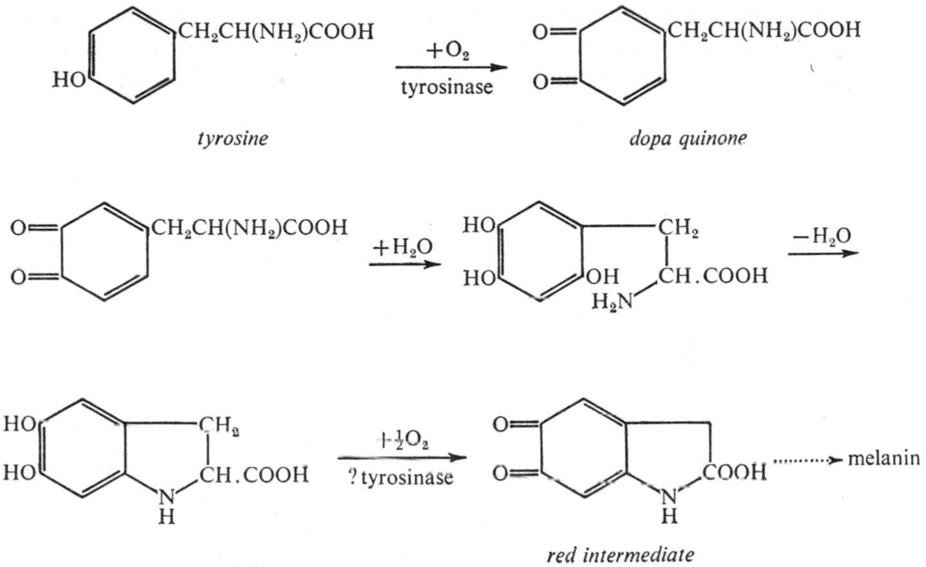

tyrosine

$+O_2$
tyrosinase

dopa quinone

$+H_2O$

$-H_2O$

$+\frac{1}{2}O_2$
? tyrosinase

········> melanin

red intermediate

in many cases of melanin formation, but other phenolic compounds too may be used. From certain insects, for example, protocatechuic (3:4-dihydroxy-benzoic) acid has been isolated as a natural substrate; others contain 3:4-dihydroxyphenylacetic acid and 3:4-dihydroxyphenyl-lactic acid. In albinism, where there is a characteristic and complete lack of melanins, tyrosinase is absent, while in piebald animals such as rabbits and guinea-pigs the dark portions of the skin contain tyrosinase, but the enzyme is absent from the

10-2

white parts. In certain insects, on the other hand, melanic pigmentation is determined by localization of the substrate rather than by that of the enzyme. Again, the dark brown or black 'ink' of the squids and octopuses consists of a fine suspension of melanin, which is very insoluble. It is elaborated in a special gland, the ink-sac, the walls of which contain tyrosinase and are at the same time rich in copper. This was for many years the only indication we had that the tyrosinase of animals is a copper protein, but we now know that mammalian tyrosinase is inhibited by reagents, e.g. cyanide, which form complexes with copper, and is released from this inhibition by Cu^{2+} though not by other metallic ions.

The first stage in the formation of melanin from tyrosine probably consists in the oxidation of tyrosine to the quinone of dihydroxyphenylalanine ('dopa quinone'). This is followed by spontaneous ring closure and by the possibly enzymic oxidation of the product to yield a red intermediate compound. Further changes, which are mainly spontaneous and include polymerization, give rise to melanin itself.

Ascorbic oxidase. Although ascorbic acid is one of the substances known to be oxidizable through the polyphenol oxidase-catechol system of plants, many plants, notably the *Brassica* family, are rich in an ascorbic oxidase. This enzyme catalyses the oxidation of ascorbic to dehydroascorbic acid:

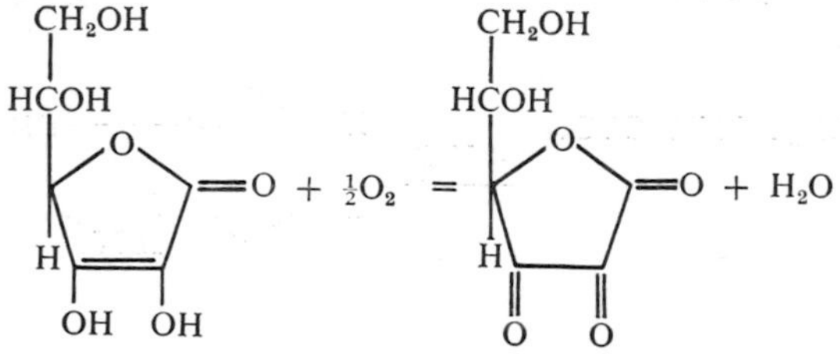

Ascorbic acid can reduce the *o*-quinonoid bodies formed by the action of polyphenol oxidase and, so long as any ascorbic acid is present in the plant or other tissue, *o*-quinones do not accumulate and discoloration, due to melanin production from such *o*-quinones (p. 132), is held in check.

The enzyme has been isolated from squash and shown to contain copper, and it is interesting to notice that a mixture of egg albumin with traces of copper ions can imitate the action of ascorbic oxidase remarkably closely.

Urate oxidase (uricase), an enzyme concerned in purine metabolism, catalyses the oxidation of uric acid to allantoin. Highly purified preparations from mammalian liver contain copper but no trace of the iron and zinc said to be present in the earlier and still impure preparations. Its action is almost certainly complex, and tracer studies carried out with ^{15}N implicate hydroxyacetylene-diureine-carboxylic acid as an intermediate:

134

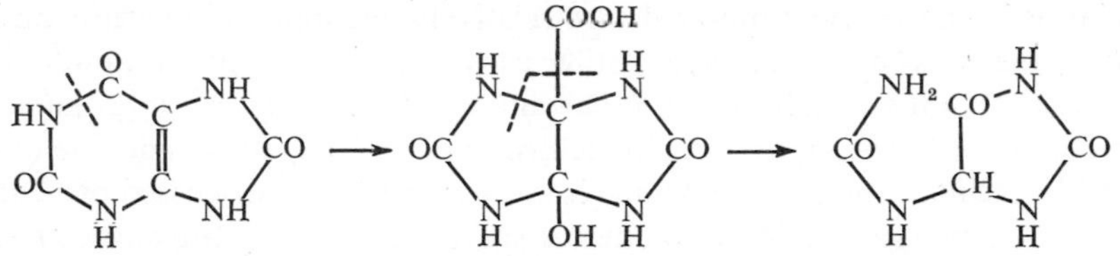

uric acid   (intermediate)   allantoin

This enzyme occurs in the livers of animals which are not uricotelic, and is present, for instance, in most mammalian livers. To this, however, there are certain curious exceptions (see p. 288).

This enzyme will catalyse the oxidation of several purines other than uric acid, but is competitively inhibited by certain others, notably by methylated derivatives of uric acid.

Cytochrome oxidase, a very important and almost universally distributed enzyme, is another example of a metalloprotein, and its prosthetic group, which contains iron, is allied to haem. It also contains copper; but we shall return to discuss this enzyme in greater detail. It is another example of an aerobic oxidase.

Aerobic dehydrogenases

D-*Amino-acid oxidase*, unlike the oxidases so far discussed, is able to utilize methylene blue as a hydrogen acceptor although its natural acceptor is molecular oxygen, which it reduces to hydrogen peroxide. This enzyme occurs in the liver and kidney tissue of mammals, and is probably present in many other animal materials. It is a curious fact that this enzyme acts only upon members of the non-natural D-series of amino-acids and is without action upon the natural L-acids. As far as the D-acids are concerned it is group-specific and attacks them all with a few exceptions, e.g. D-glutamate.

D-Amino-acid oxidase catalyses the oxidative deamination of its substrates to the corresponding α-keto-acids. This takes place in two stages, of which the first, consisting in the dehydrogenation of the substrate, is catalysed, while the second, in which the resulting imino-acid is hydrolysed, may possibly be spontaneous:

$$\underset{\substack{| \\ \text{COOH} \\ \text{D-}amino\text{-}acid}}{\overset{\text{R}}{\underset{|}{\text{CH.NH}_2}}} + O_2 = \underset{\substack{| \\ \text{COOH} \\ imino\text{-}acid}}{\overset{\text{R}}{\underset{|}{\text{C}=\text{NH}}}} + H_2O_2;$$

$$\underset{\substack{| \\ \text{COOH}}}{\overset{\text{R}}{\underset{|}{\text{C}=\text{NH}}}} + H_2O = \underset{\substack{| \\ \text{COOH} \\ \alpha\text{-}keto\text{-}acid}}{\overset{\text{R}}{\underset{|}{\text{CO}}}} + NH_3.$$

135

As far as is known, D-amino-acids are relatively uncommon in nature, and it may be that this enzyme has some other as yet unknown important biological function. That it exists, however, is certain, and it has in fact been isolated. It is a conjugated protein, the prosthetic group of which is flavin adenine dinucleotide, the mode of action of which is known and will be discussed presently.

Other deaminating oxidases include a specific *glycine oxidase* and a group-specific L-*amino-acid oxidase*, both of which appear to be flavoproteins. The powerful L-*amino-acid oxidase* of viper venom (*Vipera aspis*) also deserves a passing reference.

Xanthine oxidase is present in milk, which offers a relatively rich source from which highly concentrated preparations can fairly easily be obtained.

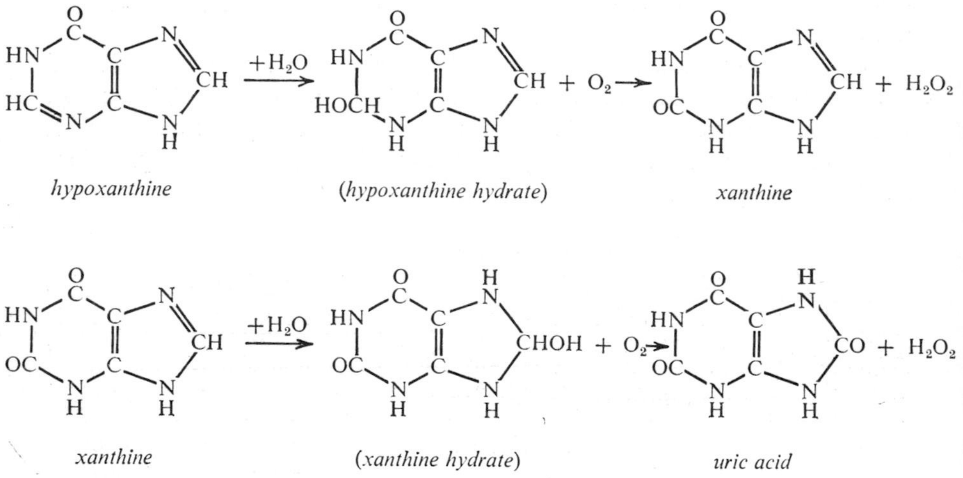

hypoxanthine   (hypoxanthine hydrate)   xanthine

xanthine   (xanthine hydrate)   uric acid

Fig. 31. Intermediate stages in oxidation of hypoxanthine and xanthine to uric acid.

The enzyme is widely though somewhat erratically distributed among animal tissues. It catalyses the oxidation of hypoxanthine to xanthine and that of xanthine to uric acid: here, as in the oxidation of aldehydes by dehydrogenation, it is probable that intermediate hydrates are formed as shown in Fig. 31.

The specificity of xanthine oxidase towards purines is very high. Its action has been tested upon numerous purines and related substances, but, apart from hypoxanthine and xanthine, no naturally occurring purines are attacked at all rapidly. The enzyme does, however, act upon certain synthetic substances such as 6:8-dihydroxypurine and upon purine itself, while it is said to be strongly and competitively inhibited by adenine.

In addition to its action upon the purines, xanthine oxidase has a second sphere of activity. It appears to be identical with the so-called Schardinger enzyme of milk, a catalyst which acts upon many aldehydes, oxidizing them to the corresponding acids:

$$\underset{aldehyde}{\overset{\displaystyle R}{\underset{\displaystyle H}{\overset{\displaystyle |}{\underset{\displaystyle |}{C}}}}=O} \quad\xrightarrow{+H_2O}\quad \underset{\substack{aldehyde\\hydrate}}{\overset{\displaystyle R}{\underset{\displaystyle H}{\overset{\displaystyle |}{\underset{\displaystyle |}{C}}}}\!\!\overset{OH}{\underset{OH}{<}}} \;+\;O_2 \quad\longrightarrow\quad \underset{acid}{\overset{\displaystyle R}{\underset{\displaystyle \|}{\overset{\displaystyle |}{C}}}\!\!\overset{OH}{<}} \;+\;H_2O_2.$$

Xanthine oxidase has been highly concentrated and, like those of D-amino-acid oxidase, the purified preparations contain flavin adenine dinucleotide, together with molybdenum and iron. The presence of these metals in flavin-containing enzymes seems to be widespread and hence the name *metallo-flavoproteins*. This particular enzyme has been studied more perhaps than any other of the metallo-flavoprotein group and there is good evidence for the belief that the hydrogen (or electron) transport chain is the following:

$$\text{substrate} \longrightarrow \text{molybdenum} \longrightarrow \text{FAD} \longrightarrow \text{iron} \longrightarrow O_2.$$

This may yet prove to be a general phenomenon in this group of enzymes.

Aldehyde oxidase of liver. Mammalian liver contains another aldehyde oxidase and a similar enzyme is present in the potato. These enzymes, like xanthine oxidase, are group-specific towards aldehydic substrates, which they oxidize to the corresponding acids; but, unlike xanthine oxidase, they have no activity towards purines. The prosthetic group contains flavin adenine dinucleotide, once more together with molybdenum.

Other important metalloflavoproteins concerned with the reduction of nitrate to nitrite have been found in a number of plants.

Glucose oxidase ('notatin') occurs in the mould *Penicillium notatum* and in some other species. It has been thoroughly purified and proves to have flavin adenine dinucleotide as its prosthetic group. The high specificity of this enzyme makes it a useful analytical tool, for it can be used for the quantitative determination of D-glucose, even in mixtures containing several sugars. The product of oxidation is D-gluconic acid, which is formed by way of the δ-gluconolactone (cf. p. 163).

Amine oxidase is a copper-containing enzyme that occurs at small concentrations in many animal tissues. It acts particularly rapidly upon compounds such as tyramine and adrenaline, and appears to be identical with the tyramine and adrenaline oxidases described many years ago. Its action upon primary amines may be described as follows:

(i) $R.CH_2NH_2 + O_2 = R.CH:NH + H_2O_2$,

(ii) $R.CH:NH + H_2O = R.CHO + NH_3$.

137

The mode of oxidation thus resembles that of the amino-acid oxidases somewhat, and it is possible that the second reaction is not catalysed but spontaneous. The enzyme also acts upon secondary and tertiary amines to form either a lower amine or else an aldehyde, together with ammonia.

An interesting point to notice here is that, while this enzyme acts upon many amines, it is inhibited by those which contain a methyl group in the α-position, for instance by ephedrine and benzedrine. Adrenaline is normally destroyed in the tissues by amine oxidase, in part at least, and its disappearance is much retarded by these drugs. Probably, therefore, these compounds owe their adrenaline-like action upon the organism to an indirect action which inhibits the oxidation of adrenaline itself, rather than to any direct action upon the tissues. The structural relationships between these substances are as follows:

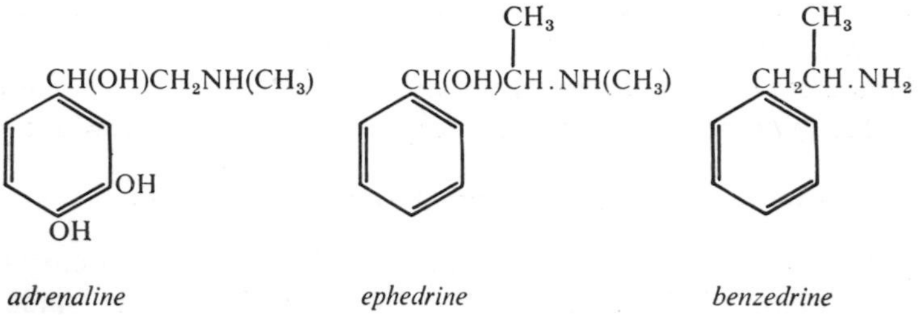

| adrenaline | ephedrine | benzedrine |

Diamine oxidase also occurs in animal tissues and is reputed to use pyridoxal phosphate as a coenzyme or prosthetic group. It catalyses the oxidation of histamine, putrescine, cadaverine and agmatine, for example, substances which can arise in the gut of animals by bacterial decarboxylation of the corresponding amino-acids. It is probable that both the amine oxidases are important in the oxidative detoxication of the poisonous amines which arise in this manner.

A *trimethylamine oxidase* has been found in cod muscle and probably plays an important part in the metabolism of trimethylamine and trimethylamine oxide. This latter product is an important substance in the metabolism of certain fishes and occurs in many other marine organisms.

The fate of hydrogen peroxide

The formation of hydrogen peroxide in the course of the *in vitro* action of aerobic dehydrogenases may be demonstrated by taking advantage of the fact that cerous hydroxide, a colourless substance, slightly soluble in water, reacts readily with hydrogen peroxide to form an insoluble, yellowish brown ceric peroxide. While there can be no doubt that hydrogen peroxide is indeed formed in isolated enzyme systems, there is little evidence that it is formed, or at any rate that it ever accumulates to any extent, in cells and tissues generally.

138

We must therefore conclude that living cells contain catalysts capable of destroying hydrogen peroxide which, as is well known, is a powerful inhibitor of many enzymes. At least three enzymes are known to participate in its destruction, namely, cytochrome c-peroxidase, peroxidase and catalase.

Cytochrome c-peroxidase has been isolated from yeast and shown to have haematin as a prosthetic group. It is probably responsible for the removal of most of the hydrogen peroxide formed in cells, but unlike other peroxidases it is strongly specific towards the substrate of oxidation, in this case the reduced form of cytochrome c. This is oxidized to the ferric form, the peroxide being at the same time reduced to water.

Peroxidase occurs in many plants, often in relatively high concentrations. Animal tissues in general appear not to contain enzymes of this kind, though a crystalline peroxidase has been isolated from milk. Horse-radish peroxidase has been shown to be a haematin compound.

Peroxidase forms a spectroscopically recognizable complex with hydrogen peroxide (p. 22), as a result of which the substrate becomes activated and capable of acting as a hydrogen acceptor for the oxidation of other substances. The classical test for peroxidase consists in adding to the suspected enzyme small amounts of hydrogen peroxide, together with guaiacum or benzidine. If peroxidase is present characteristic deep colorations develop owing to the oxidation of the guaiacum or benzidine to yield coloured derivatives. Pyrogallol is similarly oxidized through a series of intermediates to form the orange-coloured purpurogallin, which can be extracted with ether and estimated colorimetrically, a process which has been made the basis of a method for the quantitative determination of peroxidase activity. The behaviour of the peroxidase-peroxide system can be generally expressed as follows:

$$A\begin{matrix} H \\ \\ H \end{matrix} + \begin{matrix} O{-}H \\ | \\ O{-}H \end{matrix} = A + 2H_2O.$$

It is interesting to notice in passing that most haem and haematin derivatives, including even free haematin, possess a weak peroxidase activity.

Catalase occurs both in animals and plants. It has been isolated from mammalian liver and found to be a conjugated protein, the prosthetic group of which can exist in an oxidized or a reduced form, the latter being identical with the haem of haemoglobin. Catalase catalyses the splitting of hydrogen peroxide into water and molecular oxygen and, according to Keilin & Hartree, who have studied this enzyme with great thoroughness, the iron atom of the prosthetic group undergoes alternate oxidation and reduction in the process. Catalase forms spectroscopically recognizable compounds with a number of substances other than hydrogen peroxide, including cyanide, hydrogen sulphide, azide and fluoride, all of which inhibit it. Cyanide and

139

hydrogen sulphide react with the ferric form of the enzyme while azide prevents reoxidation of the ferrous form.

Catalase appears to be very specific, for it does not act upon organic peroxides such as ethyl peroxide except in very concentrated solutions. Its action upon hydrogen peroxide is very rapid however, and it has been shown that 1 mg. of pure catalase-iron can produce *ca.* 2740 l. of oxygen per hour from hydrogen peroxide, at 0° C. Expressed in molecular units this corresponds to the decomposition of no less than $4 \cdot 2 \times 10^4$ mol. of hydrogen peroxide per sec. at 0° C. by 1 mol. of catalase.

Catalase itself has a considerable peroxidase activity of its own and current opinion favours the view that this is probably more important than its catalase action. Splitting of hydrogen peroxide by catalase is only likely to take place if peroxide is formed more rapidly than it can be consumed by acting as a hydrogen acceptor under the influence of enzymes with peroxidase activity.

Mode of action of the oxidases

We saw in an earlier section that in the oxidation of biological substances three factors are involved, viz. the hydrogen donor, the hydrogen acceptor and the catalyst. In a typical oxidase system these are represented by the substrate, molecular oxygen and the oxidase respectively. The system can be analysed a little further, however, for in several cases it has been shown that the prosthetic group participates in the reaction and undergoes alternate oxidation and reduction. As a specific case we may consider the oxidation of a D-amino-acid by D-amino-acid oxidase. In this case the prosthetic group consists of flavin adenine dinucleotide (abbreviated as FAD), a substance which shows well-defined absorption in the blue region of the visible spectrum. If a D-amino-acid is added to the oxidase under strictly anaerobic conditions, the spectrum changes in a manner which indicates that the prosthetic group has undergone reduction. If oxygen is then readmitted to the system, the prosthetic group is reoxidized.

Now it is known that flavin adenine dinucleotide can be reduced by fairly powerful reducing agents, even when it is in the free condition and apart from its specific protein partner. It is not, however, reduced by amino-acids unless and until they have been activated by the oxidase. It is therefore reasonable to suppose that the oxidase protein acts as a catalyst which facilitates the transference of hydrogen from the (activated) amino-acid to the dinucleotide. Possibly, however, the protein plays no part in the second stage of the process, in which the reduced dinucleotide is reoxidized by molecular oxygen, for uncombined flavin adenine dinucleotide is spontaneously oxidized by oxygen. Tentatively, then, we may picture the action of this oxidase as a two-stage process, the protein component acting as a catalyst for the first stage, while the prosthetic group (FAD) functions as a built-in hydrogen acceptor, thus:

140

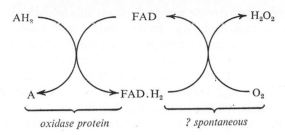

oxidase protein   *? spontaneous*

Similar considerations probably apply to other aerobic dehydrogenases in which FAD is present. How far this analysis can be extended to other flavo-protein oxidases remains to be seen, but it seems reasonable at the present time to suspect that it will prove to be of fairly general applicability.

The part played by the metallic components of metalloflavoproteins is uncertain but, in the case of xanthine oxidase, it seems likely that the scheme shown above must be expanded to read thus (p. 137);

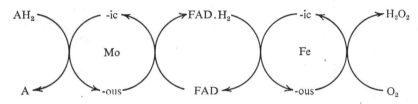

WARBURG'S RESPIRATORY ENZYME AND THE CYTOCHROME SYSTEM

Wieland's theory of dehydrogenation and the discovery by Thunberg of the dehydrogenases laid great stress on the importance of the activation of the substrates which undergo oxidation in cells and tissues. Relatively few oxido-reductase systems are able to use molecular oxygen, however, and consideration must be given to the possibility that oxygen too may be activated and thus rendered capable of acting as a hydrogen acceptor in systems that are unable to utilize molecular oxygen directly.

In a series of brilliant studies beginning in 1918, Warburg found evidence for the presence, in aerobic cells of every kind, of a catalyst, the function of which was, he believed, the activation of oxygen. His attention was drawn by the fact that iron compounds are capable of catalysing the oxidation of many different organic substances. Blood charcoal, i.e. charcoal prepared by heating blood, catalyses the oxidation of many organic substances such as the amino-acids with a simultaneous uptake of oxygen. Pure charcoal made by heating sucrose does not possess this property, and Warburg attributed the catalytic action of blood charcoal to the iron it contains.

The behaviour of blood-charcoal systems resembles that of living cells to a

141

surprising degree. Their oxygen uptake is powerfully inhibited by a number of substances known to combine with heavy metals and to inhibit cellular respi-ration. Thus cyanide at concentrations of the order of M/1000 suffices to stop 80–90% of the total respiration of most cells, and cyanide at similar con-centrations also inhibits charcoal systems very powerfully indeed. Hydrogen sulphide acts similarly. Narcotic drugs, such as urethane and the higher alcohols, also inhibit cellular respiration, but do so only at comparatively high concentrations, and they act in the same general manner upon charcoal systems as well. Warburg was led therefore to postulate the existence of a universal, iron-containing and oxygen-activating catalyst concerned in cellular respiration, to which he gave the name of *Atmungsferment*, or respiratory

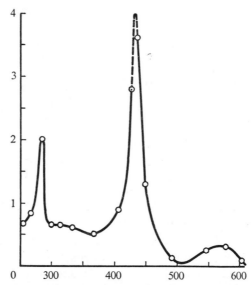

Fig. 32. Photochemical absorption spectrum of the carbon monoxide compound of the *Atmungsferment*. Ordinate: $\beta \times 10^8$. Abscissa: λ. (Plotted from data of Warburg & Negelein, 1929.)

enzyme. It is not necessary here to go into Warburg's work in great detail, but some at least of his evidence is of fundamental importance.

Most striking of all were the results of certain experiments on the inhibitory effects of carbon monoxide. Warburg found that the respiration of living cells is inhibited by carbon monoxide in the dark, but that the inhibition disappears in the light. Carbon monoxide is known to form complexes with heavy metals, including iron and copper for example, and it is characteristic of the iron complexes that they are photolabile, i.e. are dissociated by light. The copper complexes, by contrast, are not influenced by light. Here, therefore, was further evidence that the respiratory enzyme must contain iron.

Now the effectiveness of light in reversing the inhibition produced by carbon monoxide depends upon the wavelength of the light. In order to be effective, light must be absorbed. Consequently, by plotting the effectiveness of light against its wavelength, it is possible to determine the photochemical absorption spectrum of the respiratory enzyme. The resulting curve is shown in Fig. 32. It shows a sharp band at 4360 Å., and its general pattern resembles that of the haemochromogens (cf. Fig. 33). Further work enabled Warburg to determine the absolute absorption coefficient of the respiratory enzyme, and this again was in general agreement with values obtained for haemochromogens. Thus, without having ever seen his enzyme, Warburg was able to deduce not merely that it contains iron, but also that it must be a haemochromogen-like material. A great mass of other work gave general support to his conclusions and indicated that the haematin of the *Atmungsferment* is particularly closely allied to that derived from chlorocruorin (a green respiratory pigment found in the blood of a small group of annelid worms) rather than to the haematin derived from the much more widely distributed haemoglobin.

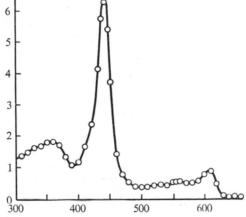

Fig. 33. Photochemical absorption spectrum of a haemochromogen (carbon monoxide compound of a chlorocruorin). Other haemochromogens give spectra of a similar pattern but with the bands in different positions. Ordinate: $\beta \times 10^8$. Abscissa: λ. (Plotted from data of Warburg, Negelein & Haas, 1930.)

In the meantime Keilin rediscovered the presence, in aerobic cells of many different kinds, of a pigment which had been described by MacMunn towards the end of the last century. MacMunn had called his pigment histohaematin or myohaematin, but it was soon forgotten. Keilin rechristened it *cytochrome*. Cytochrome, he found, was present in every kind of aerobic cell he examined, even in facultative anaerobic bacteria: indeed, the only cells in which none could be detected were strictly anaerobic bacteria. There was, moreover, a general parallel between the respiratory activities of different tissues and the amounts of cytochrome they contained.

Cytochrome is clearly recognizable by the well-defined absorption spectrum which it exhibits in the reduced form. In the oxidized form the spectrum disappears. Slight differences were noticed in the positions of the absorption bands in different tissues, but, on the whole, these differences are not greater than would be expected if the cytochromes, like the haemoglobins, are species-specific. Fig. 34 shows the absorption spectrum of reduced cytochrome as seen in the thoracic muscles of a bee, while Fig. 35 shows the

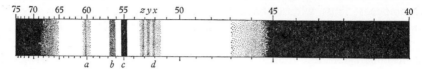

Fig. 34. Absorption spectrum of reduced cytochrome in the thoracic muscles of a bee. (After Keilin, 1925.)

positions of the four absorption bands of the reduced cytochromes in a number of different materials. That one or other of these bands is occasionally absent, and that they vary somewhat in relative intensity and in the rates at which they fade when oxygen is admitted to the system, shows that cytochrome is not a single substance but a mixture of at least three components, each being a haemochromogen-like compound with a typical two-banded spectrum. Fig. 36 shows how the spectrum of a typical cytochrome could be built up by the superimposition of the spectra of three compounds of this type.

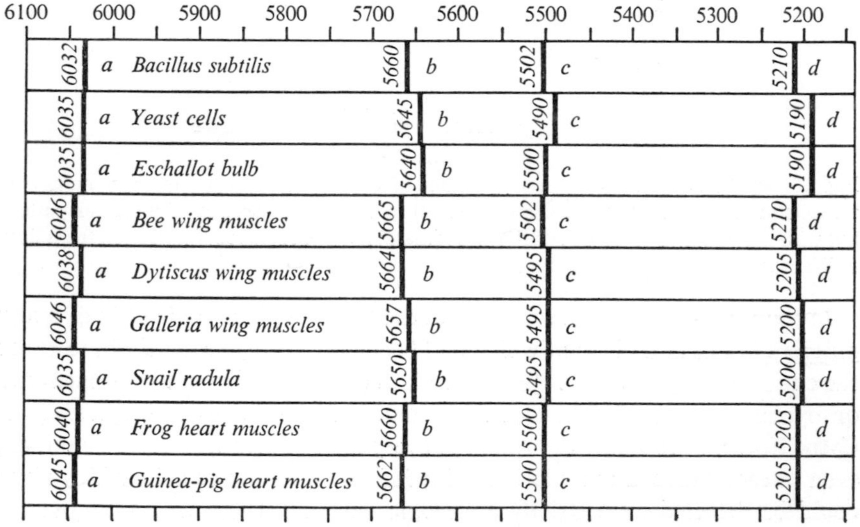

6100	6000	5900	5800	5700	5600	5500	5400	5300	5200
	a Bacillus subtilis	6032		5660 b		5502 c			5210 d
	a Yeast cells	6035		5645 b		5490 c			5190 d
	a Eschallot bulb	6035		5640 b		5500 c			5190 d
	a Bee wing muscles	6046		5665 b		5502 c			5210 d
	a Dytiscus wing muscles	6038		5664 b		5495 c			5205 d
	a Galleria wing muscles	6046		5657 b		5495 c			5200 d
	a Snail radula	6035		5650 b		5495 c			5200 d
	a Frog heart muscles	6040		5660 b		5500 c			5205 d
	a Guinea-pig heart muscles	6045		5662 b		5500 c			5205 d

Fig. 35. Absorption bands of cytochromes of various tissues.
(After Keilin, 1933.)

144

CYTOCHROME

A particularly attractive feature of much of Keilin's work is that it was possible to observe the behaviour of cytochrome within living cells when the latter were subjected to a variety of experimental conditions. Certain of these experiments were of fundamental importance. For example, a suspension of

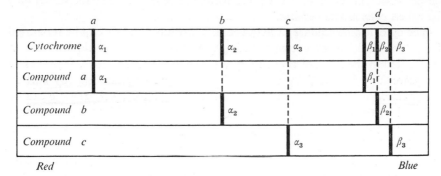

Fig. 36. Absorption bands of the components of a typical cytochrome.
(After Keilin, 1933.)

bakers' yeast shows no spectrum of cytochrome so long as oxygen is bubbled through it. If the stream of oxygen is replaced by nitrogen the spectrum of reduced cytochrome makes its appearance, and persists until oxygen is again admitted. These simple observations alone show that the cells contain systems which can reduce oxidized cytochrome and others which can reoxidize it in the presence of oxygen. Similar observations on other cells show that the existence of systems of the same kind is widespread. Thus we may write a scheme such as the following:

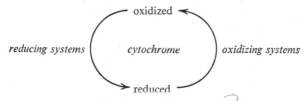

Keilin now examined the effects of respiratory inhibitors upon the behaviour of cytochrome, and found that poisons such as cyanide, hydrogen sulphide and carbon monoxide prevent the oxidation of reduced cytochrome. Narcotics, on the other hand, prevent the reduction of oxidized cytochrome. The effects of all these inhibitors can readily be studied in suspensions of yeast with apparatus no more complicated than a test-tube, a powerful lamp and a suitable spectroscope. The results just outlined indicate the presence in the cells of systems which oxidize reduced cytochrome and are inhibited by cyanide, carbon monoxide, etc., and of reducing systems that are inhibited by narcotics.

145

A few years later Keilin succeeded in isolating the *c* component of cyto-chrome from bakers' yeast. This pigment is readily reduced by mild reducing agents, but, once reduced, cannot be reoxidized merely by shaking with oxygen. Yet it is reoxidized readily enough when oxygen is admitted to a cell suspension. This argues for the presence in the cells of an enzyme which may be called *cytochrome oxidase*.

The cytochrome oxidase of living cells is insoluble and cannot be separated from the cell debris except with difficulty and by special techniques, but can be studied in the form of thoroughly washed tissue suspensions. It proves to be powerfully inhibited by cyanide, by hydrogen sulphide and by carbon monoxide. Further, as Keilin himself showed, oxidized cytochrome *c* can be reduced by the dehydrogenase systems of Thunberg, e.g. by succinate dehy-drogenase together with succinate. These dehydrogenase systems are known to be inhibited in many cases by narcotics, and are evidently responsible for the intracellular reduction of cytochrome, the systems for which, as Keilin had shown, are sensitive to the same narcotics. We may therefore summarize the position in the following manner, using heavy arrows to indicate the reactions with which the various inhibitory substances interfere:

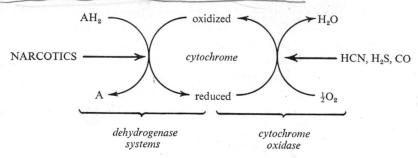

It would thus seem that we have in cytochrome the natural hydrogen acceptor of the dehydrogenase systems. Clearly, therefore, it is necessary to know something about its various components. Most is known about cytochrome *c*, which is the only soluble member of the group, the others remaining closely attached to the mitochondria. Methods have been devised for detaching the insoluble components with the aid of detergents. This has made possible the separation and identification of their prosthetic groups by conversion to the corresponding pyridine haemochromogens. The only method hitherto avail-able for their study consisted of the spectroscopic observation of their be-haviour in very finely divided and exhaustively washed tissue suspensions. Heart muscle provides a relatively rich source of these materials.

Cytochrome c consists of a protein carrying an iron-containing prosthetic group. The latter is closely allied to, and possibly identical with, the haem of haemoglobin, but is attached more firmly and in a different manner, probably through —SH groups, to its protein partner. This component is readily reduced by fairly mild reducing agents, including biological materials such as cysteine, and by relatively crude dehydrogenase preparations in the presence of their substrates. For its oxidation cytochrome oxidase is required, and the process is inhibited by respiratory inhibitors of the cyanide group but not by narcotics.

Cytochrome c_1, rather recently discovered, has an absorption spectrum and general properties very similar to those of component c.

Cytochrome b, like c and c_1, is a conjugated protein of which the prosthetic group, so far as we know, is identical with that of haemoglobin. It, too, is reducible by dehydrogenase systems, but differs from c and c_1 in that it is slowly autoxidizable, i.e. it can be slowly oxidized by molecular oxygen without the intervention of the oxidase. Its oxidation is therefore not totally abolished by cyanide, but there is a marked inhibition nevertheless, since b is oxidized more rapidly by oxygen in the presence than in the absence of the oxidase.

Cytochrome a has been found by very careful spectroscopic study to consist of two components, now called a and a_3. Both are haemochromogen-like substances, the haem of which differs from that of haemoglobin but closely resembles that of chlorocruorin, the green oxygen-carrying pigment found in the bloods of certain annelid worms. Component a resembles c and c_1 in that it is not autoxidizable and can only be oxidized through the agency of cytochrome oxidase, while a_3, which is autoxidizable, differs from all the rest in forming a compound with carbon monoxide. It is considered by many workers to be identical with cytochrome oxidase.

In addition to the iron present in its prosthetic group, a contains copper in a ratio of 1–3 atoms Cu for each atom of haem iron and there is a widespread belief that it is this copper rather than a separate a_3 that is responsible for the final reaction with molecular oxygen.

Many *other cytochrome components* are known. In many bacteria we find new components which differ from those of animal tissues and have been called a_1 and a_2. They are uniquely confined to bacteria. Special cytochromes characteristic of plants (cytochrome f for example) have been described and, in all, more than 25 different components have now been recognized. Of particular interest is the discovery that yeast contains a cytochrome, the spectrum of which resembles that of b but is not identical with it. In the intact cell the spectrum of this substance, to which the title of b_2 has been given, is not observable because it is too dilute, but the pigment has been separated, highly concentrated, and shown to form a part of the lactate dehydrogenase system of yeast. It is improbable that all the naturally occurring components of the cytochrome complex have yet been discovered, especially because some, such as b_2, are known to occur at concentrations too small to allow of direct spectroscopic observation.

The mechanism of the oxidation and reduction of the cytochromes is not certainly known. Reduction presumably involves the gain of one electron because the only observable difference between the oxidized and reduced forms lies in the valency of the iron atom, which passes with reduction from the ferric to the ferrous condition, a change which is analogous to that which takes place when methaemoglobin is reduced to haemoglobin:

$$-Fe^{3+} + e^- \rightarrow -Fe^{2+}$$

The cytochromes collectively form a chain of carriers of hydrogen (or electrons) linking the substrates undergoing intracellular oxidation with molecular oxygen. Measurements of their oxidation-reduction potentials, together with the brilliant kinetic experiments carried out by Britton Chance, indicate that the following sequence operates in intact cells and particulate suspensions alike:

$$AH_2 \cdots\!\rightarrow c_1 \longrightarrow c \longrightarrow a \longrightarrow a_3 \longrightarrow O_2.$$

The position of b however is still uncertain; probably it comes somewhere before c_1 or on a side line.

Identity of cytochrome oxidase

Oxidized cytochrome is readily reduced by many mild reducing agents such as cysteine, p-phenylene diamine, hydroquinone and the like. Keilin studied the behaviour of cytochrome oxidase in reconstructed systems in which the natural reducing systems of the cell were replaced by cysteine, while cytochrome was represented by its c component. The oxidase preparation employed consisted of a finely divided and exhaustively washed suspension of heart muscle.

Keilin found that preparations of cytochrome oxidase will catalyse the oxidation of reduced cytochrome c, and that it is closely bound to the tissue substance, and insoluble. Moreover, cytochrome oxidase is powerfully inhibited by cyanide and by hydrogen sulphide, and by carbon monoxide in the dark but not in the light.

The reader will have noticed by now that the properties of cytochrome oxidase are virtually identical with those of Warburg's respiratory enzyme, even in the effect of light upon the inhibition produced by carbon monoxide. The resemblances go even further, for both are apparently universally distributed in living cells, and both, while exceedingly sensitive to cyanide and the like, are relatively insensitive to narcotics. Keilin therefore suggested that Warburg's respiratory enzyme and cytochrome oxidase must be identical.

More recent determinations of the photochemical absorption spectrum of the carbon monoxide compound of cytochrome oxidase are shown in Fig. 37. The general form of the curve closely resembles that obtained by Warburg in his classical work on the respiratory enzyme (Fig. 32), and the main peak occurs at a similar point in both cases. Warburg's experiments were performed on yeast and on acetic acid bacteria, while Melnick, for example, used

148

heart muscle. The fact that there are distinct differences between the two may reasonably be attributed to species-specific differences between the different materials, similar to the species-specific differences between different haemoglobins, and it seems difficult to escape the conclusion that cytochrome oxidase and the respiratory enzyme are truly one and the same catalyst.

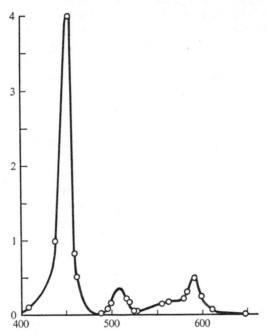

Fig. 37. Photochemical absorption spectrum of the carbon monoxide compound of cytochrome oxidase (heart). Ordinate: $\beta_\lambda/\beta_{436}$. Abscissa: λ. (Plotted from data of Melnick, 1942.)

Even if, as some believe, the systems studied by Warburg on the one hand, and by Keilin and others on the other, are not absolutely identical, it is still true that the later results demonstrate that the cytochrome oxidase activity of heart muscle is associated with a haemochromogen-like, and therefore iron-containing catalyst, just as did Warburg's work in the case of the respiratory enzyme of yeast and of acetic acid bacteria. Both, moreover, are spectroscopically similar to the haematin derivable from chlorocruorin. Probably, therefore, we are justified in concluding, with Keilin, that the respiratory enzyme is none other than cytochrome oxidase, provided that we allow for the possibility of species-specific differences between the protein components in different organisms.

That cytochrome oxidase appears to be a haemochromogen-like compound suggests that it might actually correspond to one of the numerous known components of cytochrome itself, and this does in fact seem probable. Keilin & Hartree have shown that there is a close correspondence between the properties of cytochrome a_3 and cytochrome oxidase. Each forms a compound with carbon monoxide, and the absorption curve of the CO-cytochrome a_3 complex shows maxima at 5900 and 4300 Å. The carbon monoxide compound of the respiratory enzyme of yeast shows almost identical maxima. Nevertheless, the identity of cytochrome oxidase with the a_3 component of cytochrome is not accepted by some authorities because the carbon monoxide complex of a_3 seems to be stable in the presence of light, and because some cells appear to contain neither a nor a_3, though this may merely be due to their actual presence in concentrations too small to allow of direct spectroscopic detection. There is also the serious objection that, if reduced cytochrome c is added to tissue preparations containing oxidized a_3, the two fail to react together.

To sum up, we may say that aerobic cells of many different kinds have been found to contain cytochrome components, usually a, a_3, b, c and c_1. Of these c, the only water-soluble component, not infrequently occurs in the greatest concentration. Component b is autoxidizable, if only slowly, but a, c and c_1 require the presence of cytochrome oxidase for their oxidation by molecular oxygen. This enzyme is identical with Warburg's respiratory enzyme, and in all probability is identical with cytochrome a_3 itself. All these substances occur only at small concentrations in living cells.

The cytochromes can be reduced by dehydrogenase systems and oxidized by molecular oxygen in the presence of cytochrome oxidase. If the oxidase is put out of action by the addition of M/1000 cyanide, about 80–90 % of the total respiration of most cells and tissues is abolished. It is probable, therefore, that the bulk of the whole respiration of these tissues goes on by way of cytochromes a, c and c_1. The oxidation of component b is slowed but not stopped by cyanide, and this component may therefore be responsible for a part at least of that fraction of the total respiration that is not abolished by cyanide.

We may be sure, then, that in the cytochromes we have a group of hydrogen carriers of enormous importance in cellular respiration. On the one hand they are reduced from the ferric to the ferrous condition by reacting with electrons drawn from the activated substrates of the dehydrogenases, and on the other they are oxidized by molecular oxygen. By repeated alternate oxidation and reduction, small amounts of these cytochromes are able to participate in the oxidation of relatively enormous amounts of material in living cells: their action is, in fact, that of carrier catalysts. The action of the cytochromes may be collectively expressed in the following manner, with the reservation that, in the case of cytochrome b, no oxidase is required to catalyse the reoxidation of the reduced material, though its reoxidation is much faster when the oxidase is active than when it is inhibited:

150

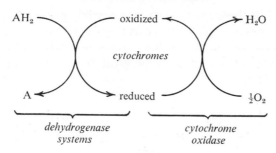

$$AH_2 \longrightarrow \quad \text{oxidized} \longleftarrow \qquad \longrightarrow H_2O$$

cytochromes

$$A \longleftarrow \qquad \longrightarrow \text{reduced} \longrightarrow \qquad \tfrac{1}{2}O_2$$

*dehydrogenase
systems*   *cytochrome
oxidase*

ACCESSORY CARRIER SYSTEMS

We have reason to believe that the bulk of the respiration of most cells is carried out by way of the cytochrome system, if only because cyanide, which inhibits cytochrome oxidase almost completely at concentrations of the order of $M/1000$, stops 80 % or more of their total respiration. We have seen already that some part of the normal respiration must be attributed to the action of *oxidases*, some of which, in the presence of appropriate substrates (e.g. the polyphenol oxidase system; p. 131), are able to act as carrier systems in their own right. But many oxidases are strongly inhibited by cyanide, and it is improbable therefore that they can account for much of the cyanide-stable fraction of the total respiration.

It is known too that certain cells and tissues contain carrier substances which, while reducible by at any rate some dehydrogenase systems on the one hand, are autoxidizable on the other. One such substance is *cytochrome* b, and it may be that a part of the cyanide-stable respiration is accomplished through this substance, though it is only slowly oxidized by molecular oxygen.

The occurrence of oxidases and of cytochrome *b* is very wide, and these substances may therefore be of general importance. But in a few special cases we know of the existence of reversibly oxidizable and reducible compounds, most of which are coloured and capable of acting naturally in much the same way as methylene blue can do artificially.

If methylene blue is added to cells previously poisoned with cyanide, their respiration is largely restored, for the dye can be reduced by the dehydro-genase systems of the cells and is spontaneously reoxidized by molecular oxygen. As the dye is autoxidizable, its reoxidation is not inhibited by cyanide. Of the naturally occurring substances capable of acting in the same manner we may refer to *pyocyanine*, a pigment produced by certain strains of *Bacillus pyocyaneus*; to *echinochrome*, which occurs in the perivisceral fluids and in the eggs of sea-urchins; and to *hallachrome*, which is found in the annelid worm *Halla parthenopaea*. The structure of echinochrome is given below, and it is interesting to notice that it is a quinonoid product. A number of natural substances with similar properties have been described, but it must be

151

realized that while these compounds may possibly play an important part in the cells and tissues in which they occur, they do not necessarily contribute to the cyanide-stable, nor indeed to the normal, respiration of cells in general, and do not therefore detract from the great and general importance of the cytochrome system.

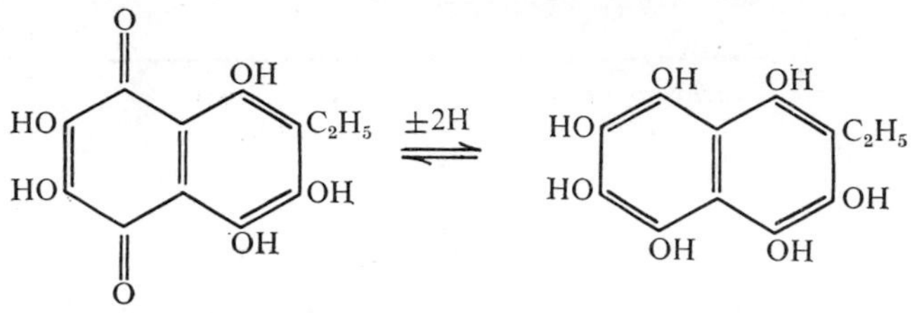

echinochrome A

In conclusion it must be pointed out that the respiration of certain cells is affected little or not at all by cyanide, and this is true of many unicellular organisms, such, for example, as *Chlorella*; the same is true of mammalian retina. In such cases it is probable either that the cytochrome system is not normally concerned in the respiration, or else that carrier substances of some other kind are present. It is, in fact, probable that cytochrome is altogether absent from some cells, but this is the exception rather than the rule.

8

OXIDOREDUCTASES:
DEHYDROGENASE SYSTEMS

DEHYDROGENASES AND CO-DEHYDROGENASES

We have already described some of the early work on the dehydrogenases, much of which was carried out by the method devised by Thunberg. An essential part of this procedure consists in washing the tissue thoroughly. It happens that some dehydrogenases are water-soluble, and any such would of course be removed by exhaustive washing. Quite apart from these, however, many of the insoluble dehydrogenases lose their power to catalyse the reduction of methylene blue as a result of thorough washing or prolonged dialysis. In such cases the activity is regained if a little boiled tissue juice is added, and it follows, therefore, that small-molecular, thermostable activators or co-enzymes of some kind are required and that these co-substances are present in the tissues.

Warburg & Christian showed that the glucose-6-phosphate dehydrogenase of red blood corpuscles requires such a coenzyme, and set themselves the task of isolating it. Starting with some 250 l. of horse blood they obtained about 20 mg. of a highly purified product. In the meantime, much interest had attached to the coenzymes of yeast juice, one of which was isolated by Euler. On analysis it appeared that these two coenzymes are closely similar in composition, since the following substances were obtained by hydrolysis:

Molecules of	Euler's yeast coenzyme = NAD	Warburg & Christian's coenzyme = NADP
Phosphoric acid	2	3
D-Ribose	2	2
Adenine	1	1
Nicotinamide	1	1

The constitution of NAD is that of a nicotinamide-adenine dinucleotide (for structural formulae of the component nucleotides see pp. 294, 296):

phosphate—D-ribofuranose—adenine
|
phosphate—D-ribofuranose—nicotinamide

This structure has been confirmed by a brilliant chemical synthesis achieved by A. R. Todd. It can be formed biologically by a reaction between nicotin-

153

amide mononucleotide and ATP, which is the pyrophosphate of adenine mononucleotide, thus:

nicotinamide mononucleotide + ATP → nicotinamide-adenine
dinucleotide (NAD) + pyrophosphate.

The position taken up by the third phosphate residue in NADP has not been certainly established but there is good evidence that this third residue is attached at position 2′ in the ribose portion of the adenosine component of NAD and can be enzymatically introduced by phosphate transfer from ATP:

$$NAD + ATP \longrightarrow NADP + ADP.$$

The behaviour of these two coenzymes has been studied by taking advantage of the fact that they show strong absorption bands in the ultra-violet (Fig. 39, p. 158). In the oxidized form they show a strong, single band at about 2600 Å., while in the reduced form the height of this band is somewhat reduced, and a second band, this time a rather broad one at about 3400 Å., makes its appearance. The sharp band at 2600 Å. can be accounted for by the known absorption spectra of adenine and nicotinamide, but the new band at 3400 Å., which appears only in the reduced forms of the coenzymes, can be accounted for by neither. It is due, as Warburg was able to show, to reduction of the pyridine ring of the nicotinamide moiety.

Working with the methiodide of nicotinamide, he found that reduction leads to the appearance of a new band at 3600 Å., so that the 3400 Å. band of the reduced coenzymes must probably be due to the presence of a reduced pyridine ring: in other words, the oxidation and reduction of NAD and NADP must take place at the pyridine ring. The nicotinamide group of the coenzymes is linked to the sugar group through its ring-bound nitrogen, and the reduction of the oxidized to the reduced form of either coenzyme is now known to take place as follows, where R represents the remainder of the molecule:

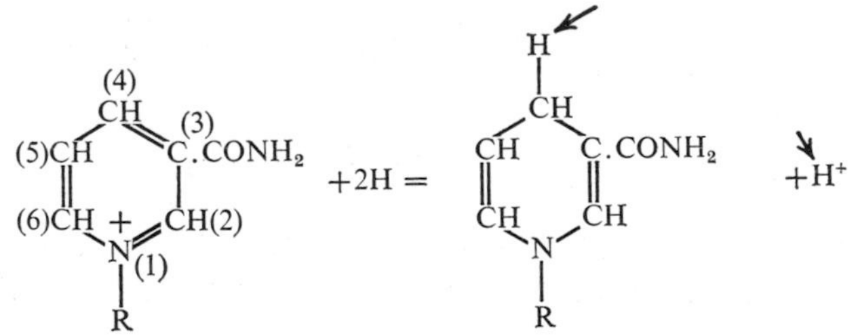

or:

$$NAD^+ + 2H = NADH + H^+$$

This is confirmed by the following observations. The oxidized form of NAD can be catalytically reduced by finely divided platinum in the presence of hydrogen and takes up 6 equivalents of the latter. If, however, it is first

154

reduced by a suitable dehydrogenase system and then treated with platinum and hydrogen, only 4 equivalents of hydrogen are taken up, indicating that the enzymatic reduction is achieved by the transference of 2 atoms of hydrogen or their equivalent. In reality the reduction is effected by the transfer of a hydride *ion*, H^-, but for the sake of simplicity we can consider the H^- as equivalent to two *atoms* of hydrogen. (See p. 124.)

This enzyme-catalysed reduction involves the insertion of one H atom at position 4. One way of proving this depends upon the oxidation of NAD by alkaline ferricyanide, which gives rise to a mixture of the corresponding 2- and 6-pyrones (Fig. 38). If the H atom at C_4 in NAD is replaced by deuterium

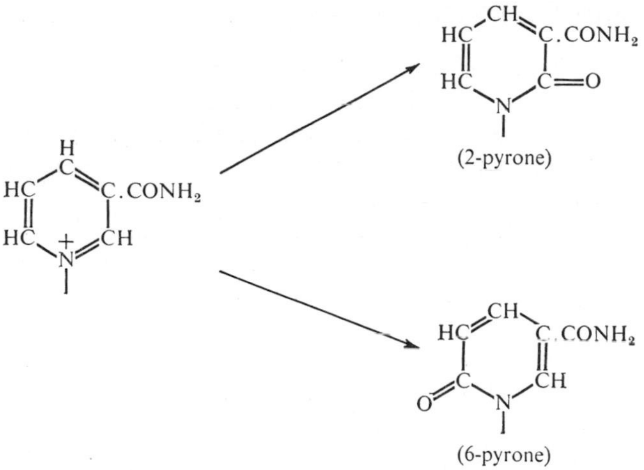

Fig. 38. Formation of pyrones from nicotinamide mononucleotide.

(*D*) and the ferricyanide oxidation is carried out we can, since H atoms could be at C_2, C_4 or C_6, argue thus:

(1) if NAD is deuterated at C_2, the 2-pyrone would not contain *D*;
(2) if NAD is deuterated at C_6, the 6-pyrone would not contain *D*;
(3) if NAD is deuterated at C_4 *both* pyrones would contain *D*.

Both pyrones do in fact contain deuterium, showing that the addition of hydrogen to the ring takes place—as we have said—at C_4.

However, the matter does not end here. If NAD is reduced with deuterium (*D*) by appropriate chemical methods we get

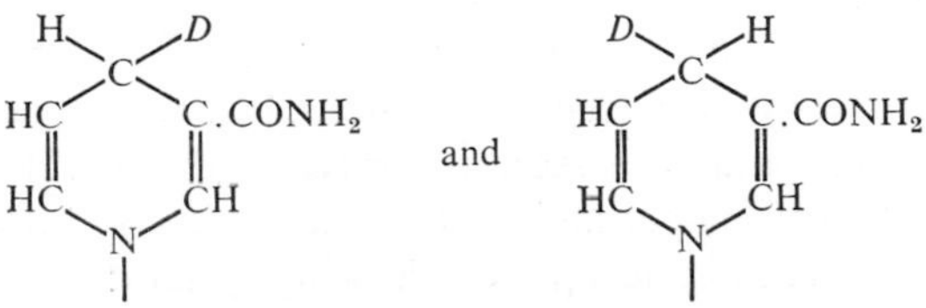

155

in equal proportions, which shows that D can lie either above or below the plane of the ring.

However, if deuterated ethanol, CH_3CD_2OH, is prepared it will reduce *oxidized* NAD^+ (which of course does not contain deuterium) in the presence of alcohol dehydrogenase:

$$NAD^+ + CH_3CD_2OH \longrightarrow NAD.D + CH_3CDO + H^+,$$

and one deuterium atom is handed on to the coenzyme. If now the reduced, deuterated coenzyme is made to react with *unlabelled* acetaldehyde in the presence of alcohol dehydrogenase, one deuterium atom appears in the alcohol:

$$NAD.D + H^+ + CH_3CHO \longrightarrow NAD^+ + CH_3CHD.COOH.$$

Pyruvate or oxaloacetate behave similarly; in both cases the coenzyme loses its deuterium and one deuterium atom turns up in the reduced product, lactate or malate as the case may be. It follows that this group of enzymes can introduce or remove deuterium or hydrogen itself to or from only one face of the ring, and not to or from both.

But if glyceraldehyde phosphate and its dehydrogenase are substituted for the alcohol dehydrogenase system, no deuterium is found in the product (α-glycerolphosphate); instead it remains in the coenzyme.

Evidently then, in biological as opposed to purely chemical systems, deuterated reduced NAD behaves differently according to whether the deuterium atom lies above or below the plane of the ring: the substrates of some dehydrogenases presumably approach from above and others from below the plane. This is a very striking and remarkable case of three-dimensional, geometrical or even geographical specificity.

The functional behaviour of NAD^+ can be studied by adding lactate and lactate dehydrogenase, for example, to a solution of the oxidized form of the coenzyme in the presence of cyanide, when the following reaction takes place:

$$CH_3CH(OH)COOH + NAD^+ \longrightarrow CH_3COCOOH + NAD.H + H^+.$$

This particular system tends towards an equilibrium that is very much in favour of lactate, and cyanide is added because it forms a cyanhydrin with the pyruvate, so that the reaction is pushed over towards the right. The reduction of the coenzyme can be followed by measuring the density of the characteristic absorption band at 3400 Å. as it develops (see Fig. 42, p. 176, for a similar example).

Clearly, therefore, the coenzyme works by acting as a hydrogen acceptor for the lactate + lactate-dehydrogenase system. If to this system we add NAD^+ and cytochrome c, the latter is reduced, provided that the dehydrogenase has not been too exhaustively purified, and by the further addition of cytochrome oxidase it is possible to build up the following system:

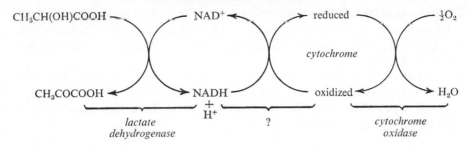

The intermediate stage marked ? also requires a catalyst, the nature of which will be discussed later, but this is present in the tissues and in any fairly impure dehydrogenase preparations extracted from them.

This system, taken as a whole, behaves in a manner parallel to that of an intact cell except, of course, that it will only oxidize lactate. Nevertheless, it is affected in the same manner and to the same extent as an intact cell by respiratory inhibitors such as narcotics on the one hand and by cyanide, carbon monoxide, etc., on the other. Lactate undergoes continuous oxidation in this system, and its oxidation is attended by the uptake of an exactly equivalent amount of oxygen.

With the isolation of NAD^+ and $NADP^+$ it became possible to classify dehydrogenases under three headings. Some can reduce methylene blue and cytochrome without the addition of any coenzyme and, according to some workers, these reduce cytochrome b in the first instance and cytochromes c_1 and c only at second hand. A second group can only reduce the c cytochromes in the presence of NAD^+, while a third group requires the presence of $NADP^+$: cytochrome b is probably not involved in the last two cases.

We have already seen that dehydrogenases are as specific towards their substrates as are other enzymes, such, for instance, as the glycosidases. We now know that they are specific also with respect to their hydrogen acceptors. Lactate dehydrogenase obtained from muscle, for instance, normally requires NAD^+ and reacts more than 100 times as fast with this as with the closely related $NADP^+$. In many cases the specificity appears to be more complete even than this; for example, glucose-6-phosphate dehydrogenase requires $NADP^+$ and cannot collaborate at all with NAD^+. In only a few cases are the two coenzymes interchangeable. But the relationships between the dehydrogenases and their substrates are more specific than those between the dehydrogenases and their coenzymes, for whereas the same coenzyme can collaborate with more than one dehydrogenase, a given substrate is activated only by the appropriate dehydrogenase and not by any other.

Lactate is not easily oxidized, and becomes easily oxidized only as a result of its union with the lactate enzyme and consequent activation. Similarly, NAD^+ undergoes a great increase in chemical reactivity in the presence of lactate dehydrogenase, with which it is able to combine. The functional

157

behaviour of the coenzyme is therefore very similar to that of the substrate, and the coenzyme might even be regarded as a second substrate. Both can combine with the dehydrogenase protein, and the Michaelis constants have been determined for both.

For convenience of reference a list of some of the more important dehydrogenases is given in Table 10 so as to show which coenzyme, if any, is required by which dehydrogenase.† Several of these enzymes have been found to contain a metal, e.g. Fe, Cu or Zn.

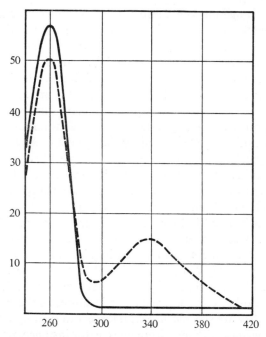

Fig. 39. Absorption spectra of oxidized and reduced NAD. Ordinates: % light absorbed. Abscissae: wavelength mμ. (After Schlenk, 1942.)

In what follows we shall refer to the three groups as *cytochrome-specific*, NAD$^+$-*specific* and NADP$^+$-*specific dehydrogenases* respectively. In the presence of their substrates, the appropriate coenzymes and other factors where necessary, all these dehydrogenases are capable of catalysing the reduction of methylene blue and of cytochrome c, provided always that they have not been too rigorously purified beforehand. In what follows the coenzymes will be denoted by NAD and NADP and their reduced forms as NAD.H$_2$ and NADP.H$_2$ respectively as a matter of convenience.

† A lengthy catalogue will be found in the *Report of the Commission on enzymes of the international union of biochemistry* (1965), Pergamon Press.

Table 10. *Coenzyme requirements of dehydrogenases*

Co-dehydrogenase required		
None	NAD	NADP
α-Glycerolphosphate[1]	α-Glycerolphosphate[2]	Glucose
Succinate	Lactate (muscle) ——	L-Glutamate
Lactate (yeast)	Malate	Glucose-6-phosphate
Choline	Triosephosphate	*iso*-Citrate
Acyl Co A	Alcohol	Malate decarboxylase
Butyryl-Co A	β-Hydroxybutyrate	Triosephosphate
	Glucose	(plants)[3]
	L-Glutamate	
	β-Hydrox-acyl-Co A	

[1] Insoluble. [2] Soluble. [3] Green parts only.

One point not so far mentioned is the existence of enzyme systems that convert NAD into NADP (p. 297), and $NAD.H_2$ into $NADP.H_2$ (p. 437). Whether or not this acts reversibly is not known.

Dehydrogenases requiring no coenzyme

α-Glycerolphosphate dehydrogenase (*insoluble*) occurs in animals, plants and micro-organisms alike. It catalyses the oxidation of D-α-glycerolphosphate to D-glyceraldehyde-3-phosphate at the expense of a suitable acceptor, e.g. methylene blue:

$$
\begin{array}{c}
CH_2O\text{\textcircled{P}} \\
| \\
CHOH + MB \\
| \\
CH_2OH
\end{array}
\rightleftharpoons
\begin{array}{c}
CH_2O\text{\textcircled{P}} \\
| \\
CHOH + MB.H_2 \\
| \\
CHO
\end{array}
$$

This enzyme is specific for the D-(+)-isomer of α-glycerolphosphate: the L-(−)-form is not attacked, nor is β-glycerolphosphate.

Succinate dehydrogenase (see also p. 172) appears to be almost universally distributed and catalyses the oxidation of succinate to fumarate:

$$
\begin{array}{c}
COOH \\
| \\
CH_2 \\
| \quad + MB \\
CH_2 \\
| \\
COOH
\end{array}
\rightleftharpoons
\begin{array}{c}
COOH \\
| \\
CH \\
\| \quad + MB.H_2 \\
CH \\
| \\
COOH
\end{array}
$$

The enzyme is insoluble and very specific. More is known about this than about any other of the group of dehydrogenases which require neither NAD nor NADP.

In the most highly purified form in which it has been obtained, succinate dehydrogenase proves to be a metalloflavoprotein, the metal present being iron. The flavin component is FAD. The enzyme has an absolute requirement

159

for inorganic phosphate to exert its catalytic effect, though the precise function of the phosphate is still uncertain.

Apart from succinate it is only known to attack 2:3-dimethylsuccinate, and is strongly and competitively inhibited by a number of other dibasic acids, notably by malonate, malate and oxaloacetate. As their formulae show, these resemble succinate rather closely and so are able to combine with the active groups of the enzyme, but yield inert products:

$$
\begin{array}{cccc}
\text{COOH} & \text{COOH} & \text{COOH} & \\
| & | & | & \text{COOH} \\
\text{CH}_2 & \text{CH}_2 & \text{CH}_2 & | \\
| & | & | & \text{CH}_2 \\
\text{CH}_2 & \text{CHOH} & \text{CO} & | \\
| & | & | & \text{COOH} \\
\text{COOH} & \text{COOH} & \text{COOH} & \\
\textit{succinic acid} & \textit{malic acid} & \textit{oxaloacetic acid} & \textit{malonic acid}
\end{array}
$$

The activity of this enzyme depends upon the presence of free —SH groups, and it is inhibited by agents which oxidize adjacent pairs of —SH groups to form —S—S— linkages. It is also inhibited by fairly high concentrations of monoiodoacetate, monoiodoacetamide and other alkylating agents, which react with and block the —SH groups irreversibly: e.g.,

$$-\text{SH} + \text{I.CH}_2\text{COOH} = -\text{S.CH}_2\text{COOH} + \text{HI}.$$

Several other dehydrogenases are similarly affected by iodoacetate, but the succinate enzyme requires unusually high concentrations for its effective inhibition.

In common with many other '—SH enzymes', succinate dehydrogenase is powerfully inhibited by many war gases, such as the arsenical smokes and vesicants, which act specifically upon —SH groups. Indeed, it has even been suggested that poisoning by oxygen may be due to oxidation of the —SH groups of sulphydryl-dependent enzymes. Enzymes of this kind can be protected against thiol-blocking agents by 2:3-dimercaptopropanol (British anti-lewisite, BAL):

$$
\begin{array}{c}
\text{CH}_2.\text{CH}.\text{CH}_2\text{OH} \\
| \quad | \\
\text{SH} \quad \text{SH}
\end{array}
$$

L-*Lactate dehydrogenase* of yeast (see also p. 172) is of special interest for two reasons, first because, unlike the lactate enzyme of muscle, it requires no coenzyme. Like succinate dehydrogenase it is a flavoprotein but flavin mononucleotide (FMN) appears to be present instead of FAD. In addition, it has been highly purified and shown to contain a conjugated protein containing haematin, which corresponds to the cytochrome b_2 of yeast (cf. p. 147). It catalyses the oxidation of L-(+)-lactate to pyruvate:

$$
\begin{array}{cc}
\text{CH}_3 & \text{CH}_3 \\
| & | \\
\text{CHOH} + \text{MB} \rightleftharpoons \text{CO} & + \text{MB.H}_2 \\
| & | \\
\text{COOH} & \text{COOH}
\end{array}
$$

D-($-$)-Lactate is not attacked by this enzyme, nor is any other β-hydroxy-acid tested. Like most dehydrogenases, the lactate enzyme of yeast can act in reverse, producing from the optically inactive pyruvate the L- but never the D-form of lactate.

Many micro-organisms other than yeast contain similar enzymes, but in some cases at least only the D-isomer of lactate can be formed and attacked.

Choline dehydrogenase, another insoluble enzyme, occurs in some but not in all animal tissues. It catalyses the oxidation of choline to betaine aldehyde:

$$\overset{+}{(CH_3)_3N}.CH_2CH_2OH + MB \rightleftharpoons \overset{+}{(CH_3)_3N}.CH_2CHO + MB.H_2.$$

Acyl-CoA dehydrogenases act upon acyl-CoA derivatives containing 4–16 carbon atoms and catalyse reactions of the following type:

$$R.CH_2CH_2CO.CoA + X \rightleftharpoons R.CH{=}CH.CO.CoA + X.H_2.$$

This introduction of an $\alpha:\beta$ double bond is the first step in the degradation of the fatty acids after they have been converted into their CoA derivatives.

There are at least two such enzymes, differing in their specificities with respect to chain length, but both are yellow flavoproteins with FAD as prosthetic group. This prosthetic group is reduced on addition of an appropriate substrate and, although in its reduced form it can reduce a number of synthetic dyes, it cannot reduce cytochrome c directly. Another factor is required (see p. 172).

Butyryl-CoA dehydrogenase resembles the fatty acyl-CoA dehydrogenases and catalyses a similar reaction but is active only for fatty acyl-CoA compounds with 4–6 carbon atoms, especially C_4. This enzyme too is a flavoprotein with FAD as prosthetic group, but the isolated enzyme is green in colour owing to the presence of copper, which may be an integral part of the enzyme. In other respects it resembles the fatty acyl-CoA dehydrogenases.

It would appear that the presence of flavin and metallic constituents in these two last enzymes absolves them from coenzyme requirements. We shall return presently to a fuller consideration of this group of enzymes (p. 172).

Dehydrogenases requiring NAD

α-Glycerolphosphate dehydrogenase (*soluble*). Animal tissues contain two α-glycerolphosphate dehydrogenases, one of which has already been described (p. 159). The soluble enzyme, unlike the insoluble, requires NAD and is specific for the L-($-$)- instead of the D-($+$)-form of α-glycerolphosphate. Methylene blue cannot be used directly as a hydrogen acceptor in this case, while the reaction product is dihydroxyacetone phosphate instead of glyceraldehyde phosphate:

$$\begin{array}{lll} CH_2O\textcircled{P} & & CH_2O\textcircled{P} \\ | & & | \\ CHOH & + NAD \rightleftharpoons CO & + NAD.H_2 \\ | & & | \\ CH_2OH & & CH_2OH \end{array}$$

Lactate dehydrogenase of muscle has been crystallized. This enzyme exists in several forms, known as isozymes, which are derived from two different basic sub-units. Like the lactate enzyme of yeast, this dehydrogenase catalyses the oxidation of L-lactate to pyruvate, again reversibly, but whereas the yeast enzyme can catalyse a direct transfer of 2H to methylene blue, that of muscle requires NAD as an immediate hydrogen acceptor:

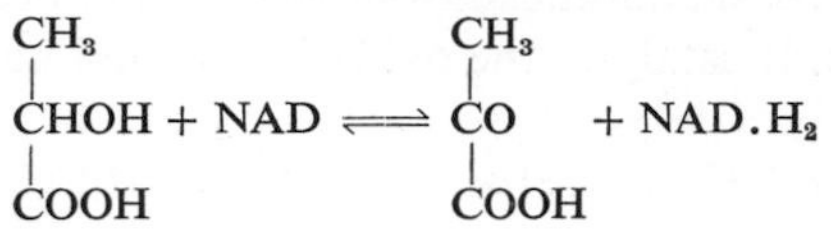

$$\begin{array}{ccc}
CH_3 & & CH_3 \\
| & & | \\
CHOH + NAD \rightleftharpoons & CO & + NAD.H_2 \\
| & & | \\
COOH & & COOH
\end{array}$$

NADP can replace NAD here but the latter reacts more than 100 times faster than the former. This system plays an important part in the metabolism of muscle and other tissues, and has been studied extensively. The equilibrium is very much in favour of the left-hand side, i.e. in favour of the hydroxy-acid, but the reaction can be forced over towards the right by adding an excess of the hydroxy-acid or, alternatively, by adding some trapping reagent, e.g. cyanide, which reacts with pyruvate to form a cyanhydrin.

Malate dehydrogenase is always found in close association with the lactate enzyme, suggesting that there may be some functional association between the two. They are not identical however. The malate enzyme catalyses the conversion of L-malate into oxaloacetate:

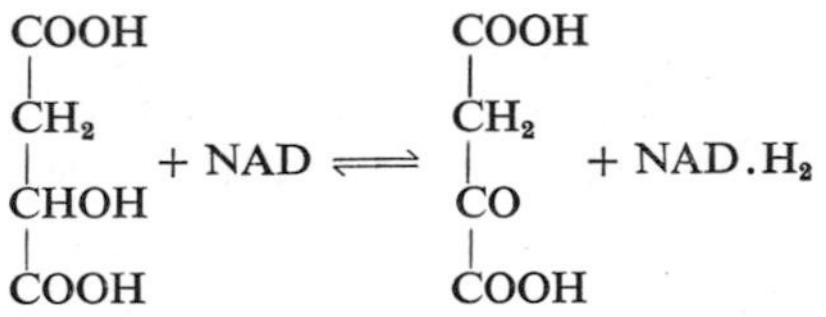

$$\begin{array}{ccc}
COOH & & COOH \\
| & & | \\
CH_2 & & CH_2 \\
| & + NAD \rightleftharpoons & | & + NAD.H_2 \\
CHOH & & CO \\
| & & | \\
COOH & & COOH
\end{array}$$

Once again the equilibrium conditions are in favour of the hydroxy-acid. Like the lactate enzyme, malate dehydrogenase can use NADP as well as NAD, but the latter is about fifteen times as active as the former.

β-Hydroxybutyrate dehydrogenase occurs in many animal tissues, especially in heart, kidney and liver. It catalyses the interconversion of L-β-hydroxy-butyrate and acetoacetate:

$$CH_3CH(OH)CH_2COOH + NAD \rightleftharpoons CH_3COCH_2COOH + NAD.H_2.$$

It has no action upon α-hydroxy- or α-keto-acids. The enzyme does not act upon β-hydroxypropionate, the next lower homologue; whether higher homologues are attacked is not known.

Alcohol dehydrogenase of yeast has been obtained in crystalline form and found to contain zinc. Like the succinate enzyme it requires the presence of its —SH groups for activity and is accordingly sensitive to iodoacetate. A somewhat similar enzyme is present in mammalian liver but not, apparently, in other mammalian tissues. The liver enzyme acts upon primary and second-ary alcohols as well to yield the corresponding aldehydes and ketones:

Primary: $\quad R.CH_2OH + NAD \rightleftharpoons R.CHO + NAD.H_2;$

Secondary:
$$R\!\!\diagdown\!\!\underset{R'}{\overset{}{CHOH}} + NAD \rightleftharpoons R\!\!\diagdown\!\!\underset{R'}{\overset{}{CO}} + NAD.H_2.$$

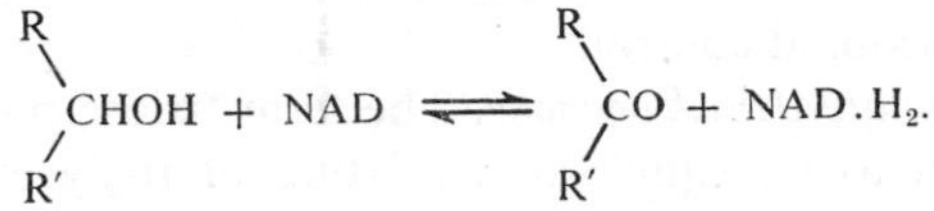

The equilibrium is in favour of the alcohol so that, in order to study the forward reactions, it is necessary to work in the presence of an aldehyde- or ketone-fixative such as sodium bisulphite.

Glucose dehydrogenase of liver catalyses the oxidation of D-glucose to the corresponding gluconic acid. In this case it is probable that the δ-lactone is formed first of all and then reacts, possibly spontaneously, with water:

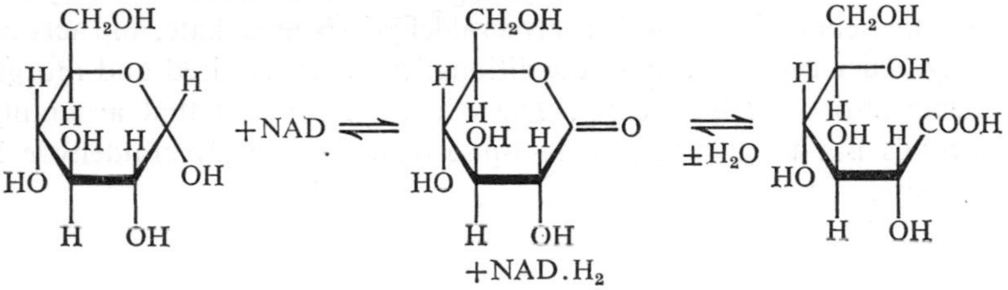

Several sugars other than D-glucose are attacked by this dehydrogenase, which has the distinction of being able to use either NAD or NADP as its hydrogen acceptor. It may be compared with the very specific glucose oxidase of *Penicillium* (p. 137).

L-Glutamate dehydrogenase is thought to catalyse the conversion of L-glut-amate to the corresponding imino-derivative, a reaction that is followed by spontaneous hydrolysis of the imino-compound to yield the corresponding α-keto-glutarate together with ammonia:

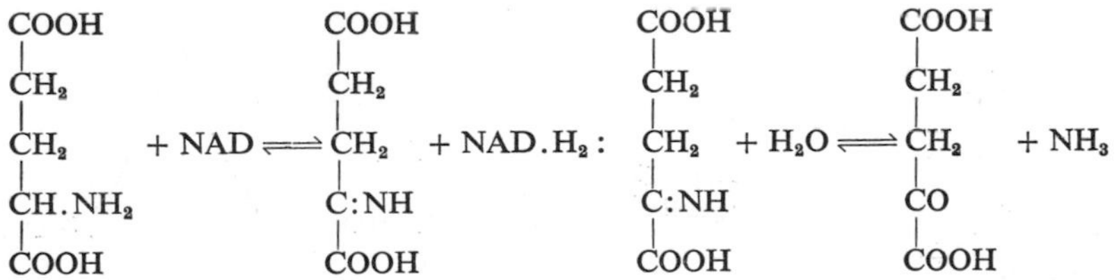

Possibly both reactions take place simultaneously.

NAD and NADP are interchangeable in this system. Both stages are revers-ible, and α-ketoglutarate can be reductively aminated by ammonia in the presence of the dehydrogenase to yield L-glutamate in liver and kidney. Similar enzymes are present in yeast and in plants, the yeast enzyme requiring NADP while that of plant tissues requires NAD.

This dehydrogenase appears to be specific for L-glutamate and has no action upon the D-isomer. However, this enzyme can also act upon L-alanine to a variable extent, the balance between the two activities being apparently subject to hormonal control.

Triosephosphate dehydrogenase. The term 'triose phosphate' as ordinarily applied refers to an equilibrium mixture of D-glyceraldehyde-3-phosphate and dihydroxyacetone phosphate:

$$\begin{array}{ccc} CH_2O\text{\textcircled{P}} & & CH_2O\text{\textcircled{P}} \\ | & & | \\ CHOH & \rightleftharpoons & CO \\ | & & | \\ CHO & & CH_2OH \end{array}$$

Attainment of this equilibrium is catalysed by triosephosphate isomerase (p. 120).

The so-called triosephosphate dehydrogenase is concerned with only one of these components, namely, with D-glyceraldehyde-3-phosphate, and acts upon this only under certain definite conditions: NAD is required and inorganic phosphate also must be present. The oxidation product that accumulates corresponds not to glyceraldehyde-3-phosphate but to glyceraldehyde-1:3-diphosphate.

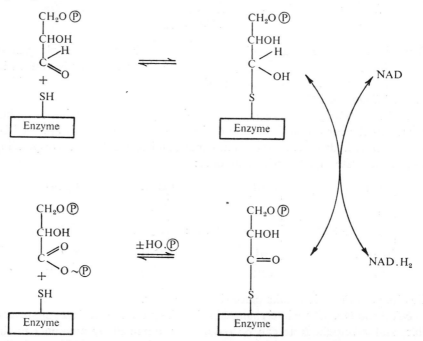

Fig. 40. Mode of action of triosephosphate dehydrogenase.

164

Enzymes similar to the triosephosphate dehydrogenase of muscle are present in yeast and in plant tissues, and are probably very widely distributed indeed. A special *NADP-specific triosephosphate dehydrogenase* occurs in the green tissues of plants and appears to be specifically concerned with photosynthesis (p. 167). The muscle enzyme is extremely sensitive to iodoacetate, from which it may be deduced that its —SH groups are required for activity and in this case there is a good deal of evidence about the part played by —SH in the reaction.

This interesting enzyme has been studied extensively, largely, in the first place, with a view to discovering the part played by inorganic phosphate. It now appears that the dehydrogenase itself has an —SH group through which it unites with its substrate (Fig. 40).

The enzyme-substrate complex loses a pair of H atoms on oxidation by NAD, to which the enzyme protein is very tightly bound, and the product is split by inorganic phosphate to yield glycerate-1:3-diphosphate. The latter contains high-energy phosphate and accordingly can react with ADP in the presence of the appropriate kinase:

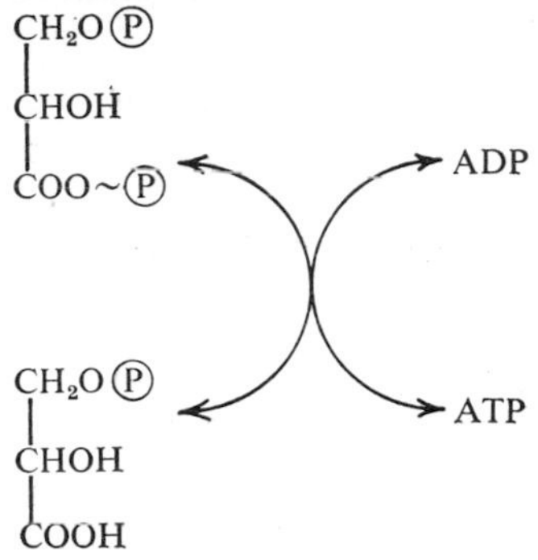

$$
\begin{array}{l}
\text{CH}_2\text{O}\,\textcircled{P}\\
|\\
\text{CHOH}\\
|\\
\text{COO}\sim\textcircled{P}
\end{array}
\qquad\rightarrow \text{ADP}
$$

$$
\begin{array}{l}
\text{CH}_2\text{O}\,\textcircled{P}\\
|\\
\text{CHOH}\\
|\\
\text{COOH}
\end{array}
\qquad\rightarrow \text{ATP}
$$

β-Hydroxy-acyl-Co A dehydrogenase plays an important part in the metabolism of fatty acids. It catalyses reactions of the following general type:

$$\text{R.CH(OH) CH}_2\text{CO}\sim\text{Co A} + \text{NAD} \rightleftharpoons \text{R.CO.CH}_2\text{CO}\sim\text{Co A} + \text{NAD.H}_2.$$

Dehydrogenases requiring NADP

Glucose and L-glutamate dehydrogenases have already been described (p. 163); mention has also been made of the NADP-specific triosephosphate dehydrogenase present in the green parts of plants.

Glucose-6-phosphate dehydrogenase occurs in red blood corpuscles and in yeast. It requires NADP and catalyses the oxidation of glucose-6-phosphate to gluconate-6-phosphate, probably by way of the δ-lactone (cf. glucose dehydrogenase):

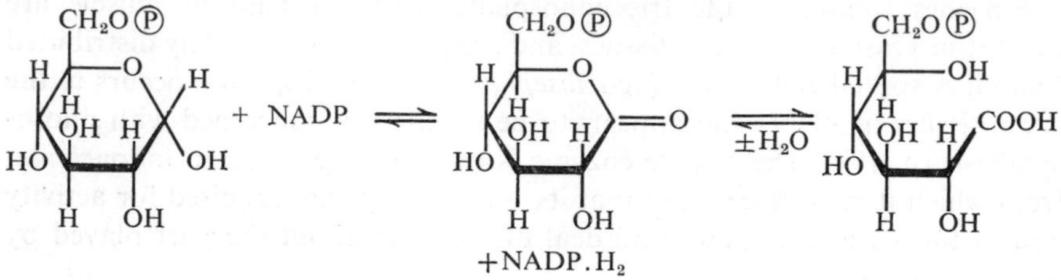

The process is reversible. This is a very specific dehydrogenase, for it has no action upon other sugar phosphates, e.g. upon fructofuranose-6-phosphate or upon fructofuranose-1:6-diphosphate, nor does it act upon glucose itself (cf. glucose dehydrogenase). The oxidation of glucose-6-phosphate to gluconate-6-phosphate is an important step in the formation of the pentose sugar, D-ribose, from hexose sources (p. 291).

iso-*Citrate dehydrogenase* is widely distributed in animal and plant tissues and is present also in many micro-organisms. It acts upon L-*iso*-citrate to produce oxalosuccinate (α-keto-β-carboxyglutarate), and the process is reversible:

$$
\begin{array}{l}
\text{COOH} \\
|\\
\text{CH}_2 \\
|\\
\text{CH.COOH} + \text{NADP} \rightleftharpoons \\
|\\
\text{CHOH} \\
|\\
\text{COOH}
\end{array}
\quad
\begin{array}{l}
\text{COOH} \\
|\\
\text{CH}_2 \\
|\\
\text{CH.COOH} + \text{NADP.H}_2 \\
|\\
\text{CO} \\
|\\
\text{COOH}
\end{array}
$$

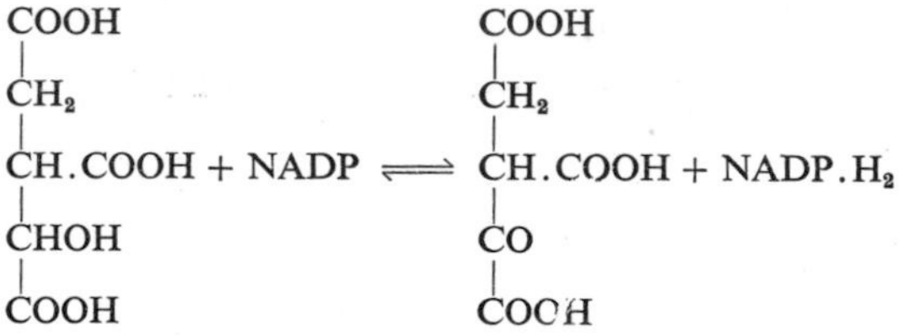

This is followed by a further reaction which involves β-decarboxylation of the product (oxalosuccinate) and gives rise to α-ketoglutarate but this second reaction, unlike the dehydrogenation, requires the presence of manganese ions. In their absence the dehydrogenation can be studied independently and there is evidence that both reactions are catalysed by one and the same enzyme-protein and perhaps take place simultaneously. Indeed this enzyme will decarboxylate added oxalosuccinate.

Perhaps this is the best place to mention a second and different *iso*-citrate dehydrogenase which is NAD-specific and, unlike the NADP-specific enzyme, has no action upon oxalosuccinate. The NADP-specific enzyme occurs mainly or wholly within the mitochondria and the NAD-specific enzyme mainly in the cell sap.

Malate decarboxylyase must be mentioned here. It differs sharply from malate dehydrogenase (p. 172) for it catalyses an oxidative decarboxylation of L-malate, using NADP as hydrogen acceptor and yielding pyruvate and carbon dioxide as products:

166

$$\begin{array}{c} \text{COO}\vdots\text{H} \\ | \\ \text{CH}_2 \\ | \\ \text{CHOH} \\ | \\ \text{COOH} \end{array} + \text{NADP} \rightleftharpoons \begin{array}{c} \text{CO}_2 \\ + \\ \text{CH}_3 \\ | \\ \text{CO} \\ | \\ \text{COOH} \end{array} + \text{NADP.H}_2$$

This enzyme thus combines the functions of a decarboxylyase and a dehydrogenase but appears nevertheless to be a single entity.

Triosephosphate dehydrogenase of plants. Plants contain several triosephosphate dehydrogenases, one of which is specific with respect to NADP. This enzyme is absent from seeds, roots and dark-grown seedlings. It appears to be strictly localized in the green tissues, in the photosynthetic activities of which it plays an important part. NADP and ATP are involved in the reaction, which can be written in abbreviated form thus;

$$\begin{array}{c} \text{CH}_2\text{O}\textcircled{P} \\ | \\ \text{CHOH} \\ | \\ \text{CHO} \end{array} + \text{H}_2\text{O} + \text{NADP} + \text{ADP} + \text{HO}.\textcircled{P} \rightleftharpoons \begin{array}{c} \text{CH}_2\text{O}\textcircled{P} \\ | \\ \text{CHOH} \\ | \\ \text{COOH} \end{array} + \text{NADP.H}_2 + \text{ATP}$$

Probably this process is as complex and involves the same intermediate stages as that catalysed by the NAD-specific enzyme (Fig. 40, p. 164).

Work on reconstructed dehydrogenase systems

Much of the work on dehydrogenases has been carried out with the aid of what are known as reconstructed systems. It is possible, for example, to build up a system containing succinate, succinate dehydrogenase, cytochrome c, cytochrome oxidase and oxygen, and such a system can 'respire', i.e. can take up oxygen and oxidize an equivalent amount of succinate to fumarate. The lactate system described on p. 157 is another example of such a system. All the reactants employed are substances which occur in nature and all can be obtained from living materials. It seems probable, therefore, that they may represent systems which actually participate in the respiration of living cells. But there are certain important criticisms that must be noted.

The fact that a given series of operations can be demonstrated in a reconstructed system is not proof positive that it takes place under biological conditions. It will be remembered, however, that the oxidation and reduction of the cytochromes can, in fact, be observed within living cells, while the presence of dehydrogenases in intact cells can be demonstrated by the Thunberg method, using a dye such as methylene blue. The discovery that the specificity of dehydrogenases is such that each can only reduce its own particular natural hydrogen acceptor shows that these substances, which do in fact occur in living cells of all kinds, must necessarily act as intermediates in processes of cellular oxidation. Thus the fact that the substances and the catalysts used in reconstructed systems *do* occur in intact cells, the fact that certain components *can* be observed at work within the cells themselves, and the incontestable facts of enzyme specificity—all these taken together make it seem improbable that we shall be led into gross error by the study of reconstructed systems, provided that the results are cautiously interpreted. But the further research has gone in this field the more complicated many apparently simple processes have turned out to be.

Intact cells, in which substances such as the cytochromes and coenzymes are present only in small concentrations, respire relatively much faster than reconstructed systems containing the same reactants at the same orders of concentration. This seems at first to suggest that the two may be fundamentally different. But there are good reasons for believing that whereas events in a reconstructed system take place in a more or less haphazard manner, the enzymes, coenzymes and other reactants of the intact cell are localized in such a way that each is present in the right place at the right time. We shall have occasion later to enlarge somewhat on this notion of intracellular organization.

The outstanding properties of the anaerobic dehydrogenases are: their *specificity towards their substrates*, their *specificity towards their hydrogen acceptors*, and the fact that, so far as our information goes, they are all *capable of acting reversibly*. In addition, they possess the usual properties of enzymes, viz. thermolability, dependence upon pH and so on. As we have seen, *three main groups can be distinguished*, the NAD- and NADP-specific types, and a third group which has no requirement for either of these coenzymes.

If now we take a dehydrogenase together with its substrate and the appropriate coenzyme, we have a system which can reduce methylene blue or cytochrome, *always provided that the enzyme preparation has not been too exhaustively purified*. But as purification is carried progressively further we find that the system eventually fails to reduce either cytochrome or methylene blue. This might be due in part to some kind of damage done to the enzyme by the methods used in its purification, but it might also mean that there must be one or more steps in the whole reaction sequence requiring catalysis by some enzyme or enzymes that are present in crude extracts but eliminated by purification. It is known, in fact, that even the so-called cytochrome-specific dehydrogenases, i.e. those requiring neither NAD nor NADP, cannot reduce the *c* cytochromes directly and that one or more additional catalysts are required to establish communication. Again, in the case of dehydrogenases that operate through NAD and NADP, there is definite evidence that additional catalysts are required to accomplish the transfer of hydrogen from the reduced coenzyme to the *c* cytochromes. The nature of these additional catalysts we shall consider in the next section.

FLAVOPROTEINS AND THE REDUCTION OF CYTOCHROME

The flavoproteins are a group of conjugated proteins which are characterized by the presence in the prosthetic group of *riboflavin*. This is a yellow substance which exhibits a strong green fluorescence even in very dilute solutions. It is identical with the flavin of milk (lactoflavin) and with that of egg-white (ooflavin), and occurs very widely indeed among cells and tissues, its importance in which may be gauged from the fact that it is a member of the B_2 group of vitamins. It is derived from the nitrogenous base 6:7-dimethyl-*iso*-alloxazine and the pentahydric sugar alcohol D-ribitol, linked together in the following manner:

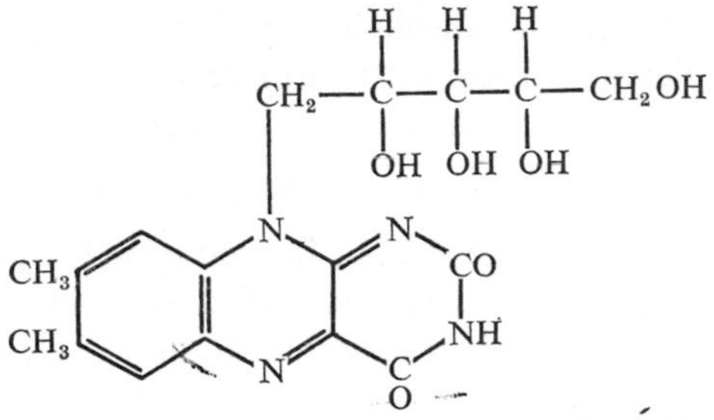

Certain points call for special comment. The substance can exist in the oxidized form shown above, but can be reduced by fairly powerful reducing agents such as dithionite ($Na_2S_2O_4$). Two atoms of hydrogen are taken up in the process, which may be described as follows:

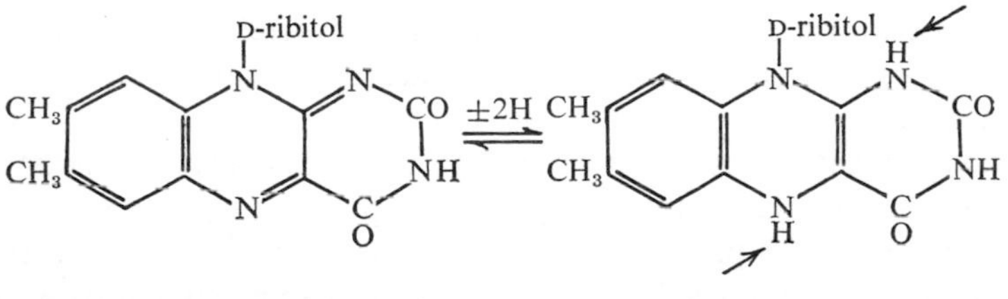

Although reagents such as dithionite are necessary to effect the reduction of the oxidized form, the reduced form is autoxidizable, i.e. it can be oxidized by shaking with molecular oxygen, the oxygen being thereby reduced to hydrogen peroxide. Another point worthy of notice is the presence in this substance not of the pentose sugar, D-ribose, but of the corresponding sugar alcohol, D-ribitol, so that the name riboflavin, suggesting as it does that the molecule contains D-ribose, is a misnomer: a better name would be ribityl-flavin but this has not found general acceptance.

Riboflavin occurs in the flavoproteins in the form of its 5'-phosphate (see p. 296), a substance which resembles a nucleotide in its general structure. Strictly speaking, however, it is not a nucleotide since, while it does contain a nitrogenous base, it does not contain a pentose sugar, but in view of its importance in cellular metabolism, in which other nucleotides also are intimately concerned, it has become common practice to refer to riboflavin phosphate as *flavin mononucleotide* (FMN) as a matter of courtesy rather than chemical exactitude. Two other important nucleotides are *adenine mononucleotide* (adenylic acid) and *nicotinamide mononucleotide*. The two latter are present in NAD and NADP, and the structures of all three mononucleotides should be compared (see pp. 294, 296).

The flavoproteins fall into two classes. In the first of these the prosthetic group consists simply of flavin mononucleotide, in the second of flavin adenine dinucleotide (FAD), i.e. the dinucleotide formed by the union of the flavin and adenine mononucleotides through their respective phosphate residues. The mononucleotide can be formed from free riboflavin at the expense of ATP by an enzyme obtained from brewers' yeast and called *riboflavin kinase*:

$$\text{riboflavin} + \text{ATP} \longrightarrow \text{riboflavin-5'-phosphate} + \text{ADP}.$$

Flavin adenine dinucleotide can then be formed from the mononucleotide by another enzymatically catalysed reaction with ATP in which the adenine nucleotide moiety of the ATP is transferred to the FMN:

$$\text{flavin mononucleotide} + \text{ATP} \longrightarrow$$
$$\text{flavin adenine dinucleotide (FAD)} + \text{pyrophosphate}.$$

The enzyme concerned is *FMN-adenylyltransferase*.

Like free riboflavin, the free mono- and dinucleotides can be reduced by dithionite, the reduced forms being autoxidizable. Similarly, when combined in the form of flavoproteins, these nucleotides can still be reduced, and when in the reduced form are sometimes though not invariably autoxidizable, according apparently as to whether they are or are not contaminated by traces of heavy metals.

We have already encountered several flavoproteins such, for example, as the flavoprotein oxidases, the prosthetic group of which acts as a built-in hydrogen acceptor for pairs of hydrogen atoms which are taken over from activated molecules of the substrate and handed on to molecular oxygen. The fact that the prosthetic group is so readily reduced by activated substrate molecules when combined with the protein component of the oxidase shows that this protein activates not only its substrate but the prosthetic group as well, just as the dehydrogenases activate their coenzymes as well as their respective substrates. Nor is this all. Many flavoproteins are now known to be associated with heavy metals, in particular Fe, Mo and Cu, and it is probable that these also play some part in the catalytic process and are activated by the protein component. In the case of the metalloflavoprotein oxidases it seems likely that the metallic constituents establish the necessary communication between the reduced flavin component and molecular oxygen. In at least one case, that of xanthine oxidase, the prosthetic group of which is FAD, molybdenum and iron are present, and here there is evidence that molybdenum acts as a carrier of H atoms (or electrons) between the substrate and the prosthetic group while the iron catalyses the reaction between the reduced FAD and molecular oxygen. The known, natural flavoproteins fall into a number of groups.

Flavoprotein oxidases (i.e. oxidases with FAD as prosthetic group) are fairly numerous. They are commonly associated with heavy metals and, as their name indicates, they can use molecular oxygen as hydrogen acceptor. Several have already been described (pp. 135–8).

Flavoprotein dehydrogenases (i.e. dehydrogenases with FAD as prosthetic groups) have been known for some years. Unable though they are to reduce molecular oxygen some of them can reduce cytochrome c. As in the case of the oxidases, the FAD of the prosthetic group acts as an intermediate carrier of hydrogen.

The two longest known of these, formerly called *diaphorases I and II*, act as specific dehydrogenases for $NAD.H_2$ and $NADP.H_2$ respectively. *They can use methylene blue as a hydrogen acceptor, but cannot reduce the cytochromes.* It was accordingly believed that an additional carrier of some kind is required to bridge the gap between the reduced form of the diaphorases on the one hand and oxidized cytochrome on the other.

During the last few years work on xanthine oxidase and the aldehyde oxidase of mammalian liver, for example, has shown conclusively that many flavoproteins are intimately associated with heavy metals. It now seems reasonably certain that the diaphorases, which act as $NAD.H_2$- and $NADP.H_2$-specific dehydrogenases, are, in the natural state, closely associated with iron in particular, but that this iron is removed by the methods formerly used for their isolation and purification in the days when the presence of heavy metals in purified enzymes was anathema. But when prepared by methods calculated to preserve the metal association, highly purified products have been obtained which are specifically reduced by the reduced form of the appropriate coenzyme, and will actively reduce cytochrome in their turn. They may therefore be regarded as *reduced coenzyme-cytochrome reductases.*

One enzyme of this kind, with Fe-association, has recently been obtained from heart muscle and, in the highly purified form, it oxidizes $NAD.H_2$ and reduces oxidized cytochrome c. Removal of the iron yields a product with virtually the same properties as diaphorase I, also isolated from heart muscle, i.e. in the absence of iron it oxidizes $NAD.H_2$ and reduces methylene blue without however reducing cytochrome c at any appreciable rate.

Perhaps it is significant that the heavy metals so far found in association with flavoproteins—iron, molybdenum, copper and perhaps manganese—are all metals that can exist in an oxidized and in a reduced form, e.g. ferric and ferrous. It may be, therefore, that the specific protein component of each of these particular flavoproteins accomplishes a threefold activation, (i) of the substrate, (ii) of the flavinadenine dinucleotide which forms the prosthetic group and (iii) of the metal(s) with which it is also associated. We may thus be justified *perhaps* in *tentatively* representing the mode of action of the coenzyme-specific dehydrogenase systems as follows:

$$AH_2 \quad NAD \quad FAD.H_2 \quad \text{-ic} \quad \text{reduced} \quad \tfrac{1}{2}O_2$$
$$metal \qquad cytochrome$$
$$A \quad NAD.H_2 \quad FAD \quad \text{-ous} \quad \text{oxidized} \quad H_2O$$

dehydrogenase   *protein component of flavoprotein*   *cytochrome oxidase*

The metals associated with flavoproteins, formerly regarded as impurities or contaminants to be removed at all costs, now appear as indispensable parts of the complete enzyme systems and this realization has, as we have just seen, enabled us to give a plausible account of the action of complete coenzyme-specific dehydrogenase systems.

The picture is more confused in the case of dehydrogenase systems requiring no coenzyme, but the recent spurt of interest in metalloflavoproteins has led to the eventual isolation of *succinate dehydrogenase* as an amber-brown coloured substance containing flavinadenine dinucleotide together with iron, possibly in the form of ferredoxin (cf. p. 321). In this case, *perhaps*, the complete system may be represented:

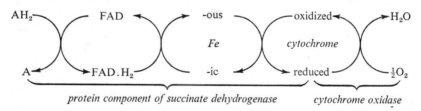

protein component of succinate dehydrogenase   *cytochrome oxidase*

In the case of the *lactate dehydrogenase of yeast* (p. 160) there is again evidence for the presence of iron, this time in haem combination. A flavin derivative is also present, apparently flavin mononucleotide, and this enzyme again reduces cytochrome.

Rather more is known about the *acyl-CoA dehydrogenases*, which are flavoproteins apparently devoid of heavy metals, and about *butyryl-CoA dehydrogenase*, also a flavoprotein. Both are reduced by an appropriate substrate, but the reduced forms are unable to reduce added cytochrome *c* directly. An additional flavoprotein, which has been called the 'electron-transporting flavoprotein' (ETF) but about which we do not yet know very much, is required to set up communication between the reduced dehydrogenases and oxidized cytochrome. For these two enzymes we may perhaps write:

COUPLED DEHYDROGENASE SYSTEMS

AH_2 — FAD — $FAD.H_2$ — oxidized — H_2O

ETF   cytochrome

A — $FAD.H_2$ — FAD — reduced — $\frac{1}{2}O_2$

protein component   *protein component of electron-*   *cytochrome*
of dehydrogenase   *transferring flavoprotein*   *oxidase*

Like the rest this must be regarded as only a tentative scheme. Much still remains to be discovered and even more new enzymes and hydrogen- or electron-carriers may still be brought to light. A quinonoid compound known as coenzyme Q or ubiquinone is one substance that now seems definitely to be implicated (p. 179). Moreover, we still have a great deal to learn about the manner in which all the catalysts concerned are arranged at their sites of action in the mitochondria or elsewhere.

REVERSIBILITY AND COUPLING OF DEHYDROGENASE SYSTEMS

If we consider the complete system involved in the oxidation of lactate in animal tissues it is clear that this oxidation is accomplished through the repeated reduction and reoxidation of a chain or series of carrier substances. A single molecule of NAD, for example, might be reduced and reoxidized, say, a thousand times, and thus contribute to the oxidation of a thousand molecules of lactate. The oxidation and reduction of these carriers takes place with great rapidity under biological conditions. The 'turn-over numbers' of oxidation catalysts, i.e. the number of times they can be reduced and reoxidized under biological conditions in 1 min. at about 30° C. range from several thousands to many millions. It is precisely because their turn-over numbers are so great that very small quantities of NAD and NADP, for example, can catalyse very large amounts of chemical change in a relatively short time. They are, in fact, catalysts, just as truly as are the enzymes with which they collaborate.

It is to be anticipated that if the supply of oxygen to the lactate system is cut off, the oxidation of lactate will cease almost immediately, since the amount of NAD available to act as hydrogen acceptor is relatively small. This is in fact true as far as the isolated or reconstructed systems are concerned. Yet many cells and tissues, including even mammalian muscle, are capable of functioning in complete absence of oxygen. It is known that metabolic oxidations can still go on in many kinds of cells under anaerobic conditions, and in this section we shall outline the mechanisms whereby these are accomplished. It is clear that, even under anaerobic conditions, the reduced coenzymes must be reoxidized in some way since, so far as we know, they are the only biological substances that can act as hydrogen acceptors for the coenzyme-specific dehydrogenases.

173

It has long been known that the lactate and succinate dehydrogenase systems of bacterial cells can be coupled together in intact cells maintained under anaerobic conditions. Lactate is oxidized to pyruvate at the expense of the reduction of fumarate to succinate, thus:

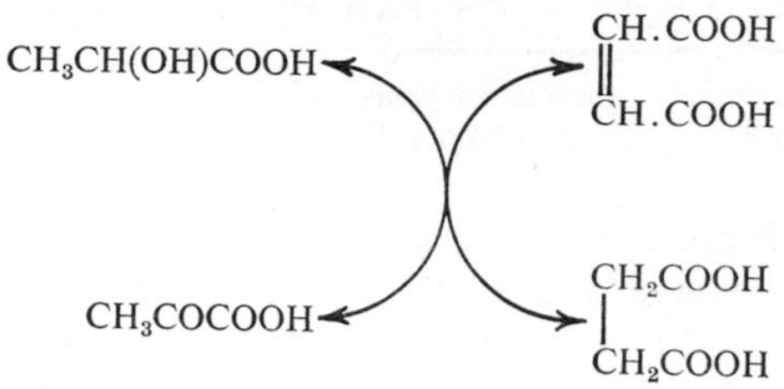

$$CH_3CH(OH)COOH \rightleftharpoons \begin{array}{l} CH.COOH \\ \| \\ CH.COOH \end{array}$$

$$CH_3COCOOH \rightleftharpoons \begin{array}{l} CH_2COOH \\ | \\ CH_2COOH \end{array}$$

At least two dehydrogenases are involved, the lactate enzyme, which catalyses the dehydrogenation of lactate, and the succinate enzyme, which catalyses the reduction of fumarate to succinate, acting in this case 'in reverse'. Neither of these requires any known coenzyme. Investigation of the mechanisms of coupled oxidation-reduction processes of this kind showed that in extracts, as opposed to intact bacterial cells, the coupling only takes place in the presence of a reversibly oxidizable and reducible compound such as methylene blue. Other substances, including cytochrome c, were also tested, but the only naturally occurring compound found to replace methylene blue as an inter-mediate hydrogen-carrier was pyocyanine, itself a reversibly oxidizable and re-ducible bacterial pigment. In the presence of methylene blue or pyocyanine it was possible to demonstrate separately the reduction of the dye by lactate in the presence of lactate dehydrogenase, and its subsequent reoxidation by fumarate in the presence of succinate dehydrogenase. When the complete system of re-actants and catalysts is taken together the conditions are such that a four-point equilibrium is finally attained, and can be modified in accordance with the usual mass-law principles. Thus we may write:

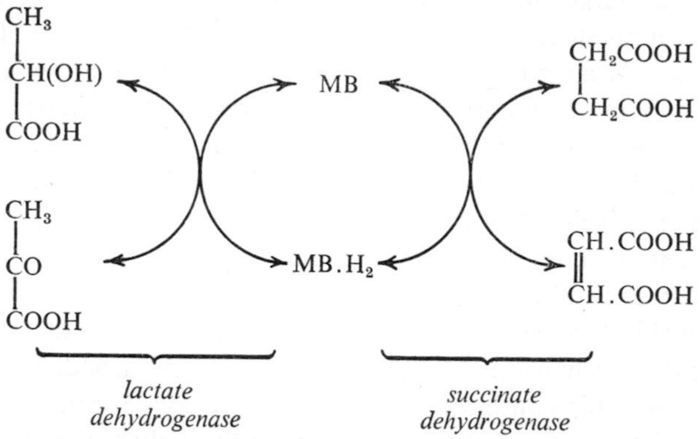

$$
\begin{array}{l} CH_3 \\ | \\ CH(OH) \\ | \\ COOH \end{array}
\quad \rightleftharpoons \quad MB \quad \rightleftharpoons \quad
\begin{array}{l} CH_2COOH \\ | \\ CH_2COOH \end{array}
$$

$$
\begin{array}{l} CH_3 \\ | \\ CO \\ | \\ COOH \end{array}
\quad \rightleftharpoons \quad MB.H_2 \quad \rightleftharpoons \quad
\begin{array}{l} CH.COOH \\ \| \\ CH.COOH \end{array}
$$

lactate dehydrogenase   *succinate dehydrogenase*

The natural intracellular carrier here is still unidentified.

174

Later work has shown that NAD and NADP can act as intermediary carriers between pairs of dehydrogenases, *provided that both enzymes are specific for the same coenzyme.* A dehydrogenase that is specific for NAD cannot be coupled to one that normally co-operates with NADP and vice versa. Many coupled pairs of coenzyme-linked dehydrogenases have now been investigated, and the triosephosphate and lactate systems of muscle, the triosephosphate and alcohol systems of yeast, and the triosephosphate and α-glycerol-phosphate systems, also of yeast, are three of the many pairs that can be linked together through NAD. The L-glutamate and glucose-6-phosphate systems of yeast can be similarly coupled together through NADP.

The part played by the coenzyme in a coupled pair can be very elegantly demonstrated by taking advantage of the absorption band at 3400 Å. which is shown by the reduced, but not by the oxidized, form of the coenzyme (Fig. 39, p. 158). Let us consider the coupling between triosephosphate dehydrogenase (see p. 164 for mechanism of action of this enzyme) and alcohol dehydrogenase (see Fig. 41).

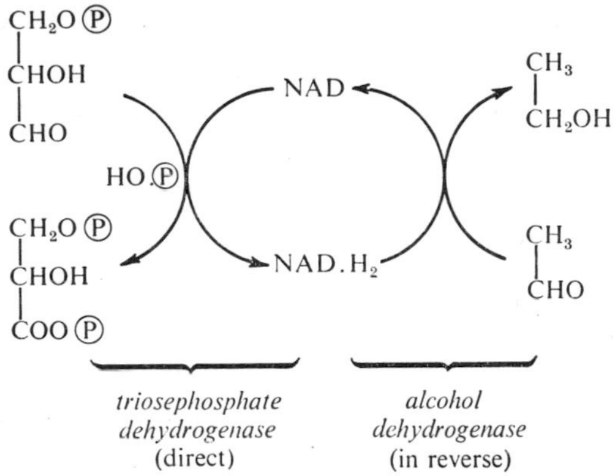

Fig. 41. Coupling of a pair of NAD-specific dehydrogenases.

If triosephosphate is taken together with the oxidized form of NAD in the presence of phosphate, no band is observable at 3400 Å. When triosephosphate dehydrogenase is added the coenzyme becomes reduced, and the progress of the reaction can be followed by measurements of the intensity of the band (Fig. 42). The coenzyme is not completely reduced because the reaction does not proceed to completion. The reaction mixture is now boiled to destroy the enzyme and then filtered or centrifuged. The intensity of the band remains unchanged, and is unaffected by the addition of acetaldehyde. If alcohol dehydrogenase is now added the second phase of the process can be observed; the coenzyme is reoxidized and the band fades.

It would be difficult to exaggerate the importance of reactions of this kind and, as we shall see later, they are a very frequent feature of metabolism

175

under anaerobic conditions, e.g. in alcoholic fermentation. Whether they take place when conditions are aerobic is difficult to say with certainty. They are still *possible* under aerobic conditions, but the rate of reoxidation of a reduced coenzyme by the next member in the oxidative chain is ordinarily so great that these reactions are probably suppressed, if not abolished altogether.

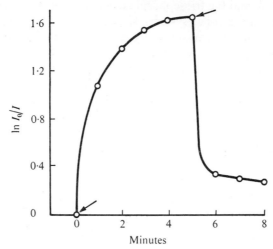

Fig. 42. Oxidation and reduction of NAD in a coupled reaction. Triosephosphate dehydrogenase added at first arrow, alcohol dehydrogenase at second; for rest of explanation see text. Ordinate: relative intensity of band at 3400Å. Abscissa: time. (Modified after Schlenk, 1942.)

We have so far been at pains to think of these as essentially reversible systems which tend towards equilibrium. This they do in isolated systems, but when, as happens under biological conditions, one or another of the reactants undergoes some further change as fast as it is formed, the system as a whole goes in one direction only, and we shall see many examples of this kind in later chapters.

INTRACELLULAR ORGANIZATION OF RESPIRATORY ENZYMES

By grinding muscle, liver or other tissue sufficiently finely in the cold, one obtains a so-called *homogenate* which will, given suitable conditions, carry out most of the metabolic activities of the original tissue. If such an homogenate is submitted to fractional centrifugation, again in the cold (see p. 198 for further details) the original whole cells can be separated into a number of different fractions, e.g. cell nuclei, mitochondria, microsomal fractions and 'cytoplasm' or cell 'sap'. Much can accordingly be learned about the intracellular localization of different enzymes by studying the enzymic activities of the various fractions.

The supernatant 'cytoplasm' remaining at the end of the fractionation procedure contains all the enzymes required for the (anaerobic) breakdown of glucose or glycogen to pyruvate or lactate, but is totally unable to oxidize these products. Indeed, this supernatant does not respire at all, because virtually all of the oxidative machinery of the cell, which has been reviewed in this and the preceding chapter, is concentrated in its mitochondria. These bodies, which contain about 30% of lipoprotein, account for from one-quarter to one-third of the total protein present, in liver tissue for example.

A fairly typical mitochondrion is, cytologically speaking, a rod-like or spiral object about 3 μ long and some 0·5 μ in diameter and stains well with Janus Green. Metabolically speaking, however, mitochondrial suspensions prepared from different sources differ somewhat in their appearance and metabolic abilities, but generally speaking they can catalyse the complete oxidation of pyruvate, the complete oxidation of fatty acids, and the capture of a large proportion of the free energy to which they gain access in the course of these katabolic processes. Many synthetic operations too can be achieved.

In recent years the interests of many enzymologists have swung rather sharply away from individual enzymes and their individual properties to those of organized enzyme systems; to intact mitochondria, by the systematic degradation of which much is being learned about the internal enzymic organization of the mitochondria themselves.

It appears, for instance, that what we may call the terminal hydrogen- or electron-transporting systems, which we have hitherto considered rather as mixtures than as an organized system of enzymes, are present in close juxta-position to each other in the mitochondria and to the systems that feed in hydrogen atoms or electrons arising from their respective substrates. By breaking down intact mitochondria in a variety of ways one might therefore hope to be able to develop as it were an 'enzyme map' of the structure of the mito-chondrion. Green, working on mitochondria prepared from beef heart muscle, has been able to separate what he calls *electron-transporting particles* (ETP) derived in all probability from the mitochondrial membranes. These possess the power to oxidize NAD.H$_2$ and succinate very rapidly indeed and to use molecular oxygen as the terminal acceptor. No co-factors, no cytochrome and nothing but the appropriate substrate need be added. It may therefore be concluded that all the necessary enzymes, coenzymes, prosthetic groups and activators are present, bound together in some way in the structure of these particles and therefore, by inference, in the mitochondria themselves. These electron-transporting particles can be broken down to smaller particular fragments in a variety of ways, the nature and activity of the fragments varying according to the method of degradation. One such case may now be described.

ETP particles contain flavin, and for each molecule of flavin there are also present 5–6 red and green haems, 30 atoms of non-haem iron and 6 atoms of copper. Further degradation of ETP by aqueous butanol, followed by differential centrifugation, gives rise to two further particulate fractions, one green in colour and containing all the copper, and the other red (see Fig. 43). The green particles have powerful *cytochrome oxidase activity* and contain, in addition to copper, a haem the spectral absorption bands of which correspond closely with those of *cytochromes* a *and* a_3.

The red particles have high succinate dehydrogenase activity and considerable dehydrogenase activity towards reduced NAD, but molecular oxygen can no longer be utilized: cytochrome *c* must now be added, and is reduced.

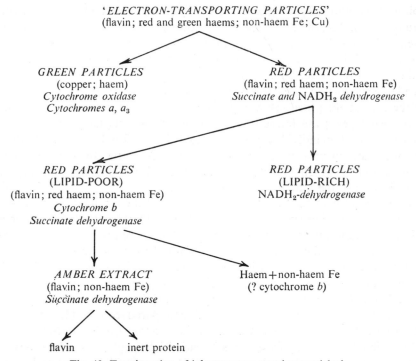

Fig. 43. Fractionation of 'electron-transporting particles'.

Further fractionation of the red particles yields two further red fractions, one of which is rich in lipid material and has considerable activity with respect to reduced NAD. The other fraction is poor in lipids and oxidizes reduced NAD only slowly but has intense succinate dehydrogenase activity.

The succinate dehydrogenase fraction contains flavin, a good deal of non-haem iron, and a red haem with absorption bands corresponding to those of *cytochrome* b. Further splitting with alkali at pH 11 dissolves out an amber-brown substance which carriers all the succinate dehydrogenase activity with it, leaving behind an inactive particulate residue which contains the haem and some of the non-haem iron. The succinate dehydrogenase contains *flavin* and some non-haem *iron* but no haem, and can catalyse the dehydrogenation of succinate if cytochrome *c* is added as hydrogen acceptor; the ability to use molecular oxygen was lost much further back in the fractionation, it will be remembered.

Finally, the succinate enzyme can be still further degraded and separated into a protein component and the flavin, and when this is done the succinate dehydrogenase activity is completely lost. Neither cytochrome *c* nor even the usual synthetic dyes can now be reduced in the presence of succinate.

This is merely one example; a specimen of the kind of work that is going on on an ever increasing scale at the time of writing. But it bristles with problems. For example, although cytochrome *c* can replace molecular oxygen as the hydrogen acceptor for various partial degradation products of the original ETP, and although in the earlier parts of this discussion we have come to think of cytochrome *c* as a particularly important carrier of hydrogen or electrons in the cell, no trace of this substance can be detected in ETP, despite the fact that these particles manage somehow or other to carry hydrogen from succinate or reduced NAD all the way through to oxygen. Clearly there is still much to be learned about the respiratory reaction chains.

One point may be emphasized at once. Mitochondria and particulate materials prepared from them are very rich in lipoproteins and as yet we know little about the functions of these substances. These lipids and the lipid-soluble substances associated with them must be expected to play some part in the oxidation-reduction systems and one such compound, coenzyme Q (ubiquinone), is already implicated. It is a quinonoid substance:

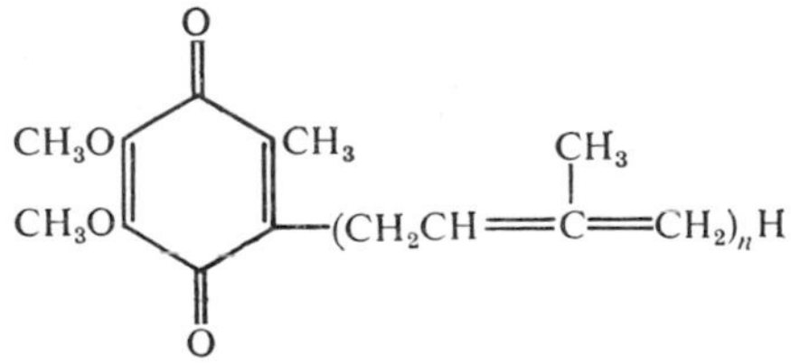

Several forms exist with *n* values from 6 to 10.

Coenzyme Q undergoes oxidation and reduction when $NAD.H_2$ or succinate are being oxidized. If ETP are extracted with acetone, succinate dehydrogenase activity is removed, but the activity can be restored by adding coenzyme Q and cytochrome *c*. Neither of these additives alone restores the succinate dehydrogenase activity. It seems likely therefore that if coenzyme Q actually plays a part in the respiratory chains it presumably operates at some point after succinate dehydrogenase and before the *c* cytochromes.

Green's ETP represent one of a probably large number of mitochondrial fragments which, when we know them all and can fit them properly together, will complete the jig-saw enzyme map of the mitochondrion. Significant and important the first few pieces certainly are and the completion of the picture is a task which biochemists have before them today. Fulfilment of this ambition will be one of the greater triumphs of modern biochemistry.

TISSUE RESPIRATION AND THE CAPTURE OF FREE ENERGY

We have so far neglected to consider the significance of tissue oxidations as sources of free energy for the organism. We know that, in one or two cases, the dehydrogenation of a substrate leads at once to the appearance of a new high-energy phosphate group in the product of oxidation. The first reaction of this kind to be investigated in detail was the oxidation of glyceraldehyde phosphate to glycerate-3-phosphate under the influence of triosephosphate dehydrogenase. The complete process involves several stepwise reactions, one of which entails the reduction of NAD, which are shown on p. 164. As an overall result, one new molecule of ATP is formed from ADP for each molecule of triosephosphate oxidized to glycerate-3-phosphate. The free energy of the new $\sim$ ⓟ has its origin in the intrinsic energy of the starting material, and the new $\sim$ ⓟ itself presumably arises as a consequence of a redistribution of energy within the molecule brought about by dehydrogenation and the resulting reorganization of the molecular structure.

A new $\sim$ ⓟ also appears in reactions in which an α-keto-acid, such as pyruvate, undergoes oxidative decarboxylation according to the following overall general equation:

$$R.CO.COOH + NAD + ADP + HO.ⓟ \longrightarrow R.COOH + CO_2 + NAD.H_2 + ATP.$$

Details of the intermediate reactions are given on p. 389.

Both of these energy-yielding reactions can now be studied with the aid of purified enzymes in fairly simple systems. But if reactions of this kind are allowed to take place in more complex systems, systems in which the 2H removed by dehydrogenation can be passed on through the appropriate co-enzyme (NAD in both cases), flavoprotein and cytochrome to molecular oxygen, *more than one new high-energy phosphate always appears and more than one molecule of ATP is formed.* These additional ATP molecules arise in the course of hydrogen transfer along the respiratory carrier chain.

A classical case was studied as early as 1939 by Belitzer & Tsibakowa who found that the oxidation of succinate, fumarate and citrate by minced frog muscle is attended by esterification of inorganic phosphate. They determined the so-called P/O ratio (mols. phosphate esterified per atom of oxygen consumed) and obtained figures of about 2. Since this phosphate fixation could not be accounted for by phosphorylation at the substrate level it must, they concluded, be associated with the concomitant processes of oxidation. Indeed, oxidative processes in general are now known to be attended by the esterification of phosphate and the resulting synthesis of ATP and, since these self-

180

same oxidative processes are mainly confined to the mitochondria, the latter are the principal generating stations of energy for ATP synthesis.

When ATP is formed in association with oxidative processes in general we speak of *oxidative phosphorylation* as distinct from *substrate phosphorylation* such as takes place in the oxidation of triosephosphate to glycerate-3-phosphate or the oxidative decarboxylation of pyruvate to yield acetate and CO_2. The difference is not a matter of mere terminology for, unlike *substrate* phosphorylation, *oxidative* phosphorylation is inhibited by dinitrophenol, gramicidin and aureomycin among other substances. A comparable process known as photophosphorylation has also been observed in the isolated choroplasts of spinach.

Since the synthesis of ATP in oxidative phosphorylation is evidently in some way coupled to the oxidative processes themselves, dinitrophenol and the other inhibitory substances are said to uncouple the two processes and are commonly referred to as *uncoupling reagents*.

The yield of high-energy phosphate can be determined by allowing the oxidation to proceed in the presence of ADP and estimating the ATP formed. Alternatively, in the presence of hexokinase and glucose, traces of ADP or ATP can be used to transfer the newly generated $\sim$ ℗ to glucose, yielding glucose-6-phosphate, which may then be estimated. Such methods always give low results, however, because most tissue preparations contain powerful ATP'ases which tend to decompose ATP as fast as it is formed. Losses through this channel can nevertheless be minimized by the use of finely divided and thoroughly washed tissue, making use of the hexokinase reaction and working in the presence of fluoride, which is a powerful inhibitor of ATP'ase activity. Quantitative results have been sought in experiments by several groups of investigators.

In one such experiment β-hydroxybutyrate was used as the substrate and between 2 and 3 molecules of ATP were formed for each atom of oxygen consumed. Now since it is impossible to deal in fractions of a high-energy phosphate group, and since allowances have to be made for losses of ATP under the experimental conditions, this corresponds to a minimum yield of 3 $\sim$ ℗ for each molecule oxidized (P/O ratio = 3). We do not know exactly where these new $\sim$ ℗ are formed nor do we yet know the mechanisms that lead to their generation. Similar experiments point to a P/O ratio of 4 for the oxidative decarboxylation of pyruvate or α-ketoglutarate, but one $\sim$ ℗ arises by substrate phosphorylation in oxidative decarboxylation itself so that, once again, *the transmission of* 2H *from the reduced coenzyme and along the oxidizing reaction chain leads to the generation of* 3 *new high-energy phosphate groups*.

Now as we pass along the chain of carriers from the coenzyme, to cytochrome and finally to molecular oxygen, there is a loss of free energy at each stage. The magnitudes of these losses can be approximately calculated from the known free energies of the carriers themselves, which can be determined by measurements of their respective oxidation-reduction potentials. This is not the place to enter into a discussion of these potentials and the methods used for their measurement:† we shall merely quote some approximate data which have been obtained in this way (Table 11). The total loss of free energy between NAD and molecular oxygen amounts in all to some 52,000 cal. per g.mol., which is sufficient to account for the formation of a number of molecules of ATP from ADP for each pair of hydrogen atoms (or electrons) carried from $NAD.H_2$ to oxygen.

Table 11. *Approximate standard free energies of some respiratory carriers*

	$-\Delta F°$ (cal.)	Change of $\Delta F°$ for the step
NAD	4,000	
Flavoprotein	15,000	$-11,000$
Cytochrome c	31,000	$-16,000$
$\frac{1}{2}O_2$	56,500	$-25,500$

Now the first stage in the oxidation of β-hydroxybutyrate, the transference of 2H from substrate to NAD, entails a loss of only a few thousands of calories of free energy, too small an amount to allow of the formation of a high-energy phosphate group. Since, however, the complete process yields 3 such groups, it follows that these must arise at later stages in the oxidative reaction sequence. The data of Table 11 show that one new molecule of ATP (= *ca.* 10,000 cal. per g.mol.) might be generated at the stage between NAD and the flavoprotein (reduced-NAD-cytochrome reductase), and a second between the flavoprotein and cytochrome c, leaving a third to be produced in the reoxidation of the reduced cytochrome. It has in fact been experimentally demonstrated that the oxidation of reduced cytochrome c by cytochrome oxidase is attended by oxidative phosphorylation with P/O = 1. This was done by adding ascorbic acid and cytochrome c to liver preparations; the other product of the reaction, dehydroascorbic acid, is not oxidized by the preparation used. Other experiments have shown that the two further molecules of ATP are formed when $NAD.H_2$ is oxidized by cytochrome c, but we do not know the mechanisms that lead to their generation. Other experiments have shown that the yield of ATP in the oxidation of succinate to fumarate is only 2, but here no coenzyme is involved. In this connexion it is surely significant that highly purified succinate dehydrogenase has an absolute requirement for inorganic phosphate.

With regard to the mechanisms involved, which under certain conditions can be made to operate 'in reverse', it is perhaps permissible to speculate a

† An excellent and most lucid account is given in M. Dixon's *Multienzyme Systems* (Cambridge, 1949).

little, bearing in mind the fact that a common feature of substrate phosphorylation is that the dehydrogenated substrate is not immediately liberated but remains bound to the responsible enzyme or other catalyst (see the case of triosephosphate dehydrogenase on p. 164). If now we represent a pair of adjacent members of the respiratory chain by AH_2 and B, and let C be the enzyme or cofactor required for their interaction, then perhaps:

$$\text{(i) } AH_2 + B + C \rightleftharpoons A \sim C + BH_2,$$
$$\text{(ii) } A \sim C + HO℗ + ADP \rightleftharpoons A + C + ATP.$$

The uncoupling effect of dinitrophenol might depend on an inhibition of (ii), leading to an alternative hydrolytic breakdown of $A \sim C$ thus:

$$\text{(iii) } A \sim C + H_2O \longrightarrow A + C.$$

This, of course, is purely hypothetical at present and the nature of the substance 'C' is the subject of much contemporary research.

Although present data are still preliminary and will undoubtedly require modification in the light of future work, it seems certain nevertheless that they represent minimal rather than maximal values because of the more or less serious losses of ATP encountered in experiments of the kind carried out in this field. In the meantime they may be used to arrive at approximate estimates of the minimal yields of ATP associated with a variety of other processes, and an example may not be out of place here.

As we shall see in a later chapter, the production of one molecule of pyruvate from glucose leads to the production of one molecule of ATP by *substrate phosphorylation* and of three more by *oxidative phosphorylation* if conditions are aerobic. This, a total yield of 4 ATP, is a relatively unrewarding process compared with the subsequent oxidation of pyruvate itself as the following considerations make clear.

Today we have much information about the intermediate stages in the oxidation of pyruvate to water and carbon dioxide, for which the following overall reaction may be written:

$$CH_3COCOOH + 5O \longrightarrow 3CO_2 + 2H_2O.$$

The following table summarizes the estimated ATP yields of the intermediate reactions.

Reaction	Primary H-acceptor	Estimated yield of ATP
Pyruvate → acetate + CO_2	NAD	4
Acetate + oxaloacetate → citrate	None	−1
iso-Citrate → oxalosuccinate	NADP	3
α-Ketoglutarate → succinate	NAD	4
Succinate → fumarate	Cytochrome	2
Malate → oxaloacetate	NAD	3
	Total	15

Thus the minimal yield, estimated on the basis of the data considered here, corresponds to 15 mols. ATP per g.mol. of pyruvate oxidized, of which 13

183

arise by oxidative and only 2 by substrate phosphorylation. This compares more than favourably with the yields of ATP from fermentation (1 mole ATP per mole of pyruvate) even when it is aerobic (4 moles of ATP per mole of pyruvate).

If, as we have every reason to believe, ATP is the primary immediate energy source for most if not all kinds of biological work it is obvious enough that oxidative phosphorylation is a process of most fundamental importance and interest. Since, moreover, the principal site of oxidative phosphorylation lies in the mitochondria, it is to these intracellular particles above all that we must look to find the principal power houses of the living cell.

PART II
METABOLISM

9

METHODS EMPLOYED IN THE INVESTIGATION
OF INTERMEDIARY METABOLISM

GENERAL PRINCIPLES

A great variety of methods is available for the study of metabolic processes and it is necessary to have some idea of their applicabilities and limitations. No attempt will be made in this chapter to compile a list of all the methods that are or have been used, nor to discourse upon modern instrumention, which nowadays covers the whole range from test-tubes to computers, but rather to consider those methods which are or have been most usually employed; nor shall we attempt here to discuss preparative and analytical methods. These can best be learned by experience in the laboratory.

Ideally, of course, metabolic experiments would be carried out under completely normal conditions, but this is seldom possible. The normal organism under normal conditions is a system which ingests certain materials and excretes others, and the conversion of ingesta into excreta proceeds very smoothly. In order to discover the pathways through which metabolism proceeds it is usually necessary therefore to interfere with normal processes in some way so as to encourage the formation and accumulation of intermediary products, or else to study the organism piece by piece. In the majority of metabolic experiments, therefore, an element of more or less serious physiological or pathological abnormality is necessarily introduced.

In an intact, normal animal, to take a specific example, we cannot obtain much information about the metabolism of proteins by straightforward investigation of nitrogenous substances entering and leaving the organism. If proteins are fed to a mammal the bulk of the ingoing protein-nitrogen emerges again in the form of urea, or in a bird in that of uric acid. Very little more can be discovered. How the nitrogen is detached from the protein, and how it is built up into urea in the one case and into uric acid in the other, we cannot discover without taking the animal more or less to pieces. If, however, we take a mammal from which the liver has been removed, it will survive for some days provided that proteins are withheld from the diet. If a protein meal is given, however, the animal quickly dies. Urea is no longer formed; instead, ammonia appears in the blood and the urine. Furthermore, the presence of unusually large amounts of amino-acids in the blood shows that these products of digestion are being absorbed in the usual manner; the free ammonia arises from amino-acids by deamination, chiefly in the kidneys.

Deamination therefore takes place in the hepatectomized animal. But whereas ammonia set free by deamination is converted into urea in the normal animal, urea production ceases with hepatectomy. It follows, therefore, that the synthesis of urea must probably be accomplished in the liver, and further evidence regarding the mechanisms involved in that synthesis is therefore to be sought by studying the liver.

An alternative method of approach is that of feeding the intact animal with substances, e.g. amino-acids, isotopically 'labelled' with heavy nitrogen (^{15}N), in place of some or all of the normal nitrogen. The isotopic form is chemically indistinguishable from ordinary nitrogen and will, we may therefore anticipate, suffer the same metabolic fate. But heavy nitrogen can be recognized and estimated by use of the mass spectrometer, and its trail through the organism can therefore be traced. Whereas in the former procedure we used abnormal animals provided with normal foods, we are now using normal animals provided with modified foods. These, in fact, represent the two fundamental methods of approach to the problems of intermediary metabolism in the intact animal. As a rule both methods are used. But even when the broad, main lines of metabolism have been traced out, there still remains the task of analysing them into their separate steps and stages so that, starting with a whole, intact organism, we find ourselves studying organs, tissues, tissue extracts, subcellular particles such as mitochondria and microsomes, groups of enzymes and even single, highly purified enzymes, as the work of analysis proceeds.

In general it must be pointed out that all the methods at present available for metabolic studies, whether in whole cells or in whole organisms, are liable to lead to erroneous conclusions on account of abnormalities introduced by the experimental conditions. The fact that such-and-such a reaction can be demonstrated in a given animal preparation or tissue extract is no guarantee that the reaction is one which normally takes place in the intact organism. Evidence from no one source should be regarded as absolutely conclusive, no matter how convincing it may appear: in every case the evidence obtained by one method should be checked against that obtained by another with different inherent limitations.

STUDIES ON NORMAL ORGANISMS

Feeding experiments on normal, intact animals led to the discovery of vitamins. Diets deficient in vitamin A lead to abnormalities in the retina, lens and cornea of the eye; deficiency of thiamine (B_1) to polyneuritis; deficiency of D to rickets and, in extreme deficiency, to osteomalacia; and so on. Here, however, we are more concerned with the normal metabolic functions of the various vitamins than with the clinical consequences of vitamin deficiencies, but nutritional experiments of this kind cannot, of course, be ignored in the present context.

Apart from work of this sort, the chief method in which normal, intact organisms are used consists essentially in the administration, by feeding, injection or otherwise, of the material the metabolism of which is to be investigated, followed by examination of the tissues and excreta for possible intermediate or end products.

If we assume that a given compound A undergoes conversion through a series of intermediates, B, C, D, etc., to yield in the end a product X, it will in many cases be possible to detect, isolate and identify X among the excreta. But even this seemingly simple procedure is beset with pitfalls. The substance X, we will suppose, is found in the urine. Unless the urine is analysed very soon after it has been passed, or else is treated with some suitable preservative, there is every prospect of heavy bacterial contamination which may transform X into some other substance or substances. The same danger arises much more acutely in the case of faecal analysis for here, even before the faeces are voided at all, they will already have been incubated for some hours in the presence of a massive bacterial population, and under conditions which are about optimal for bacterial growth. To attain complete faecal sterility is virtually impossible even in these days of antibiotics and, in consequence, faecal analysis plays a relatively small part in most metabolic studies. This particular difficulty has been overcome in some cases by opening the upper regions of the intestine to the surface by surgical operation, but the animal cannot then be regarded either as intact or as strictly normal.

The possibility of bacterial involvement is of particular importance in work on herbivorous animals, most of which maintain in their alimentary tract vast populations of symbiotic micro-organisms upon which they rely for the digestion of much of their food. Any substance fed to such an animal will have to run the gamut of these symbionts before it reaches the blood stream of their host and may be more or less extensively modified in the process. In experiments upon herbivorous animals, such as the rabbit, it is therefore advisable to inject the material to be studied, whether subcutaneously, intravenously or otherwise. For direct feeding experiments carnivorous animals such as dogs and cats are usually preferred, since the incovenience of injection can thus be avoided although even here there are still some bacterial factors to be reckoned with. And this fact alone raises further complications, for herbivorous animals and carnivores may differ considerably in their metabolism. It is necessary therefore to be on one's guard against the temptation to argue from a carnivore to a herbivore, and from one species to another.

We have so far considered only the end-products of metabolism. Several methods may be used to discover the nature and identity of the intermediates B, C, D, etc. One possible procedure is to administer massive doses of A. In a reaction sequence $A \rightarrow B \rightarrow C \rightarrow D \rightarrow$ and so on, the rate of the process as a whole will be limited by that of the slowest reaction in the chain, say $D \rightarrow E$. If the system is overloaded by giving massive doses of A, D will tend to accumulate and may therefore appear in the excreta, blood or tissues, in any or all of which it may be sought. But the administration of massive quantities of A may have consequences that we do not anticipate. It is possible that A may be transformed along more lines than one, so that, when the concentration is high, unusual or even abnormal side-products begin to accumulate. These may be discovered and mistaken for normal intermediates. Again, if D, which in the ordinary way is converted into E as fast as it is formed, attains any appreciable concentration in the tissues, it, too, may be converted into abnormal side-products which may once more be mistaken for normal intermediates. It is usual, therefore, to administer suspected intermediates in fresh feeding or injection experiments and see whether they yield the same products as normally arise from A itself.

Another possibility is that A may normally undergo conversion into several products, B_1, B_2, B_3 and so on, one of which, say B_2, is excreted without further change. Since the only apparent intermediary we shall detect in such a case is B_2, we may be misled into believing that the whole of A is normally transformed into B_2. It is necessary that these various possibilities should be kept in mind in the interpretation of results obtained in feeding or injection experiments: provided they are realized and that due allowance is made for them, valuable information can usually be obtained. Experiments of this kind have done yeoman service in the past and will doubtless continue to do so in the future.

A further method in which intact, normal animals are used involves the chemical alteration of the substance A in such a manner that it and its products can more easily be detected and recognized. Thus Knoop, in his classical experiments on the oxidation of fatty acids, introduced a phenyl group into the terminal position of the fatty chain and was able then to find aromatic derivatives in the urine of animals to which these ω-phenylated fatty acids had been administered; ω-phenyl valeric acid, for example, gave rise to hippuric acid when given to dogs:

$$C_6H_5 . CH_2CH_2CH_2CH_2COOH \longrightarrow C_6H_5 . CO—HN . CH_2COOH.$$

It was already known that the administration of benzoic acid to dogs gives rise to the appearance of hippuric acid in the urine, and Knoop was therefore able to conclude that phenyl valeric acid is converted into benzoic acid by the animal's tissues.

This method is open to a number of serious objections. First, we cannot assume that if we modify the starting material we shall not alter its fate in the organism, nor, secondly, can we assume that by feeding abnormal materials we shall not induce some completely new series of reactions which, in the ordinary way, play no important part in metabolism.

Valuable information has nevertheless been gained in the past from work of this kind, and the method has its present-day counterpart in the use of *isotopes*, such as deuterium, tritium, heavy nitrogen, radioactive carbon, phosphorus, sulphur and so on, as 'tracers'. These isotopes are not chemically distinguishable from the normal elements, and it may reasonably be supposed therefore that their metabolism will follow strictly normal lines. The isotope method is used extensively at the present time and is an enormous advance on substitution methods of the kind used by Knoop and others among the earlier workers. Many examples of the use of isotopes will be found in these pages.

STUDIES ON ABNORMAL ORGANISMS

Organisms that are intact but suffering from some pathological derangement of metabolism sometimes offer valuable experimental material. Certain very special metabolic abnormalities occur spontaneously, though rarely for the most part, and are usually genetic in origin. Albinos, for example, are devoid of the enzyme tyrosinase, and may be used in studies of certain aspects of the metabolism of the aromatic amino-acids. The metabolism of tyrosine goes astray in a number of other curious genetic freaks, notably in alcaptonuria, a disorder in which the urine becomes dark brown or black when allowed to stand exposed to the air. The blackening is due to the oxidation of a diphenol, homogentisic acid, which arises from the aromatic amino-acids.

Cases of alcaptonuria were studied in an early attempt to decide whether amino-acids undergo oxidative or hydrolytic deamination. Homogentisic acid is no longer excreted if aromatic amino-acids are excluded from the diet. It is reasonable therefore to suppose that any substance which lies on the route between tyrosine and homogentisic acid will, if administered to an alcaptonuric deprived of aromatic amino-acids, give rise to a renewed excretion of homogentisic acid. Now if tyrosine were hydrolytically deaminated the deamination product would be *p*-hydroxyphenyl-lactate: if oxidatively, the first product would be *p*-hydroxyphenylpyruvate. Both these substances were accordingly prepared and separately administered to human patients suffering from alcaptonuria. It was then found that whereas *p*-hydroxyphenylpyruvate was almost quantitatively converted into homogentisic acid, little or none was formed from the corresponding hydroxy-acid. The relationships of these substances are shown in Fig. 44. It was concluded that the deamination of tyrosine, and by inference that of other amino-acids, is an oxidative rather than a hydrolytic process.

The danger-points in this argument are, first, the supposition that because *p*-hydroxyphenylpyruvate yields homogentisic acid it necessarily lies on the pathway tyrosine → homogentisic acid: it might equally well form homogentisic acid by some independent and possibly abnormal route. Secondly, even if we discount the first objection and take it as established that tyrosine is in fact deaminated with production of the corresponding keto-acid, it is exceedingly dangerous to assume that amino-acids other than tyrosine also

191

undergo oxidative deamination, if only because, in alcaptonuria, the metabolism of tyrosine itself is seriously deranged.

Particularly important among the pathological conditions of which advantage has been taken is the state of diabetes. Spontaneous diabetes, diabetes induced by surgical removal of the pancreas or by injections of alloxan or of the growth hormone of the anterior pituitary and the pseudo-diabetes induced by injection of the drug phlorrhizin, have all been put to service, especially

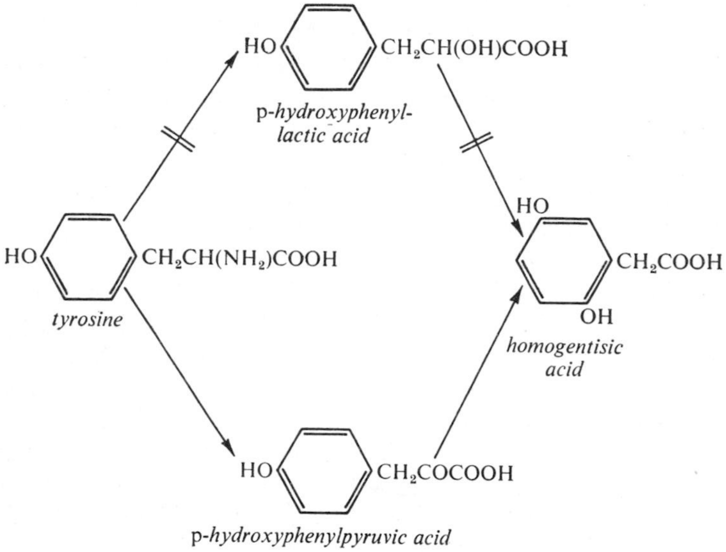

Fig. 44. Formation of homogentisic acid from tyrosine.

in studies of the metabolism of lipids and carbohydrates. These animal preparations have two important features in common. First, carbohydrate metabolism is profoundly deranged and glucose, instead of being stored in the liver in the form of glycogen, is eliminated in the urine. Secondly, there is a large-scale excretion of the so-called acetone or ketone bodies. These compounds, acetoacetate and β-hydroxybutyrate, together with acetone, are formed from fatty sources. If a diabetic or phlorrhizinized animal is maintained on a constant diet, a steady rate of excretion of glucose and acetone bodies can be established. If now substances such as alanine, lactate, glycerol and the like are administered, an increased output of glucose ensues, indicating that these substances give rise to or replace carbohydrate in the organism. Other compounds, such as butyrate and acetate, together with a number of amino-acids, increase the output of acetone bodies. The diabetic or phlorrhizinized animal is thus useful as a device which allows us to detect the formation of carbohydrate and lipid materials from substances of other kinds.

192

Particularly important among the surgical preparations is the hepatectomized animal. Total removal of the liver is a difficult operation and the subjects do not survive for more than a few days. An alternative procedure is to establish an Eck's fistula, i.e. to by-pass the liver by leading the portal blood directly into the inferior vena cava. The liver plays a leading part in many metabolic processes, and in its absence these are thrown out of gear or even stopped altogether. Intermediary products tend to pile up and commonly appear in the urine. Mention has already been made of one such case: ammonia produced by the deamination of amino-acids is normally converted into urea by mammalian liver, and into uric acid by the liver of birds, but these synthetic operations cease with removal of the liver or establishment of an Eck's fistula, amino-acids and ammonia accumulating instead. This tells us that urea and uric acid are formed from ammonia and that their synthesis takes place in the liver, but gives no indication of the mode of synthesis. It does, however, serve to show what particular organ we must study in order to elucidate the rest of the story.

The hepatectomized animal is of particular value on account of the central metabolic role of the liver, but pancreatectomized, adrenalectomized, hypophysectomized, thyroidectomized and other preparations have also been employed, especially in attempts to discover the parts played by hormones in the regulation and control of metabolic processes. In all such cases the preparation is abnormal in certain known respects, but it is necessary to realize that processes other than those which we know to be deranged may also be thrown out of gear. Confirmation of results obtained by one method should therefore always be sought, and usually is sought, with the aid of other methods and different preparations.

STUDIES ON PERFUSED ORGANS

It is often possible to study the metabolic activities of a particular organ by providing it artificially with an independent circulation. The organ to be perfused may either be left *in situ* in the animal, or may be removed and kept under conditions that approximate as closely as possible to those which it enjoys under normal physiological conditions. The circulating medium may be the animal's own blood or blood drawn from another individual of the same species; alternatively, it is possible to use certain physiological salines which we shall discuss presently. The necessary head of pressure may be obtained by means of mechanically operated pumps arranged to imitate the action of the heart, and many types of 'artificial lungs' have been devised for oxygenation of the medium.

The method of artificial perfusion is open to a number of serious objections. It takes time to establish the artificial circulation, and the animal must of course be anaesthetized during the operation. This means that, for the first few minutes, the organ is liable to be influenced by temporary deprivation of oxygen as well as by the anaesthetic. The choice of anaesthetic has therefore to receive careful consideration. How much damage may be done by the temporary shortage of oxygen is not easy to ascertain but, given speed and skill, damage from this cause can be minimized. If blood is used for the perfusion it is necessary to add an anti-coagulant such as heparin, and the possible action of this upon the organ has also to be reckoned with. Care must be taken to ensure that the perfusion medium is kept at the right pressure, temperature, pH and so on, and that it shall be kept well oxygenated, but all these are largely technical matters which can be dealt with, given experience and skill on the part of the operator.

There are, however, other objections that cannot so easily be countered. As long as the organ enjoys its normal blood supply it is exposed to nervous and hormonal influences which cannot as yet be exactly reproduced outside the animal. Thus the liver, a favourite object of study by the perfusion method, plays a central part in the metabolism of carbohydrates, and this, as we know, is profoundly affected by a number of hormones, notably by insulin, adrenaline and certain pituitary and adrenocortical hormones. When the liver is removed from the body the influence of these is withdrawn, and this may be expected to result in abnormal metabolic behaviour. Thus, from the moment at which the normal circulation is replaced by the experimental perfusion system, the organ is exposed to conditions which are already abnormal, and probably become progressively more abnormal as time goes on. It is always difficult to be certain that the whole of the organ is actually being perfused, so that parts may be moribund or dead long before the whole. Nevertheless, for the first hour or two, a skilfully manipulated preparation behaves in a manner which approximates fairly closely to normal, and results obtained with such preparations commonly find confirmation by other techniques.

The general procedure, following the successful establishment of the artificial circulation, consists in adding substances the metabolism of which is to be studied to the circulating medium, samples of which are withdrawn from time to time for analysis. Perfusion experiments may be done with liver, muscle, heart, kidney and so on. Large animals are usually preferred since large size facilitates the operative procedure, though it makes considerable demands upon laboratory accommodation and costly technical assistance at the same time.

As well-known examples of the successful employment of this method we may refer once more to the classical observation that isolated, perfused dog liver synthesizes urea from added ammonia, goose liver producing uric acid by contrast; and, in addition, to the work of Embden, Friedmann and others on the formation of ketone bodies from fatty acids. Liver perfusion in particular has regained in recent times some of the popularity which it at one time enjoyed.

194

USE OF PHYSIOLOGICAL SALINES

Perhaps the most important single contribution ever made to physiology and biochemistry was the discovery in the early 1880's by Sidney Ringer that appropriately balanced solution of the chlorides of sodium, potassium and calcium can maintain the action of the perfused hearts of frogs and tortoises. Subsequent work has shown that with the aid of slightly more complex media the heart-beat even of warm-blooded animals can be maintained for many hours or even days. There is reason to think that, of all the multifarious constituents of mammalian blood, many are specialized features of secondary importance, the ionic constituents alone being absolutely fundamental and essential. Given a supply of well-oxygenated physiological saline at the appropriate temperature, pH and osmotic pressure, it seems that the fundamental physiological requirements even of mammalian tissues can be fulfilled.

Table 12. *Composition of mammalian blood serum and* Krebs's *physiological saline*

	Mammalian serum (averages)	Krebs's physiological saline
Na^+	*c.* 320	327
K^+	22	23
Ca^{++}	10	10
Mg^{++}	2·5	2·9
Cl^-	370	454
PO_4^-	10	11
SO_4^-	11	11·4
HCO_3^-	54 vol. %	54 vol. %
CO_2 (at 38° C.)	2·5 vol. %	2·5 vol. %
pH	7·4	7·4

All concentrations are in mg. per 100 ml., excepting bicarbonate and carbon dioxide, which are expressed as vol. CO_2 per 100 ml. *Glucose* (0·2 %) is also added before use.

Numerous salines have been introduced for specific purposes. They can be used, for example, to replace blood in perfusion experiments, and were at one time used clinically on a large scale to make up the blood volume after severe haemorrhage. Solutions containing the chlorides of sodium, potassium, calcium and magnesium, together with small amounts of phosphate, are suitable for many purposes, and are best buffered with bicarbonate and carbon dioxide. Glucose is often added to provide 'food' for the tissues. The ionic composition of one such saline is given, side by side with that of the mammalian serum it is designed to imitate, in Table 12; this particular medium has been used extensively in work involving the use of tissue slices.

USE OF TISSUE SLICES

The somewhat messy method of perfusion, which makes considerable demands upon the surgical skill of the experimenter, has been largely abandoned in favour of the use of tissues in the form of thin slices. Provided that certain conditions are fulfilled, these slices will survive for some hours, apparently in a manner that approximates closely to the physiological, and are simple to prepare and manipulate. The size of the average cell is such that, although many cells are inevitably damaged when the tissue is sliced, the proportion of damaged to undamaged cells is very small, while the debris of those that are damaged can be removed fairly completely by washing. Provided that the organ to be used is removed, sliced and washed rapidly, we can obtain small fragments of virtually normal tissue. Their removal from the normal blood supply of course implies that they are removed also from the influence of the animal as a whole, just as is the case in perfusion experiments, but whereas an unknown and often considerable proportion of the cells in a perfused organ may well be in a poor if not actually moribund condition, washed tissue slices contain relatively few cells that are appreciably injured.

Certain conditions must be fulfilled in the preparation and use of these tissue slices. It is usually convenient to have fragments one or two centimetres square. Their thickness must be such that the cells in the middle of the slices, which can acquire oxygen only by inward diffusion from the medium, do not suffer from lack of oxygen. Usually, therefore, the medium is kept in equilibrium with an atmosphere containing 2·5–5 % of carbon dioxide for buffering purposes, and 97·5–95 % oxygen. If such a gas mixture is used the slices must be not more than 0·3 mm. thick. Satisfactory slices can fairly easily be cut free-hand with a sharp razor blade.

The usual procedure is as follows. Suitable vessels are filled with the appropriate saline, through which the gas mixture is bubbled, the whole being gently shaken in a thermostatic bath at body temperature. Other samples of the warmed medium are prepared for washing the slices. The animal is killed, by a blow on the neck for example, and the organ required is rapidly removed and placed on clean absorbent paper moistened with warm saline. Slices are rapidly cut and placed at once in the warm, oxygenated saline until enough have been accumulated. They are next washed two or three times and then transferred to the main vessels. The substances to be studied are added and the apparatus is gently shaken for a suitable period, during or after which samples of the medium are withdrawn, deproteinized and analysed.

This method is open to several of the criticisms that apply to the perfusion technique, though others are obviated. No anaesthetic is necessary, and a small animal such as a rat or a guinea-pig will supply enough material for a number of experiments. It is possible that the cells may behave abnormally as a result of their exposure to the high partial pressure of oxygen required to ensure adequate oxygenation of the deeper layers of cells. The method has, however, found wide favour. On account of the small size of the tissue fragments, the method is very suitable for the application of manometric methods which, as is well known, can be used for a very wide range of measurements and estimations on the micro- or semimicro-scale.

USE OF BREIS, HOMOGENATES, SUBCELLULAR PARTICLES AND EXTRACTS

The analysis of a complex series of metabolic events into its component reactions usually provides evidence for the participation of a number of enzymes and accessory catalysts, and for a complete analysis the discovery and identification of the function of each of these is required. To obtain this information it is necessary to separate the enzymes one from another, to destroy some enzymes and preserve others, or in some other way so to disrupt the cellular organization that intermediate products can be discovered. Sometimes this can be done by the use of *specific inhibitors* known to inactivate particular enzymes; in other cases '*trapping' reagents* can be employed to fix particular intermediates. More usually it is necessary to extract the enzymes, though it sometimes suffices to mince or grind the tissue. The resulting *minces* and '*breis*' contain all the enzymes of the original material, but the normal spatial relationships between them are destroyed by disruption of the cellular architecture.

In recent years the use of *homogenates* has found wide favour. These are prepared by grinding the chopped tissue very finely indeed in a special mill constructed of glass, nylon or teflon and consisting of an outer tube and an inner, closely fitting, mechanically driven pestle. The grinding surfaces are roughened by previous grinding with carborundum powder. Complete disintegration of the cells can be easily and rapidly achieved with a good homogenizer while sub-cellular particles as large as or smaller than the mitochondria can be preserved intact by having an appropriate amount of play between the pestle and the mortar. All the operations are normally carried out at a low temperature and water, buffer, 0·25 M sucrose or isotonic saline are used as suspension media. The homogenates contain all the enzymes of the original tissue, and by dilution, which lowers the concentrations of all the substrates, coenzymes and other co-factors, the metabolism of the homogenates can be reduced to a very low level. If cytochrome c and AMP, ADP or ATP are then added, the homogenate will oxidize many added substrates. Intermediate products can frequently be caused to accumulate by working in the presence of specific inhibitors.

As early as 1913 Warburg observed that the oxidative processes going on in extracts of broken cells are virtually confined to the larger 'granules' liberated from the cells. Now the nuclei, mitochondria and other particulate

14-2

cell-constituents which are present intact in homogenates can be separated one from another by differential centrifugation.

A usual procedure is as follows, all the operations being conducted at 0–2° C. A 10 % homogenate of the tissue is prepared using 0·25 M sucrose as the suspension medium. The product is centrifuged at about 500 g for 10 min.; this removes cell debris and cell nuclei. The supernatant is centrifuged for a further 10 min. at 8000 g to sediment the mitochondria, which are re-suspended, usually at 20 % on the basis of the weight of the original tissue, and put through the fractionation procedure once again. The use of sucrose solutions is designed to prevent osmotic swelling and bursting on the part of the mitochondria.

After removal of the mitochondria the microsomes can be sedimented by centrifuging for a further 30 min. at 20,000–50,000 g leaving the cytoplasmic enzymes in the supernatant solution. Suspensions of cell nuclei too can be obtained but their separation from general cell debris presents a number of special problems. Methods are also available for preparing suspensions of other particulate cell inclusions, e.g. melanin particles, nucleoli and Golgi apparatus.

Washed suspensions of mitochondria have been much used in studies of oxidative metabolism since they contain the enzymes involved in cellular respiration and require only to be fortified with cytochrome c, ATP, ADP or AMP and the necessary coenzymes, usually Mg^{2+} and inorganic phosphate. NAD and NADP are firmly bound in the mitochondria and are not removed by washing.

The addition of ADP has been used in studies of oxidative phosphorylation, i.e. the synthesis of ATP which is associated with oxidative metabolism. The presence of ATP'ases, which split ATP as fast as it is formed, led to many failures in early experiments but these enzymes are strongly inhibited though not, unfortunately, totally inactivated, by fluoride. A great step forward was made with the introduction of a simple method of purifying yeast hexokinase. If the latter is added together with glucose, only traces of ATP are necessary to phosphorylate the glucose, the residual ADP being rephosphorylated by the mitochondria and so on. The total synthesis of ATP can then be determined in terms of glucose-6-phosphate formed.

Many tissue enzymes can be extracted with water or saline, freed from the general cell debris by filtration or centrifugation, and later purified. Of the soluble enzymes some tolerate precipitation with acetone at or below 0° C. and enzymes of this kind can be extracted with aqueous media, precipitated by means of ice-cold acetone, and then extracted again with water or saline from the resulting 'acetone powders'.

A fine example of the usefulness of whole, *cell-free extracts* is found in the case of yeast juice, which is prepared by macerating the cells with sand and kieselguhr (a siliceous earth) and squeezing the mass in a suitable press. Many of the enzymes extracted in this way require coenzymes, which can be removed by *dialysis*, and much of our present knowledge of fermentation has been gained by the use of dialysed yeast juice, often with supplementary tools in the form of *selective inhibitors*. Similarly, the enzymes involved in glycolysis can be obtained in solution by aqueous extraction of minced muscle for example.

In the end it is often necessary to have recourse to *purified enzymes*. Finely divided tissue is allowed to stand with ice-cold water or isotonic potassium chloride, for example, with or without gentle mechanical stirring. After centrifugation of the extract to remove insoluble cell debris it is possible to purify many enzymes by fractional precipitation, fractional adsorption, chromatography and other more specialized procedures, so that highly concentrated preparations are obtained. By further rigorous purification, the details of which vary according to the nature of the enzyme, crystalline preparations are obtainable in many cases. It is well to remember, however, that proteins in general—and enzymes are no exception to the rule—are very prone to the formation of mixed crystals while still far from being chemically pure. Many important enzymes, e.g. cytochrome oxidase and succinate dehydrogenase, are insoluble however. Until recently they were usually studied in thoroughly washed and suitably fortified suspensions of finely minced or homogenized tissue. Often it is possible to coax water-insoluble enzymes into solution by the use of detergents, and in more recent times a number of insoluble, particulate enzymes have been isolated by systematic degradation of washed mitochondria, and a broad description of the separation of cytochrome oxidase, reduced NAD-cytochrome reductase and succinate dehydrogenase from mitochondria will be found on pp. 176–9.

Detailed knowledge of the processes catalysed by single enzymes often helps us to analyse into several stages a process which seems at first sight to be a single metabolic operation, and information gained by the study of 'built-up' or reconstructed systems, comprising several enzymes and their appropriate accessory catalysts, may give valuable indications of the manner in which the individual stages are organized in the metabolic whole. We shall come across numerous examples of this kind, but for the moment the reader may be reminded of the use of reconstructed systems in the study of the dehydrogenases (p. 167).

USE OF ISOTOPES

In biochemical as in many other kinds of research the now ready availability of isotopes has been of immense benefit. Indeed they have largely deposed many of the more classical techniques we have discussed in the earlier parts of this chapter. *Isotopic elements differ not at all in their chemical or metabolic properties but are detectable by differences in their physical properties.*

Deuterium (heavy hydrogen, ^2H) is twice as heavy as 'ordinary' hydrogen, since, while the nucleus of the latter contains only 1 proton, the heavy isotope contains 1 proton and 1 neutron and so has twice the atomic mass. Similarly, while normal carbon, ^{12}C, has 6 nuclear protons and 6 neutrons, there also exist isotopic carbons ^{13}C and ^{14}C, the nuclei of which contain 1 and 2 additional neutrons respectively. The nuclei of these two isotopes are unstable and consequently radioactive.

Heavy isotopes can readily be detected and estimated in mixtures by means of the mass spectrometer; radioactive isotopes can be similarly handled with the liquid scintillation counter. Failing all else, heavy hydrogen can be detected by burning it or substances containing it, collecting the water formed and measuring its physical characteristics, since 'ordinary' water and heavy water differ in a number of ways; e.g.:

	'Ordinary' water	'Heavy' water
Freezing point (° C.)	0	3·82
Boiling point (° C.)	100	101·42
Specific gravity	1·000	1·1074

By accurate measurements of one or more of these properties it is possible to determine the amount of heavy water present in a mixture with ordinary water.

The radioactive carbons and many other radioactive isotopes can readily be detected by the mere fact of their radioactivity, and quantitative determinations can be made by counting the particles they emit. But ^{12}C, ^{13}C and ^{14}C all have the same number (6) of orbital electrons, and it is the number of these orbital electrons that determines the chemical properties and therefore the metabolic behaviour of the element and its compounds.

Of the stable isotopes perhaps ^{2}H (deuterium), ^{18}O and ^{15}N have been most widely used, while among those that are radioactive, ^{3}H (tritium), ^{14}C, ^{35}S and ^{32}P have perhaps found most employment in the past, but there has been a great increase in the number and variety of isotopes available in recent times, expensive though they are.

Many examples of the use of isotopes will be found in this book, but for sake of example we may take the synthesis of creatine, which is represented in Fig. 45. Reaction (1) takes place in the kidney and reaction (2) in the liver. If kidney slices are incubated with arginine together with glycine labelled with ^{15}N, the glycocyamine formed carries heavy nitrogen in the corresponding position. Similarly, arginine with ^{15}N in one of its own amidine nitrogen atoms yields a glycocyamine with heavy nitrogen in the corresponding position in its amidine group. If now glycocyamine is incubated with liver slices and trideuteromethionine, i.e. methionine, the methyl group of which contains 3 deuterium (^{2}H) atoms, the creatine formed has a trideutero-methyl group.

This is but one example. Isotopic substances have been used in feeding experiments, in experiments involving tissue slices, homogenates, mitochondrial suspensions, purified enzymes and so on.

200

In addition to their use in metabolic studies isotopes have been extensively used in analytical procedures. Suppose, for example, that we have a protein hydrolysate and wish to know how much of each of several amino-acids it contains. Before the days of chromatography and the development of Moore

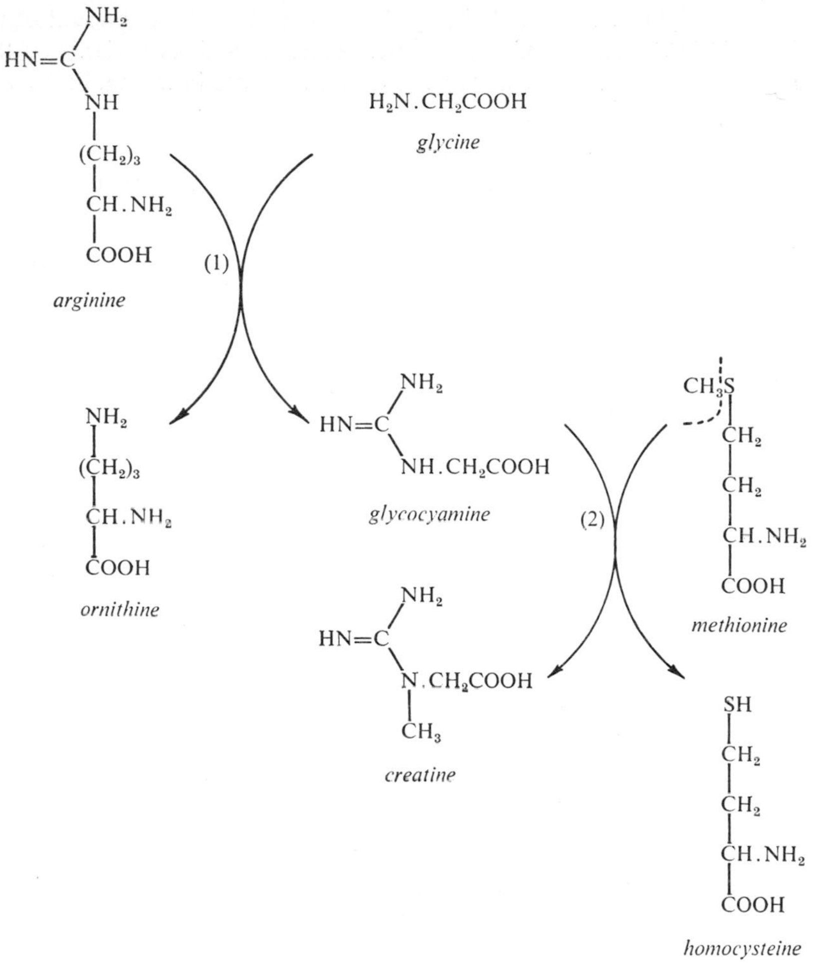

$$HN=C\begin{smallmatrix}NH_2\\NH\end{smallmatrix}$$

(CH₂)₃

CH.NH₂

COOH   (1)

arginine

H₂N.CH₂COOH

glycine

NH₂

(CH₂)₃

CH.NH₂

COOH

ornithine

$$HN=C\begin{smallmatrix}NH_2\\NH.CH_2COOH\end{smallmatrix}$$

glycocyamine   (2)

$$HN=C\begin{smallmatrix}NH_2\\N.CH_2COOH\end{smallmatrix}$$

CH₃

creatine

CH₃S

CH₂

CH₂

CH.NH₂

COOH

methionine

SH

CH₂

CH₂

CH.NH₂

COOH

homocysteine

Fig. 45. Biosynthesis of creatine as demonstrated by isotopes
(see text for full explanation).

& Stein's automatic amino-acid analyser, the quantitative separation and purification of these amino-acids by normal chemical methods was tedious and exceedingly difficult, and losses were apt to be high. Today we can, however, use the method of *isotope dilution*.

An isotopically labelled sample of one of the amino-acids in which we are interested is prepared and its specific activity is measured. A known amount is then added to a known amount of the hydrolysate. A specimen of the amino-acid in question is now isolated; there is no need to isolate the whole. The isolated sample will contain the isotope, and the ratio of its specific activity to that of the sample originally added will be equal to the ratio between the weight of the sample isolated and that of the total amount of the amino-acid in the sample of hydrolysate. Several ingenious variants on this method of isotope dilution have been devised for use in metabolic as well as in analytical procedures.

10

FOOD, DIGESTION AND ABSORPTION

Living organisms can be broadly divided into two groups. Some, like the green plants, only require to be provided with simple inorganic materials from which, with the aid of energy drawn from the external world, they can accomplish the synthesis of everything required for their life, growth and reproduction. Others, like the animals, can only live and reproduce if provided with complex, energy-rich, organic materials, collectively designated as food. These two groups of living organisms are known as *autotrophes* and *heterotrophes* respectively.

Predominant among autotrophic organisms are the *green plants*, which are able to fix and utilize the energy of solar radiation. This is brought to bear, in a manner which is now fairly well understood, upon the synthesis of complex energy-rich materials; and, as raw materials for the synthesis, carbon dioxide, water, salts, and some simple source of nitrogen such, for instance, as ammonia or nitrate are all that is necessary. The key substance in photosynthesis, chlorophyll, finds counterparts in the specialized bacterial pigments upon which *photosynthetic bacteria* rely for a comparable fixation of solar energy. In the remaining group of autotrophic organisms, the *chemosynthetic bacteria*, energy is not obtained from the sun but by harnessing the chemical energy of some inorganic process such as the oxidation of ammonia to nitrite or nitrate, the oxidation of hydrogen sulphide to elemental sulphur, or that of ferrous compounds to the ferric state. The autotrophes are in every case competent to synthesize all the structural, catalytic and storage materials they need for growth, maintenance and reproduction: everything their life requires can be produced from the simplest of starting materials, the necessary energy being collected from the external world.

Heterotrophic organisms stand in sharp contrast to the autotrophes, for not even the most versatile of heterotrophic forms can live except by exploiting the industry and synthetic ingenuity of other organisms. Only by fermenting, oxidizing, or in some other way degrading complex organic material can the heterotrophes obtain the energy required to maintain themselves. It may therefore be said that all heterotrophes require 'food', that is to say, oxidizable or fermentable material by the breakdown of which free energy can be released and harnessed for locomotion, chemical synthesis and other energy-consuming processes.

Many heterotrophic forms of life such, for example, as the *free-living bacteria* and *yeasts*, can live and reproduce in very simple media. Apart from water, salts and some simple source of nitrogen, they need only to be provided with 'food' in the form of some fairly simple organic compound such, say, as lactate. Given these substances the free-living bacteria can accomplish *de novo* the synthesis of everything their life requires. But many micro-organisms are more exacting. The presence of certain particular compounds in the habitual environment of a given strain can lead in the end to the loss within that strain of the ability to synthesize the substances in question. Thus many milk-souring bacteria, such as are cultivated for the manufacture of cheese, cannot live or multiply except in media containing riboflavin. Free-living organisms can synthesize this important substance for themselves, but the cultivated milk-sourers have lived for so many generations in milk, which is a fairly rich source of riboflavin, that their ability to synthesize it has been lost for lack of employment. For these organisms, riboflavin has become an indispensable accessory food factor; in other words, a vitamin or essential growth factor.

The nutritional requirements of many micro-organisms have been carefully investigated over many years, and there is now abundant evidence that some degree of synthetic disability is a common feature among them. Yeasts which have been nursed and pampered in vineyards and breweries require the provision of a number of the factors that free-living forms can make for themselves, while among bacteria and protozoa many forms, including numerous highly pathogenic strains and species, have been found to have nutritional requirements that are very exacting indeed. Their loss of synthetic ability seems to be a stepwise process, for certain bacteria, given β-alanine or nicotinamide, can synthesize pantothenic acid or NAD respectively, but in other cases it is necessary that these more complicated substances should be given intact; even the ability to join together the constitutional fragments has been lost. The work of Beadle, Tatum and their collaborators and followers on mutant strains of the bread-mould, *Neurospora crassa*, revealed many examples of step-by-step loss of synthetic ability (pp. 310–13). This loss of synthetic ability has gone even further in some micro-organisms than it has in animals, for some of them are unable even to synthesize haematin.

By contrast with green plants, or even with free-living micro-organisms, *animals* are very exacting creatures indeed. In addition to water, salts, and 'food' in the sense of energy-yielding organic substances, animals of every kind need to be provided with certain amino- and fatty acids, and with a number of the other indispensable accessory food factors collectively known as *vitamins*. These include thiamine, riboflavin and nicotinamide, all three of which are constituents of essential coenzymes. This implies the inability of animals to synthesize many of the tissue constituents and catalysts which they require. More recently evidence has come forward to point to catalytic roles for most of the B vitamins including, in addition to those already mentioned, pyridoxal and biotin, together with folic, lipoic and pantothenic acids (see Table 13 for some examples).

Table 13. *Catalytic functions of some members of the B vitamins*

	Compound	Present or concerned in	Formula see p.
B_1	Thiamine (B_1)	Co-carboxylase; coenzyme of oxidative de-carboxylation, transketolation and trans-aldolation	302
B_2 *group*	Nicotinamide (niacin)	NAD and NADP (co-dehydrogenases)	154
	Riboflavin (B_2)	Prosthetic groups of flavoproteins	169
	Pyridoxal (B_6)	Prosthetic groups of transaminases, decarb-oxylases, serine and threonine dehydratases	100
	Pantothenic acid	Coenzyme A	241
	Biotin	Co-factor involved in fixation of CO_2	118
	Folic acid	THF, one-carbon metabolism (B_{12}), etc.	108
	Inositol ⎱ Choline ⎰	Phospholipid metabolism	— 112
	Carnitine (insects)	Complexes with acetyl-CoA	112
	Cobalamine (B_{12})	Transcarboxylation	—
	Lipoic acid	Oxidative decarboxylation	388

Many examples of the functions of these compounds can be located in the text by consulting the index.

These general notions help greatly in the interpretation of the food relationships that exist between living organisms of different kinds. In the words of Charles Elton, 'Animals are not always struggling for existence. They spend most of their time doing nothing in particular. But when they do begin, they spend the greater part of their lives eating. The primary driving force of all animals is the necessity of finding the right kind of food and enough of it.' The 'right kind of food' is largely determined, of course, by the animal's ability to capture and kill other organisms, but on the chemical side we can say that the 'right kind of food' is that which provides the eater with the energy requirements of its kind and, at the same time, with whatever special materials, amino-acids, vitamins and the like, are essential to the species in consequence of its synthetic disabilities.

In any natural animal and plant community we can trace out what are known as *food chains*. A food chain typically begins with green plants, which are exploited by herbivorous animals and these, in their turn, by carnivores. These become the prey of larger and more powerful carnivores and so on until, in the end, we arrive at an animal so large and powerful that it has virtually no enemies except, perhaps, that ubiquitous animal, man. Always in these food chains the starting-point is with autotrophic organisms, most usually with green plants or photosynthetic algae. Herbivorous animals rely at first hand,

and carnivores at second or third hand, upon the autotrophes for supplies of the numerous accessory food factors which they require, as well as for a sufficiency of complex, energy-yielding organic foodstuffs. Synthesized by autotrophic organisms, especially by the green plants, and gathered together in the first instance by herbivorous beasts, these essential materials are passed stage by stage along the food chains.

The same general ideas are also valuable in the interpretation of food relationships of other kinds. *Parasitism*, for example, presents many problems which a thorough knowledge of the nutritional requirements of parasitic organisms might go far towards solving. We have already seen that the requirements of a given organism are liable to be influenced by the availability of particular substances in the environment to which the organism is accustomed, quite apart from the general physico-chemical properties of the habitat. This sort of process can be carried to great lengths. Many micro-organisms are now confined to particular habitats and have, in fact, become absolutely dependent, i.e. parasitic, upon those habitats because they have lost the ability to synthesize one after another of a few or many essential substances and can therefore only survive in environments in which all those substances are to be found ready-made. Possibly the same will prove to be true of other parasites, such, for instance, as the tape-worms and round-worms that are such a common feature of the intestinal fauna of animals of every kind.

Another important type of nutritional association is *symbiosis*. A cow may harbour large numbers of parasitic worms in addition to the multitude of symbiotic micro-organisms that inhabit its rumen, but the relationships between the cow and the worms on the one hand, and between the cow and its symbionts on the other, are very different. The cow acts virtually as a food-collecting machine for both groups of organisms, but gets nothing in return from its parasitic inhabitants except illness in some cases. The symbionts, however, reward their host by breaking down cellulose and other cow-indigestible materials, from which they produce short-chain fatty acids which the cow can utilize. Similar arrangements are found in herbivores of many kinds, from cows to cockroaches. But the host member of the pair stands to gain yet further rewards for hospitality rendered. Some at least of the symbionts can synthesize from very simple materials all the amino-acids and vitamins that they themselves require. These compounds become incorporated in the first instance into the substance of the symbionts, but these organisms are not immortal. When they die and undergo eventual autolysis, or digestion by the host's enzymes, their essential amino-acids and vitamins become available to the host, at any rate in part. There can be little doubt that some herbivores depend largely upon their intestinal flora and fauna for supplies of essential accessory food materials, though it may be doubted whether supplies from these sources are ever sufficient by themselves.

Provided that their somewhat exacting nutritional requirements are ful-

filled and that sufficient energy-yielding substances are available, heterotrophic organisms such as the mammals are capable of dismantling their food
materials and rearranging the component parts in a very versatile manner.
From its food proteins, for instance, an animal can build up species-specific
protein characteristic of its own tissues and secretions; carbohydrates and
even complex lipids can be produced from the deaminated residues of superfluous amino-acids; carbohydrates can be converted into fats and so on.
Herbivorous animals can lay down both carbohydrate and fat at the expense
of the short-chain fatty acids produced by the exertions of their intestinal
symbionts.

The first steps in this direction consist in the digestion and absorption of the
food, and these are followed by storage and eventual metabolism of the ingested materials. In human feeding *cookery* plays an important part. Cooking
denatures the food proteins and bursts the granules of natural starch, thus
facilitating their attack by the digestive enzymes. By *digestion* we mean the
hydrolytic breakdown of food materials, which consist pre-eminently of relatively large molecules, into simpler compounds which a given organism can
absorb and then use to build up its own tissues and food reserves. This definition is one that can be widely interpreted, for the food may be the food eaten
by an animal, on the one hand, or, on the other, it may comprise the materials
provided in a seed or an egg for the embryonic development of a plant or an
animal. A seedling plant is as heterotrophic as any animal until it reaches the
daylight, produces chlorophyll for itself, and can begin photosynthetic activities on its own account.

The seeds of plants contain considerable reserves of organic foodstuffs
from which new plants develop. At or before the time of germination, enzymes are present which may be said to have digestive functions, since they
serve to dismantle the food materials into simpler components which the
young plant then oxidizes or rearranges in its own characteristic manner. For
example, the seeds of the castor-oil plant, *Ricinus*, are rich in oils and contain
a powerful lipase: barley, which is rich in starch, contains powerful α- and
β-amylases and a maltase at germination and so on.

Among animals, digestion may be accomplished in either of two main ways.
The food may be phagocytically ingested and then intracellularly digested, or
it may undergo extracellular digestion before being absorbed. Often both
mechanisms are used side by side in one and the same animal. Intracellular
digestion is probably more primitive than its extracellular counterpart, for
phagocytosis is only possible for particles up to a certain order of size. That
extracellular digestion arose as an adaptation to the necessity of breaking up
relatively large food masses prior to absorption seems very likely, and it is
probably significant in this respect that the peptidases involved in the extracellular digestion of proteins in the mammals appear to be qualitatively and
quantitatively homospecific with the intracellular peptidases or kathepsins.

In some animals it is possible to observe what seems to be a transitional process that is neither entirely intracellular nor wholly extracellular. In certain platyhelminth worms, for instance, the gut is lined with cells endowed with considerable amoeboid activity. When food is taken, these cells absorb water from it and swell up, sending out processes which form a syncytial network that fills the gut cavity and enmeshes the food mass. Digestion takes place within the syncytium, which is later withdrawn. In certain other animals, however, a syncytium is formed but withdrawn before digestion has proceeded very far, and digestion continues even in its absence. Here we have as it were a half-way house between intra- and extracellular digestion, for it would appear that the function of the syncytium has been discharged once the enzyme granules it contains have been discharged.

In many organisms, notably among the protozoa and sponges, phagocytosis followed by intracellular digestion is the only mechanism available for assimilation, but in other phyla it is not uncommon to find both the intracellular and extracellular modes of digestion used together. It usually appears in cases such as these that if an animal is carnivorous its extracellular enzymes are those which act upon proteins; if it is herbivorous the extracellular enzymes are those that act upon carbohydrates. Thus among the coelenterates, which are mainly carnivorous, peptidases are secreted by the walls of the coelenteron while non-protein materials are digested intracellularly. Similarly, the only extracellular digestive enzymes found among lamellibranch molluscs, which are almost exclusively herbivorous, are the amylases of the so-called crystalline style.

The general disintegration of the food mass that results from the action of extracellular enzymes is usually facilitated by mechanical movements of the walls of the digestive cavity, so that the mass is eventually reduced to a particulate dispersion fine enough to allow of phagocytosis, and digestion is subsequently completed within the cells. But when we come to animals as complex and as highly specialized as the mammals we find that digestion is entirely extracellular. Indeed, the only remnant of the phagocytic systems which are so important in many invertebrates is that found in the wandering scavenger cells of the reticulo-endothelial system.

Relatively little is known about the digestive processes of invertebrates, but the very large literature of the subject indicates, in a general manner, that animals as a whole are equipped with enzymes competent to break down lipids, proteins and carbohydrates into their simpler constituents. Peptidases, lipases and carbohydrases have been detected, either in the extracellular digestive juices or in the cells of the digestive glands themselves, in a very large number of cases. It is not always possible however to be sure that, because a protein-splitting enzyme is demonstrably present in the cells of a digestive gland, it necessarily has a digestive function.

The nature of the chemical operations involved in digestion appears to be substantially the same in all kinds of animals, whether digestion is intra- or extracellular. Most is known about these processes and the enzymes which catalyse them in mammals, and the ensuing description of digestion relates mainly to these animals.

It is not unusual to think of digestion as a process which is divisible into a series of nicely defined steps, each of which leads to equally nicely defined products. We find in the text-books an abundance of statements to the effect that pepsin digests proteins thus far and no farther, that trypsin digests them farther to another definite point, and so on. While it is perfectly true that each enzyme, taken by itself, will carry out certain perfectly definite operations and cease acting when these have been accomplished, it must be realized that digestion is not carried out in this manner. Food that has passed into the small intestine of a mammal, for example, is exposed to the simultaneous activities of all the pancreatic and intestinal enzymes, and its digestion is not separable into a series of discrete steps and stages but is, rather, a continuous process. The fact that the digestive secretions of animals have been resolved into a number of individual catalysts, each of which can be studied separately, has tended somewhat to encourage the step-by-step outlook on digestion as a whole. Perhaps the best way to check this tendency is to think of the enzymes involved in digestion, not as a mere collection of catalysts, but rather as an organized *system* of catalysts, so ordered and regulated as to carry out a long and complex but nevertheless continuous chain of processes.

The products of digestion form the raw materials for the processes of *metabolism*, a general term used to cover all the chemical changes going on in the cells and tissues of living organisms. These changes may result in a chemical simplification of the starting material, in which case we speak of *katabolism*, or in an increase of chemical complexity, when we speak of *anabolism* (cf. Chap. 3).

Living organisms of every kind appear to be able to accomplish anabolism at the expense of katabolism, but the ability to carry out anabolic changes at the expense of external energy is the prerogative of autotrophic organisms. Thus, when we are studying processes of katabolism we should have constantly in mind the question, how much energy becomes available to the organism? We may also ask in what form it becomes available, and how it is converted into chemical, mechanical, electrical, thermal or osmotic work as the case may be. Similarly, when anabolic changes are being considered, we must inquire whence and in what form the necessary free energy is forthcoming, and how it is transferred from its metabolic source to its intramolecular destination. These are important questions; questions which, moreover, have remained practically unanswered and unanswerable until comparatively recent times, but at last we have a few clear indications on which we are beginning to found a new knowledge of biological energetics. A new and biologically applicable thermodynamics has still to be developed however.

Before going on to consider metabolism in detail it is desirable to examine the phenomena of digestion, taking our information mainly from the mammals. This task we shall attempt in the rest of this chapter.

DIGESTION AND ABSORPTION OF PROTEINS

Saliva contains no proteolytic enzyme, and the first phase of digestion takes place in the stomach under the influence of pepsin. Pepsin, it will be remembered, is secreted in the form of an enzymatically inactive precursor, pepsinogen (pro-pepsin). This is activated by the hydrochloric acid of the gastric juice, which provides at the same time an acid medium in which the pH is about optimal for the action of pepsin. The latter, which acts more rapidly upon denatured than upon native proteins, opens up certain particular peptide links in its substrates but whether or not it is able to complete its work depends a good deal on the consistency of the gastric contents. As soon as these have become liquid they are forced through the pyloric sphincter, whether peptic digestion has been completed or not.

After its passage through the pylorus the partially digested food mass, or chyme, is mixed with the pancreatic juice and the bile. Taken together, these secretions contain about enough free alkali to neutralize the acid that has come through from the stomach, and the pH of the intestinal contents is brought nearly to neutrality.

Trypsinogen (pro-trypsin) and chymotrypsinogen (pro-chymotrypsin), activated by enteropeptidase and by trypsin respectively, yield trypsin and chymotrypsin. These enzymes continue the process of hydrolytic disintegration begun by pepsin and open up more, but different, peptide links, to produce peptide fragments much smaller than the original food proteins. Few free amino-acid molecules are produced at this stage. Carboxypeptidases, contributed by the pancreatic juice, and aminopeptidases, secreted by the intestine, take up the task of degrading the polypeptide fragments inwards from the ends, liberating amino-acid molecules one at a time until, when the dipeptide stage is reached, the substrates pass out of their range of specificity and into that of the dipeptidases of the intestinal secretions, and these complete the digestion. Eventually, therefore, the amino-acids which enter into the composition of the food proteins are set free, absorbed into the portal blood stream, and carried away into the general circulation by way of the liver.

In the past there has been considerable discussion as to whether protein foodstuffs are, in fact, completely broken down into their constituent amino-acids before being absorbed. Some favoured this view, while others believed that so long as the protein has been reduced to some soluble form such, for example, as a mixture of peptones, the function of the digestive enzymes has been satisfactorily discharged. There is now evidence in plenty to show that this view is erroneous. In the first place it is unlikely that peptones

210

are absorbed as such because, if peptones are injected into the blood stream of mammals, a condition known as 'peptone shock' results, but nothing comparable follows the consumption of a protein meal. Abel, using an ingenious technique known as vividiffusion, showed many years ago that amino-acids, but no protein fragments of larger size, can be detected in the blood leaving those regions of the gut from which absorption takes place. This he did by leading off the emergent blood through a series of collodion tubes immersed in warm physiological saline, and returning it then to the circulation. After this performance he was able to isolate several amino-acids from the saline medium and to detect the presence of a number of others by chemical tests, but no trace of products more complex than the amino-acids were detected. There is, moreover, a large increase in the concentration of free amino-acid nitrogen in the blood while absorption is taking place in a normal animal. It may also be argued, though perhaps the argument savours a little of teleology, that animals do in fact possess a series of enzymes capable of carrying digestion right through to the free amino-acid stage, and that these enzymes would hardly have been perpetuated in the course of evolution unless they were of some use, i.e. survival value, to the organism. There is also the telling fact that certain students acquired considerable fame for themselves by consenting over considerable periods to the replacement of their dietary protein by mixtures of purified amino-acids without, however, appearing any the worse for the experience.

Food proteins, then, are not absorbed by an animal as such. For one thing, cow-protein or sheep-protein is not the same as man-protein, and if a man is to rely on the cow or the sheep to keep up his own levels of man-protein, the protein he eats has to be completely dismantled by his digestive enzymes and the component amino-acids must then be rearranged by the man's own enzymic apparatus to produce his own kind of protein, but very little is known even today about the mechanisms involved in the absorption of these vitally important amino-acids.

DIGESTION AND ABSORPTION OF CARBOHYDRATES

Few animals are equipped with enzymes capable of attacking cellulose, although this polysaccharide plays a very large part in the nutrition of herbivorous animals. In these creatures the task of digesting cellulose is usually delegated to vast hordes of symbiotic micro-organisms (p. 74), and the useful products of their activity consist in the main of short-chain fatty acids. The mechanisms involved in this degradation are complex, if only because many different kinds of micro-organisms are concerned.

Like cellulose, the so-called hemicelluloses (xylans, arabans, mannans, galactans, etc.) and fructofuranosans (such as the levans of grasses and the inulins of the Jerusalem artichoke and other Compositae) are not digestible

211

by the enzymes of most animals, though they can be handled by symbiotic micro-organisms and probably yield products similar to those from cellulose.

In the insoluble granular form in which it occurs in nature, starch is very resistant to digestion and one of the functions of cooking is the disruption and partial solubilization of the starch granules. The digestion of starch and glycogen is initiated by salivary amylase but, unless the eater follows the precept of Mr Gladstone and chews each mouthful of food quite an unbelievable number of times, little digestion takes place in the mouth. The food, more or less intimately mixed with saliva, is swallowed and passes on into the stomach. Although the optimal pH for salivary digestion lies very near to neutrality, the secretion of the strongly acid gastric juice does not put a sudden end to salivary digestion because it takes time for the acid to penetrate into the food bolus. The consistency of the food mass is therefore an important factor. Eventually, however, the free acid of the gastric contents reduces the pH to a value at which the salivary amylase is inactive and is actually destroyed, but in the meantime starch and glycogen alike have been at least partly broken down to yield maltose, together with some maltotriose and, if digestion is not yet complete, an assortment of dextrins.

The gastric juice itself contains no carbohydrase, but a notable concentration of free hydrochloric acid is present and contributes something to the digestion of carbohydrates containing fructofuranose units. Fructofuranosides such as sucrose, inulins and levans are hydrolysed with great ease and rapidity by warm, dilute mineral acids, and it is probable, therefore, that substances such as these undergo at any rate a partial hydrolysis during their stay in the stomach. The hydrolytic activities of the hydrochloric acid are cut short when the chyme passes through the pyloric sphincter and into the duodenum, where it encounters the strongly alkaline pancreatic juice and bile. Here the pH rises nearly to neutrality, and under these conditions the amylase of the pancreatic juice has almost its optimal activity. This enzyme finishes the work begun by the salivary amylase, and the conversion of starch and glycogen into maltose is completed. Some free glucose is produced at the same time.

Maltose, however, is only a transitory product, for it is rapidly hydrolysed under the influence of an α-glucosidase, the so-called 'maltase' of the intestinal juice. This secretion also contains a powerful β-fructofuranosidase, which completes the hydrolysis of sucrose, and a β-galactosidase, 'lactase', that deals with lactose. Ultimately, therefore, the digestible carbohydrates of the food are resolved into their constituent monosaccharides and absorbed in this form. It is improbable that appreciable quantities of di- or higher saccharides are absorbed because, as is known from injection experiments, disaccharides present in the blood stream are largely excreted unchanged, and it is only in exceptional and probably abnormal cases that disaccharides appear in the urine, though lactosuria is common during pregnancy and lactation.

212

The rates of absorption of different monosaccharides vary much more widely than might have been expected in view of the fact that all hexoses have the same molecular weight, while that of the pentoses is not very different. It follows, therefore, that the absorption of sugars from the gut cannot be explained in terms simply of diffusion. The same conclusion follows from the fact that glucose, for example, can be absorbed from very strong solutions, and therefore against large osmotic gradients. Many experiments have been made to discover what mechanisms are involved in the absorption. One theory after another has been proposed to account for the intestinal absorption of the sugars but no satisfactory explanation of the known facts has so far been propounded, let alone proved.

The rate of absorption can be determined by opening up an experimental animal, such as a rat, and introducing a known amount of the sugar to be studied into a loop of intestine, previously tied off at both ends. The animal is kept for a known length of time and the contents of the intestinal loop are then removed and analysed. The amount of sugar absorbed is found by difference. While experiments carried out on these lines show that galactose and glucose are absorbed much more rapidly than other sugars, they are open to criticism. There is, in the first place, a definite possibility that direct damage may be done to the gut, and it is also possible that the anaesthetic that must necessarily be used may interfere with the normal processes of absorption. But other methods of investigation are possible and yield substantially the same results. Cori, working on unanaesthetized rats which he fed by stomach-tube, obtained the results shown in Table 14. Other workers, have obtained substantially the same figures, though rather different ratios have been found in different animal species.

Table 14. *Absorption of monosaccharides from the small intestine of rats*

Sugar	(*After* Cori) Relative rate of absorption
D-Glucose	100
D-Galactose	110
D-Fructose	43
D-Mannose	19
L-Xylose	15
L-Arabinose	9

Pentoses are absorbed at about the same rate as indifferent substances such as sodium sulphate. Galactose and glucose are absorbed so much more rapidly than the rest that a special mechanism of some kind must be deemed to be involved in their case. Verzár showed that the selective absorption of glucose and galactose can be abolished by adding iodoacetate or phlorrhizin

15-2

to the contents of tied-off intestinal loops, and that absorption in the intact animal is much delayed by previous injection of these drugs. Both phlorrhizin and iodoacetate are known to be powerful inhibitors of fermentation and glycolysis, and in both these processes phosphorylation is known to play a fundamental part. Verzár therefore regarded his results as evidence that phosphorylation is involved in the absorption of sugars from the intestine. This argument assumes that the experimental results obtained with iodo-acetate are valid, and this is questionable because the concentrations of iodo-acetate used were high enough to cause sloughing of the mucosa from the gut wall in some of the experiments.

The most probable explanation of the action of iodoacetate lies in the fact that, for energetic reasons and whether it involves phosphorylation or not, the absorption of glucose probably requires the participation of ATP. The provision of ATP depends in turn upon glycolysis and oxidation, several steps in which are powerfully inhibited by iodoacetate. The interference of iodo-acetate in the normal absorption of glucose is probably due, therefore, to inhibition of energy-yielding metabolism. The only verdict we can give on the phosphorylation theory is one of not proven and, indeed, it is clear today that absorption is a more complex affair than it formerly appeared.

A point of interest may be added in passing. It is well known that phlor-rhizin also abolishes the reabsorption of glucose from the urine by the cells of the kidney tubule, as well as its specific absorption from the small intestine. Phlorrhizin, like iodoacetate, is a powerful inhibitor of glycolysis and oxida-tion, and the facts suggest that the intestine and the renal tubule alike carry out the work of absorption by essentially similar mechanisms, whether phos-phorylation is involved or not. Another interesting drug is cetramide; this substance not only inhibits the absorption of sugars but actually reverses the flow so that the levels of glucose in the blood and in the gut contents become identical.

DIGESTION AND ABSORPTION OF FATS

Saliva contains no lipase and, while the presence of a lipase in the gastric juice has been reported in a number of cases, the activity of the alleged gastric lipase at the pH of the gastric contents is such that it can be of little importance in digestion. Most authors believe that it must be pancreatic lipase that has regurgitated from the small intestine. But although no appreciable digestion takes place in the stomach, the fats of the food are warmed and softened, if not actually liquefied. When presently the chyme is somewhat forcibly squirted into the duodenum there is a marked tendency for the fat to become emulsified, a tendency which is emphasized by the presence of the bile salts contributed by the liver. The commonest representatives of this important group of substances are conjugated derivatives of cholic and other bile acids with glycine and taurine (p. 216). They are remarkable for their property of very greatly reducing the surface tension at fat/water interfaces, and for this reason they not only facilitate emulsification but tend to stabilize an emulsion once it has been formed.

Until the beginning of the century it was generally believed that finely emulsified fat can be absorbed without previous digestion, but it appears that the bile salts alone cannot produce a sufficiently fine dispersion. The essential conditions for absorption are, according to Frazer, that the particles shall be less than $0.5\,\mu$ in diameter and that they shall be negatively charged.

Frazer and his colleagues attempted to find natural emulsifying agents which could produce the degree of emulsification required for direct absorption under physico-chemical conditions similar to those which prevail in the small intestine. The substances studied included bile salts, cholesterol, a free fatty acid (oleic) and a monoglyceride (glyceryl monostearate), separately and in various combinations. Only with bile salts + fatty acid + monoglyceride was it found possible to obtain the necessary degree of dispersion.

But not all the food fat is absorbed in the emulsified form. Animals possess powerful lipolytic enzymes, which probably would not have survived unless they were useful to their possessors. The pancreatic and intestinal lipases act only on fats in the emulsified form. Now, the action of pancreatic lipase upon ordinary neutral fats disengages only the terminal fatty acids from their combination with glycerol, giving a mixture of free fatty acids with monoglycerides and these, in the presence of bile salts, are precisely the materials required for the emulsification of the remaining unhydrolysed fat. It may therefore be concluded that at least a part of the fat of the food undergoes digestion before being absorbed, and that the products of partial digestion, together with the bile salts, facilitate the emulsification of the remainder, which is then absorbed without previous digestive hydrolysis. In this case, therefore, digestion and absorption are closely interconnected.

Bile salts play a large and very important part in the digestion of lipid materials by facilitating their emulsification and so presenting the digestive lipases with a larger surface upon which to attack their substrates. Pancreatic lipase only acts on emulsified fats, and it is difficult to reconcile with these the further fact that bile salts are not essential for digestion. In experimental animals in which the bile duct has been ligated, or in human subjects in whom the bile duct is occluded, e.g. by the presence of gall stones, fat is still digested, as is attested by the presence of free fatty acids in the faeces. Again, if a fat such as olive oil is introduced into a tied-off loop of intestine, with or without the addition of bile salts, no absorption takes place, but if a small amount of lipase is also added, the contents of the loop undergo digestion. But the oleic acid liberated is only absorbed if bile salts were introduced into the loop at the outset. Thus bile salts play some important part in the absorption of free fatty acids, as well as in that of unhydrolysed fat.

This is usually attributed to the so-called hydrotropic action of the bile

215

salts, i.e. their ability to form water-soluble complexes with fatty acids. This effect can be demonstrated readily enough by adding a solution of sodium glyco- or taurocholate to an aqueous emulsion of a fatty acid. If enough bile salt is added, the emulsion becomes water-clear. The ability of these complexes to pass through a membrane will therefore be determined by the relative proportions of fatty acids and bile salts. If the fatty acid:bile salt ratio is low the particles will be small, and their ability to pass through the intestinal barrier will be expected to be proportionately greater.

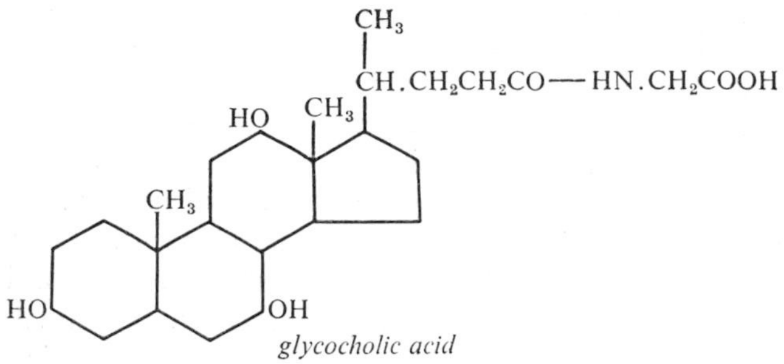

glycocholic acid

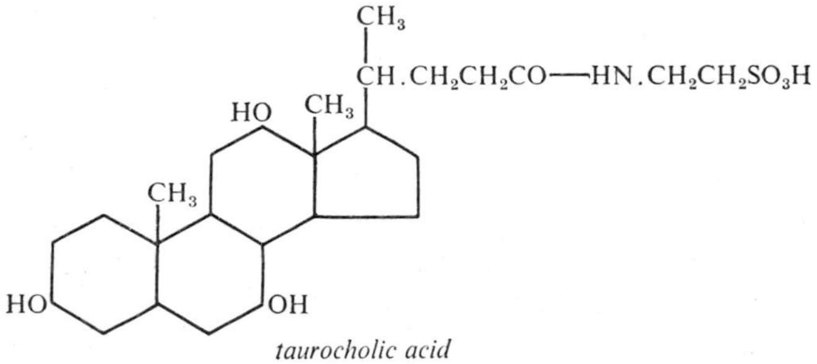

taurocholic acid

Fig. 46. Structures of some common bile salts.

Many experiments were carried out by Verzár and his colleagues on the absorption of fatty acids, but in discussing them it is well to remember that, at the time, Verzár himself was of the opinion that fats must be fully hydrolysed before they can be absorbed. In his experiments use was again made of tied-off intestinal loops. Oleic acid and bile salts were introduced into the loops in all the experiments, together with the other substances indicated in Table 15, which is taken from his work. Neither glycerol nor phosphate alone leads to any acceleration of absorption, but if both are present together the rate increases two or three times. This suggests the possible formation of some

216

Table 15. *Absorption of fatty acids* (*after* Verzár)

Contents of intestinal loop	Oleic acid absorbed in 6 hr.
Oleic acid + bile salt	29·3
Oleic acid + glycerol + bile salt	24·9
Oleic acid + phosphate + bile salt	10·3
Oleic acid + phosphate + glycerol + bile salt	48·9
Oleic acid + glycerol phosphate + bile salt	72·7
Oleic acid + glycerol phosphate + bile salt + iodoacetate	0

compound of fatty acids with glycerol and phosphate, i.e. of a lecithin-like substance.

Like the absorption of sugars, that of fatty acids is inhibited by phlorrhizin and by iodoacetate, probably for the same reasons as is the absorption of glucose. Perhaps phosphorylation is involved, for the phospholipid content of the blood is higher during the absorption of a fatty meal than at any other time, while the phospholipid content of the intestinal lymph rises from a resting level of about 2·2 to about 7·5 mg. % while fat absorption is taking place. Furthermore, if an animal is fed with fat that has been 'labelled' with iodine or with heavy hydrogen, iodized or deuterated phospholipids can be recovered from the gut mucosa while absorption is in progress. It remains true, however, that the great bulk of the food fat reappears after absorption in the form of so-called 'neutral' fat.

It appears, then, that a part of the food fat is digested by the pancreatic and intestinal lipases, and that the products of partial digestion, aided by bile salts, serve to emulsify the remainder so finely that it can pass into the cells of the mucosa without previous hydrolysis. How much of the total fat undergoes digestion and how much is absorbed directly is not certainly known.

Fat which is absorbed in the emulsified condition passes into the cells of the gut wall, in which it can be observed in the form of minute droplets which stain with dyes such as Sudan III and have in fact all the histochemical characteristics of neutral fat. From the cells these droplets, the chylomicrons, make their way into the lacteals and hence, through the lymphatic system, the thoracic duct, and then through the liver and presently into the adipose tissues by way of the blood stream, where they are responsible for the condition of post-absorptive lipaemia. Free fatty acids pass into the intestinal mucosa, perhaps in the form of water-soluble complexes with bile salts, and here they appear, at any rate in part, to be resynthesized to ordinary neutral fat. These too pass into the lymph, for in experiments in which isotopically labelled pentadecanoic acid was administered to experimental animals, the isotope was almost quantitatively recovered in the effluent from a cannula tied into the thoracic duct. Other experiments with isotopically labelled acids have shown that the shorter chain acids—up to 8 carbon atoms in length—are largely absorbed by way of the blood stream, perhaps because they are appreciably (8 C) or very (2–3 C) soluble in water.

Here, as in the field of carbohydrate absorption, many theories have been put forward one after the other. None has so far been completely convincing nor completely proved, but the foregoing account seems best to accord with known facts.

11

GENERAL METABOLISM OF AMINO-ACIDS

Proteins constitute an indispensable article of food for all animals. We have abundant direct evidence from feeding experiments that the usual laboratory animals require supplies of certain, but not of all, amino-acids, not only for growth while the animal is young, but also for the maintenance of normal physiological condition during adult life. Amino-acids can therefore be classified as essential or non-essential according as they must be or need not be supplied in the food.

The *essential amino-acids* are only to be found in proteins. Protein food-stuffs are therefore indispensable. In most and probably all herbivores an indirect, secondary source of essential amino-acids is found in the tissue proteins of symbiotic micro-organisms inhabiting the gut but, although the value of this supplement may be considerable in some cases, we do not know if it is ever sufficient alone. Probably it is not. Evidence regarding the amino-acid requirements of invertebrate animals, unfortunately, is scanty. What information we have relates mostly to insects, and does not give grounds for supposing that their ingenuity in amino-acids synthesis is any greater than our own. Tryptophan, lysine and probably histidine seem to be essential for animals of every kind.

The naturally occurring amino-acids can be classified as in Table 16. Under the heading of essential amino-acids are some that can be replaced by other

Table 16. *Nutritional status of amino-acids*

	Essential	
Non-essential	Irreplaceable	Replaceable
Glycine	Threonine	
Alanine	Valine	
Serine	Leucine	
Aspartate	isoLeucine	
Glutamate	Methionine	Cysteine: cystine
Proline	Phenylalanine	Tyrosine
Hydroxyproline	Histidine	
Arginine	Tryptophan	
	Lysine	
	Arginine	

members of the essential group; thus tyrosine can be formed from phenyl-alanine, while cysteine can be produced if methionine is available. Provided that enough phenylalanine is available to discharge the essential and characteristic functions of phenylalanine itself, *and* to produce at the same time enough tyrosine to fulfil those of tyrosine, tyrosine itself need not be provided. Tyrosine, however, cannot discharge the functions of phenylalanine. Glycine, aspartate and glutamate, by contrast, need not be provided at all; these the organism can make for itself from non-protein materials, but of all the amino-acids that enter into the composition of the tissue and other proteins, there are less than ten which the animal organism can produce by its own resources.

The position of *arginine*, which figures as essential and as non-essential alike, calls for special comment. Adult rats remain alive and healthy, and young rats grow, on diets wholly devoid of arginine. But the growth of young rats on an arginine-deficient diet can be accelerated by the administration of that amino-acid. It is therefore probable that, while the growing animal can evidently synthesize arginine to some extent, it cannot do so fast enough to keep pace with the requirements of optimal growth. A similar effect has been observed in chicks, which require but cannot synthesize *glycine*. The case of *histidine* is also somewhat peculiar for it is thought that man, in contra-distinction to other mammals, has no histidine requirement. It may be, indeed, that important substances other than amino-acids are synthesized slowly enough to limit the growth-rate of young organisms.

Amino-acids, essential and non-essential alike, are required for numerous purposes. Quite apart from the special products to which particular individual amino-acids give rise, new tissue proteins must be synthesized, damaged or wasted tissues must be repaired or replaced, and normal supplies of enzymes and hormones must constantly be maintained. The formation of noradren-aline, adrenaline and the thyroid hormones, for example, makes essential the provision of tyrosine (or phenylalanine), to which they are closely related and from which they are in all probability produced, as witness their respective formulae (Fig. 47). The production of another hormone, insulin, imposes particularly heavy demands upon the supplies of essential amino-acids, for it is a polypeptide containing roughly 16% leucines, 8% phenylalanine, 12% tyrosine, 4% histidine, 3% threonine, 2% lysine and 12% cystine.

Provided that the food proteins contain enough of all the amino-acids that are essential, the organism can probably make good any deficiency of the rest, although, in the ordinary way, it is not likely to be called upon to do so. The essential amino-acids are, on the whole, the least common, so that an adequate intake of these in the form of protein is more or less inevitably attended by an adequate intake of the rest. It is necessary, however, that the requisite amino-acids should be presented to the organism together, all at one time, as they are if an adequate meal of protein is taken. Amino-acids that are not more or less immediately incorporated into proteins or otherwise utilized

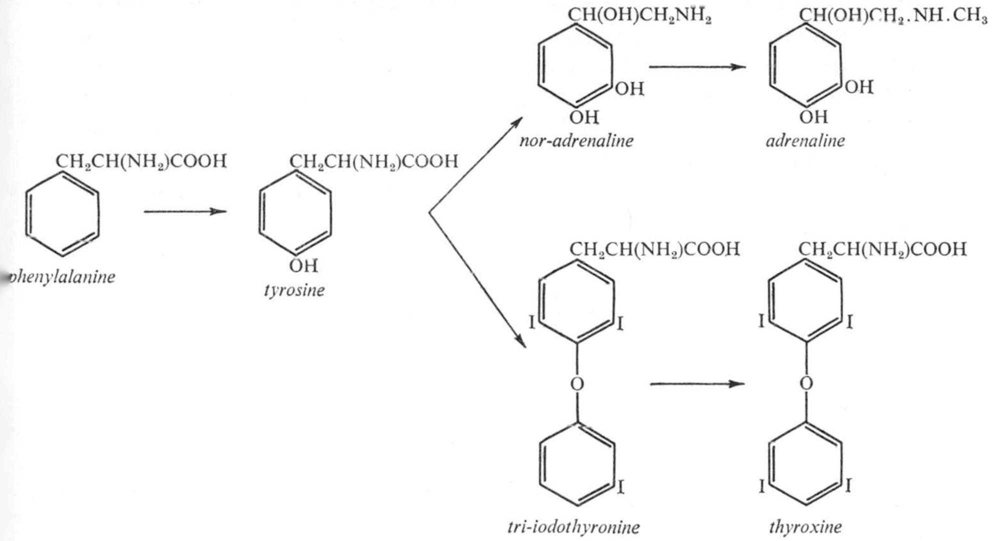

Fig. 47. Relationships between aromatic amino-acids and some
adrenal and thyroid hormones.

undergo deamination, be they essential or non-essential, and are therefore
lost for purposes of protein synthesis. This has been demonstrated by feeding
mixtures of purified amino-acids simultaneously and at different times.

The elaboration of hormones, enzymes and other special products still goes
on even during starvation, when it can only be done at the expense of the
tissue proteins. Prolonged deprivation of protein therefore leads to emaciation
and eventually to death. For some time before death ensues there is a small,
fairly constant, daily excretion of nitrogen, the magnitude of which may be
taken as an index to the amount of protein being broken down for processes
essential to the functioning of the body machine. Death itself is heralded by a
sudden extreme rise in the rate of nitrogenous excretion, known as the 'premortal rise', and this begins when, the available carbohydrate and fat reserves
of the tissues having been exhausted, the organism is left with only its tissue
proteins as a source of energy, so that a large-scale degradation of tissue
protein begins.

On a diet that contains very little protein it is possible for the daily intake
of protein nitrogen to lie below the output of urinary nitrogen. So long as
output exceeds intake the organism, on balance, is the loser, and the deficit of
nitrogen is withdrawn from the tissues. If, however, the protein allowance
is gradually increased, a point is eventually reached at which intake just
suffices to balance output. The organism is then said to be in a state of nitrogen
balance or, *nitrogenous equilibrium*. The amount of protein required just to
attain this equilibrium condition in a given individual is therefore a measure

221

of the *minimum protein requirement* of that individual, and since proteins are among the most expensive articles of food, this is a matter of economic as well as academic interest. Many workers have accordingly investigated the minimum protein requirements of the human organism, and the results obtained have varied very widely. Rubner and his colleagues put it at about 100–120 g. protein per diem for an average man, whereas Chittenden, using himself as the experimental animal, found that he could satisfy his personal requirements with only some 30–35 g. per diem, his health improving as a result of the experiment. These large differences do not, as might at first appear, merely reflect differences in individual requirements, but differences rather in the chemical nature of the food proteins chosen. These proteins must supply enough of the essential amino-acids, and no amount of protein, however great, that fails to accomplish this can suffice to establish nitrogenous equilibrium. Animal proteins are, on the whole, much richer than plant proteins in terms of their content of essential amino-acids, and it follows that smaller amounts of protein are required when meat, fish, eggs, cheese, milk and the like are chosen (e.g. by Chittenden) than when the food selected consists largely of cereals and pulses (e.g. by Rubner). The maize protein, zein, is notoriously deficient in tryptophan and in lysine, both of which are essential, and if zein is taken as the sole protein of a diet, nitrogenous equilibrium can never be established, no matter how much of it is consumed. Gelatin is similarly deficient in tryptophan and in phenylalanine, and, like zein, is a protein of 'poor biological value'.

The primary function of protein food is to supply the amino-acids needed for the growth, repair and general maintenance of the structural and catalytic machinery of living cells. If, as is commonly the case, the proteins of the food provide more amino-acid units than are required for the discharge of these primary and very specific functions, the excess can be degraded and made to subserve the secondary and less specific function of providing fuel for the machine. As we shall see presently, proteins and amino-acids are not normally stored to any appreciable extent in the normal adult organism. Nitrogen retention on a significant scale is only observed during periods of tissue growth, during childhood and pregnancy, for example, or during periods of protein replacement, as during convalescence after a wasting disease or after protein starvation. The non-nitrogenous residues of surplus amino-acids are retained and serve to contribute to the stores of 'energy-producing' materials, i.e. carbohydrates and fats.

If a meal of protein is administered to a phlorrhizinized or diabetic animal an increased output of glucose and of acetone bodies is observed. Part of the protein must therefore be considered as convertible into carbohydrate derivatives and part into fatty metabolites. If the amino-acids are administered individually it is found that some are *glycogenic*, i.e. give rise to glycogen, which appears as glucose in the urine, while others are *ketogenic*, giving rise to ketone or acetone bodies. The known fates of the amino-acids are summarized in Table 17. It will be noticed that certain amino-acids, including some of the essential group, give rise to both glucose and to ketone bodies.

Table 17. *Fates of amino-acids administered to a diabetic or phlorrhizinized dog*

Glycogenic	Ketogenic	Both
Glycine	Leucine	*iso*-Leucine
Alanine		Lysine
Serine		Phenylalanine
Threonine		Tyrosine
Cysteine, cystine		
Methionine		
Valine		
Aspartate		
Glutamate		
Arginine		
Histidine		
Tryptophan		
Proline		
Hydroxyproline		

Note. These results have mostly been confirmed by methods of isotopic labelling.

If the reader will refer again to Tables 16 and 17 (pp. 219 and above) several interesting points will be noticed. First, *all the non-essential amino-acids are glycogenic*, which probably indicates that their conversion into carbohydrate is a reversible operation. *Of the essential amino-acids only a few are glycogenic*, and the fact that they are essential presumably indicates that the reaction chains through which they are transformed into carbohydrate include some irreversible step or steps. Thus carbohydrate can contribute to the synthesis of some amino-acids, but not of all. *All the ketogenic amino-acids are essential.* They give rise to fatty metabolites, but the fact that they are essential, i.e. cannot be synthesized by the animal organism, shows that their conversion into ketone bodies is not a reversible performance. Thus fat, unlike carbohydrate, contributes little or nothing towards the synthesis of amino-acids or of protein. These general points may be summarized as follows:

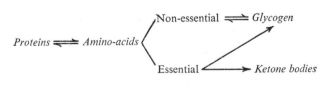

FATE OF α-AMINO-NITROGEN

Neither glucose nor the ketone bodies contain nitrogen. It follows, therefore, that, at an early stage in their metabolism, the amino-acids suffer the removal of their characteristic α-amino-groups. In a typical mammal such as a dog, this α-amino-nitrogen ultimately appears in the urine in the form of urea. In birds, snakes and lizards, by contrast, the final end-product is uric acid, while in most aquatic animals ammonia is excreted instead.

The urine of a dog starved of protein contains very little urea, but if a protein meal is taken, urea production soon begins and an equivalent amount of nitrogen is eliminated in the form of urea within 24 hr. or thereabouts. Now we are already aware that the food proteins are broken down by digestive peptidases to yield the component amino-acids, and that it is in this form that the food proteins are actually absorbed into the blood stream. We have therefore to discover how, where, and in what form the α-amino-nitrogen is detached from the amino-acid molecules, and how urea is elaborated from the primary nitrogenous product.

A partial answer to these questions is obtained by studying a hepatectomized animal or, alternatively, an animal with an Eck's fistula. An animal of this kind will survive for some days, but dies quickly if it is allowed to eat protein. At death, unusually large amounts of amino-acids are found in the blood, but neither the blood nor the urine contains any urea in the case of a dog, or uric acid in that of a bird. Instead, ammonia is present, and ammonia poisoning is one of the causes of death. These observations show (a) that the amino-groups of the amino-acids are probably split off in the form of ammonia, and (b) that the conversion of this ammonia into urea or uric acid, as the case may be, normally takes place only in the liver. The latter conclusion is confirmed by liver-perfusion experiments, by experiments on liver slices and by isotopic labelling with ^{15}N for example. Again, if small concentrations of ammonia are perfused through a surviving liver or shaken with liver slices, urea is formed in the case of dog liver, while if a bird's liver is used, the addition of ammonia leads to the production of uric acid.

We shall discuss these processes separately, dealing first with the removal of the amino-groups, a process which is known as deamination.

DEAMINATION

On paper, the deamination of amino-acids with production of ammonia might be accomplished in either of two ways, both of which have been considered. It might be a hydrolytic (equation (1)) or an oxidative process (equation (2)). Very little experimental evidence has ever been adduced in favour of hydrolysis as the mode of deamination in animal tissues, though it is known that hydrolytic deamination takes place in some bacteria. The vast

$$(1) \qquad \underset{\underset{COOH}{\diagdown}}{\overset{\overset{NH_2}{\diagup}}{R.CH}} +H_2O = \underset{\underset{COOH}{\diagdown}}{\overset{\overset{OH}{\diagup}}{R.CH}} +NH_3.$$

$$(2) \qquad \underset{\underset{COOH}{\diagdown}}{\overset{\overset{NH_2}{\diagup}}{R.CH}} +\tfrac{1}{2}O_2 = \underset{\underset{COOH}{\diagdown}}{\overset{\overset{O}{\diagdown\!\diagdown}}{R.C}} +NH_3.$$

bulk of evidence relating to animal metabolism is in favour of oxidative deamination.

The most convincing work on this problem was that carried out by Krebs, who made use of the tissue-slice technique. Slices of various rat tissues were shaken under physiological conditions of temperature, pH, etc., in the presence of various amino-acids. After an hour or two the reaction mixture was deproteinized and the corresponding α-keto-acids were sought and found by taking advantage of the fact that they form very insoluble, characteristic 2:4-dinitrophenylhydrazones. Among mammalian tissues only liver and kidney deaminate amino-acids at all rapidly, and these tissues use more oxygen when they are deaminating than when they are not. Surviving liver and kidney slices deaminate both the common, naturally occurring L-series and the much rarer 'non-natural' D-series of amino-acids, but not all amino-acids are attacked at the same rate.

To obtain strictly quantitative evidence in favour of equation (2) is more difficult. If liver tissue is used, the ammonia set free by deamination disappears because it is converted more or less completely into urea, but this difficulty can be obviated by the use of kidney slices, which do not form urea. But in liver and kidney alike, the other product of deamination, the α-keto-acid, is liable to be further metabolized, by oxidative decarboxylation in the first instance. This process, Krebs found, can be prevented by the addition of arsenious oxide. Working therefore with kidney slices and in the presence of arsenite, he was able to demonstrate that, for every molecule of ammonia produced, an extra atom of oxygen was consumed and a molecule of the corresponding α-keto-acid formed. Essentially the same results were obtained with extracts prepared from acetone powders of kidney tissue (Table 18).

Now pulp preparations of liver and kidney alike act upon both the L- and the D-series of amino-acids, again in an oxidative manner, but as soon as the pulp is appreciably diluted its ability to attack the naturally occurring L-acids disappears. The enzyme responsible for the deamination of the D-series, however, is resistant to dilution, and powerful preparations of the D-amino-acid oxidase can be made by extracting fresh, finely divided kidney tissue with water or buffer, centrifuging to remove tissue debris, and treating the clear

225

Table 18. *Oxidative deamination of amino-acids by kidney extract*
(*after* Krebs)

Amino-acid added	Mol. O_2 : NH_3 : keto-acid
D-L-Alanine	1 : 1·94 : 1·83
D-L-Valine	1 : 2·08 : 2·20
D-L-*nor*Leucine	1 : 1·85 : 1·85
D-L-Leucine	1 : 2·42 : 2·28
D-L-Phenylalanine	1 : 2·17 : 1·85

extract with 10 vol. of ice-cold acetone under ice-cold conditions. By filtering off the resulting acetone powder and drying it carefully, a stable preparation can be had which retains its activity for some weeks, and from which active enzyme solutions can be made by extraction with water or phosphate buffer. Preparations made in this way deaminate all the amino-acids of the D-series with three exceptions: glycine, D-glutamate and D-lysine. A specific oxidase was later discovered which deals with glycine, but there is reason to think that lysine is never deaminated at all (p. 229). The D-amino-acid oxidase has been extensively concentrated and finally isolated, and a more detailed description of its nature and properties will be found on pp. 135, 140.

So far, however, no evidence had been obtained about the enzyme or enzyme systems involved in the deamination of the 'natural' L-series of amino-acids. Numerous attempts were made to obtain enzyme preparations which would act upon these naturally occurring amino-acids, but for a number of years only one such enzyme was known.

This enzyme, which occurs in liver and kidney, is completely specific with respect to L-glutamate. It requires either NAD or NADP, and is, in fact, a typical, coenzyme-specific dehydrogenase, now known as L-*glutamate dehydrogenase*. Unlike that of D-amino-acid oxidase its action is freely reversible.

With the more recent discovery and eventual isolation of an L-amino-acid oxidase from rat liver and kidney, we are able to account for the deamination of the majority of L-amino-acids, for the L-oxidase resembles the D-enzyme in being a true oxidase. It is a group-specific enzyme and attacks all the mono-amino-mono-carboxylic amino-acids except glycine and those that contain a hydroxyl group, but has no action upon the diamino- or dicarboxylic acids. How important this enzyme may be seems uncertain, for even in the rat it acts relatively feebly.

Whether it is catalysed by the D- or by the L-amino-acid oxidase, or by L-glutamate dehydrogenase, *deamination is always oxidative and takes place in two possibly simultaneous stages*. In the first a pair of hydrogen atoms is transferred to an appropriate hydrogen acceptor and the corresponding α-imino-acid is probably formed as a transient intermediate:

$$(1) \quad R.\overset{\displaystyle NH_2}{\underset{\displaystyle COOH}{CH}} = R.\overset{\displaystyle NH}{\underset{\displaystyle COOH}{C}} \quad +2H \text{ (to acceptor)}.$$

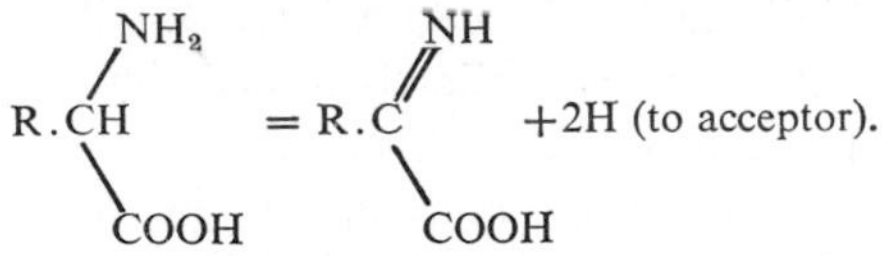

The imino-acid reacts, apparently spontaneously, with water to yield the α-keto-acid, together with ammonia.

$$(2) \quad R.\overset{\displaystyle NH}{\underset{\displaystyle COOH}{C}} + H_2O = R.\overset{\displaystyle O}{\underset{\displaystyle COOH}{C}} + NH_3.$$

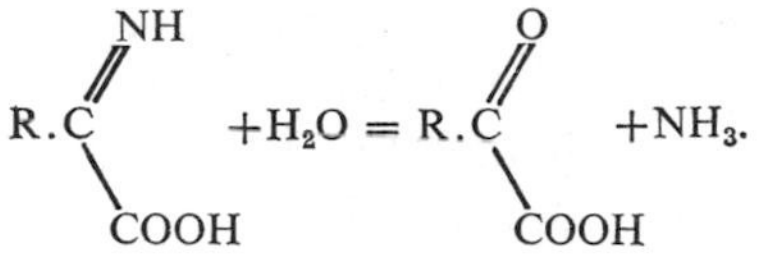

Special non-oxidative mechanisms are involved in the deamination of the hydroxy-acids, serine and threonine (p. 114), but again imino-acids are probably formed as transient intermediates.

TRANSDEAMINATION

In addition to deaminating enzymes, animal and plant tissues possess catalytic mechanisms which carry out the transference of amino-groups from amino-acids to certain α-keto-acids. If L-glutamate and pyruvate are added together to chopped liver or muscle tissue, the α-amino-group of the glutamate is in part transferred to the pyruvate, so that α-ketoglutarate and alanine are formed. The system tends towards an equilibrium which can be approached equally from either side, and the process is referred to as 'transamination'. The enzymes concerned are called transaminases or *aminotransferases*.

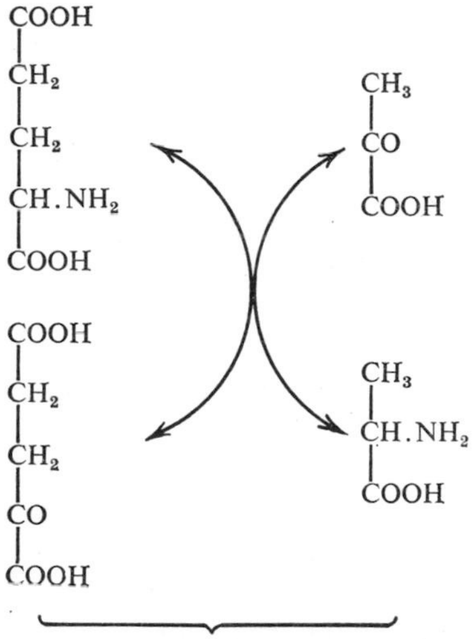

COOH
|
CH_2
|
CH_2
|
CH.NH_2
|
COOH

CH_3
|
CO
|
COOH

COOH
|
CH_2
|
CH_2
|
CO
|
COOH

CH_3
|
CH.NH_2
|
COOH

transaminase

Enzymes of this kind seem to be very widely distributed in plant and animal tissues alike, and it is worth while to notice that, unlike the deaminating enzymes, they are not confined to liver and kidney among animal tissues, but are present also in brain, muscle, and heart, for example. They are specific, moreover, towards amino-acids of the natural L-series. One group of transaminases is specific towards aspartate and the corresponding α-keto-succinate (oxaloacetate), and another towards glutamate and the corresponding α-ketoglutarate. The glutamate enzymes seem to predominate in animal tissues and also in many plants. (At this point the reader should refer back to the earlier section on transamination (pp. 99–102).)

Braunstein suggested that the deamination of L-amino-acids in general might involve the transaminases. It was already known that α-ketoglutarate is a common metabolite, arising as it does from carbohydrate as well as from protein sources. Under the influence of the glutamate transaminases the α-amino-groups of any incoming amino-acid could, it was suggested, be transferred to α-ketoglutarate to yield glutamate which then, under the influence of the L-glutamate dehydrogenase of liver or kidney, would undergo deamination, α-ketoglutarate being regenerated and ammonia set free:

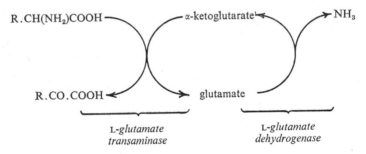

$$\text{R.CH(NH}_2)\text{COOH} \qquad \text{α-ketoglutarate} \qquad \text{NH}_3$$

$$\text{R.CO.COOH} \qquad \text{glutamate}$$

L-*glutamate transaminase* L-*glutamate dehydrogenase*

Catalytic concentrations of α-ketoglutarate would be all that is necessary, if Braunstein's suggestion is correct, to catalyse the deamination of large amounts of L-amino-acids in tissues containing both the glutamate transaminases and the corresponding dehydrogenase. This same scheme could also explain why the dilution of a liver or kidney pulp leads, as Krebs had already discovered, to the disappearance of its ability to deaminate the L-series of amino-acids: dilution would mean dilution of the α-ketoglutarate and glutamate which play the part of essential carriers or coenzymes in the Braunstein scheme.

It has now been shown that this system can participate in the deamination of amino-acids in general and that pyridoxal phosphate plays the part of co-enzyme, or 'co-transaminase' (cf. p. 101). Taken together with L-amino-acid oxidase this system enables us to account for the deamination of practically all the naturally occurring amino-acids, and it seems quite certain that the transamination machinery is the more important of the two.

Special emphasis must be placed upon the reversibility of transamination, for this process appears to be involved both in the breakdown of amino-acids and in their synthesis from non-protein sources. Since transamination is a reversible process, it follows that new amino-acid molecules can be synthesized at the expense of glutamate, provided that the appropriate α-keto-compounds are available. At least three such α-keto-derivatives arise in the course of carbohydrate metabolism, viz. α-ketoglutarate, oxaloacetate and pyruvate. Glutamate can be synthesized from α-ketoglutarate through the reversed action of L-glutamate dehydrogenase, and can then act as an amino-group donor for the synthesis by transamination of aspartate and alanine from oxaloacetate and pyruvate respectively. These mechanisms thus enable us to account for the production from non-protein sources of at any rate three of the relatively few amino-acids that animal tissues are capable of synthesizing for themselves.

There can be little doubt that, on account of its reversibility, transamination plays a part of fundamental importance in the anabolism as well as in the katabolism of nitrogenous compounds. Schoenheimer and his co-workers kept a group of young rats on a low protein diet and added heavy ammonium citrate to their food. The idea was to see whether the animals could utilize the heavy nitrogen, ^{15}N, for the synthesis of nitrogenous compounds, and a low protein diet was used in order to increase the chances of its utilization. Later the animals were killed and creatine, glutamate, aspartate, histidine, arginine and lysine were isolated from the carcasses and analysed for heavy nitrogen. With a single exception in the case of lysine, all these compounds were found to contain significant quantities of ^{15}N, showing that the heavy ammonia had, in fact, been used. The fact that ^{15}N had found its way into all these substances serves to show (a) that ammonia can be used for the synthesis of amino-acids, and (b) that the α-amino-group of a given amino-acid is not simply a static feature of the amino-acid molecule, but that it must be undergoing constant, dynamic interchange with other amino-groups. That lysine fails to acquire ^{15}N suggests that this particular amino-acid never undergoes biological deamination at all. Again, the ^{15}N of arginine was found to be located almost entirely in the amidine part of the molecule, suggesting that arginine, and probably ornithine too, therefore, do not undergo deamination. The close structural relationship between ornithine and lysine seems to be in accordance with this view. Schoenheimer's work provides important circumstantial evidence for the central importance of transamination in amino-group metabolism, and serves also to underline the importance of its reversibility.

A reversible system in metabolism is not merely to be regarded as something that goes one way at one minute and the other way the next, but as a process that goes on simultaneously in both directions at one and the same time, so that the process as a whole is one that is essentially dynamic. But, as we have

16-2

already had several occasions to mention, degradation and synthesis, balanced though they may be, often proceed by different routes (cf. interconversion of glucose and glycogen, pp. 93–7).

STORAGE OF AMINO-GROUPS

It will be realized that if amino-acid synthesis is to proceed in living tissues, amino-groups must be forthcoming. These might be taken from incoming amino-acids by transamination or, alternatively, ammonia itself might be utilized, if available. But ammonia is a very toxic metabolite and, since non-essential amino-acids can be produced by animals, even under conditions of protein starvation, it follows that ammonia or amino-groups must in some way be stored in the tissues in an innocuous form. Further evidence that such a storage is possible is found in the fact that any tendency towards acidaemia is counteracted in the mammal by the excretion of a more than usually acid urine, the excess acid present being neutralized more or less extensively by ammonia produced in the kidney. This process of ammonia production can still go on in animals temporarily deprived of protein foodstuffs.

No clue to the manner of this storage was found until, in the course of his experiments on deamination, Krebs observed that, of the ammonia produced by the deamination of added amino-acids, a part sometimes failed to put in an appearance. The disappearance of ammonia in liver tissue could be attributed to its conversion into urea, and it might have been thought that some of this urea is later broken up by a tissue urease to furnish ammonia when the latter is required for amino-acid synthesis or for ammonia production by the kidney. In fact, however, there is no evidence for the occurrence of urease in mammalian tissues, although this enzyme is known to be present in the tissues of numerous invertebrates. If 'heavy' urea is administered to a mammal, preferably by injection so as to avoid possible bacterial intervention, the heavy nitrogen is quantitatively eliminated in the urine, still in the form of urea.

Even in the case of kidney tissue, which forms no urea, Krebs found that substantial amounts of ammonia could disappear on occasion and accordingly set himself the task of discovering its fate and showed that, in fact, ammonia can react with glutamate to form glutamine. This is now known to be an endergonic reaction involving the participation of ATP (see p. 98). It is known too that many tissues, including even blood, contain considerable quantities of glutamine, and it would seem that glutamine can be synthesized by many tissues at times when ammonia is available, and broken down again by glutaminase, a widely distributed hydrolytic enzyme, to furnish ammonia or —NH_2 groups when required.

Glutamine has long been known as an important constituent of plant materials, as also has the corresponding amide of aspartic acid, asparagine. The function of the latter as a storage depot for ammonia had been demonstrated many years earlier in studies of etiolated seedlings of the tree-lupin, *Lupinus luteus*. If the protein-rich seed of this plant is allowed to germinate in the dark, a seedling develops which is devoid of chlorophyll. Its development fairly soon comes to an end, and a study of the nitrogen distribution in the plantlet shows that some 80 % of the protein-nitrogen of the original seed has been converted into asparagine. The remaining 20 % has been transferred to the new tissue proteins of the seedling. If the seedling is now allowed access to light, asparagine begins to disappear and new protein is synthesized at the expense of the asparagine-nitrogen.

These changes can be accounted for if we suppose, as is undoubtedly true, that the processes of photosynthesis can give rise in one way or another to α-keto-acids corresponding to all the amino-acids required for protein synthesis. It is necessary also to suppose that transamination in the plant allows the transfer of amino-groups from aspartate to any normal α-keto-acid.

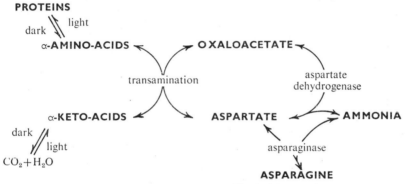

Fig. 48. Nitrogen metabolism in seedlings of *Lupinus luteus*

In the etiolated seedling, oxaloacetate arises in the course of carbohydrate breakdown and can function as an amino-group receptor with respect to amino-acids formed by hydrolysis of the seed proteins. This allows the deaminated residues of the amino-acids to be utilized as fuel at a time when the organism is as yet unable to draw upon solar energy through chlorophyll and photosynthesis. The aspartate formed by transamination can react with more ammonia, produced by deamination or transdeamination of amino-acids, to yield asparagine, and this process can continue as long as the protein reserves hold out, subject only to the limitation imposed by the fact that the tissues, even of an etiolated seedling, contain some protein. The whole cycle of events can be summarized as in Fig. 48.

When presently the plant gains access to light, chlorophyll makes its appearance and photosynthesis begins. This leads to the production of α-keto-acids, among other substances and, as the concentration of these begins to increase, they begin to undergo transamination at the expense of aspartate so that new

231

amino-acids are formed and participate then in protein synthesis. The utilization of aspartate and the consequent fall in the concentration of the latter leads, in its turn, to the breakdown of asparagine. This gives rise to more aspartate, which can be used for transamination, and to ammonia, which can be employed in direct amination of more of the α-keto-acids, yielding more amino-acids for protein synthesis. Alternatively, the free ammonia can be used to synthesize more aspartate from oxaloacetate, the latter arising as an intermediate in the oxidative breakdown of carbohydrate.

Asparagine and aspartate seem, therefore, to play in this plant a part parallel to that of glutamine and glutamate in many other plants and in animal tissues. Aspartate and its amide have long interested plant biochemists, and with good reason; it was brought again into prominence by claims that the primary amino-acid synthesized by the symbiotic root-nodule bacteria of leguminous plants is aspartate itself.

Reference has frequently been made in this chapter and elsewhere to the synthesis of proteins but there are good reasons why a consideration of this challenging problem should be held over until a later stage in our deliberations.

FATE OF DEAMINATED RESIDUES

The α-keto-acids left after removal of the amino-group from the parent amino-acids commonly undergo conversion into intermediates on the metabolic pathways either of fats or of carbohydrates. As a general rule the first step in this direction consists in *oxidative decarboxylation* which, as we shall see, is a complex operation (p. 381 *et seq.*) but may be summarized for present purposes in the following manner:

$$R.CO.COOH + \tfrac{1}{2}O_2 \longrightarrow R.COOH + CO_2.$$

This process, incidentally, is strongly inhibited by arsenite.

Certain amino-acids give rise to glucose if administered to diabetic or phlorrhizinized animals, or to glycogen in the livers of starving animals; others yield acetoacetate and the other ketone bodies, while a few give rise to both. These conclusions can be conclusively verified by administering isotopically labelled amino-acids to normal animals and seeking out the isotope in the fat or glycogen stores of the body.

If alanine is administered to a diabetic animal, all three of its carbon atoms are transformed into glucose. To state categorically that the extra glucose excreted has been formed from the alanine is not strictly accurate, for it may only be that the deaminated residues of the amino-acids were metabolized in place of glucose so that an equivalent quantity of glucose or glycogen was spared the fate of oxidation. Nevertheless, if the deaminated material replaces glucose or glycogen it presumably does so because it joins the metabolic pathways of glucose and glycogen, so that, for all practical purposes, it suffices to say that the amino-acid is convertible into glucose or glycogen.

Glucose and glycogen are formed from amino-acids by way of reaction chains which are often of considerable length. Many simple organic substances are known to be convertible into glucose or glycogen, and in some cases, e.g. lactate, pyruvate and glycerol, we know precisely what the intermediate stages are.

In the next chapter we shall trace, as far as we can, the routes of conversion of amino-acids into substances which, like propionate, lactate and pyruvate, are known carbohydrate-formers or into compounds which, like acetate and acetoacetate, are precursors of fatty acids. The rest of the stages are common to the metabolism of carbohydrates or fatty acids respectively, and will be considered later.

The first step in the transformation of the amino-acids consists in the deamination or transdeamination of the amino-acids, the mechanisms of which we have already discussed. For convenience of reference the known modes of deamination of the naturally occurring L-acids are summarized in Table 19. It may be noticed that there is serious doubt whether lysine, ornithine and arginine undergo biological deamination (p. 229). It is known that serine and

Table 19. *Deamination of amino-acids of the naturally occurring L-series in animal tissues*

Amino-acid	L-Amino-acid oxidase	Specific enzyme
Glycine	−	Glycine oxidase
Alanine	+	.
Serine	−	Serine dehydratase
Threonine	−	Threonine dehydratase
Cysteine	+	.
Methionine	+	.
Valine	+	.
Leucine	+	.
isoLeucine	+	.
Aspartate	−	L-Aspartate dehydrogenase (plants)
Glutamate	−	L-Glutamate dehydrogenase
Arginine	−	.
Ornithine	−	.
Lysine	−	.
Phenylalanine	+	.
Tyrosine	+	.
Tryptophan	+	.
Histidine	+	.

Most and probably all can be *transdeaminated* except arginine, ornithine and lysine.

Note. The use of a negative sign indicates only that the amino-acid is not at present known to be deaminated by L-amino-acid oxidase.

threonine too can be atypically deaminated by non-oxidative enzymes present in rat liver (p. 114), known as *serine dehydratase* and *threonine dehydratase*, respectively. Both require pyridoxal phosphate as co-factor.

BACTERIAL ATTACK AND DETOXICATION

Bacteria of many kinds possess powerful and specific amino-acid decarboxy-lyases, and although these do not fall within the scope of this book they must be mentioned here because they have a considerable impact on animal meta-bolism. The alimentary canal of animals is always densely populated with a variety of micro-organisms at some region or other, and the activities of the members of this population lead to the production of materials which may then be absorbed by the animal itself, the extent of this absorption depending upon the region in which microbial activity takes place. In herbivorous animals, of course, micro-organisms play a very large and important part in the digestion of the food, but even in animals that do not delegate their digestive operations to symbiotic bacteria, the metabolism of micro-organisms has still to be taken into account.

The bacterial inhabitants of the intestine have access to the food, or to the products of digestion of the food, among which amino-acids are included; and certain bacteria can decarboxylate one or more of these to produce the corresponding amines:

$$R.CH(NH_2)\overset{...}{COO}H = R.CH_2NH_2 + CO_2.$$

All of these bacterial decarboxylyases probably use pyridoxal phosphate as a coenzyme or prosthetic group. Some of the products are intensely poisonous and, when they are absorbed into the animal's blood stream, undergo what is usually known as *detoxication*. This is accomplished by oxidation, reduction, acetylation, methylation, or in some other manner. Furthermore, by pro-gressive bacterial degradation of the side-chains of acids containing aromatic rings, poisonous phenolic substances are formed, and these, like the amines, undergo detoxication in the animal body. We shall deal individually with individual cases in the next chapter, and we shall also comment upon the parts played by the amino-acids themselves, since several of them act as important detoxicating agents.

In addition to amino-acids, many other substances play a part in detoxica-tion, and these may be briefly mentioned at this point. Acetylation (p. 104) is a common fate among aromatic substances containing $-NH_2$ groups, and the administration of aniline, substituted anilines, sulphonamide drugs and the like, is followed by their excretion, partly or wholly in the acetylated form. Amines are commonly oxidized, yielding harmless aldehydes or acids, by the amine and diamine oxidases of the tissues (p. 137). Phenolic substances

234

are frequently excreted in conjugation with glucuronic acid or with sulphuric acid, the latter being probably derived from the sulphur-containing amino-acids, cysteine and cystine.

This book does not contain a special chapter on detoxication or, as it is often termed, protective synthesis; it has seemed more suitable to the author to dispense with such a chapter since a large part of the subject can be dealt with under the special functions of individual amino-acids, while the foregoing remarks, together with a judicious use of the index, will fill in most of the gaps that will remain when the present chapter has been studied. It is worth while in passing to note that the terms 'detoxication' and 'protective synthesis' are misnomers. Foreign, extraneous substances are sometimes caught up in the metabolic machinery and treated as if they were normal metabolites. Often their toxicity is reduced thereby, but cases are known in which precisely the reverse happens. Sodium fluoroacetate is probably not very toxic *per se*, but becomes built up into fluorocitrate, a powerful inhibitor of the oxidation of carbohydrate, fat and protein alike.

12

SPECIAL METABOLISM OF THE AMINO-ACIDS

INTRODUCTION

In addition to the 20 amino-acids that enter into the composition of proteins in general, the tissues of green plants contain a very considerable number of special amino-acids of unknown function which do not become incorporated into proteins. They will not be considered here. Further, although a great deal is now known about the biosynthesis, even of the essential amino-acids in micro-organisms of various kinds, it has seemed best to deal here mainly with amino-acid metabolism as it relates to animals. However, to give some satisfaction to the curious, a table showing some of the known sources from which newly synthesized amino-acids arise is appended to this chapter.

In such a context as this it would not be possible, even if it seemed desirable, to present more than a sample of all the information that is available today in a single chapter but for a very much more detailed exposition the reader may be referred to the excellent 2-volume work of Meister.†

OPTICAL ISOMERISM AND NOMENCLATURE

With the single exception of glycine, the simplest member of the group, all the amino-acids which occur in proteins are optically active. As a rule only one of the possible optical antipodes occurs naturally on the large scale and this is usually referred to as the 'natural' form, even though the other antipode may also occur on occasion. Similarly among the sugars, one isomer always preponderates in natural materials, the other being encountered rarely if at all.

The simple monosaccharides can all be regarded as derivatives of glyceraldehyde, which is itself optically active and is, in fact, a triose sugar *conventionally* formulated thus:

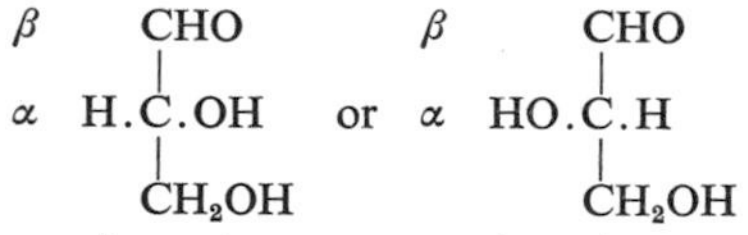

$$\begin{array}{cccccc} \beta & CHO & & \beta & CHO \\ \alpha & H.C.OH & or & \alpha & HO.C.H \\ & CH_2OH & & & CH_2OH \\ & \textit{dextro-isomer} & & & \textit{laevo-isomer} \end{array}$$

Tetroses, pentoses, hexoses and higher monosaccharides can be regarded as being derived from glyceraldehyde by the interposition of additional groups, usually $>CHOH$, between the α-carbon and the reducing group.

† A. Meister, *Biochemistry of the Amino-acids* (2nd ed.) (Academic Press: New York & London, 1965).

236

One form of glyceraldehyde is dextro- and the other laevo-rotatory towards polarized light and in accordance with classical practice, these were formerly described as *d*- and *l*-glyceraldehyde respectively. Now the stereochemical arrangement about the α-carbon atom of *dextro*-glyceraldehyde is present also in the preponderant, i.e. the 'natural', forms of all naturally occurring sugars and because of this structural resemblance, which extends to ketoses as well as to aldoses, it became usual in biochemical circles, where structure is very much more important than optical rotation, to refer to them as *d*-sugars. This practice was however not universally adopted in more purely chemical circles and thus led to considerable confusion. Fructose, for example, is a *d*-sugar in biochemical parlance but, being laevo-rotatory, remained *l*-fructose to the chemists, and is still sometimes even referred to as 'laevulose'.

In an endeavour to overcome this difficulty the prefixes -(+)- and -(−)- were introduced to signify dextro- and laevo-rotation respectively, *d*- and *l*- being retained in their purely structural signification. Thus the 'natural' form of glucose became *d*-(+)-glucose and 'natural' fructose became *d*-(−)-fructose. But this system, too, failed to find universal respect. Accordingly *d*- and *l*- have now been abandoned, and replaced by D- and L- by general agreement (at any rate between Great Britain and the United States of America) and we shall use this nomenclature here. *These small capital letters have purely structural significance and give no indication whatsoever of optical rotation.* If it is necessary to specify rotation as well the supplementary symbols, -(+)- or -(−)-, or the supplementary prefixes, *dextro*- or *laevo*-, are employed in addition to D- and L-.

Returning now to the sugars, we recall their structural relationships to what is now officially recognized as D-glyceraldehyde and refer to the 'natural' forms as D-sugars. L-Sugars do occur in natural products but are relatively rare. By the same token the 'natural' amino-acids are all members of the L-series; D-amino-acids, though they do occasionally occur, as constituents of the cell walls of some bacteria for example, are uncommon. This time, however, the standard of reference is not glyceraldehyde, a triose sugar, but the amino-acid, serine. Actually, however, D-serine is structurally related to D-glyceraldehyde and L-serine to L-glyceraldehyde:

$$
\begin{array}{cc}
& \text{COOH} \qquad \beta \qquad \text{CHO} \\
& \quad | \qquad\qquad\qquad\quad | \\
\alpha \quad \text{H}_2\text{N.C.H} \qquad \alpha \quad \text{HO.C.H} \\
& \quad | \qquad\qquad\qquad\quad | \\
\beta \qquad \text{CH}_2\text{OH} \qquad\qquad \text{CH}_2\text{OH} \\
\end{array}
$$

L-*serine*   L-*glyceraldehyde*

Now it is conventional to use D- and L- with reference to the spatial relationships prevailing about the *α-carbon atom* of any optically active substance, but unfortunately the α-position is differently defined in different groups of compounds. In amino-acids the position of reference is the carboxyl group,

237

so that *the immediately adjacent carbon atom is the α-carbon*. In carbohydrates, however, reference is made to the carbon atom most remote from the reducing group, so that *the α-carbon is the most remote but one from the reducing group*.

This difference of convention has led to difficulties in some cases, notably with the amino-acid, threonine, so named because it is structurally related to the tetrose sugar, D-threose. Regarded as a sugar derivative, the 'natural' form of threonine can be formulated as follows:

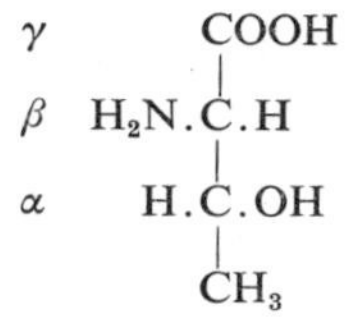

$$
\begin{array}{ll}
\gamma & \text{COOH} \\
\beta & \text{H}_2\text{N.C.H} \\
\alpha & \text{H.C.OH} \\
& \text{CH}_3 \\
\end{array}
\qquad
\begin{array}{ll}
\gamma & \text{CHO} \\
\beta & \text{HO.C.H} \\
\alpha & \text{H.C.OH} \\
& \text{CH}_2\text{OH} \\
\end{array}
$$

threonine 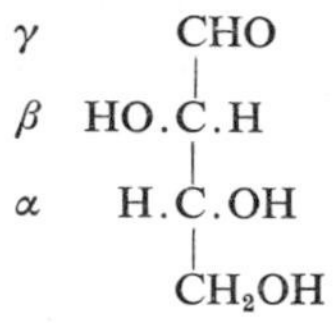   D-*threose*

If we consider threonine as a sugar derivative it must be called D-threonine because the configuration at the α-carbon atom is the same as that in D-threose. But regarded as an amino-acid, threonine must be described thus:

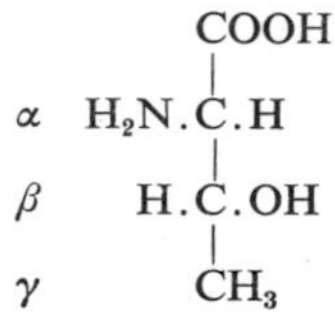

$$
\begin{array}{ll}
& \text{COOH} \\
\alpha & \text{H}_2\text{N.C.H} \\
\beta & \text{H.C.OH} \\
\gamma & \text{CH}_3 \\
\end{array}
\qquad
\begin{array}{ll}
& \text{COOH} \\
\alpha & \text{H}_2\text{N.C.H} \\
\beta & \text{CH}_2\text{OH} \\
\end{array}
$$

threonine 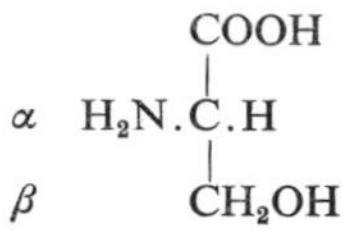   L-*serine*

and from this point of view it must be called L-threonine because the configuration at the α-carbon atom is the same as in L-serine. The difference here is simply due to the use of two different conventions.

Cases like this do not often arise but, when they do, order can be produced out of the apparent chaos by using subscripts to the usual D- and L-. Where a structure is referred back to that of glyceraldehyde, to which all carbohydrates are referred in the end, $_g$ is used; where the structure is referred back to that of serine we use $_s$. Thus the 'natural' form of threonine can be written as D_g-threonine or as L_s-threonine, and the two are synonymous.

SPECIFIC METABOLISM

GLYCINE, $\text{H}_2\text{N.CH}_2\text{COOH}$

(non-essential: glycogenic), is the simplest of the naturally occurring amino-acids and is the only member of the group that does not contain an asymmetric carbon atom. It gives rise on deamination by a *glycine oxidase* to

238

GLYCINE

glyoxylate, which in turn can be oxidized to *formate* but there is reason to think that glycine oxidase is identical with D-amino-acid oxidase:

$$\begin{array}{ccc} \underset{\mathrm{COOH}}{\overset{\mathrm{CH_2NH_2}}{|}} & \longrightarrow & \underset{\mathrm{COOH}}{\overset{\mathrm{CHO}}{|}} \end{array} \quad \xrightarrow[-2\mathrm{H}]{+\mathrm{H_2O}} \quad \mathrm{H.COOH + CO_2}$$

Formate itself reacts with tetrahydrofolate (THF) to give *formyltetra-hydrofolate* by the transformation of which the formyl group (Fig. 28, p. 108) can be reduced to a hydroxymethyl or methyl residue, any or all of which may then be transferred to appropriate acceptor molecules. Formyl groups are used in the synthesis of purines; hydroxymethyl groups in the interconversion of glycine and serine (below); methyl groups serve to convert *S*-adenosylhomocysteine into *S*-adenosyl methionine, which in its turn acts as a methyl donor in many reactions. These methyl groups give rise to those of choline and creatine for example and those of many other methylated compounds (pp. 110–12).

We do not know how glycine undergoes conversion to carbohydrate in the animal body, but it is interesting to notice that there is a considerable delay between the administration of glycine to a diabetic or phlorrhizinized animal and the ensuing excretion of glucose in the urine. We do know however that glycine can be transformed into *serine,* which is also glycogenic. This transformation takes place through the agency of hydroxymethyl-tetrahydrofolate thus:

$$\begin{array}{ccc} \mathrm{THF{-}CH_2OH} & & \mathrm{THF.H} \\ + & \longrightarrow & + \\ \underset{\mathrm{COOH}}{\overset{\mathrm{CH_2NH_2}}{|}} & & \underset{\underset{\mathrm{COOH}}{|}}{\overset{\mathrm{CH_2OH}}{\underset{}{|}}}\mathrm{CH.NH_2} \end{array}$$

The reaction is catalysed by a specific *serine-hydroxymethyl*-THF-*transferase.* Glycine can also be formed from *threonine* (p. 243).

Glycine gives rise to two methylated derivatives, *sarcosine,* a monomethyl glycine which can be formed by the biological methylation of glycine itself, and *glycine betaine,* in which three methyl groups are present. The distribution and functions of these substances are described elsewhere (p. 112). Glycine also enters into the formation of *creatine* (p. 201). And in all these cases the methyl groups arise, not directly from methyl-THF, but through *S*-adenosyl methionine (p. 111).

$$(\mathrm{CH_3})\mathrm{H_2N^+.CH_2COO^-} \qquad (\mathrm{CH_3})_3\mathrm{N^+.CH_2COO^-} \qquad \mathrm{H_3N^+.CH_2COO^-}$$
sarcosine   glycine betaine   glycine (zwitterion)

Experiments with isotopic glycine have shown that this amino-acid enters into the formation of uric acid and other *purines* (p. 282) and into that of

239

porphyrins too. Glycine is also present in certain *bile salts* formed by the union of glycine with cholic and other bile acids (p. 216).

It occurs again in combination in the peculiar tripeptide *glutathione*, which has the constitution of γ-glutamyl-cysteinyl-glycine: the peculiarity here lies in the fact that it is the γ- and not the α-carboxyl group of the glutamate unit that is engaged in the peptide linkage:

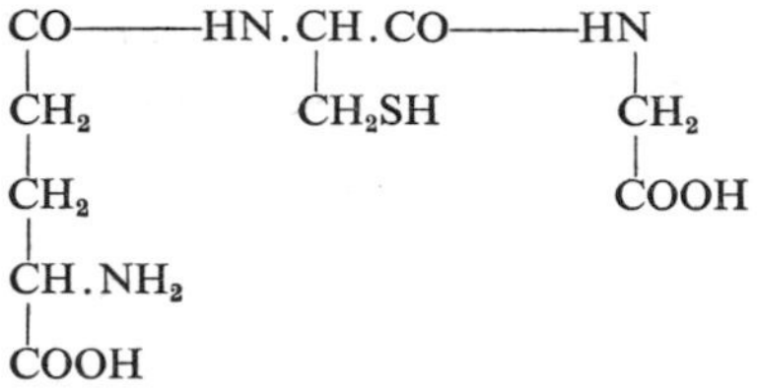

$$CO\text{——}HN.CH.CO\text{——}HN$$

$$\begin{array}{ccc} CH_2 & CH_2SH & CH_2 \\ CH_2 & & COOH \\ CH.NH_2 & & \\ COOH & & \end{array}$$

γ-glutamyl———cysteinyl———glycine

Finally, glycine plays a part in the *detoxication* of aromatic acids such, for instance, as benzoic and phenylacetic acids, and also in that of nicotinic acid. These acids are excreted in the urine in the form of conjugates with glycine:

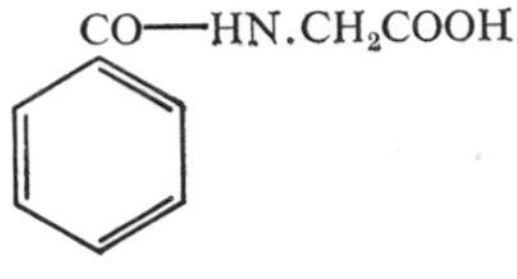

$$CO\text{—}HN.CH_2COOH$$

$$CO\text{—}HN.CH_2COOH$$

hippuric acid 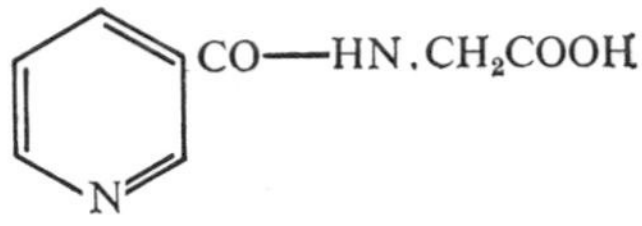   *nicotinuric acid*

ALANINE, $CH_3CH(NH_2)COOH$

(non-essential: glycogenic), is the simplest of the optically active amino-acids. Two forms exist, one of which has the same spatial configuration as D-serine and is called D-alanine, while the other, with the spatial configuration of L-serine, is known as L-alanine; the latter is the predominant or 'natural' form in biological structures and systems.

It is an interesting fact that, while all the naturally occurring amino-acids can exist in both the D- and the L-forms, the D-isomers occur only rarely in nature; so rarely, in fact, that the prefix L-, which refers to the common forms, has been omitted throughout this book except where it is necessary for the sake of clarity. Chemical synthesis of the amino-acids ordinarily yields racemic products, i.e. mixtures of the D- and L-forms. Biological synthesis, on the other hand, yields only the L-form in the ordinary way because of the stereochemical specificity of the enzymes that take part.

Another important general feature of the amino-acids found in proteins and simpler peptides is that they are all members of the α-series, i.e. the amino- and carboxylic groupings involved in peptide formation are both attached to

240

α-carbon atoms. To this rule there are but few exceptions. As we have just seen, there is an exception in the case of glutathione, a natural tri-peptide in which a γ-carboxyl residue is involved in peptide formation. An exception of another kind is found in the occurrence of β-*alanine*,

$$\overset{\beta}{H_2N.CH_2}\overset{\alpha}{CH_2}COOH.$$

β-Alanine occurs in a member of the B_2 group of vitamins, viz. *pantothenic acid*, a constituent of coenzyme A (p. 297):

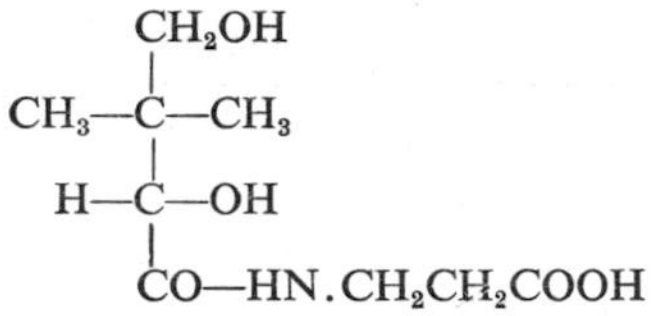

$$
\begin{array}{c}
CH_2OH \\
| \\
CH_3-C-CH_3 \\
| \\
H-C-OH \\
| \\
CO-HN.CH_2CH_2COOH
\end{array}
$$

pantothenic acid

It is found again in a pair of related histidine derivatives,

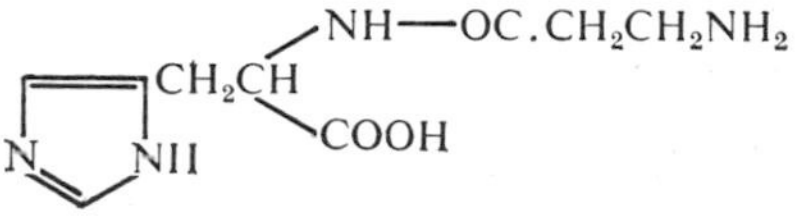

carnosine   anserine

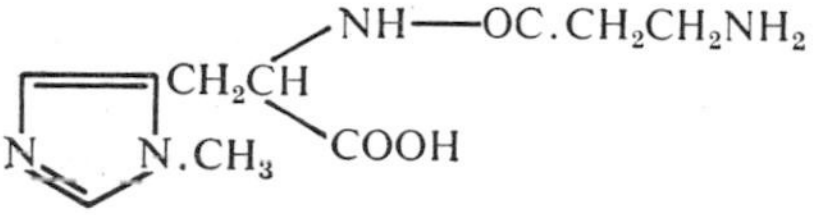

Carnosine (β-alanylhistidine), is something of a biological curiosity, being derived as it is from the rare β-amino-acid, β-alanine. It is, therefore, not a typical dipeptide, since all such are derived wholly from α-amino-acids, but it has been obtained from the muscles of representatives of all classes of vertebrate animals.

Anserine a methylated carnosine, is also a derivative of β-alanine. It was first isolated from the muscle of the goose, *Anser*, from which it received its name. Carnosine and anserine often occur side by side in one and the same tissue. The function of these peculiar bases is quite unknown. How they are formed in animal tissues is still unknown, though anserine can arise by the transmethylation of carnosine. Carnosine itself can replace the essential amino-acid histidine in the diet of rats, from which we may conclude that carnosine can be hydrolysed in the tissues. A specific carnosinase is in fact known to occur in the rat while a corresponding anserinase has been found in fish muscle.

α-*Alanine*, the normal isomer, yields pyruvate on deamination or trans-deamination. This product is known to lie on the direct line of breakdown and synthesis of glucose and glycogen.

241

$$\textit{SERINE,} \quad \begin{array}{l} CH_2OH \\ | \\ CH.NH_2 \\ | \\ COOH \end{array}$$

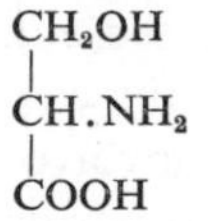

(non-essential: glycogenic), is one of the less common amino-acids, and not much is known about its metabolism. It is deaminated anaerobically by an enzyme, *serine dehydratase* (p. 114), which requires pyridoxal phosphate as coenzyme and is present in cell-free extracts of liver, to yield pyruvate, a known glucose-former:

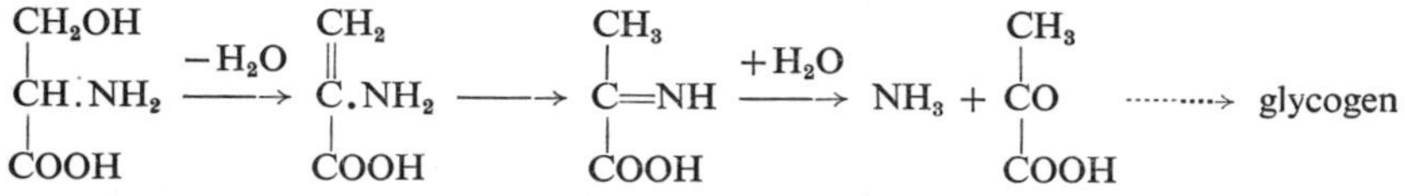

$$\begin{array}{ccccccc}
CH_2OH & & CH_2 & & CH_3 & & CH_3 \\
| & \xrightarrow{-H_2O} & || & \longrightarrow & | & \xrightarrow{+H_2O} & | \\
CH.NH_2 & & C.NH_2 & & C=NH & & NH_3 + CO \quad \cdots\cdots\rightarrow \text{ glycogen} \\
| & & | & & | & & | \\
COOH & & COOH & & COOH & & COOH
\end{array}$$

Probably the first of these reactions is catalysed by the enzyme, the remainder as far as pyruvate being perhaps spontaneous.

Serine itself can give rise to and be formed from *glycine* (p. 239) and can undergo decarboxylation to form *ethanolamine*:

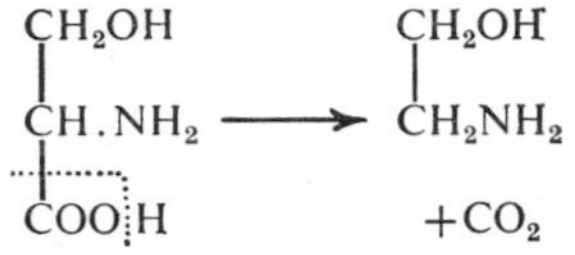

$$\begin{array}{ccc}
CH_2OH & & CH_2OH \\
| & & | \\
CH.NH_2 & \longrightarrow & CH_2NH_2 \\
| & & \\
COOH & & +CO_2
\end{array}$$

In addition, it has been shown that serine can exchange its hydoxyl group for the sulphydryl of homocysteine to yield *cysteine* (p. 244).

Serine has been isolated from acid hydrolysates of phosphoproteins such as casein in the form of the difficultly hydrolysable *serine phosphate*, suggesting that the organically bound phosphorus, which is a characteristic constituent of casein and other phosphoproteins, may find attachment to the protein molecule through the hydroxylic side-chains of serine residues. Indeed, serine or serine phosphate appear to be implicated in the active centres of many enzymes.

$$\textit{THREONINE,} \quad \begin{array}{l} CH_3 \\ | \\ CHOH \\ | \\ CH.NH_2 \\ | \\ COOH \end{array}$$

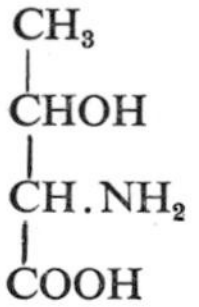

(essential: glycogenic), like serine, is a hydroxylic amino-acid, but its functional importance is still uncertain. Threonine is deaminated by a specific *threonine dehydratase* which, like serine dehydratase, requires pyridoxal phosphate as coenzyme, and gives rise to propionate and thence to glycogen. Conversion of the propionate into carbohydrate is an elaborate process which involves ATP, coenzyme A, biotin and vitamin B_{12} (cf. p. 119) among other things.

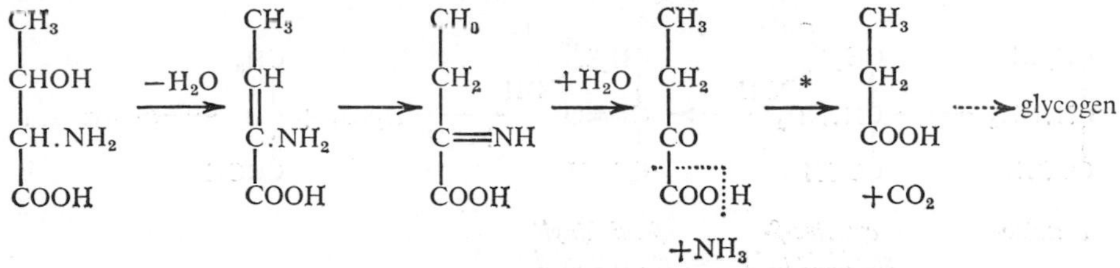

$$\begin{array}{c} CH_3 \\ | \\ CHOH \\ | \\ CH.NH_2 \\ | \\ COOH \end{array} \xrightarrow{-H_2O} \begin{array}{c} CH_3 \\ | \\ CH \\ \| \\ C.NH_2 \\ | \\ COOH \end{array} \longrightarrow \begin{array}{c} CH_0 \\ | \\ CH_2 \\ | \\ C{=}NH \\ | \\ COOH \end{array} \xrightarrow{+H_2O} \begin{array}{c} CH_3 \\ | \\ CH_2 \\ | \\ CO \\ | \\ COOH \\ +NH_3 \end{array} \xrightarrow{\ *\ } \begin{array}{c} CH_3 \\ | \\ CH_2 \\ | \\ COOH \\ +CO_2 \end{array} \cdots\cdots> glycogen$$

Threonine is one of the few essential amino-acids that are glycogenic, and it seems probable that, if the reactions suggested here correspond to reality, the reason why threonine cannot be synthesized in the body is that oxidative decarboxylation (*), a general reaction to which all α-keto-acids are liable, is irreversible under the conditions obtaining in animal tissues.

Finally, threonine can be split enzymatically to yield *glycine* and acetaldehyde:

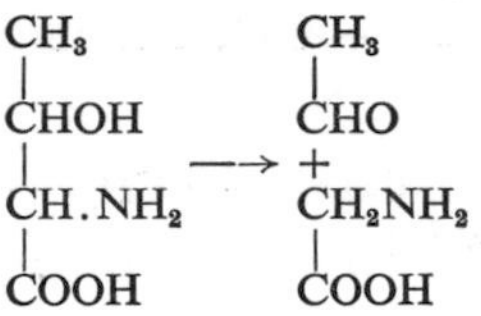

$$\begin{array}{c} CH_3 \\ | \\ CHOH \\ | \\ CH.NH_2 \\ | \\ COOH \end{array} \longrightarrow \begin{array}{c} CH_3 \\ | \\ CHO \\ + \\ CH_2NH_2 \\ | \\ COOH \end{array}$$

The uncommon amino-acid, *α-aminobutyric*, can be formed by transamination of the α-ketobutyrate formed by deamination of threonine.

$$\textit{CYSTEINE,}\ \begin{array}{c} CH_2SH \\ | \\ CH.NH_2 \\ | \\ COOH \end{array} \qquad \text{and}\qquad \textit{CYSTINE,}\ \begin{array}{cc} CH_2S{-}{-}S.CH_2 \\ | \qquad\qquad | \\ CH.NH_2 \quad CH.NH_2 \\ | \qquad\qquad | \\ COOH \quad\ \ COOH \end{array}$$

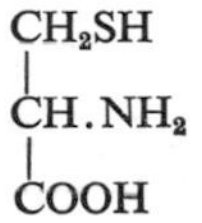

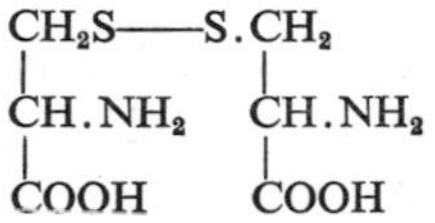

(essential but replaceable by methionine: glycogenic), are the commonest sulphur-containing amino-acids. Cystine can be reduced to form free cysteine and the latter, under the influence of *cysteine desulphydrase*, a lyase, loses its sulphydryl group and gives rise to pyruvate and thence to glycogen:

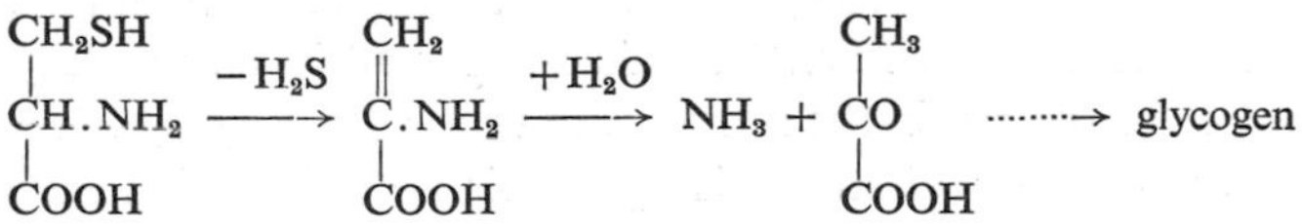

$$\begin{array}{c} CH_2SH \\ | \\ CH.NH_2 \\ | \\ COOH \end{array} \xrightarrow{-H_2S} \begin{array}{c} CH_2 \\ \| \\ C.NH_2 \\ | \\ COOH \end{array} \xrightarrow{+H_2O} NH_3 + \begin{array}{c} CH_3 \\ | \\ CO \\ | \\ COOH \end{array} \cdots\cdots> \text{glycogen}$$

The desulphurase reaction is anaerobic but under aerobic conditions the H_2S set free can be oxidized to sulphate. It is also known that *cysteine sulphinic acid* can be formed by oxidation of the —SH of cysteine and can itself undergo transamination with α-ketoglutarate yielding glutamate and *β-sulphinyl-pyruvate*. The latter can then be split enzymatically into pyruvate and inorganic sulphate:

$$
\begin{array}{c}
\text{CH}_2\text{SH} \\
| \\
\text{CH.NH}_2 \\
| \\
\text{COOH}
\end{array}
\longrightarrow
\begin{array}{c}
\text{CH}_2\text{S}{<}^{\text{O}}_{\text{OH}} \\
| \\
\text{CH.NH}_2 \\
| \\
\text{COOH}
\end{array}
\longrightarrow
\begin{array}{c}
\text{CH}_2\text{S}{<}^{\text{O}}_{\text{OH}} \\
| \\
\text{C}{=}\text{O} \\
| \\
\text{COOH}
\end{array}
\xrightarrow[+\text{H}_2\text{O}]{}
\text{H}_2\text{SO}_4 +
\begin{array}{c}
\text{CH}_3 \\
| \\
\text{CO} \\
| \\
\text{COOH}
\end{array}
\text{----}\!\!\rightarrow \text{glycogen}
$$

| cysteine | cysteine-β-sulphinic acid | β-sulphinyl pyruvic acid | |

That dietary cysteine and cystine can be replaced by *methionine* argues that they can be derived from it, and this is indeed the case. If methionine containing radioactive sulphur is administered to animals, radioactive sulphur can be recovered in cysteine and in cystine isolated from the tissues. Cysteine does not arise directly from methionine but from its demethylated product, *homocysteine*. If sliced rat liver is incubated with homocysteine in the presence of *serine*, cysteine is formed by a transthiolation reaction in which an —SH group changes place with an —OH. An intermediate reaction complex, *cystathionine*, is formed:

$$
\begin{array}{cc}
\text{CH}_2\text{SH} & \text{CH}_2\text{OH} \\
| & | \\
\text{CH}_2 & +\text{CH.NH}_2 \\
| & | \\
\text{CH.NH}_2 & \text{COOH} \\
| & \\
\text{COOH} &
\end{array}
\xrightarrow[]{-\text{H}_2\text{O}}
\begin{array}{cc}
\text{CH}_2\text{S}\!\!-\!\!\!-\!\!\!-\!\!\text{CH}_2 \\
| \qquad | \\
\text{CH}_2 \quad \text{CH.NH}_2 \\
| \qquad | \\
\text{CH.NH}_2 \quad \text{COOH} \\
| \\
\text{COOH}
\end{array}
\xrightarrow[]{+\text{H}_2\text{O}}
\begin{array}{cc}
\text{CH}_2\text{OH} & \text{CH}_2\text{SH} \\
| & | \\
\text{CH}_2 & +\text{CH.NH}_2 \\
| & | \\
\text{CH.NH}_2 & \text{COOH} \\
| & \\
\text{COOH} &
\end{array}
$$

| homocysteine | serine | cystathionine | homoserine cysteine |

Two enzymes are involved, one in the synthesis and the second in the breakdown of cystathionine. The first of these (cystathionine synthetase) is apparently identical with *serine dehydratase* while the second (cystathionase) is identical with *homoserine dehydratase*. Pyridoxal phosphate acts as a coenzyme for both of these enzymes.

Cysteine and cystine are particularly important because of the ease with which a *pair of* —SH *groups* of cysteine can be oxidized to give the —S—S— bond of cystine, and vice versa. This property extends to many compounds into the composition of which cysteine enters. Thus it is present in *glutathione*, which we may represent as G.SH. This compound is very readily oxidized, e.g. by molecular oxygen in the presence of traces of heavy metals, to give the oxidized form, G.S—S.G, and it is upon this behaviour that the functional importance of glutathione appears to depend.

Linkages of the —S—S— type also play an important part in the intramolecular structure of *hair keratin* and other scleroproteins. Hair contains about 7·3% of cystine and it is believed that —S—S— bonds are formed between adjacent molecular fibres, and that the tensile strength and other mechanical properties of the hair fibre as a whole are largely due to these linkages. Cross linkages of the —S—S— type also form an integral part of the structure of many enzymes and other proteins; they occur, for example, in ribonuclease and in many proteolytic enzymes.

Many other enzymes, however, require free —SH groups. If these are oxidized to give —S—S— bonds, catalytic activity disappears but can usually be restored by the addition of reduced glutathione. Although this is not by any means a universal property of enzymes, certain dehydrogenases and other enzymes are reversibly inhibited by mild oxidation and irreversibly by iodoacetate, iodoacetamide and many other alkylating reagents. Iodoacetate, for example, reacts with and blocks —SH groups in the following manner:

$$-SH + I.CH_2COOH = HI + -S.CH_2COOH.$$

A considerable number of war gases (cf. p. 160), including lachrymators and vesicants, also act by blocking —SH groups which can, however, be 'protected' or reactivated by certain dithiols, in particular by British anti-lewisite ('BAL'; 2:3-dimercaptopropanol):

$$\underset{\underset{SH}{|}\quad\underset{SH}{|}}{CH_2.CH_2.CH_2OH}$$

Like glutathione these act by reducing —S—S— bonds to —SH.

Cysteine is the mother substance of *taurine*, a compound which is very widely distributed, often in remarkably large amounts (p. 103). It is found in the bile of many vertebrates as a conjugant in the *bile salts* (p. 216). Taurine is known to arise from *cysteic acid* through the agency of a specific cysteic acid decarboxylyase, one of the few 'straight' decarboxylyases known to occur in animal tissues (liver and kidney). Cysteic acid itself is believed to be formed from cysteine by oxidation, most probably by way of *cysteine sulphinic acid*. That the reaction takes place cannot be seriously doubted, for the administration of methionine containing radioactive sulphur can be followed by the isolation, not only of radioactive cysteine and cystine from the tissues, but by that of radioactive taurine also. The formation of taurine is therefore believed to take the following course:

$$\underset{\text{cysteine}}{\underset{\underset{COOH}{|}}{\overset{CH_2SH}{\underset{CH.NH_2}{|}}}} \longrightarrow \underset{\substack{\text{cysteine-}\beta\text{-}\\\text{sulphinic}\\\text{acid}}}{\underset{\underset{COOH}{|}}{\overset{CH_2S{\overset{\overset{O}{\diagup}}{\diagdown}}_{OH}}{\underset{CH.NH_2}{|}}}} \longrightarrow \underset{\substack{\text{cysteic}\\\text{acid}}}{\underset{\underset{COOH}{|}}{\overset{CH_2SO_3H}{\underset{CH.NH_2}{|}}}} \overset{-CO_2}{\longrightarrow} \underset{\text{taurine}}{\underset{\underset{CH_2NH_2}{|}}{\overset{CH_2SO_3H}{|}}}$$

By decarboxylation of cysteine itself *β-mercapto-ethylamine* is formed; this is a constituent of coenzyme A (p. 297).

The oxidation of cysteine to cysteic acid probably represents an early step in the oxidative degradation of cysteine and cystine, the sulphur of which ultimately appears in the urine of mammals in the form of *inorganic sulphate*. Perhaps cysteine is also the source of the so-called *ethereal sulphates* which appear in the urine following the absorption of phenolic substances into the body: their conjugation with sulphuric acid is one of several devices involved in the *detoxication* of phenols. Examples of ethereal sulphates are the following:

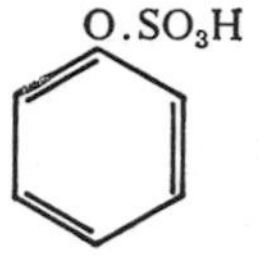

*phenolsulphuric
acid*

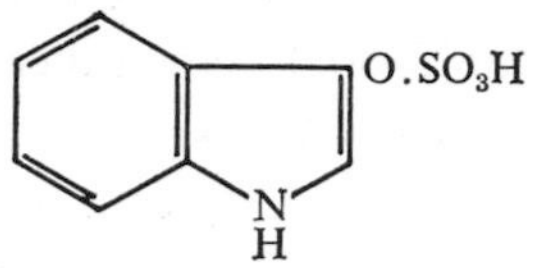

*indoxylsulphuric acid
(urinary indican)*

Unlike cysteic acid and taurine, these ethereal sulphates are true sulphates and not sulphonic acids: the uncommon but direct carbon-to-sulphur linkage characteristic of the sulphur-containing amino-acids is absent from the ethereal sulphates.

Cysteine also contributes to the formation of *mercapturic acids*, which are products of detoxication of certain aromatic substances. A classical example of this is seen in the fate of bromobenzene administered to dogs. This substance is eliminated in conjugation with *N*-acetylcysteine in the form of *p*-bromophenyl-mercapturic acid, a remarkable achievement on the part of an animal that is never likely to meet bromobenzene except through the medium of the laboratory. Naphthalene also is converted in part into a mercapturic acid:

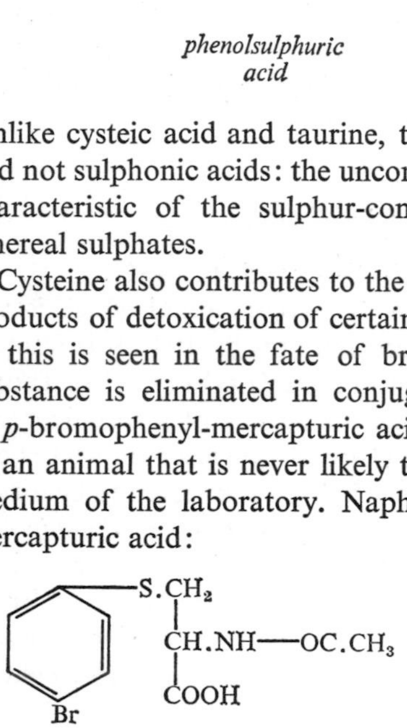

p-bromophenyl-mercapturic acid

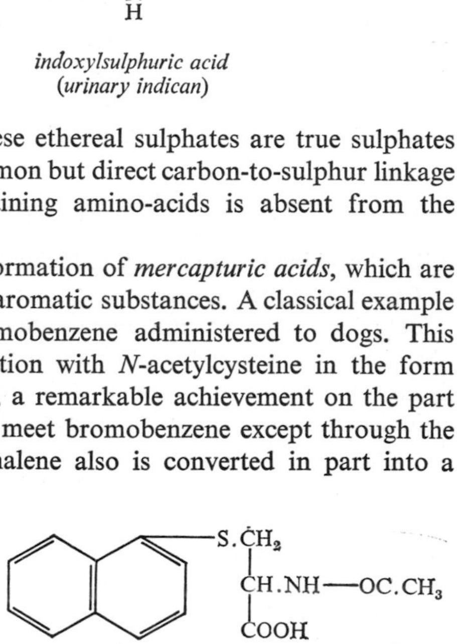

naphthyl-mercapturic acid

In addition to the inorganic and ethereal sulphates, mammalian urine contains a third fraction, the so-called '*neutral sulphur*'. This comprises a mixed bag of sulphur compounds, including traces of thio-alcohols (mercaptans) and mercapturic acids. Not much is known about this fraction, but there is no doubt that much of it arises from the sulphur-containing amino-acids. There exists, moreover, a rare and abnormal condition known as *cystinuria*, in which the neutral sulphur fraction is very large and consists mainly of cystine itself. The administration of cysteine or cystine to a cystinuric does not, however, increase the output of urinary cystine, showing that the cystine excreted is not directly derived from that of the food.

246

$$\textit{METHIONINE,} \quad \begin{array}{l} S.CH_3 \\ | \\ CH_2 \\ | \\ CH_2 \\ | \\ CH.NH_2 \\ | \\ COOH \end{array}$$

(essential: glycogenic). Its main known function, which it discharges in the form of S-adenosyl-methionine (p. 111), is that of a *biological methylating agent*. The primary source of methyl groups is the methylated derivative of tetrahydrofolic acid (Fig. 28, p. 109) but, once transferred to S-adenosyl-homocysteine to form S-adenosyl-methionine, these methyl groups enter into numerous methyltransferase reactions. Thus S-adenosyl-methionine completes the synthesis of creatine by transferring a methyl group to glycocyamine. Very probably it is the methylating agent involved in the *detoxication* of pyridine and certain pyridine derivatives, including nicotinic acid, for pyridine itself is excreted in the form of N-methylpyridine, and nicotinic acid partly as trigonelline. There is evidence that methionine, again in the form of its S-adenosyl-derivative, supplies methyl groups for the synthesis of adrenaline (p. 257) and sarcosine (p. 239).

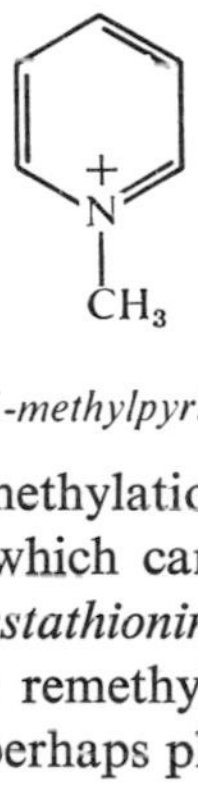

N-*methylpyridine* 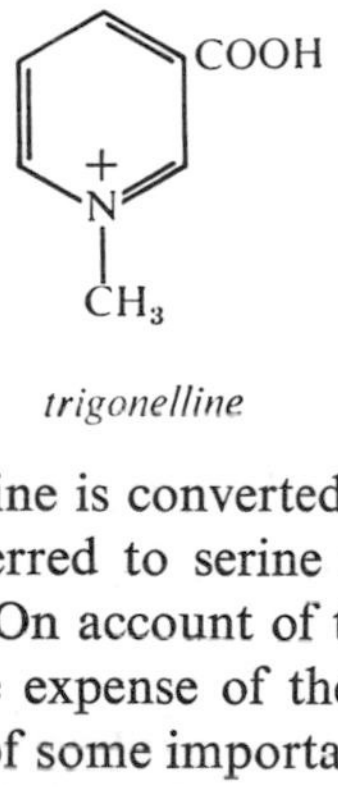   *trigonelline*

By undergoing demethylation, methionine is converted into *homocysteine*, the —SH group of which can be transferred to serine in the synthesis of *cysteine* by way of *cystathionine* (p. 244). On account of the ease with which homocysteine can be remethylated at the expense of the methyl groups of choline, methionine perhaps plays a part of some importance in the metabolism of *phospholipids* by converting ethanolamine into choline and vice versa (p. 111). Homocysteine can be remethylated at the expense of choline, glycine betaine and possibly of sarcosine, but not at that of creatine.

$$\textit{VALINE,} \quad \begin{array}{c} CH_3 \quad CH_3 \\ \diagdown \diagup \\ CH \\ | \\ CH.NH_2 \\ | \\ COOH \end{array}$$

(essential: glycogenic), was formerly believed to be ketogenic, in common with the other branched-chain amino-acids. Relatively little is known about the metabolism of this and the other amino-acids containing branched chains. The conversion of valine into glycogen has been studied extensively, largely with the aid of isotopes and it now appears that, contrary to expectation, there is no removal of one of the methyl groups. The first two steps follow the usual pattern—deamination and oxidative decarboxylation—leading to iso-butyryl-CoA. This coenzyme A plays a major part in the subsequent reactions;

$$
\begin{array}{ccccccc}
\underset{\displaystyle\underset{\displaystyle\underset{\displaystyle\text{COOH}}{|}}{\overset{\displaystyle\text{CH}}{|}}}{\overset{\displaystyle CH_3 \quad CH_3}{\diagdown\diagup}} &
\longrightarrow &
\underset{\displaystyle\underset{\displaystyle\underset{\displaystyle\text{COOH}}{|}}{\overset{\displaystyle\text{CO}}{|}}}{\overset{\displaystyle CH_3 \quad CH_3}{\diagdown\diagup}} &
\xrightarrow{-CO_2} &
\underset{\displaystyle\underset{\displaystyle\text{CO.CoA}}{|}}{\overset{\displaystyle CH_3 \quad CH_3}{\diagdown\diagup}} &
\xrightarrow{-2H} &
\underset{\displaystyle\underset{\displaystyle\text{CO.CoA}}{|}}{\overset{\displaystyle CH_3 \quad CH_2}{\diagdown\diagup}} \quad \xrightarrow{+H_2O}
\end{array}
$$

(CH.NH₂ form on left, CO form second; third is *iso-butyryl-CoA*)

$$
\begin{array}{ccccccc}
\underset{\displaystyle\text{CO.CoA}}{\overset{\displaystyle CH_3 \quad CH_2OH}{\diagdown\diagup CH}} &
\xrightarrow{-CoA} &
\underset{\displaystyle\text{COOH}}{\overset{\displaystyle CH_3 \quad CH_2OH}{\diagdown\diagup CH}} &
\xrightarrow{-2H} &
\underset{\displaystyle\text{COOH}}{\overset{\displaystyle CH_3 \quad CHO}{\diagdown\diagup CH}} &
\xrightarrow[-2H \;\; +CoA]{+H_2O} &
\underset{\displaystyle\text{CO.CoA}}{CH_3.CH} \xrightarrow{B_{12}}
\end{array}
$$

methylmalonyl-CoA

$$
\underset{\displaystyle\underset{\displaystyle\underset{\displaystyle\underset{\displaystyle\text{CO.CoA}}{|}}{\text{CH}_2}}{\text{CH}_2}}{\text{COOH}} \longrightarrow
\underset{\displaystyle\underset{\displaystyle\underset{\displaystyle\underset{\displaystyle\text{COOH}}{|}}{\text{CH}_2}}{\text{CH}_2}}{\text{COOH}} \dashrightarrow
\underset{\displaystyle\underset{\displaystyle\text{COOH}}{\text{CO}}}{\text{CH}_3} \dashrightarrow \text{glycogen}
$$

succinyl-CoA

Succinyl-CoA is certainly glycogenic but methylmalonyl-CoA is an intermediate in the synthesis of fatty acids, as we shall see later on.

$$
\textit{LEUCINE} \quad
\underset{\displaystyle\underset{\displaystyle\underset{\displaystyle\underset{\displaystyle\text{COOH}}{|}}{\text{CH.NH}_2}}{\underset{\displaystyle\text{CH}_2}{|}}}{\overset{\displaystyle CH_3 \quad CH_3}{\diagdown\diagup CH}}
$$

(essential: ketogenic). Its conversion into ketone bodies proceeds through the reactions shown on page 249. Both products lie directly on the lines leading to the synthesis of fatty acids and coenzyme A again plays a fundamental part in the process.

248

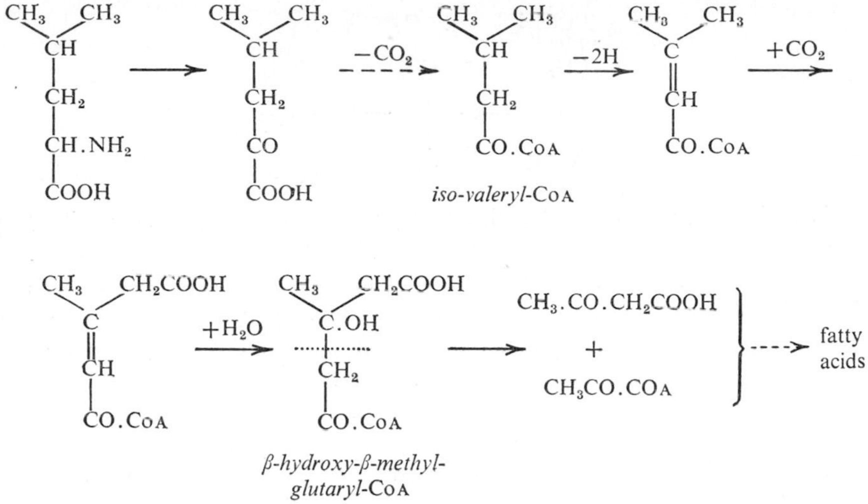

$$\begin{array}{c} CH_3 \quad CH_3 \\ \diagdown \diagup \\ CH \\ | \\ CH_2 \\ | \\ CH.NH_2 \\ | \\ COOH \end{array} \longrightarrow \begin{array}{c} CH_3 \quad CH_3 \\ \diagdown \diagup \\ CH \\ | \\ CH_2 \\ | \\ CO \\ | \\ COOH \end{array} \xrightarrow{-CO_2} \begin{array}{c} CH_3 \quad CH_3 \\ \diagdown \diagup \\ CH \\ | \\ CH_2 \\ | \\ CO.C_OA \end{array} \xrightarrow{-2H} \begin{array}{c} CH_3 \quad CH_3 \\ \diagdown \diagup \\ C \\ \| \\ CH \\ | \\ CO.C_OA \end{array} \xrightarrow{+CO_2}$$

iso-valeryl-CoA

$$\begin{array}{c} CH_3 \quad CH_2COOH \\ \diagdown \diagup \\ C \\ \| \\ CH \\ | \\ CO.C_OA \end{array} \xrightarrow{+H_2O} \begin{array}{c} CH_3 \quad CH_2COOH \\ \diagdown \diagup \\ C.OH \\ | \\ CH_2 \\ | \\ CO.C_OA \end{array} \longrightarrow \begin{array}{c} CH_3.CO.CH_2COOH \\ + \\ CH_3CO.COA \end{array} \left.\begin{array}{c} \\ \\ \\ \end{array}\right\} \dashrightarrow \begin{array}{c} \text{fatty} \\ \text{acids} \end{array}$$

β-hydroxy-β-methyl-
glutaryl-CoA

iso-*LEUCINE*, $\begin{array}{c} CH_3 \\ | \\ CH_2 \quad CH_3 \\ \diagdown \diagup \\ CH \\ | \\ CH.NH_2 \\ | \\ COOH \end{array}$

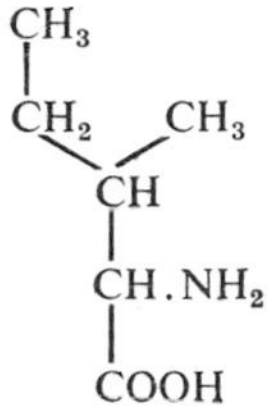

(essential: glycogenic *and* ketogenic), might be converted into glycogen by mechanisms similar to those postulated for valine, but would of course require the removal of one more carbon atom. Coenzyme A again plays a large part and the eventual products are a molecule each of acetyl-CoA and pro-pionyl-CoA. Of these the former is lipogenic and the latter glycogenic.

ASPARTIC ACID, $\begin{array}{c} COOH \\ | \\ CH_2 \\ | \\ CH.NH_2 \\ | \\ COOH \end{array}$

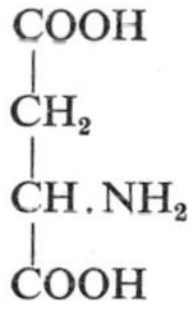

(non essential: glycogenic), yields oxaloacetate on deamination or transde-amination. This product is somewhat unstable and undergoes slow, sponta-neous β-decarboxylation under physiological conditions of temperature and pH. The liver contains an enzyme, *oxaloacetate decarboxylyase*, which catalyses this reaction, of which the product, pyruvate, is known to give rise freely to glucose and glycogen:

249

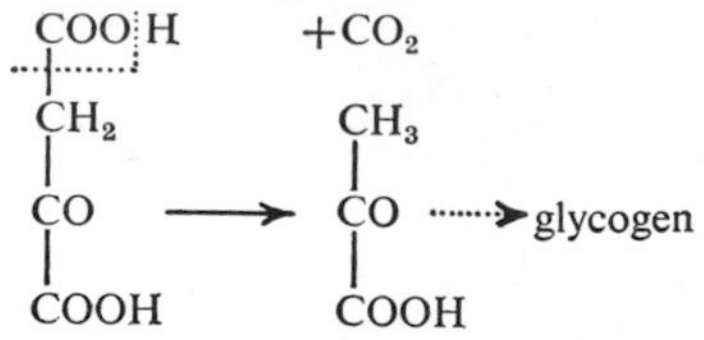

$$\begin{array}{ccc}
\underset{}{COO\vdots H} & +CO_2 \\
\mid & \\
CH_2 & CH_3 \\
\mid & \mid \\
CO & \longrightarrow & CO \dashrightarrow glycogen \\
\mid & \mid \\
COOH & COOH
\end{array}$$

This rather rare process of β-decarboxylation contrasts sharply with the oxidative decarboxylation that is characteristic of α-keto-acids in general.

Aspartate can react with ammonia in the presence of ATP and under the influence of a synthetase to form *asparagine*, which is itself non-essential and glycogenic. This system plays an important part in the *storage of amino-groups* in the tissues of many plants (p. 231). Aspartate also plays a central part in *transamination and transdeamination* in some animal and plant tissues and plays a part also in *ureogenesis* by donating its amino-group to citrulline to yield arginine (p. 276). It plays a somewhat similar part as an amino-group donor in the synthesis of certain nucleotides (p. 285).

Finally, by a reaction with carbamoyl phosphate, aspartate gives rise to *carbamoyl aspartate* in the presence of *aspartate carbamoyltransferase*:

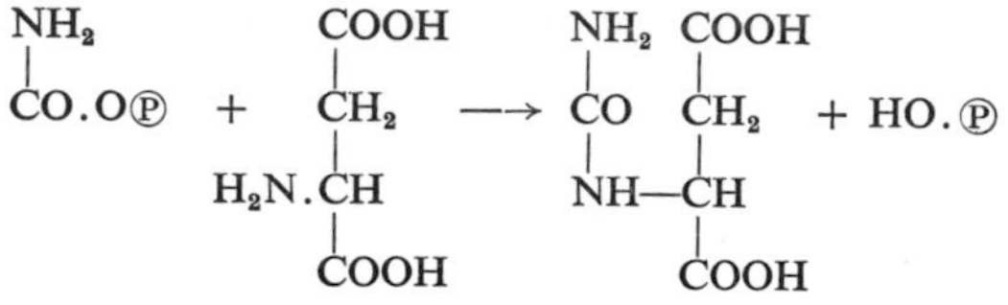

$$\begin{array}{ccccc}
NH_2 & & COOH & NH_2 & COOH \\
\mid & & \mid & \mid & \mid \\
CO.O\circledP & + & CH_2 & \longrightarrow & CO & CH_2 & + HO.\circledP \\
& & \mid & & \mid & \mid \\
& & H_2N.CH & & NH—CH \\
& & \mid & & \mid \\
& & COOH & & COOH
\end{array}$$

The product, also known as *ureidosuccinate*, is the starting material for pyrimidine synthesis (p. 290). Carbamoyl-aspartate can also be formed by the transference of a carbamoyl group to aspartate from *citrulline* (p. 103).

$$\begin{array}{ll}
GLUTAMIC\ ACID, & COOH \\
& \mid \\
& CH_2 \\
& \mid \\
& CH_2 \\
& \mid \\
& CH.NH_2 \\
& \mid \\
& COOH
\end{array}$$

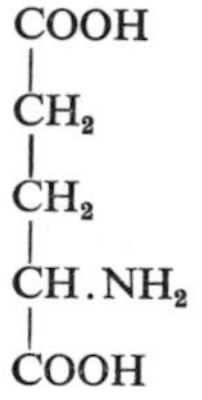

(non-essential: glycogenic), yields α-ketoglutarate on deamination. The product, an α-keto-acid, undergoes oxidative decarboxylation in the usual way, giving rise to succinate by way of succinyl-Co A. Succinate itself is converted into fumarate by dehydrogenation, catalysed by succinate dehydrogenase; and fumarate, under the influence of fumarase, takes on water to give malate. Malate is dehydrogenated in its turn by the action of malate dehydrogenase and NAD, to give oxaloacetate. The latter, on β-decarboxylation, yields pyruvate and hence gives rise to glycogen:

250

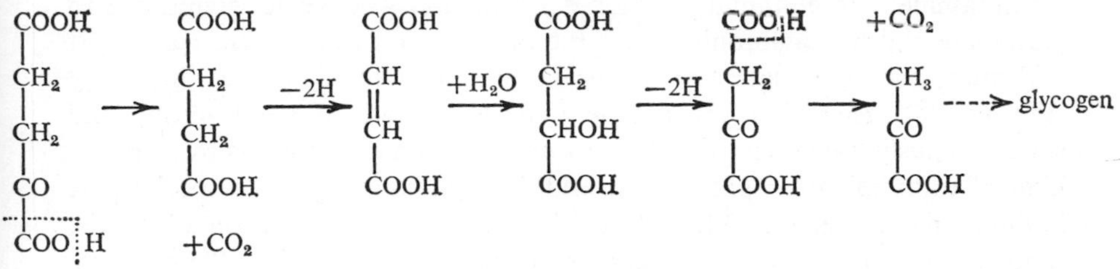

$$\begin{array}{c}\text{COOH}\\|\\\text{CH}_2\\|\\\text{CH}_2\\|\\\text{CO}\\|\\\text{COO}\vdots\text{H}\quad +\text{CO}_2\end{array}\rightarrow\begin{array}{c}\text{COOH}\\|\\\text{CH}_2\\|\\\text{CH}_2\\|\\\text{COOH}\quad +\text{CO}_2\end{array}\xrightarrow{-2\text{H}}\begin{array}{c}\text{COOH}\\|\\\text{CH}\\\|\\\text{CH}\\|\\\text{COOH}\end{array}\xrightarrow{+\text{H}_2\text{O}}\begin{array}{c}\text{COOH}\\|\\\text{CH}_2\\|\\\text{CHOH}\\|\\\text{COOH}\end{array}\xrightarrow{-2\text{H}}\begin{array}{c}\text{COO}\vdots\text{H}\\|\\\text{CH}_2\\|\\\text{CO}\\|\\\text{COOH}\end{array}\rightarrow\begin{array}{c}+\text{CO}_2\\\text{CH}_3\\|\\\text{CO}\\|\\\text{COOH}\end{array}\cdots\!\!\to\text{glycogen}$$

Glutamate can also give rise to and be formed from *proline* (Fig. 49). The reactions involve glutamic-γ-semialdehyde and reduced NAD or NADP:

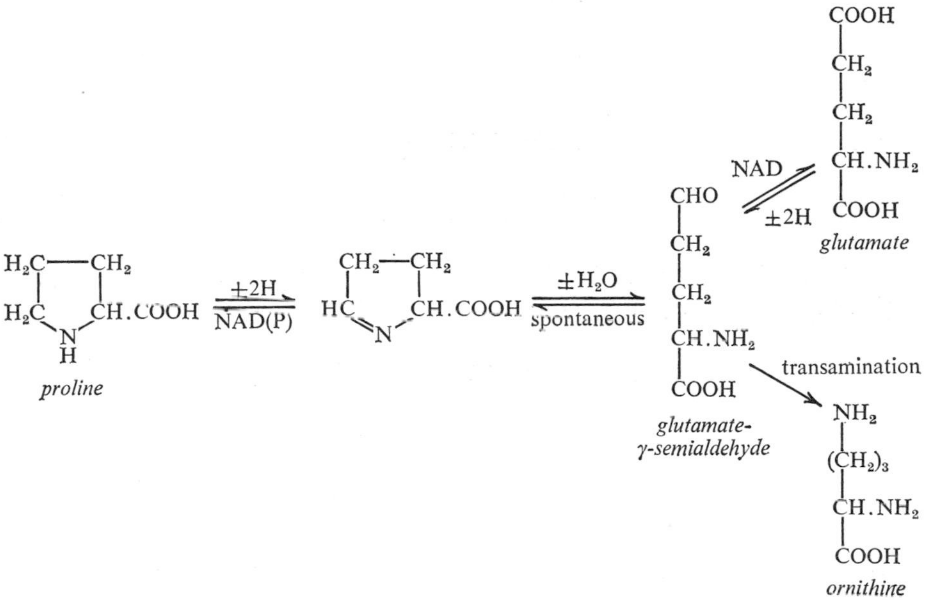

Fig. 49. Interconversion of glutamate, proline and ornithine.

The existence of a *glutamate-α-decarboxylyase* in animal tissues accounts for the formation of γ-*aminobutyrate*, which occurs in brain tissue and elsewhere:

$$\text{H}_2\text{N}.\text{CH}_2\text{CH}_2\text{CH}_2\text{COOH}.$$

Like β-alanine this is a somewhat uncommon and thoroughly atypical amino-acid and perhaps it is worth suggesting that β-alanine itself (p. 241) might conceivably arise by the action of an analogous enzyme upon aspartate.

Glutamate together with glutamate dehydrogenase plays a central part in *transamination and transdeamination*, and is a constituent of *glutathione*, while its amide, *glutamine*, occurs in considerable quantities in many plant and animal tissues, in which it serves as a *store of amino groups*.

Glutamine, like glutamate, is non-essential and glycogenic, giving rise to glutamate and free ammonia under the influence of *glutaminase* (p. 83). It is formed from glutamate and ammonia by an endergonic process involving ATP (p. 98). There is evidence too that it can lose its amido-group and simultaneously transfer its α-amino-group to certain α-keto-acids by a special kind of transamination process. In addition, glutamine plays a special part in ammonia storage (p. 230) and also in the *detoxication* of aromatic acids, though only among Primates, according to present information. Phenylacetic acid is excreted in the form of a conjugate with glycine by most mammals, but in man and the chimpanzee it gives rise to phenyl-acetylglutamine:

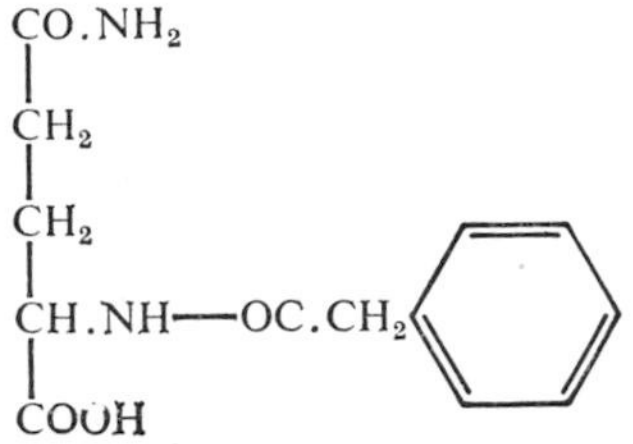

$$CO.NH_2$$
$$|$$
$$CH_2$$
$$|$$
$$CH_2$$
$$|$$
$$CH.NH\!-\!\!-OC.CH_2\langle\ \rangle$$
$$|$$
$$COOH$$

phenylacetylglutamine

Glutamine is also heavily involved in *uricogenesis* (pp. 283–5) while *N*-acetyl-glutamate acts catalytically in *ureogenesis* (p. 276).

$$ARGININE,\quad HN\!=\!\!C\!\!<\!\!\begin{array}{l}NH_2\\NH\end{array}$$
$$|$$
$$(CH_2)_3$$
$$|$$
$$CH.NH_2$$
$$|$$
$$COOH$$

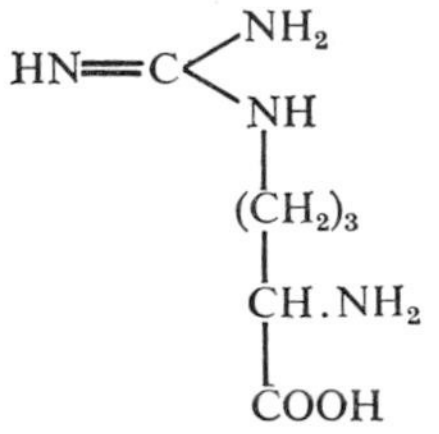

('half-essential', see p. 220: glycogenic), can lose its amidine group under the hydrolytic influence of *arginase* to yield *ornithine*, which is also glycogenic. There are indications that arginine does not normally undergo deamination (p. 229).

Arginine can also part with its amidine radical by participating in group-transfer reactions, i.e. by *transamidination* (p. 102). It plays a vital part in the synthesis of urea by the so-called *ornithine cycle* of Krebs; the mechanisms leading to its resynthesis from ornithine are considered elsewhere (pp. 271–7). Arginine occurs much more extensively in invertebrates than among verte-brates, for it forms the guanidine base of the commonest of the *invertebrate phosphagens* (p. 370): in addition, arginine is the parent substance of creatine and probably of some at least of the peculiar guanidine derivatives found in invertebrate tissues such, for example, as *taurocyamine, octopine* and *agmatine* (pp. 102–3).

$$CITRULLINE, \quad O{=}C\begin{array}{l} \diagup NH_2 \\ \diagdown NH \end{array}$$

$$\begin{array}{c} | \\ (CH_2)_3 \\ | \\ CH.NH_2 \\ | \\ COOH \end{array}$$

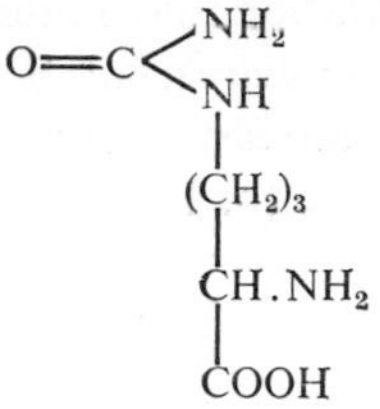

(non-essential: glycogenic), does not enter into the composition of proteins, but occurs as an intermediary between ornithine and arginine in the '*ornithine cycle*' (p. 276). It is itself formed by transcarbamoylation of ornithine at the expense of carbamoyl phosphate but can hand on its carbamoyl group to aspartate, yielding *carbamoyl aspartate*, the starting material for pyrimidine synthesis (p. 290).

$$ORNITHINE, \quad NH_2$$

$$\begin{array}{c} | \\ (CH_2)_3 \\ | \\ CH.NH_2 \\ | \\ COOH \end{array}$$

(non-essential: glycogenic), has not been isolated from protein hydrolysates but arises from the action of arginase upon arginine and participates in the '*ornithine cycle*'. It yields *citrulline* by accepting a carbamoyl group (—CO.NH$_2$) from carbamoyl phosphate (see p. 276). There is some doubt whether it undergoes biological deamination (cf. lysine, below) but it has been shown in feeding experiments that deuterium-labelled ornithine can give rise to glutamate and thence, presumably by cyclization of the γ-semialdehyde, to proline: the process seems to be reversible (see Fig. 49), and both products are glycogenic.

Among birds, ornithine discharges a special function in the *detoxication* of aromatic acids such as benzoic: the latter is excreted in the urine of birds in the form of dibenzoylornithine, or *ornithuric acid*:

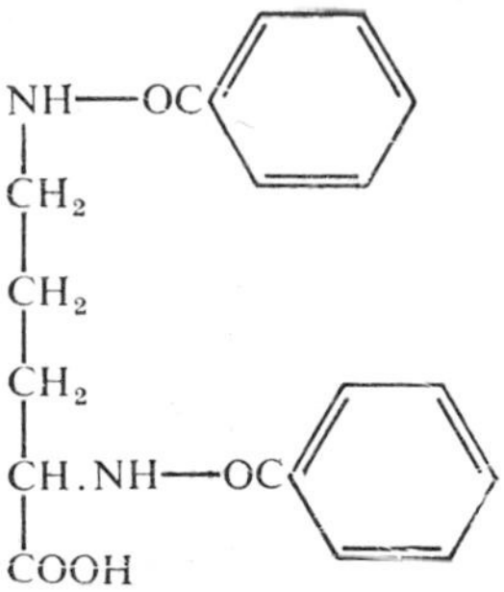

ornithuric acid   *putrescine*

By bacterial decarboxylation in the intestine and elsewhere, ornithine can give rise to the toxic diamine *putrescine*. Small amounts of this substance are absorbed by animals and are *detoxicated,* probably undergoing oxidation at the hands of diamine oxidase (p. 138).

$$LYSINE, \quad \begin{array}{c} NH_2 \\ | \\ (CH_2)_4 \\ | \\ CH.NH_2 \\ | \\ COOH \end{array}$$

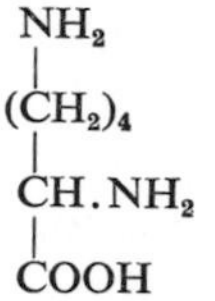

(essential: glycogenic *and* ketogenic), is the next higher homologue of ornithine. It is believed not to undergo deamination in the body (p. 229) but is metabolically degraded in other ways. The details are known today and the following are among the reactions concerned:

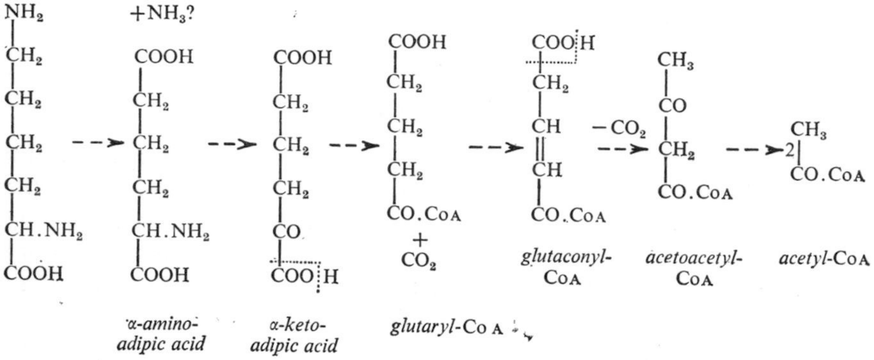

Very little is known about the specific functions which, as an essential amino-acid, it must be presumed to discharge in the organism. Under the influence of bacterial enzymes it can yield *cadaverine*,

$$H_2N(CH_2)_5NH_2,$$

which, like putrescine, is highly toxic and can be oxidatively detoxicated by diamine oxidase (p. 138).

Mention must also be made of the uncommon amino-acid *mesodiaminopimelic acid* which undergoes simple decarboxylation to give lysine:

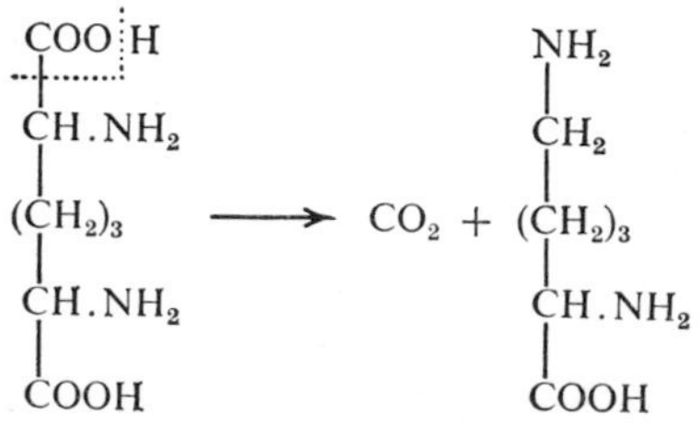

PHENYLALANINE, $CH_2CH(NH_2)COOH$

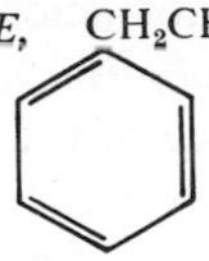

(essential: glycogenic *and* ketogenic), and

TYROSINE, $CH_2CH(NH_2)COOH$

OH

(replaceable by phenylalanine: glycogenic *and* ketogenic). It may be presumed that phenylalanine is irreversibly convertible into tyrosine (p. 220), by oxidation. These important amino-acids are known to be both glycogenic and ketogenic and it is considered that the ring must be opened in the process; the side-chain contains only three carbon atoms whereas four are required for the production of acetoacetate. Work with isotopically labelled phenylalanine and tyrosine indicates that the α- and β-carbon atoms of the side-chain and two more from the ring furnish the four carbon atoms of acetoacetate, and there is evidence that the following are the principal reactions involved:

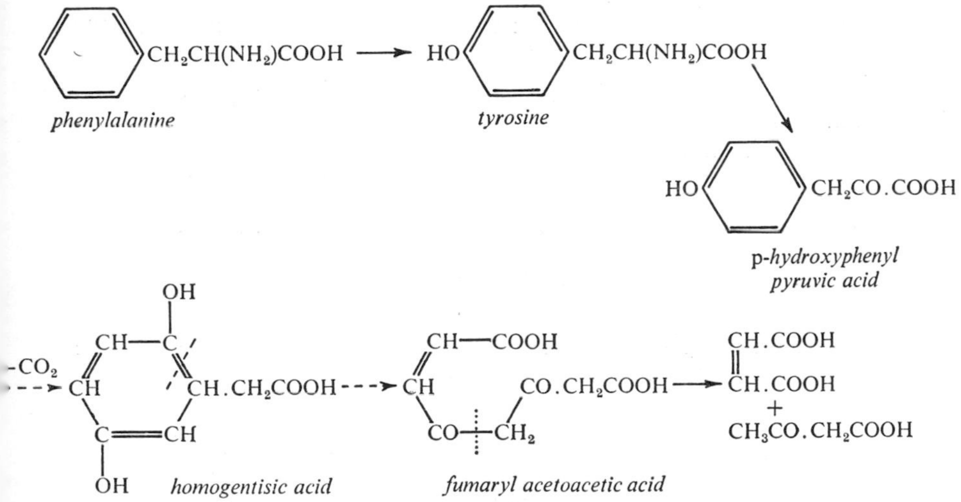

$CH_2CH(NH_2)COOH$ $\longrightarrow$ HO $CH_2CH(NH_2)COOH$

phenylalanine   *tyrosine*

HO $CH_2CO.COOH$

*p-hydroxyphenyl
pyruvic acid*

$$-CO_2 \dashrightarrow$$

homogentisic acid $\dashrightarrow$

$$\begin{array}{c} CH—COOH \\ \parallel \\ CH \\ CO—CH_2 \end{array}$$

$$CO.CH_2COOH \longrightarrow \begin{array}{c} CH.COOH \\ \parallel \\ CH.COOH \\ + \\ CH_3CO.CH_2COOH \end{array}$$

fumaryl acetoacetic acid

Fumarate gives rise to pyruvate by way of malate and oxaloacetate in the usual way; the acetoacetate is, as usual, lipogenic. Enzymes capable of catalysing all of these reactions have been found in liver preparations of various kinds.

255

The metabolism of the aromatic acids goes astray in a group of interesting hereditary '*inborn errors of metabolism*' presumably because of the deficiency, defect or absence of one or more of the numerous enzymes concerned in the normal metabolism of these aromatic amino-acids and brought about by genetic mutations. A peculiar form of mental deficiency known as *imbecillitas* (*oligophrenia*) *phenylpyruvica* owes its name to the curious fact that the urine of those afflicted regularly contains small amounts of phenylpyruvic acid. In *albinism* the enzyme tyrosinase is completely lacking, and the dark-coloured *melanic pigments* are characteristically absent from the skin, hair, eyes and other usual situations. A few cases of *tyrosinosis* have been reported, the

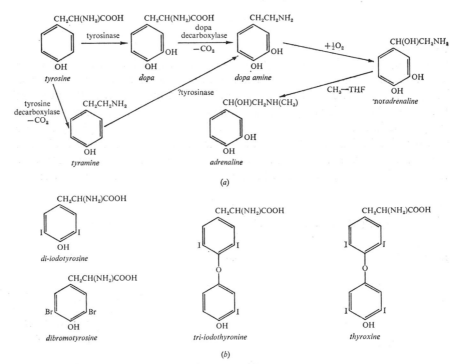

Fig. 50. (*a*) Possible pathways of synthesis of adrenaline and *nor*adrenaline. (*b*) Some halogenated derivatives of tyrosine.

urine in this disorder containing traces of tyrosine as a regular feature. *Alcaptonuria* is another disorder in which the metabolism of the aromatic amino-acids appears to be blocked; *homogentisic acid* is formed in the usual way but fails to be further metabolized and is consequently to be found in the urine (p. 255).

Phenylalanine and tyrosine are of particular importance in animal metabolism as the parent substances of at least four hormones, *nor*adrenaline,

adrenaline, tri-iodothyronine and *thyroxine*. This has been confirmed by the use of isotopes. *Tyramine* can be formed from tyrosine by the action of a weak, specific tyrosine decarboxylyase that occurs in the kidney and liver of mammals, and may possibly be an intermediate in the elaboration of adrenaline. If so, it must presumably then be attacked by tyrosinase to yield the amine corresponding to dihydroxyphenylalanine ('dopa'). It is known, however, that dopa can be formed from tyrosine directly by the action of tyrosinase and, moreover, that there exists a specific dopa decarboxylyase which, acting upon dopa itself, could yield the same amine once more. The known stages on the route to adrenaline, and the possible step just described may be summarized as in Fig. 50. The *N*-methyl group of adrenaline is transferred from *S*-adenosyl-methionine, for administration of methyl-labelled (^{14}C) methionine leads to the production of *N*-methyl-labelled adrenaline in the adrenal medulla.

Tri-iodothyronine and *thyroxine* are heavily iodinated derivatives of tyrosine but little is known about their biological formation. They both occur together with *di- and tri-iodotyrosine*, in the thyroid tissue of vertebrates. Di-iodotyrosine, together with the corresponding *dibromotyrosine*, has also been described as a constituent amino-acid of the skeletal protein material of a coral, *Gorgonia*, for which reason these halogenated tyrosines are sometimes called iodogorgoic and bromogorgoic acids respectively.

Phenylalanine and tyrosine give rise to a series of toxic products when submitted to bacterial attack. These include phenol, *p*-cresol, tyramine and phenylethylamine. The amines are probably oxidatively *detoxicated* by amine oxidase, and the phenols by conjugation, usually with sulphuric acid.

HISTIDINE, ⌐────⌐CH$_2$CH(NH$_2$)COOH

N⤸ ⤸NH

(essential for many animals but apparently not for man; glycogenic). Apart from its presence in proteins this amino-acid occurs in combination with *β*-alanine in the dipeptide *carnosine* (p. 241), large amounts of which are present in the muscles of vertebrates of most kinds, though invariably absent from invertebrate tissues. The analogous compound *anserine* (p. 241) is similarly distributed and also contains *β*-alanine, but histidine is here replaced by its 1-*N*-methyl derivative. The methyl group arises by transference from *S*-adenosyl-methionine.

Animal tissues, especially the liver, are known to contain enzymes which open the imidazole ring through a series of reactions. The first step leads to the formation of *urocanic acid* and is catalysed by a specific *histidase*. Subsequent reactions lead through a long series of intermediates to glutamate and *α*-ketoglutarate, both of which are, of course, glycogenic.

257

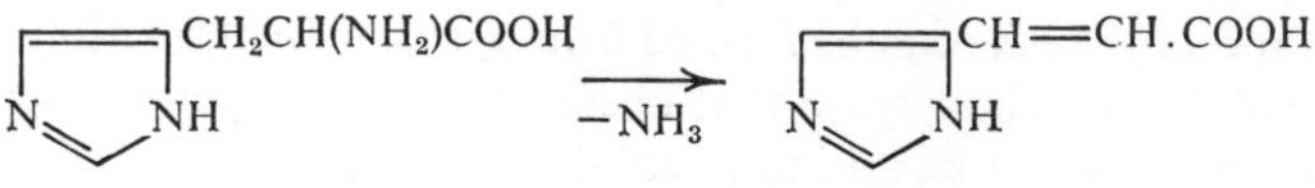

CH$_2$CH(NH$_2$)COOH $\xrightarrow{-NH_3}$ CH=CH.COOH

urocanic acid

Histidine is the mother substance of *histamine*, being attacked by a specific *histidine decarboxylyase*, traces of which are present in liver and kidney. Histamine can also be produced by bacterial activity in the intestine and elsewhere, and is *detoxicated* by a histamine oxidase, probably identical with the diamine oxidase of animal tissues, though its distribution is a little peculiar.

TRYPTOPHAN, CH$_2$CH(NH$_2$)COOH

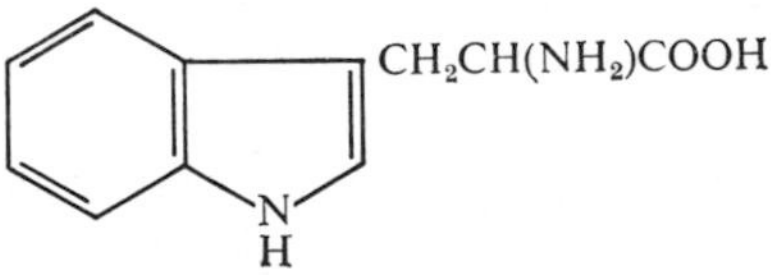

(essential: glycogenic). This amino-acid gives rise, when administered in fairly large doses, to the excretion of *kynurenic acid*. The product is formed by way of *kynurenine* (see Fig. 51 *a*).

There is evidence that kynurenine may be further metabolized to yield *nicotinic acid* in the animal; this is the only known case in which a vitamin can be produced from an essential amino-acid. 3-Hydroxyanthranilic acid, which can be formed from kynurenine, is known to be a precursor of nicotinic acid and the intermediate stages include those shown in Fig. 51 *b*. It is interesting to notice that quinolinic and nicotinic acids are formed as the corresponding nucleotides.

Bacterial decarboxylation of tryptophan leads to the formation of a poisonous amine, *tryptamine*, which is probably destroyed by an amine oxidase, like other amines. Further degradation of the side-chain by bacteria is also possible and leads to the formation of a pair of foul-smelling compounds, *indole* and *skatole*; these are said to be largely responsible for the odour of faeces. Indole undergoes bacterial conversion into the corresponding alcohol, *indoxyl*, which, if absorbed, is *detoxicated* by conjugation with sulphuric acid and excreted in the urine in the form of the corresponding ethereal sulphate:

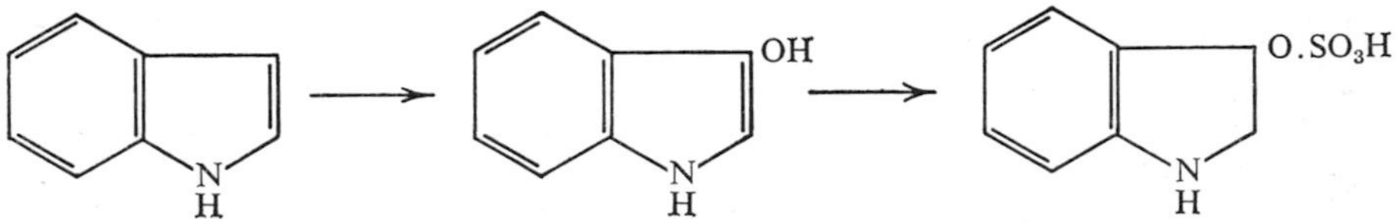

indole   *indoxyl*   *indoxylsulphuric acid (urinary indican)*

258

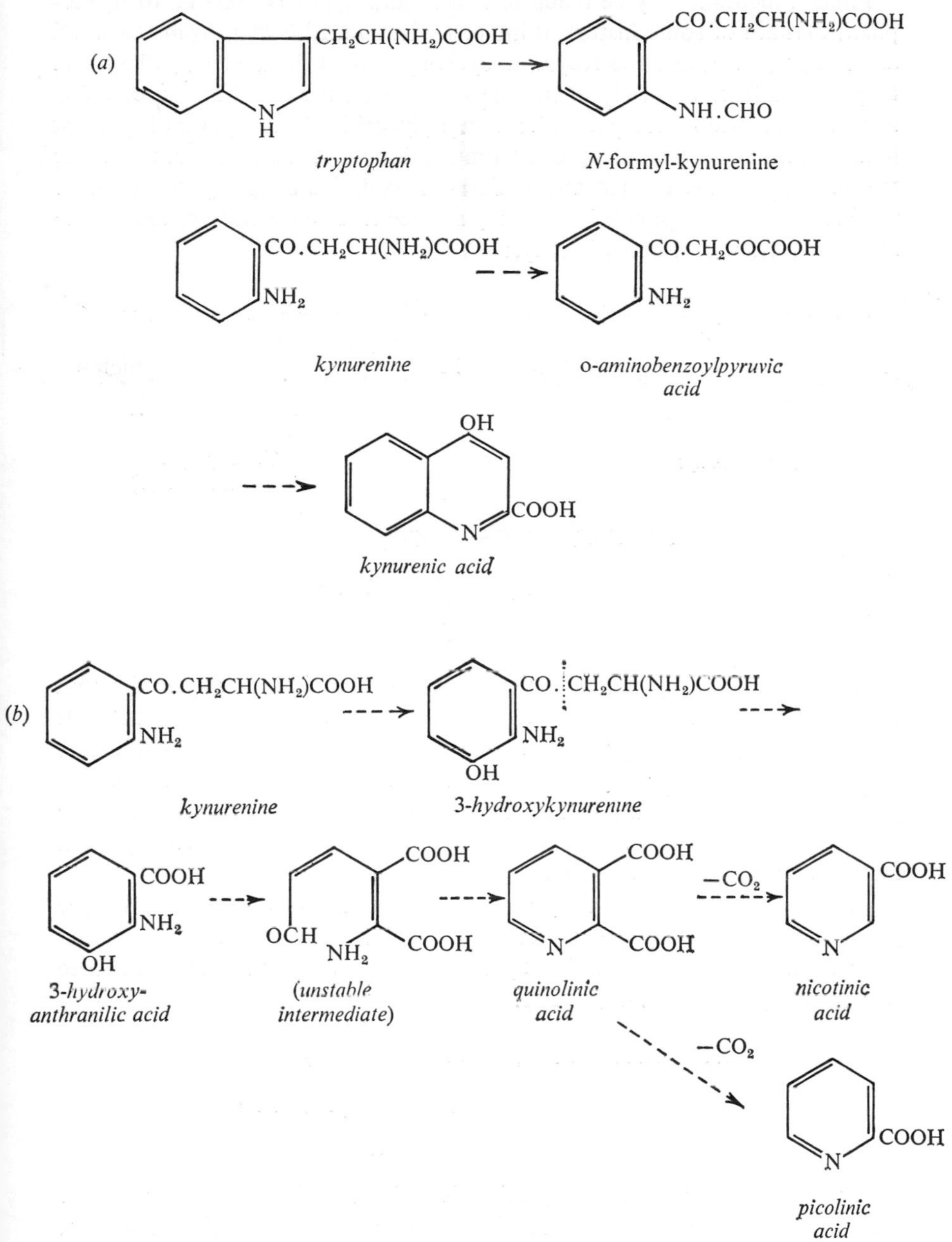

Fig. 51. Some pathways of tryptophan metabolism. (*a*) Formation of kynurenine
and kynurenic acid. (*b*) Formation of nicotinic and picolinic acids.

Finally, mention may be made of two natural pigments related to trypto-phan, that are of some historical interest, viz. *natural indigo*, from the woad and indigo plants, and the Royal or *Tyrian purple* of the ancients, which can be prepared from a variety of marine gastropods, the classical source being *Murex* spp. Natural indigo arises from a β-glycoside of indoxyl that is present in the plant juices and undergoes decomposition when the tissues are bruised; free indoxyl is thus formed and undergoes oxidative coupling in the air to yield indigo. Tyrian purple is similarly formed from a derivative, thought to be a mercaptan, of 4-brom-indoxyl:

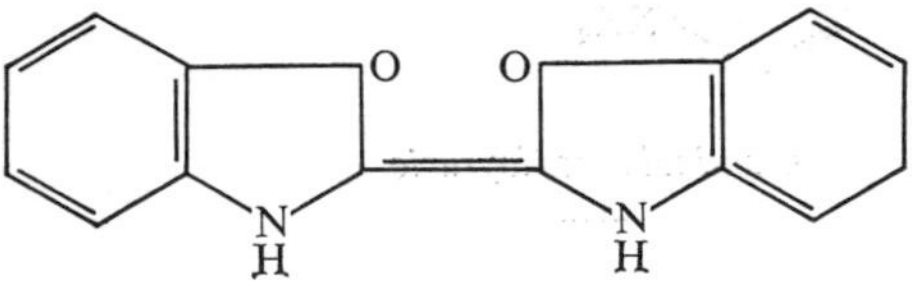

natural indigo

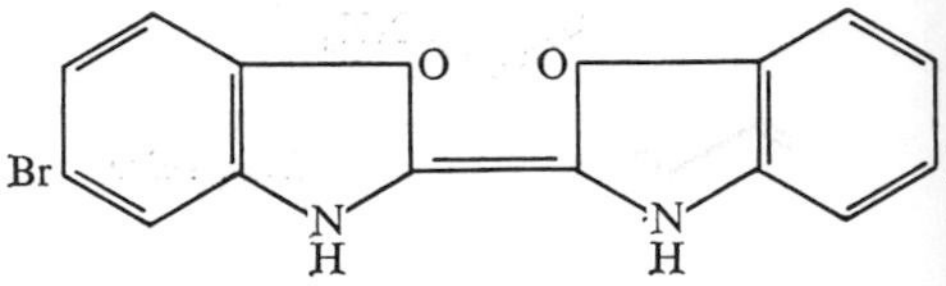

Tyrian purple
(4-4′-dibromindigo)

PROLINE, H$_2$C———CH$_2$
H$_2$C   CH.COOH
N
H

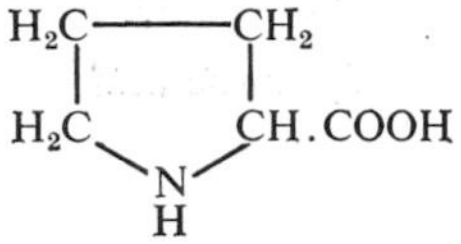

(non-essential: glycogenic) is, strictly speaking, an imino- rather than an amino-acid. Little is known about its metabolism, apart from the fact that *proline oxidase*, which is present in kidney tissue, can open the ring and give rise to glutamate. This may perhaps be a preliminary to the deamination of proline. The ring opening is not a hydrolytic but essentially an oxidative process. Since glutamate is glycogenic, the formation of glucose and glycogen from proline can be understood.

There is evidence that this process can in some way be reversed since glutamate can act as a precursor of proline itself. There is also evidence that proline can arise from ornithine by way of glutamic-γ-semialdehyde: these reactions appear in Fig. 49, p. 251.

HYDROXYPROLINE, HO.CH———CH$_2$
H$_2$C   CH.COOH
N
H

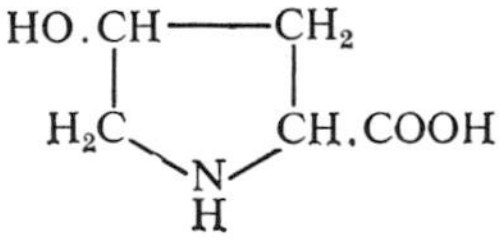

(non-essential: glycogenic), like proline, is an imino-acid. It is converted into carbohydrate in the somewhat round about manner shown in the flow-sheet. It is attacked, perhaps by proline oxidase or some similar enzyme, to give *γ-hydroxyglutamate-γ-semialdehyde*, which occurs in nature and is known to be glycogenic.

260

HO.CH——CH₂   proline   HO.CH——CH₂   +H₂O   HO.CH——CH₂   →
| | oxidase? | | −2H | | transamination
CH₂ CH.COOH   CHO CH.COOH   COOH CH.COOH
  \N/ | |
   H   NH₂   NH₂

semialdehyde   *γ-hydroxyglutamate*

HO.CH——CH₂   CHO   CH₃
| | → | + CO
COOH C.COOH   COOH |
       ‖   COOH
       O

α-keto-γ-hydroxyglutarate   *glyoxal*   *pyruvic acid*

Although hydroxyproline arises from proline, as it does in all probability, there seems to be no return route; perhaps the γ-hydroxyl group is an insuperable obstacle.

BIOSYNTHESIS OF AMINO-ACIDS

We have already pointed out in these pages, and shall do so again later on, that it is easy enough to account for the synthesis of a number of the few amino-acids that are 'non-essential' i.e. can be synthesized as far as animals are concerned. For example, pyruvate, oxaloacetate and α-ketoglutarate all arise in the course of metabolism generally and require only L-glutamate dehydrogenase and the appropriate aminotransferases to produce alanine, aspartate and glutamate respectively. The formation of several others can also be accounted for in fairly simple terms. But the synthetic ability of animals is far inferior to that of green plants and micro-organisms, for green plants and free-living yeasts and bacteria alike can synthesize each and every one of the amino-acids that ordinarily enter into the structure of proteins; the plants, in fact, produce a large number of other rather exotic amino-acids which play no part in protein structure and as yet have no known function.

We know relatively little about the synthesis of the essential amino-acids as far as the plants are concerned but in bacteria, on the other hand, a great deal more is known. The biosynthetic pathways in these unicellular organisms are seldom simple; usually they are very devious and complicated and we shall have occasion to mention only one or two of these pathways later on in any detail, but in another context. However it may be worth while to look at Fig. 72 on p. 312 at this point.

This book is not concerned particularly with bacterial metabolism and so a detailed study of the pathways followed in amino-acid synthesis by bacteria and the like must be considered to be outside our present terms of reference, but in order that the reader may at least be able to see what the raw materials are for amino-acid synthesis in general, these have been collected together and appear in Table 20. Some will already be familiar, especially in the case of those amino-acids that are 'non-essential' (marked with an asterisk. Where that symbol is in brackets the case is exceptional.

       18-2

Table 20. *Biosynthesis of the amino-acids*

Amino-acid	Atom	Source
Glycine(*)		? Serine
Alanine*	All carbon	Pyruvate
Serine*	All carbon	Probably $\textcircled{P}$-glycerate; glycine
Threonine	All carbon	Aspartate (via semialdehyde and homoserine)
Cysteine* (replaceable)	All carbon	Serine
	—SH	Sulphide, thiosulphate
Methionine	C-1 to C-4	Aspartate (via semi-aldehyde and homoserine)
	—SH	Cysteine
	—CH$_3$	Methyl-THF
Valine	C-1, C-2, C-4	Pyruvate
	C-3, C-4′	Pyruvate-2, pyruvate-3
Leucine	C-3, C-4, C-5, C-5′	Pyruvate
	C-1, C-2	Acetate
Isoleucine	C-1, C-2, C-5, C-6	Threonine
	C-3, C-4	Pyruvate-2, pyruvate-3
Aspartate*	All carbon	Oxaloacetate
Glutamate*	All carbon	α-Ketoglutarate
Arginine(*)	C-1 to C-5	Glutamate to ornithine
	Guanidino H$_2$H—C$\diagdown$	CO$_2$, NH$_3$ (via carbamyl phosphate)
	Guanidino HN=	Aspartate
Lysine	All carbon	*meso*-Diaminopimelate
meso-Diaminopimelate	C-1 to C-4	Aspartate (via siemaldehyde)
	C-5 to C-7	Pyruvate
Phenylalanine	Same as tyrosine	As for tyrosine
Tyrosine (replaceable)	Side chain	$\textcircled{P}$-enolpyruvate
	Ring	Erythrose-4-phosphate + $\textcircled{P}$-enolpyruvate
Histidine	Imidazole N-1, C-2	N-1, C-2 of adenylic acid
	Imidazole N-3	Amide of glutamine
	Remaining C	Ribose-5-phosphate
Tryptophan	Side chain	Serine
	Benzene ring	As for tyrosine
	C-2, C-3 of indole	C-1, C-2 or ribose-5-phosphate
	N of indole	Amide of glutamine
Proline*	All carbon	Glutamate
Hydroxyproline*	All carbon	Glutamate (via proline)

Note. The numbering of the C atoms begins with the α-carboxyl group in every case, e.g.

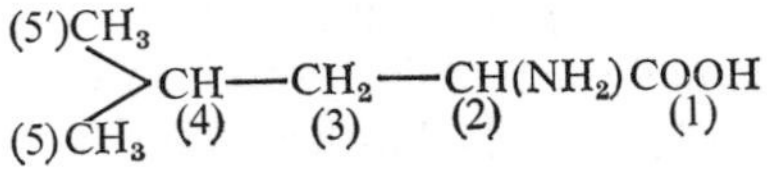

$$\begin{array}{c} (5')CH_3 \\ \diagdown \\ (5)CH_3 \end{array} CH - CH_2 - CH(NH_2)COOH$$
$$\qquad\quad (4) \quad\;\; (3) \qquad (2) \qquad (1)$$

(leucine)

262

13

EXCRETORY METABOLISM OF PROTEINS AND AMINO-ACIDS

NATURE OF THE NITROGENOUS END-PRODUCTS

The great bulk of all the nitrogen entering a typical animal arises from the α-amino-nitrogen of its food proteins. In mammals, at any rate, this is split off by deamination or transdeamination in the form of ammonia. Small quantities of nitrogen also arise from other sources, from the deamination of aminopurines and aminopyrimidines for example, but the great mass originates in the food proteins.

The tissues have little capacity to store proteins or amino-acids as such, but considerable amounts of nitrogen can undoubtedly be stored in animal tissues in the form of the amide-N of glutamine. The storage capacity in animals is, however, small compared with the average daily turnover of protein and amino-acid nitrogen. In plants, which have no excretory apparatus, larger quantities of nitrogen can be stored, either in the form of asparagine or glutamine or both (p. 231), according to the species, but in animals, which possess efficient excretory machinery, the superfluous nitrogen is excreted and the excreta of animals, especially the urine, contain a variety of nitrogenous substances of varying degrees of complexity.

Only among invertebrates do we find significant amounts of amino-acid nitrogen being excreted as such, and in certain invertebrate groups as much as 20–30% of the total nitrogen ingested may be excreted in the form of unchanged *amino-acids*. Whether this is due simply to leakage of amino-acids from the body fluids of these animals, or whether it indicates some sort of metabolic disability we do not know. In marine invertebrates, at any rate, it is certain that the surface membranes are permeable to water, to ions and to small molecules, and it is therefore possible that amino-acid molecules may be lost by diffusion to some extent.

Up to the present we have dealt mainly with the nitrogenous metabolism of mammals, chiefly because so much more is known about them than about any other group of animals. But there is reason to think that animals of every kind possess digestive enzymes capable of dismantling their food proteins completely to yield the component amino-acids, which can then be reassembled to produce the appropriate species-specific proteins. Although an excretion of unchanged amino-acids is observed among many invertebrates, the greater part of the ingested nitrogen is excreted in the form of ammonia even among these animals, especially if they are aquatic. It is probable that

263

all animals deaminate or transdeaminate at any rate the greater part of their incoming amino-acids with production of ammonia. Whether the deaminating machinery is always the same or even of the same general kind we do not at present know, but such evidence as there is points to a large-scale production of ammonia in all animals.

Now ammonia is a very toxic substance. Just how toxic it is can be appreciated if we consider some experiments carried out by Sumner, who injected purified urease into rabbits. The blood of the rabbit contains a small amount of urea and, from this urea, ammonia is formed by the hydrolytic action of the enzyme. Sumner's animals died as soon as the concentration of ammonia in the blood rose to about 5 mg. per 100 ml., i.e. about 1 part in 20,000, a very high order of toxicity indeed. Death occurred before any change in the pH of the blood could be detected, and it is therefore probable that the toxicity of ammonia is due to some specific property of the ammonium ion rather than to the basicity of ammonium hydroxide. Although foreign proteins often are toxic it is extremely improbable that death was due to any toxic properties of the enzyme itself, for urease was also injected into birds, the blood of which does not normally contain urea, and in this case the animals were unharmed; but fatal results followed the injection of urease together with urea.

If we examine the excreta of many different kinds of animals, representing as many different phyla and classes as possible, we find that among the nitrogenous substances present, some one compound always predominates. Over and above the traces of assorted odds and ends such as creatine, purines, betaines and the like, we find either *ammonia, urea or uric acid* accounting as a rule for two-thirds or more of the total nitrogen excreted. In a few special cases, some compound other than these predominates, but such cases are rare and, in fact, animals as a whole may be divided rather sharply into three groups according as their main nitrogenous excretory product is ammonia, urea or uric acid. These three groups are respectively said to be *ammoniotelic, ureotelic* and *uricotelic*. This raises several important problems. First we must inquire why some animals are content to excrete their waste ammonia unchanged, and why others convert ammonia, the primary product of deamination, into secondary products in the form of urea and uric acid. Then we must ask why it is that, among animals that do elaborate these secondary products, some produce urea and others uric acid. Finally, we have to inquire into the mechanisms whereby these ultimate end-products are synthesized.

The nature of the predominant end-product in any particular case seems to be conditioned by the nature of the habitual environment of the particular organism, and the known facts are best explained on the supposition that *the conversion of ammonia to other products is an indispensable adaptation to limitation of the availability of water.*

If we consider the invertebrates first of all it may be said at once that they

264

fall into a very large group of ammonioteles on the one hand, and a much smaller group of uricoteles on the other. Aquatic invertebrates, almost without exception, are ammoniotelic. Ureotelism seems not to have been much developed as an excretory device by members of the invertebrate phyla, while uricotelism is found only among terrestrial representatives of groups which, like the insects and the gastropod snails, have succeeded in colonizing the dry land. Animals living in water have at their disposal a comparatively vast reservoir into which they can discharge waste ammonia, a relatively diffusible substance, without running any grave risk of being poisoned by their own excrement. Terrestrial invertebrates, on the other hand, are often hard put to it to find enough water for their essential needs, and the impossibility of disposing of ammonia fast enough to avoid toxaemia is overcome by the biological conversion of ammonia, a very soluble and highly poisonous material, into the insoluble and relatively innocuous uric acid. Terrestrial woodlice, however, appear to be ammoniotelic, and in their case adaptation to terrestrial existence seems to have been achieved, not by uricotelism, but by an over-all reduction in protein metabolism. The daily turnover of nitrogen here is of the order of only 10 % of that of related marine and fresh-water species, and this suppression of protein metabolism seems to be the simplest and probably the most primitive device for combating the dangers of ammonaemia.

The general picture appears particularly clearly among the vertebrates. Taking the fishes first, it must be remembered that they fall into two main classes, the teleosts, or bony fishes, and the elasmobranchs, or cartilaginous fishes. Each of these classes is well represented in fresh and in salt water alike. To appreciate their position in the matter of water supply must involve a short digression.

Invertebrates. The lives of aquatic animals are complicated by a factor which terrestrial creatures like ourselves have little cause to appreciate. The fact that an animal is aquatic does not necessarily mean that it enjoys an unlimited supply of water. Among *marine invertebrates* the membranes bounding the body surface are, as a rule, more or less permeable to water and to small molecules. Ammonia formed in the body of such an organism can therefore escape comparatively readily by diffusion into the external environment, especially if the organism or its immediate environment is mobile. In *fresh-water invertebrates*, however, the boundary membranes are much less permeable: indeed, the main part of the body surface is impermeable to salts, and not very permeable even to water. This impermeability is important because most cells and tissues can only survive in the presence of considerably higher concentrations of salts than are present in the surrounding fresh water, and surface impermeability is one of several devices that serve to prevent leakage of salts out of the animal. But even so, the animal not only lives in water, it *breathes* in water, and this means that, in certain organs specialized for the purposes of respiration, the animal's blood comes into very close proximity to the surrounding water. In these respiratory organs, oxygen is taken into the blood and carbon dioxide eliminated, and the membranes of these organs have necessarily to be freely permeable to these dissolved gases. It seems that the necessary degree of permeability to dissolved gases is inseparable from an appreciable degree of permeability to water. In the respiratory organs of a fresh-water animal, then, we find membranes that are permeable to water, though impermeable to salts; they are, in fact, approximately semipermeable. On the outer side of these membranes there is fresh water, which is virtually free of dissolved salts, and on the other lies the animal's blood, which contains on the average about 1 % of dissolved salts. For this reason there is a considerable osmotic force tending to drive water into the animal from outside. The entry of this water leads to dilution of the salts of the blood, but these animals possess elaborate salt-absorbing and excretory organs which enable them to turn out the unwanted water, while maintaining at the same time a constant internal salinity. A fresh-water invertebrate may therefore be pictured as having a constant, osmotically-driven current of water passing through its body. Any ammonia formed in the cells and tissues can diffuse into the blood of such an animal and be carried away with the superfluous water in the form of a copious but very dilute urine.

Teleosts. The position is substantially the same for fresh-water teleosts. Ammonia formed in the tissues escapes rapidly and readily by way of the urinary system, and here, as in aquatic invertebrates, there is no danger of toxaemia due to the accumulation of ammonia. For *marine teleosts*, however, the position is considerably more difficult. The gill membranes and the mucuous membranes of the mouth are appreciably permeable to water, as they are in the fresh-water forms. But sea water contains about 3 % of salts as against the 1 % or thereabouts present in the blood, and the osmotic flow of water in this case is therefore *away* from the fish, instead of *towards* it. Marine teleosts, therefore, although they inhabit a watery medium, are nevertheless poorly supplied with water. They lose water constantly to their environment and are liable to die of desiccation, unlike their fresh-water relatives, whose lives are constantly imperilled by the imminent threat of flooding.

The nitrogenous excretion of marine teleosts has been investigated in a very ingenious experiment devised by Homer Smith. A wooden box is divided into two compartments by a watertight rubber dam pierced by a hole large enough to fit closely round the belly of a fish. The animal is placed in the apparatus, which is filled with sea water, in such a way that the head and the gills are accommodated in one compartment and the tail and the excretory aperture in the other. After a suitable time, samples of water from either compartment are withdrawn and analysed for nitrogenous compounds, and it then appears that 80–90 % of the total nitrogen excreted is found in the forward compartment and must therefore have been excreted by way of the gills, only a small part of the whole being evacuated by way of the kidneys. About two-thirds of all the nitrogenous material excreted consists of ammonia, indicating that, although the fish has a comparatively poor water supply, it can, nevertheless, dispose of most of its ammonia by diffusion across the gill membranes, without previously converting it into any less noxious nitrogenous compound. The remaining third is not present in the form of urea, nor yet as uric acid; indeed, it defied identification for some time, but turned out in the end to be *trimethylamine oxide* $(CH_3)_3N \rightarrow O$. This is a practically neutral, soluble and innocuous material.

There is no record of the occurrence of significant amounts of trimethylamine oxide in the tissues or in the excreta of fresh-water teleosts, though its presence in the tissues and excreta of marine forms has been abundantly confirmed. The possibility that it represents a detoxicated form of ammonia has therefore to be considered.

Elasmobranchs. The elasmobranch fishes—dog-fishes, sharks, rays, skates and chimeroids—present considerably more complex problems. Marine elasmobranchs produce and retain within their bodies large amounts of urea. Retention of urea in the blood and tissue fluids of these fishes is possible because the gills are impermeable to urea, while the kidney possesses a specialized mechanism that can control the loss of urea from the body by reabsorption of urea from the renal tubules. Enough is always retained to keep up a concentration of 2–2·5% of urea in the blood; over and above this concentration, urea is excreted, and the elasmobranch fishes are, in fact, ureotelic. The presence of so much urea in the blood raises the total osmotic pressure of the blood to a level somewhat higher than that of the surrounding sea water, and these fishes therefore escape the constant loss of water which threatens the existence of the marine teleosts. Instead of losing water to their environment, they constantly receive an osmotically driven supply of water from the sea. By resorting to ureotelism, therefore, marine elasmobranchs are not only protected against toxaemia due to ammonia but, by retaining some of the urea they produce, now find themselves in a very favourable position as regards water supply.

Like marine teleosts, the *marine elasmobranchs* excrete a part of their waste nitrogen in the form of trimethylamine oxide. This suggests that they have at some time in the past experienced the same osmotic difficulties as confront the marine teleosts of the present day, but that they faced them by making urea, which is even less toxic than trimethylamine oxide, and, by retaining enough of it in their tissues, turned the osmotic gradient to their advantage instead of their detriment.

The *fresh-water elasmobranchs* are believed to be descended from their marine cousins, which they resemble in being ureotelic, although in their case the amount of urea retained is only of the order of 1%. Thus, even among the fishes, an essentially aquatic group, we find ureotelism already well developed.

No discussion of the fishes would be complete without some mention of the air-breathing Dipnoi, or *lung-fishes*. These creatures inhabit swamps and rivers in tropical regions. During the hot season the water dries up, and the lung-fishes shut themselves up in cocoon-like structures in the mud to wait inanimate until the rains come. As long as water is available, these fishes behave like fresh-water teleosts and are essentially ammoniotelic. But during the period while they lie dormant and cut off from the water, they switch over to ureotelism and, when the rains come and the rivers fill again, almost their first act on emerging is to excrete a mass of urea that has accumulated during their aestivation. This phenomenon is a particularly interesting one, since it constitutes a test case of the validity of our general hypothesis—that the detoxication of ammonia is essentially an adaptation to restriction of water supply.

Going on now to the *Amphibia*, the frogs, toads, newts and the rest, we are in the company of animals most of which are able to spend longer or shorter periods away from the water. We should expect, in the terms of our hypothesis, that no animal could live long away from water without exposing itself to the hazards of ammonia-poisoning and, indeed, that the colonization of dry land could hardly have been begun until some mechanism had been evolved by means of which ammonia could be detoxicated. The Amphibia would be expected, then, to be either ureotelic or uricotelic. We can get some very interesting evidence here by studying the humble tadpole. Tadpoles are aquatic and ammoniotelic. Later, the tadpole gradually undergoes the metamorphosis that changes it from a wholly aquatic animal into a true amphibian and, at the same time precisely, it also undergoes a gradual chemical metamorphosis and abandons ammoniotelism in favour of ureotelism. The adult frog, in common with the adult forms of most other Amphibia, is ureotelic, and it seems as though, in the course of its development, it recapitulates some of the essential features of its evolutionary past.

There are however among the amphibia a number of species that have secondarily returned to fresh water, which they never leave. One of these, *Xenopus laevis*, has been studied in considerable detail. So long as it is kept in water it is essentially ammoniotelic, but still produces a good deal of urea as evidence of its evolutionary ancestry. Like the lung-fishes, *Xenopus* can aestivate during the dry season, and specimens kept out of water and in only a moist atmosphere cut down their production of ammonia and manufacture more urea instead. In this respect, therefore, their chemical behaviour resembles that of the lung-fish.

The rest of the vertebrates are generally considered as having evolved from some primitive kind of amphibian stock which, we may reasonably suppose, must at one time have been ureotelic. Leading from the Amphibia we find two main, diverging lines of evolution, one leading to the mammals and the other to the reptiles and the birds. There exists among the reptiles one ancient group, the Chelonia (tortoises and turtles) which are of rather particular evolutionary interest as a transitional group, at any rate from the chemical

point of view. Probably the original chelonian stock was terrestrial or am-
phibious, but today we find wholly aquatic, semi-aquatic and wholly terrestrial
species. Table 21 presents some data relating to the nitrogen excretion of some
of these animals. Those which are *terrestrial* today but favour marshy surroundings are essentially ureotelic; little ammonia or uric acid is produced.
In *wholly aquatic* species a significant degree of ureotelism is still retained—
evidence again of evolutionary ancestry—but the ratio of urea-N to ammonia-N
is much lower, just as it is in *Xenopus*, suggesting that these forms tend to
revert towards an ammoniotelic habit. But in the *desert-living and wholly
terrestrial forms* we find that, although urea formation still persists, though
on a much reduced scale, the bulk of the total N is now excreted in the form
of uric acid.

Table 21. *Nitrogen excretion of some chelonian reptiles*
(*Averages after* V. Moyle)

| | % of total N as | | |
Habitat	Ammonia	Urea	Uric acid
Wholly aquatic	20–25	20–25	5
Semi-aquatic	6	40–60	5
Wholly terrestrial:			
Hygrophilous	6	30	7
Xerophilous	5	10–20	50–60

Other dry-living reptiles, the Sauria (snakes and lizards), together with
the birds, have already abandoned ureotelism in favour of uricotelism. The
mammals on the other hand have continued in the amphibian manner and
clung to the more primitive ureotelic habit.

The case of the Chelonia cries out for further study, but the position is
much clearer in the other great ureotelic group, the *mammals*. There still
remain a few egg-laying mammals, e.g. *Echidna*, whose eggs are incubated
always in wet situations. Not much is known about the water relationships of
these eggs, but it has nevertheless been established that *Echidna* is ureotelic.
Little is known about the marsupials, but the rest of the mammals undergo
embryonic development in intimate contact with the maternal circulation.
Food materials travel from the maternal blood stream across the placenta to
the embryo, and waste products can likewise travel back across the placental
barrier to be excreted by the maternal kidneys. The mammalian embryo, with
the entire water resources and the excretory apparatus of the maternal organism at its disposal, has no need to do otherwise than remain ureotelic and
excrete its waste nitrogen by proxy.

Conditions are very different in the eggs of certain tortoises and all *snakes,
lizards and birds*, and Joseph Needham was the first to point out the impor-

269

tance of the conditions of embryonic as compared with adult life. These eggs are laid with a supply of water barely sufficient to see them through development, and no more is to be had, apart from metabolic water formed by the oxidation of food reserves as development proceeds. These eggs are surrounded by tough membranes or hard shells which are practically impermeable to water. In such a system the production of ammonia could be nothing short of disastrous. Urea would be a more suitable end-product if only because it is relatively harmless, but apart from the elasmobranch fishes, no organisms are known that can stand up to more than a very mild uraemia without more or less serious disturbances of normal physiological function. Needham has calculated that, if the waste nitrogen actually produced during the embryonic development of the chick was converted into and retained as urea, the resulting uraemia, by human standards, would be sufficient to give the embryo a bad headache at the very least. 'In which case', as Needham says, 'natural selection would hardly have preserved it for our entertainment.' Embryos which develop in these closed-box, or 'cleidoic', eggs had their problems solved by the conversion of waste ammonia, not into urea, but into uric acid, and the habit of uricotelism which they acquire during embryonic existence persists into, and throughout, their adult life. Whereas urea is a very soluble compound, the excretion of which requires a comparatively liberal supply of water, uric acid, which is almost equally innocuous, is exceedingly insoluble and can simply be dumped in the solid form. It is carried away from the embryo proper and deposited in a little membranous bag, the allantois, the contents of which include solid nodules of uric acid at the end of development.

To summarize, we may make the following statements. Among invertebrates, the ammoniotelic type of metabolism is found in aquatic animals; ureotelism appears to be little employed, while uricotelism is confined to organisms that have become adapted to life under terrestrial conditions. Among vertebrates, ammoniotelism is confined to animals that are entirely aquatic and, even among these, ureotelism and perhaps trimethylamine oxide formation have been exploited by certain fishes which, though aquatic, experience considerable shortage of water. With the conquest of the land, ureotelism appears to have been generally adopted and is still found today among the Amphibia. It has been retained by the mammals, including such aquatic mammals as whales, seals and porpoises, whose embryos have the water of the maternal blood-stream at their disposal. The dry-living reptiles, together with the birds, have abandoned ureotelism in favour of uricotelism, a change which is associated with embryonic development under the conditions of acute water shortage implied by the cleidoicity of the egg. Thus the detoxication of ammonia by conversion to urea or uric acid appears in every case to be intimately associated with limitations of water supply. A tabulated form of summary is given in Table 22.

Table 22. *Nitrogen excretion of vertebrates in relation to water supply*

Group	Environment	Water supply	$(CH_3)_3N \to O$	NH_3	Urea	Uric acid
Pisces:						
Teleostei	FW	Abundant	−	+	−	−
	SW	Poor	+	+	−	−
Elasmobranchii	FW	Abundant	−	−	+	−
	SW	Good	+	−	+	−
Dipnoi	FW	Abundant	−	+	−	−
	T*	None	−	−	+	−
Amphibia:						
Urodela	FW	Abundant	−	+	−	−
Anura: Tadpole	FW	Abundant	−	+	−	−
Frog	FW/T	Good/poor	−	−	+	−
Reptilia:						
Chelonia	FW	Abundant	−	+	+	−
	FW/T	Good/poor	−	−	+	−
	T	Poor	−	−	(+)	+
Sauria	T	Poor	−	−	−	+
Avas	T	Poor	−	−	−	+
Mammalia	T	Poor	−	−	+	−

FW = fresh water; SW = sea water; T = terrestrial.
* During aestivation.

SYNTHESIS OF THE END-PRODUCTS: UREA

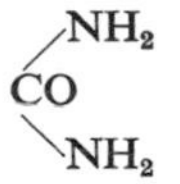

Most of the early work on the biological synthesis of urea was, not unnaturally, carried out on mammalian materials. It has long been known that hepatectomy in the dog leads to cessation of urea production, an observation that points to the liver as the sole seat of urea synthesis in the mammalian organism. Work with perfused mammalian liver confirmed this conclusion, for a synthetic formation of urea from added ammonia was readily demonstrated. Little was known about the mechanism of the synthesis for many years, although several alternative theories were expounded and exploded.

The existence of a urea-producing enzyme in mammalian liver was suspected at the end of the last century, following the discovery that, if liver tissue is allowed to autolyse, urea is produced. This urea originates in arginine, set free by autolysis of the tissue proteins, under the influence of a hydrolytic enzyme which was called *arginase*, and which catalyses the following reaction:

271

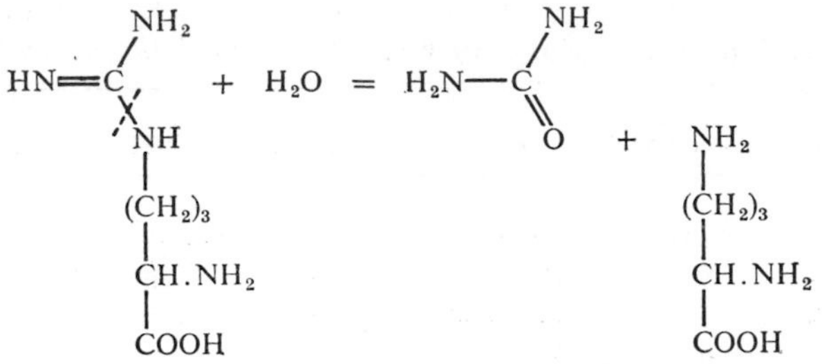

$$\underset{\substack{\text{arginine}}}{\overset{\displaystyle NH_2}{HN\!=\!\underset{\substack{NH \\ | \\ (CH_2)_3 \\ | \\ CH.NH_2 \\ | \\ COOH}}{C}}} + H_2O = \underset{\substack{\text{urea}}}{\overset{\displaystyle NH_2}{H_2N\!-\!\underset{O}{C}}} + \underset{\substack{\text{ornithine}}}{\overset{\displaystyle NH_2}{\underset{\substack{(CH_2)_3 \\ | \\ CH.NH_2 \\ | \\ COOH}}{}}}$$

It was Clementi who first drew attention to the striking fact that, while *arginase occurs in high concentrations in the liver of ureotelic animals*, it is present in traces at most in the liver of those which are uricotelic (Table 23). It was already clear that arginase could be held responsible for the production of some of the urea excreted by mammals, inasmuch as arginine occurs in considerable amounts in most proteins. Arginine arising from the food proteins could therefore account for some, though not by any means for all of the urea formed.

Table 23. *Distribution of arginase in liver and kidney of vertebrates*
(*Data according to* Clementi)

Class	Species	Liver	Kidney
Mammalia	Dog	+	−
	Ox	+	−
	Pig	+	−
	Guinea-pig	+	−
	Rat	+	−
	Monkey	+	−
	Man	+	−
Aves	*Gallus domesticus*	−	+
	Columba livia	−	+
	Turtur turtur	−	+
	Fringuilla cloris	−	+
Reptilia:			
Chelonia	*Emys europae*	+	−
Sauria	*Lacerta agilis*	−	−
	Anguis fragilis	−	−
	Coronella austriaca	−	−
Amphibia	*Rana esculenta*	+	−
	Rana temporia	+	−
Pisces:			
Elasmobranchii	*Torpedo ocellata*	+	.
	Raia clavata	+	.
Teleostei	*Perca fluviatilis*	+	−
	Abramis brama	+	−
	Barbus fluviatilis	+	−

Later, Krebs, using the tissue-slice technique, showed that surviving slices of rat liver can convert added ammonia into urea. Moreover, when amino-acids are added to the slices, the ammonia set free by deamination undergoes approximately quantitative conversion into urea. Different amino-acids are deaminated at different rates and the rate of urea formation from amino-acids therefore varies from one to another. In one case, however, the rate of synthesis exceeded all expectation. It was to be anticipated that when *orni-thine* was used in the presence of added ammonia, not more than one molecule of urea could be formed from each molecule of ornithine itself, one N-atom arising from the α- and a second from the δ-amino-group. In fact, however, ten or more molecules of urea were formed for each molecule of ornithine added. This suggested that ornithine must act catalytically in the synthesis of urea from added ammonia.

Following up this clue, the effect of arginine was also investigated, and again it was found that the rate, as well as the amount of urea synthesized from added ammonia, was far greater than could be accounted for except in terms of catalysis. Krebs therefore concluded that *ornithine* must react with ammonia and carbon dioxide to form *arginine* which, under the influence of the liver *arginase*, could give rise to urea, ornithine being regenerated and used over and over again as a carrier.

The formation of arginine from ornithine must evidently be a complex process, and Krebs sought to discover possible intermediate compounds. It was already known that there exists in the water-melon, *Citrullus vulgaris*, a substance which, chemically speaking, lies midway between ornithine and arginine. This substance, known as *citrulline*, has been tested and evidence has been obtained that, like ornithine and arginine, citrulline can act catalytic-ally in the synthesis of urea from added ammonia, though the results obtained with citrulline are usually less dramatic.

Krebs therefore proposed the so-called 'ornithine cycle', the outline of which is illustrated in Fig. 52. This scheme explains the catalytic behaviour of ornithine, citrulline and arginine and, involving as it does the participation of arginase as an integral part of the system, provided for the first time a rational basis for the empirical rule enunciated by Clementi, that arginase is always pre-sent in the liver of ureotelic animals but absent from those that are uricotelic.

Kreb's original work on rat liver has been extended to the liver of other ureotelic animals, and it has now been shown that ornithine and citrulline act catalytically in the synthesis of urea by the livers of other mammals, of a tor-toise, of frogs and toads and also in marine elasmobranchs. Bird and snake livers, as would have been anticipated, form no urea when ammonia and orni-thine or citrulline are provided: uric acid is formed instead. Moreover, such small amounts of arginase as are present in the livers of these uricotelic animals appear to differ in general properties and specificity from the liver arginase of ureoteles.

273

The case of the fishes (Table 24) calls for some special comment. Teleostean fishes are characteristically ammoniotelic rather than ureotelic, although their livers contain arginase. Much of the rest of the ornithine cycle mechanism is lacking, however, and no urea is produced when ornithine is provided, to-

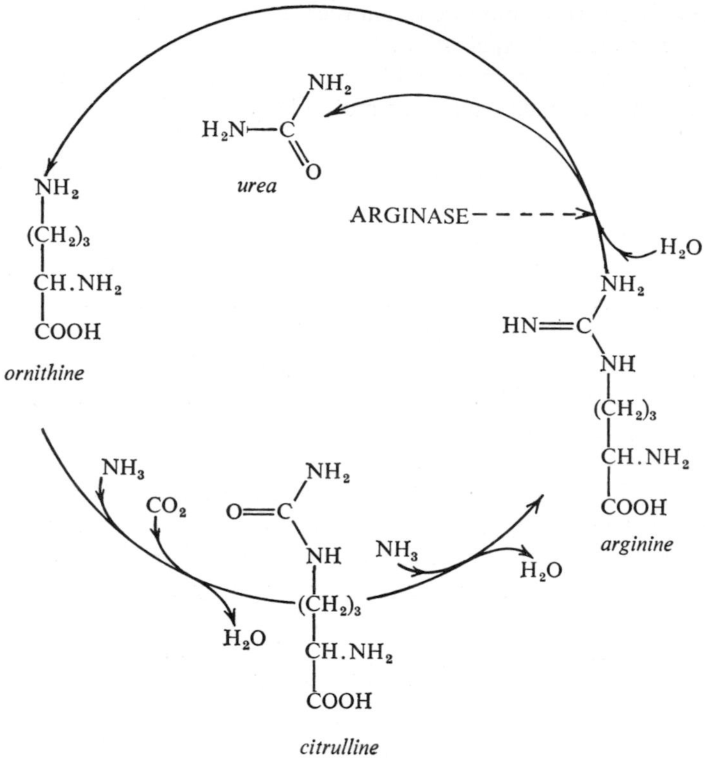

Fig. 52. The ornithine cycle.

Table 24. *Arginase contents of various tissues (arbitrary units)*
(*Data from* Hunter & Dauphinee)

Species	Liver	Kidney	Heart	Muscle
Cat	1280	2·7	.	.
Rabbit	369	0·9	.	.
Hen	0	.	.	.
Pigeon	0	18·3	.	.
Mud-turtle	14·2	0	.	.
Dog-fish	319	31	109	2·2
Herring	181	7	8·8	0·9
Other teleosts	8–110	1–5	.	.

gether with ammonia and lactate. The elasmobranchs, by contrast, are ureotelic and here, unlike other ureotelic organisms, arginase is present in considerable concentrations in practically every part of the body. The study of urea synthesis in elasmobranch tissues such as liver is difficult since, to ensure survival of the tissue, it is necessary to work in a medium which resembles elasmobranch blood, and the latter already contains a high proportion of urea. There is now, however, every reason to believe that all the reactions of the ornithine cycle go on in elasmobranchs. That arginase is present all over the elasmobranch body suggests that the synthesis of urea may not be wholly confined to the liver, as it is in other ureoteles, and it is probably significant that, whereas hepatectomy in the dog is followed by cessation of urea production, the hepatectomized dog-fish maintains its normal high level of uraemia.

As far as the invertebrates are concerned there is, as has been stated, little reason to believe that ureotelism has been much exploited. Arginase, however, is present in many such organisms, usually in traces, but occasionally in concentrations as high as or even higher than those in mammalian livers. One might suppose that here, as among the teleostean fishes, the *tour de force* necessary for the acquisition of the rest of the ornithine cycle has not been achieved, or, if perhaps it was present in the remote past, it has been lost over the course of evolutionary time. It is also conceivable that the earliest of all of the ancestral fishes were ureotelic, that ureotelism was retained by elasmobranchs and dipnoans, but lost in other groups. Much more needs to be done if this rather unorthodox notion is to be seriously entertained. However, specialization by loss is, after all, not an unknown phenomenon (p. 288).

The mechanisms that underlie the ornithine cycle have now been worked out in considerable detail. If homogenates of mammalian liver are incubated with ammonia, bicarbonate, ornithine, N-acetyl-glutamate and ATP, urea is slowly formed and citrulline accumulates. The second stage of the process, the conversion of citrulline to urea, requires Mg^{2+} or Mn^{2+} and, if these are added, urea is more rapidly formed and citrulline no longer accumulates.

Formation of citrulline from ornithine. This first stage in the synthesis of urea can be demonstrated in the presence of ammonia, carbon dioxide (as bicarbonate), ATP and ornithine itself in the presence of liver mitochondria. These substances react together in some at present unknown manner to produce *carbamoyl phosphate*, the latter reacting with ornithine to form citrulline (see Fig. 53), under the influence of an *ornithine carbamoyltransferase*. Catalytic amounts of N-acetyl-glutamate, which is demonstrably present in and has actually been isolated from mammalian liver, are also required, but the part played by this glutamate derivative is not known.

The synthesis of carbamoyl phosphate probably involves several steps and several enzymes and in the meantime the collective title of *carbamoyl phosphate synthetase* is applied to the whole catalytic system. The formation of carbamoyl phosphate requires the presence of mitochondrial enzymes, but

275

the subsequent reaction between carbamoyl phosphate and ornithine can be demonstrated in the presence of a soluble enzyme preparation obtained from liver mitochondria. (It is worthy of note that citrulline can also transfer its carbamoyl group to aspartic acid, yielding carbamoyl-aspartate. The latter plays a fundamental part in pyrimidine synthesis (p. 290).)

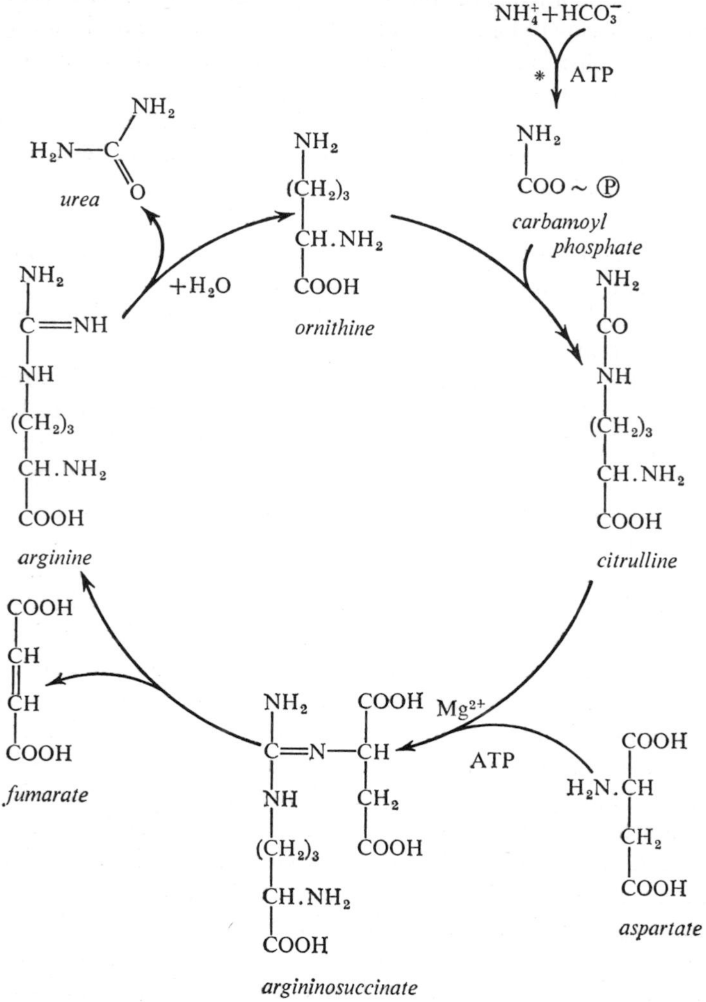

Fig. 53. Intermediate reactions in ureogenesis.
(*Traces of N-acetyl-glutamate are also required.)

Formation of arginine from citrulline. The conversion of citrulline to arginine can take place in extracts prepared from acetone liver powders and again requires ATP. Magnesium ions are also necessary. It has been established that citrulline reacts with *aspartate* to form a complex, *argininosuccinate*, which is then decomposed to give arginine and fumarate, and has been isolated. (It is worth mentioning that a few genetically abnormal children are known who excrete large quantities of argininosuccinate in their urine in spite of the fact that their urea output is apparently normal.)

Three enzymes are required here, namely *ornithine carbamoyltransferase, argininosuccinate synthetase* and *argininosuccinate lyase*. The reactions appear to be those shown in Fig. 53. The products of the reaction, arginine and fumarate, have been positively identified but it is not known what part ATP plays in the process.

The final stage, the fission of arginine by arginase, is well known. Evidently the synthesis of urea is a complex process and one, moreover, that calls for the expenditure of a good deal of energy (2 mols. ATP per mol.), so that the animal has to pay a fairly heavy price for the advantages it can derive from ureotelism. In passing it is interesting to note that carbamoyl phosphate is also produced by certain bacteria from which, in fact, it was first isolated. In these organisms only one, instead of two molecules of ATP are used in the formation of each molecule of urea.

In conclusion it should be pointed out that, while a virtually cast-iron case has been made out for the existence of this cyclical system it does not necessarily follow that alternative ureogenic systems may not also exist. But although such alternative mechanisms have frequently been postulated there is little evidence as to their nature or to their existence in reality.

SYNTHESIS OF THE END-PRODUCTS: TRIMETHYLAMINE, $(CH_3)_3N$, AND TRIMETHYLAMINE OXIDE, $(CH_3)_3N \rightarrow O$

These have been obtained from many animal sources, vertebrate and invertebrate alike. As a rule the oxide occurs in quantities that completely overshadow those of the free base and, in general, it seems that trimethylamine arises from its oxide through bacterial action. The characteristic odour of dead marine fishes, which incidentally is not observable in fresh-water species, is largely due to free trimethylamine formed by the action of putrefactive bacteria upon trimethylamine oxide present in the tissues. Perfectly fresh fish muscle contains traces at most of the free base but free trimethylamine has long been known as a trace constituent of mammalian urine, and it has also been detected in human menstrual blood.

Perhaps the most striking fact that has emerged in connexion with trimethylamine oxide is that, while it is present in a great variety of marine animals, it seems never to occur on a substantial scale among fresh-water organisms. Its function in the fishes has been studied a good deal. That it occurs only in marine species suggests that it might play a part analogous to that of urea in marine elasmobranchs in the regulation of the osmotic pressure of the blood, for it resembles urea in being nearly neutral and relatively innocuous. Elasmobranch bloods contain about 100–120 mM per kg., as compared with about 330–440 mM of urea. It accounts therefore for 20–25 % of that part of the total osmotic pressure that is not due to salts. Further, the concentration of trimethylamine oxide in the urine of marine elasmobranchs is only about one-tenth as great as that present in the blood, so that this substance, like urea, is actively retained by the elasmobranch kidney. Its apparently active reabsorption by the kidney adds weight to the idea that it plays a significant part in osmotic regulation in these fishes.

Among teleosts, however, matters are different. The trimethylamine oxide content of fresh cod muscle, for example, has been estimated at about 20 mM per kg., only about one-fifth of the amount present in elasmobranch muscle. Furthermore, there is no reason to believe that the substance is retained in this case, for it is one of the major nitrogenous constituents of the excreta of many marine teleosts. It is therefore less likely that trimethylamine oxide plays any significant osmotic role in marine teleosts; it is far more probable that it is essentially an excretory product but, while undoubtedly it is exogenous in part, it may possibly represent a detoxicated form of ammonia.

Feeding experiments have been carried out with young salmon to elucidate the origin of this compound. These fishes are euryhaline and were kept at first in fresh water on a diet of fresh, minced ox liver, when no trimethylamine oxide was produced. They were then transferred to sea water, still on the same diet, but still no trimethylamine oxide appeared until food containing preformed trimethylamine oxide was given in place of the ox liver. This would seem to indicate that this compound is wholly exogenous in origin, but the excretion in some cases of nearly 50 % of the total nitrogen in the form of trimethylamine oxide by some marine fishes suggests that this substance must have a synthetic origin in these fishes, to some extent at least. Probably a part originates as such in the food, for many marine organisms, including the small crustaceans which form an important article of diet for many fishes, contain substantial amounts of trimethylamine oxide, which is taken over by the feeder and excreted unchanged.

Another possible source is *glycine betaine*. This substance occurs abundantly in some animals and in many plants. It is known that if cows are fed on sugar-beet residues, which are a rather rich source of glycine betaine, trimethylamine oxide appears in the milk. But although betaine gives rise to trimethylamine oxide when administered to cows, animal tissues do not, in general, appear to be capable of converting betaine into trimethylamine oxide, though trimethylamine itself is oxidized by mammalian tissues to yield the oxide, and a trimethylamine oxidase has been discovered in fish muscle. One can do little more at present than guess at the extent to which such a conversion can take place in fishes, and in any case there remains to be explained the excretion of trimethylamine oxide by marine yet not by fresh-water fishes.

It is conceivable that trimethylamine itself might first be formed by the biological methylation of ammonia, and subsequently oxidized to the oxide in animal tissues; mechanisms for methylation are known (p. 110), and it is known too that trimethylamine can be oxidized to its oxide in the tissues of some animals.

SYNTHESIS OF THE END-PRODUCTS: URIC ACID

Uric acid is just as important an excretory product among uricotelic animals as is urea in those that are ureotelic. Hepatectomy in the dog is followed by cessation of urea production, and hepatectomy in the goose is similarly followed by cessation of the synthesis of uric acid. In both of these cases hepatectomy leads to the accumulation of ammonia and amino-acids in the blood, and it follows that, just as urea is synthesized from ammonia in the liver of the dog, so too is uric acid synthesized from ammonia in that of the bird. Early experiments with perfused goose liver confirmed this conclusion, and showed convincingly that the uric acid excreted by birds can be formed from ammonia.

The suggestion was made quite early on that uric acid might be synthesized by way of urea, but Krebs, using liver slices, was able to confirm the old observation that ammonia can be built up into uric acid by bird liver, but could find no indication that any synthesis takes place from urea. Again, if isotopically labelled urea is injected into birds, no isotope appears in the uric acid excreted.

Further knowledge was gained by taking advantage of the peculiar fact that, whereas the livers of most birds, e.g. the domestic hen and the goose, form uric acid from added ammonia, that of the pigeon does not yield uric acid but a precursor of some kind which, under the influence of an enzyme present in the kidney, undergoes conversion into uric acid itself. Krebs and his co-workers, working with liver slices, collected enough of the intermediary body for identification, and showed that it is hypoxanthine:

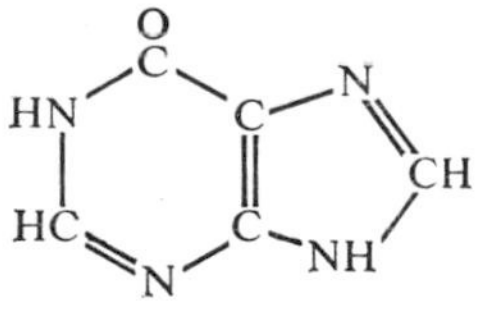

hypoxanthine

The reason for this interesting dislocation of uricogenesis in the pigeon is that, whereas the liver of most birds contains a xanthine dehydrogenase, that of the pigeon does not. In this particular bird, xanthine dehydrogenase is present in the kidney, and hypoxanthine formed in the liver is passed on to the kidney and there catalytically oxidized to uric acid, presumably by way of xanthine.

279

In passing it is interesting to notice here a case of evolutionary divergence. The hypoxanthine and xanthine oxidizing systems of birds and reptiles appear to be dehydrogenases and differ from the classical mammalian xanthine oxidase in that they are unable to use molecular oxygen as their hydrogen acceptor (cf. p. 136) although they can reduce methylene blue.

The administration to birds of isotopically labelled ammonia, glycine, formate and CO_2, followed by systematic degradation of the uric acid excreted, made it possible to specify the origins of the atoms which go to make up the purine ring. The results are summarized in Fig. 54:

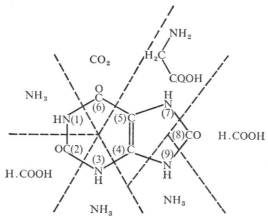

Fig. 54. Origin of the constituent atoms in purine synthesis.

The evidence provided by these experiments is borne out by work with homogenates and particle-free extracts of pigeon liver. All the evidence suggests, however, that the N atoms at (1), (3) and (9) are not directly introduced by molecules of ammonia as such but rather by transference from substances formed from ammonia such, for example, as glutamate or glutamine.

Important evidence about the intermediate steps was obtained with the aid of acetone powder extracts prepared from pigeon liver. Given ATP these preparations will synthesize hypoxanthine from ammonia, carbon dioxide, glycine and formate, but do so only if glutamine or glutamate is added, together with ribose-1-phosphate or a precursor of ribose-1-phosphate. The synthesis then proceeds according to the following overall equation:

$$3NH_3 + CO_2 + \text{glycine} + 2 \text{ formate} \longrightarrow \text{hypoxanthine.}$$

The requirement for glutamate or glutamine indicates that one or other of these substances is required to provide the N atoms at (1), (3) and (9), while that for ribose-1-phosphate suggests that synthesis proceeds through ribosidic intermediates. In experiments in which the formate used was labelled with ^{14}C, not only hypoxanthine but the corresponding nucleoside (inosine) and nucleotide (inosinic acid) also were isolated and all were shown to contain the label. It seems, therefore, that the equation can be expanded to read as follows:

$$3NH_3 + CO_2 + \text{glycine} + 2 \text{ formate} + \text{ribose-1-}\textcircled{P} \longrightarrow$$
$$\text{inosinic acid} \longrightarrow \text{inosine} \longrightarrow \text{hypoxanthine} + \text{ribose-1-}\textcircled{P}.$$

Thus all the evidence available to date indicates that the first-formed purine is hypoxanthine.

New light was thrown upon the problem with the discovery in certain bacteria of a hitherto unknown substance, 5-*amino-4-iminazolecarboxamide*:

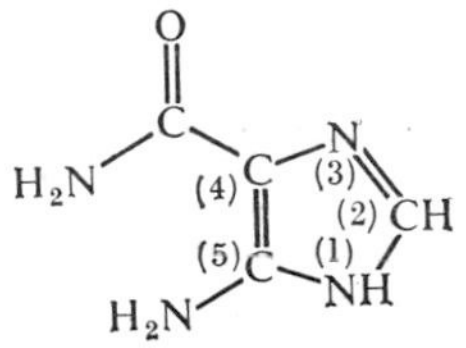

This compound, which accumulates in the medium of sulphonamide-sensitive bacteria grown in the presence of sulphanilamide, could, at any rate on paper, react with formaldehyde or formate to give hypoxanthine by ring closure.

Now it has long been known that sulphanilamide owes its bactericidal or bacteriostatic properties to its structural similarity to *p*-aminobenzoic acid, a similarity close enough to enable it to compete with *p*-aminobenzoic acid, which is itself an essential growth factor or vitamin for sulphonamide-sensitive bacteria.

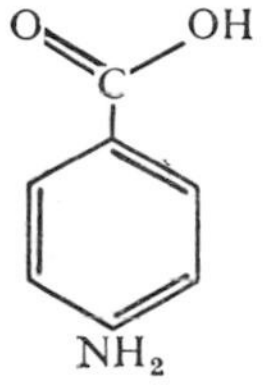

p aminobenzoic acid 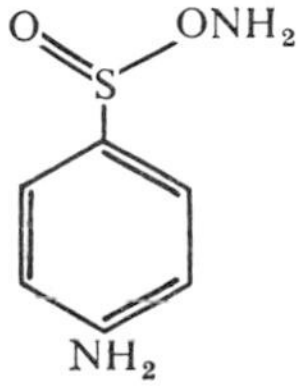   sulphanilamide

p-Aminobenzoic acid is required for the synthesis of *folic acid* (formula on p. 108), an essential vitamin for animals which, even if given *p*-aminobenzoic acid, cannot synthesize this rather elaborate compound for themselves. On the other hand, sulphonamide-sensitive bacteria, which need folic acid as much as animals do, can synthesize it if *p*-aminobenzoate is provided, but the presence of sulphanilamide checks the synthesis by competition with this essential raw material and so slows or stops the growth, multiplication and eventually the life of the organism.

In its reduced form, *tetrahydrofolic acid* (THF_4 or TFH_4), it can combine with formate, which arises in a variety of metabolic processes (pp. 108, 109 for example), and produces a formyl derivative under the influence of an ATP-dependent *formyl-tetrahydrofolate synthetase*:

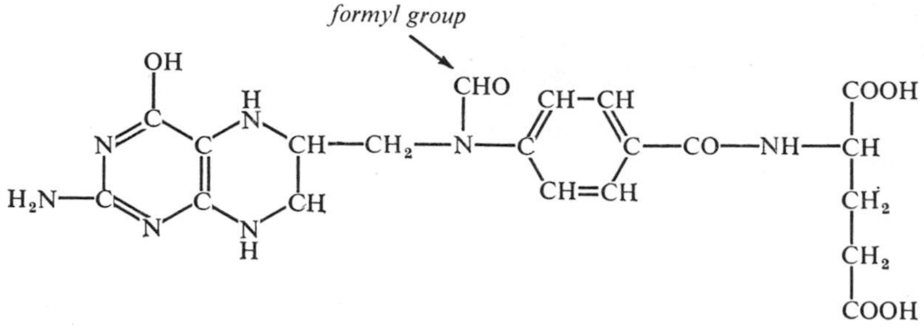

formyl group

OH
|
H_2N—C ... CHO ... COOH

formyl-tetrahydrofolic acid

Formyl-tetrahydrofolate acts biologically as an acceptor, carrier and donor of formyl groups and will in fact react with 5-amino-4-iminazolecarboxamide under the influence of a pigeon liver enzyme, thus providing the additional carbon atom required for completion of the second of the two purine rings:

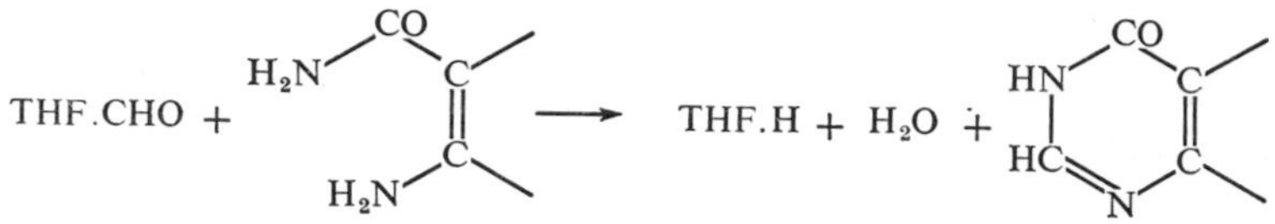

$$THF.CHO + \quad \longrightarrow \quad THF.H + H_2O + $$

(*hypoxanthine*)

This interesting compound is similarly involved in another earlier reaction in uricogenesis.

On administration to pigeons 5-amino-4-iminazolecarboxamide gives rise to hypoxanthine, the liver providing the necessary formyl group by way of THF_4, and when labelled with ^{14}C gives rise to labelled hypoxanthine if incubated with homogenized pigeon liver. Considerations involving the distribution of the label indicate, however, that the compound does not lead directly to hypoxanthine but that a *ribonucleotide derivative is formed as an intermediate*. This ribonucleotide then, and only then, reacts with formyl-THF_4 and produces the hypoxanthine ribonucleotide, inosinic acid.

Once the origin of the atoms of the purine ring had been determined it became possible to manufacture hypotheses, to test them experimentally, and so to build up step by step a scheme to account for purine formation from the known starting materials. A key discovery had already paved the way for these new advances. A reaction that takes place between ATP and ribose-5-phosphate, which can be produced from the 1-phosphate by phosphoglucomutase (see p. 295), leads to the formation of a pyrophosphate derivative of the 5-phosphate. The enzyme concerned is called *ribosephosphate pyrophosphokinase* and the product, usually abbreviated as PRPP, is *5-phosphoribityl-1-pyrophosphate*.

Starting from PRPP and glutamine, hypoxanthine formation has now been traced through a somewhat lengthy series of intermediate reactions. Enzymes capable of catalysing all these reactions have been found in pigeon liver and the various products isolated, though information about the cofactors and possible intermediates is not yet complete. All the postulated intermediates are convertible into uric acid in the pigeon and ATP is required at almost every one of the intermediates reactions.

(1) The initial reaction takes place between PRPP and the amide N of glutamine, (this and the ensuing reactions are summarized in Fig. 55):

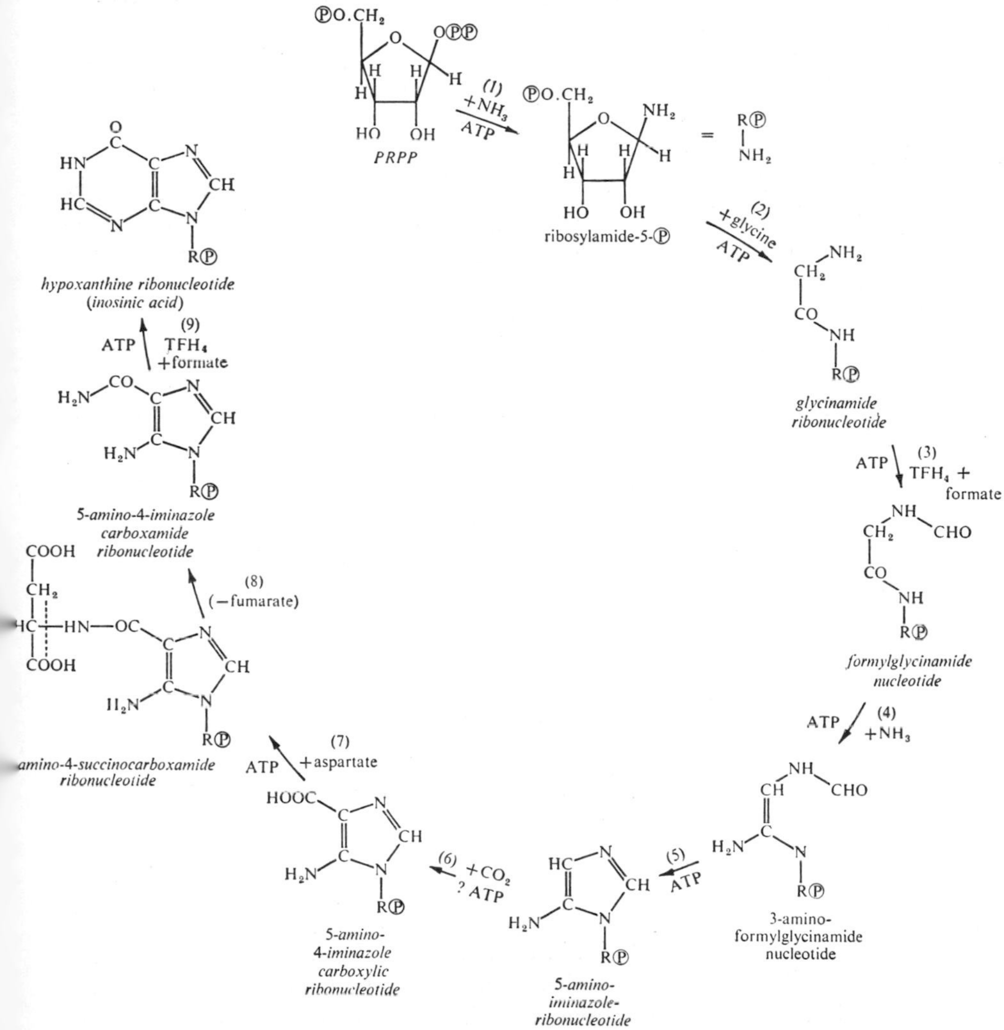

PRPP

ribosylamide-5-$\textcircled{P}$

hypoxanthine ribonucleotide
(inosinic acid)

(9) ATP, TFH$_4$ +formate

5-amino-4-iminazole carboxamide ribonucleotide

(8) (−fumarate)

amino-4-succinocarboxamide ribonucleotide

(7) ATP +aspartate

5-amino-4-iminazole carboxylic ribonucleotide

(6) +CO$_2$? ATP

5-amino-iminazole-ribonucleotide

(5) ATP

3-amino-formylglycinamide nucleotide

(4) ATP +NH$_3$

formylglycinamide nucleotide

(3) ATP, TFH$_4$ + formate

glycinamide ribonucleotide

(2) +glycine ATP

(1) +NH$_3$ ATP

Fig. 55. Intermediates in nucleotide synthesis (inosinic acid) (TFH$_4$ ≡ THF).

283

(2) The product, represented here as $℗R\text{-}NH_2$, now reacts first with glycine and then with formyltetrahydrofolate to give

(3) *Glycinamide-ribosyl-5-phosphate* (glycinamide ribonucleotide) followed by *formylglycinamide-ribosyl-5-phosphate* (formylglycinamide ribonucleotide).

(4) In the next step an amino-group is transferred, again from glutamine, followed by ring closure (5).

(6) A carbon dioxide molecule is now introduced yielding *5-amino-4-iminazole carboxylic acid ribonucleotide*.

(7) This product reacts with aspartate, forming *5-amino-4-succinocarboxamide ribonucleotide*.

(8) In the next reaction fumarate is split off by a specific lyase and the product is the ribonucleotide of *5-amino-4-iminazole carboxamide*, and the latter, by a further reaction with formate, in the form of formyltetrahydrofolate, produces *inosinic acid* (9).

One outstanding problem remains; that is the mechanism whereby CO_2 is introduced in reaction (6); conceivably this is 'active CO_2', i.e. the CO_2 of a biotin~CO_2 complex (cf. p. 118). Finally it should be noticed that since ATP is required at practically every step in the process of uricogenesis, uric acid is a very expensive commodity in terms of free energy.

These conclusions are summarized in Fig. 56 and are superficially similar to those in Fig. 54. However the numbers here indicate the order in which the various atoms are incorporated and correspond to the numbers of the equations outlined in the preceding paragraphs.

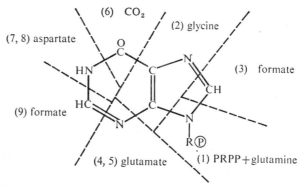

Fig. 56. Order of reactions in nucleotide synthesis (cf. Fig. 55 for details).

Two other important nucleotides, adenylic and guanylic acids, are formed from inosinic acid and give rise in their turn to ATP, which is of enormous importance throughout metabolism as a whole, and GTP, which has a part to play in a number of reactions, including the formation of adenylic acid itself. The processes leading to the formation of these other nucleotides are summarized in Fig. 57. Free hypoxanthine itself is not converted into either adenine or guanine but its nucleotide, inosinic acid, can be transformed into the corresponding nucleotides of adenine and guanine, *adenylic* and *guanylic acids* respectively, by soluble enzymes present in bone marrow. The pathways of these interconversions shown in the figure seem probable in view of present evidence. R. Ⓟ represents ribose-5-phosphate as before.

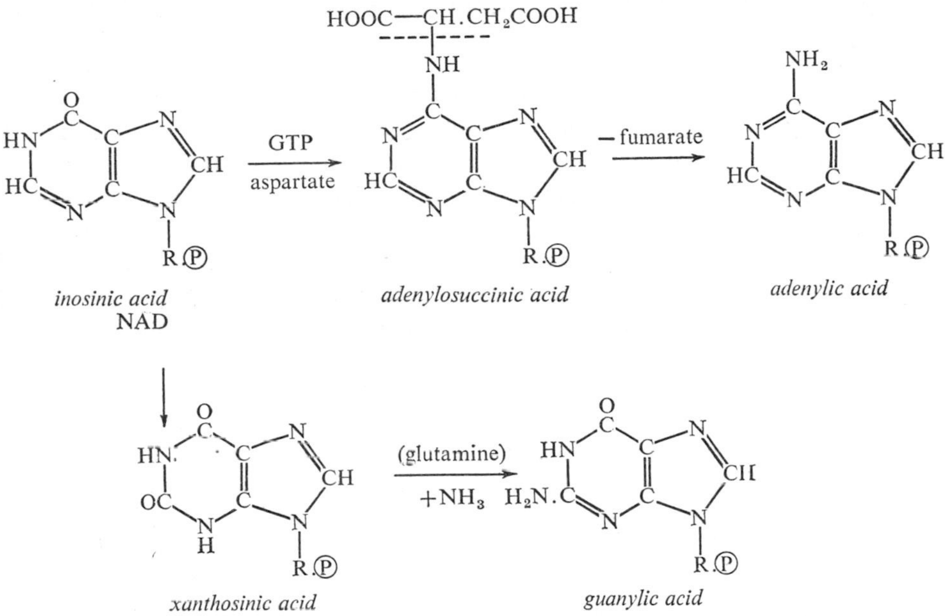

Fig. 57. Formation of adenylic and guanylic acids from inosinic acid.

All three of these important bases, hypoxanthine, adenine and guanine can give rise to uric acid in ureotelic and uricotelic animals alike and we shall go on presently to take up the question of purine metabolism.

Removal of the ribose-5′-phosphate can be effected by hydrolytic removal of the 5′-phosphate followed by a transribosylation reaction:

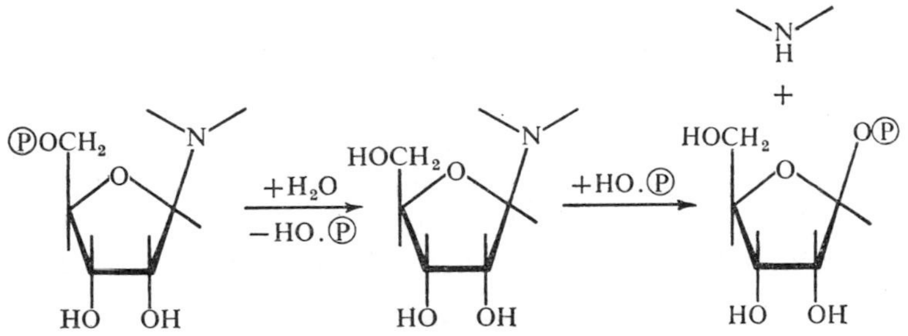

to give the free purine bases.

So much for uricogenesis in the birds. Little work has been done on urico-genesis in their evolutionary antecedents, the reptiles. As far as the uricotelic invertebrates are concerned, a good deal of work has been done, but the results are, at best, inconclusive. The outlook in this field has been much prejudiced by the view, now abandoned in so far as it affects uricotelic verte-brates, that urea is an intermediary in the synthesis. The best attitude to adopt at the present time is one of ignorance.

METABOLISM OF PURINE BASES

Adenine (6-aminopurine) occurs in small amounts in the free state and has been isolated from many vertebrate and invertebrate tissues. It also arises by the breakdown of nucleic acid. On the synthetic side it gives rise to adenosine, adenylic acid (AMP), ADP and ATP, and is degraded by hydrolytic deamina-tion to *hypoxanthine* (6-hydroxypurine) under the influence of a hydrolytic purine deaminase called adenine aminohydrolase (adenase). Adenase appears to be widely but somewhat erratically distributed and has been detected in the tissues of numerous animal species.

Guanine (2-amino-6-hydroxypurine), like adenine, occurs in nucleic acid, but although GTP occasionally replaces ATP as a phosphate-plus-energy donor, there is no reason to think that it plays as large a part in cellular metabolism as does adenine. It is an exceedingly insoluble substance and is deposited in crystalline form in special cells, known as iridocytes, in many animals, where it is responsible for the beautiful iridescence of many fishes, for example. Guanine has a special and apparently unique function among spiders, where it replaces uric acid as the predominant product in nitrogenous excretion. In spite of a popular belief to the contrary, the spiders constitute a group that is morphologically quite distinct from the insects: they demonstrate their independence chemically, too, by excreting guanine in place of uric acid. Guanine, if anything, is even less soluble than uric acid, and contains one amino-group per molecule over and above the four ring-bound nitrogen atoms of the purine ring. Evidently, therefore, guanine is well qualified to take over the excretory functions of uric acid.

Like adenine, guanine is degraded by hydrolytic deamination by a guanine aminohydrolase (guanase) and yields *xanthine* (2:6-dihydroxypurine). Hypo-xanthine and xanthine then undergo serial oxidation under the influence of xanthine oxidase or the corresponding dehydrogenase to give *uric acid*. These metabolic relationships may be summarized as shown in Fig. 58. In passing, it should be noticed that all the hydroxypurines (often known as 'oxypurines') are tautomeric substances which readily undergo transforma-tion at the

$$-N{=}C(OH){-} \rightleftharpoons -NH{-}CO{-}$$

groupings. In the scheme shown in Fig. 58, only the *enol* forms are given for the sake of clarity, except in the case of uric acid, in which the keto form is believed to predominate.

The distribution of adenase, guanase and xanthine oxidase among animal tissues is very erratic. It is said, for example, that man and the rat possess no adenase, though the enzyme is common elsewhere, while the tissues of the

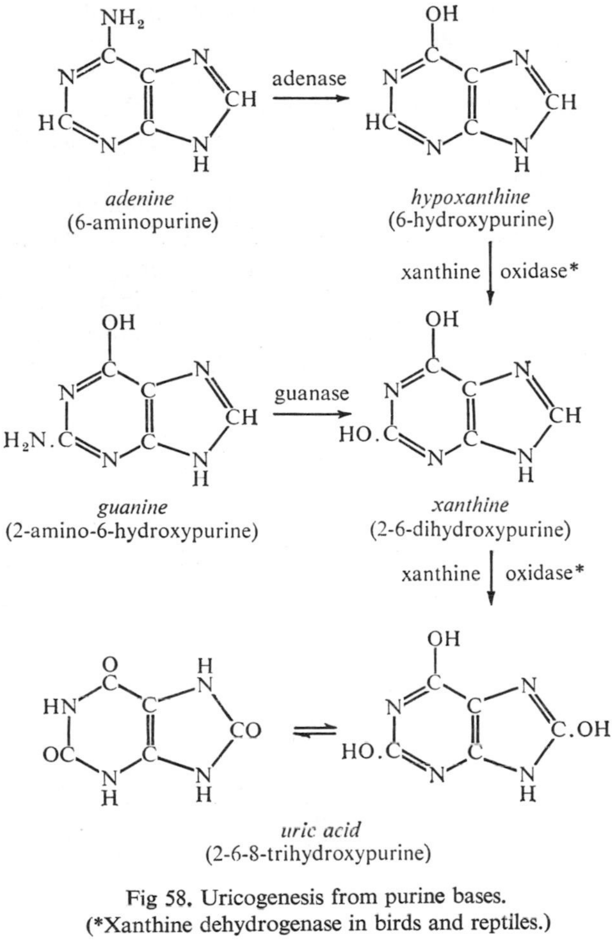

Fig 58. Uricogenesis from purine bases.
(*Xanthine dehydrogenase in birds and reptiles.)

embryonic pig are stated to contain guanase, in contradistinction to those of the adult, which do not. Again, the xanthine oxidase of mammals is thought to be replaced by a dehydrogenase that is present in the liver of most birds, e.g. goose and domestic fowl, but is absent from that of the pigeon.

287

METABOLISM OF URIC ACID

Even though it is often only an intermediate product, uric acid may arise from purine bases in animals of any kind, whether they are ammoniotelic, ureotelic or uricotelic. The biosynthesis of the purine ring system has already been discussed (pp. 279–85). Uricotelic animals, as we have seen, convert the bulk of their waste nitrogen into uric acid, but the amounts of uric acid that arise from purine metabolism itself are always relatively small, accounting perhaps for about 5 % of all the nitrogen excreted.

Uric acid is excreted without further chemical manipulation by uricotelic animals, but in most other forms it is more or less extensviely degraded before being excreted (see Fig. 59). The first stage in the process of uricolysis consists in the oxidation of uric acid itself to the more soluble substance allantoin, under the influence of *urate oxidase* (urico-oxidase). This takes place in all mammals apart from man and the higher apes (Primates), while the Dalmatian coach-hound is peculiar among dogs in that it excretes only a small part of its total purines in the form of allantoin. It is a strange fact that the liver of this dog is nevertheless fairly rich in urate oxidase; one possible explanation of this paradox is that the renal threshold for uric acid appears to be abnormally low in this animal, so that uric acid escapes very rapidly into the urine before the liver enzyme has had time to oxidize it all.

In other mammals, however, allantoin is excreted in place of uric acid, and uricolysis stops at this stage. In most other non-uricotelic animals, allantoin is further degraded to yield allantoic acid, thence to urea, and finally even to ammonia, though in many animal groups the complete set of uricolytic enzymes is lacking. The stages involved in the complete process are summarized in Fig. 59, together with the names of the enzymes concerned and some indications of their distribution.

It is worthy of note that, like adenase, guanase and xanthine oxidase, the uricolytic enzymes are very erratically distributed among animals. In particular, it is interesting to notice that, with the evolution of more complex forms of life, the tendency, as far as purine metabolism is concerned, has been to lose old enzymes rather than to acquire new ones, a fact which is amply illustrated by Fig. 59.

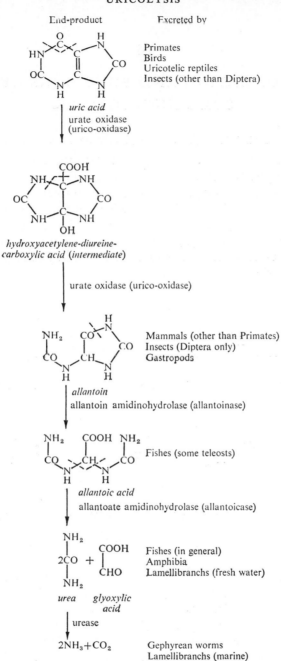

Fig. 59. End-products of purine metabolism. (After Florkin and Duchateau.)
(Trivial names are in brackets, systematic names are not.)

14

FORMATION AND FUNCTIONS OF SOME NUCLEOTIDES

FORMATION OF THE NITROGENOUS BASES

Purines. We have already given some attention to the biosynthesis of hypoxanthine, the parent member of the group of purines found in the nucleic acids (pp. 279–85). Free hypoxanthine itself is not converted directly into either adenine or guanine but its nucleotide, inosinic acid, can be transformed into the corresponding nucleotides of adenine and guanine, *adenylic* and *guanylic acids* respectively, by soluble enzymes present in bone marrow. The pathways of these interconversions are shown in Fig. 57 (p. 285), where R℗ represents ribose-5-phosphate.

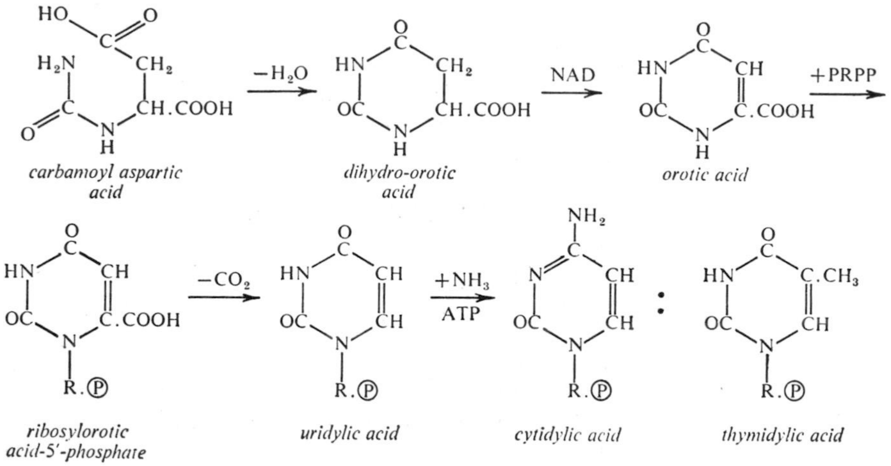

carbamoyl aspartic acid $-H_2O$ dihydro-orotic acid   NAD   orotic acid $+PRPP$

ribosylorotic acid-5′-phosphate $-CO_2$ uridylic acid $+NH_3$ / ATP   cytidylic acid   :   thymidylic acid

Fig. 60. Biosynthesis of pyrimidine nucleotides.

Pyrimidines. Much has now been learned about the biosynthesis of the pyrimidine bases. Of central importance is *orotic acid*, which arises from carbamoyl aspartate (see Fig. 60). The formation of the nucleotide in this case (cf. p. 283) takes place only after the ring system has been completed.

By a reaction between orotic acid and PRPP (p. 295) the corresponding nucleotide is formed and can then be converted into *uridylic acid*, the nucleotide of uracil, by decarboxylation.

The synthesis of cytidine nucleotide (*cytidylic acid*, CMP) has also been demonstrated. The starting materials include uridine triphosphate (UTP), which can be formed by a reaction between uridylic acid (UMP) and ATP,

$$ATP + UMP \rightleftharpoons AMP + UTP.$$

Ammonia, which is not replaceable by glutamine or glutamate nor by the corresponding aspartate compounds, is also required. The resulting product, cytidine triphosphate, can then react with AMP:

$$CTP + AMP \rightleftharpoons CMP + ATP.$$

The formation of the third important pyrimidine nucleotide, *thymidylic acid*, is less clearly understood; presumably the starting material is uridylic acid and possibly a folic acid derivative supplies the additional carbon atom of its methyl group, if not the intact methyl group itself.

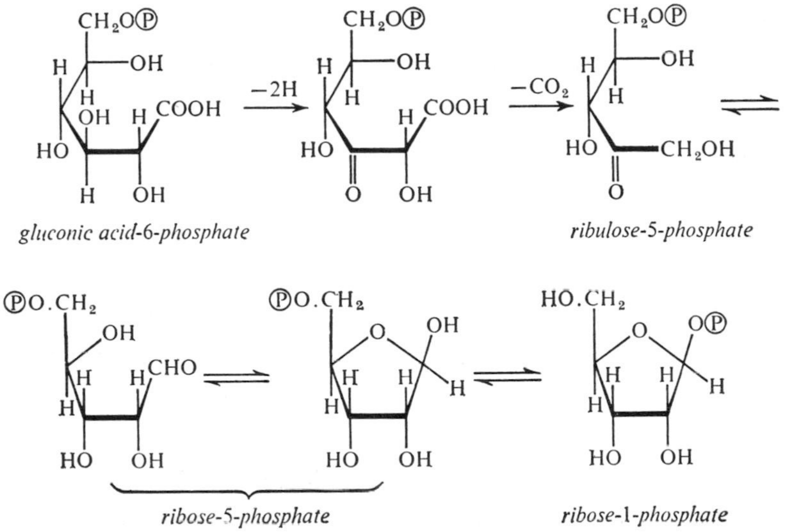

Fig. 61. Formation of ribulose- and ribose-5-phosphates from gluconate-6-phosphate.

BREAKDOWN AND FORMATION OF NUCLEOSIDES AND NUCLEOTIDES

Origin of the pentose sugars

The naturally occurring nucleosides and nucleotides are N-β-glycosides either of D-ribofuranose or D-2-deoxyribofuranose, two very unstable sugars the occurrence of which appears to be restricted to this particular group of compounds and their derivatives. D-Ribose can arise from gluconate-6-phosphate formed by the action of glucose-6-phosphate dehydrogenase and NADP upon glucose-6-phosphate (p. 165). The reactions involved are summarized in Fig. 61.

The first reaction in the sequence destroys the asymmetry at carbon 3 of the gluconic acid and carbon 1 is then removed by decarboxylation, yielding the 5-phosphate ester of the ketopentose *ribulose*. The conversion of this into the corresponding aldopentose is catalysed by *ribose-5-phosphate isomerase*. The cyclized form of the aldose, ribofuranose-5-phosphate, is finally brought into equilibrium with the 1-phosphate, a riboside; the enzyme concerned has been called *phosphoribomutase* and may possibly be identical with phospho-glucomutase (p. 123).

Ribulose-5-phosphate can also be formed from xylulose-5-phosphate by the isomerizing enzyme *ribulose-5-phosphate epimerase* (p. 121). Xylulose-5-phosphate and ribose-5-phosphate also arise by the transketolation of glyceraldehyde-3-phosphate and sedoheptulose-7-phosphate (p. 105).

Less is known about the production of the corresponding derivatives of 2-deoxyribose. The latter could perhaps be formed from acetaldehyde and glyceraldehyde-3-phosphate by a specific *transketolase*:

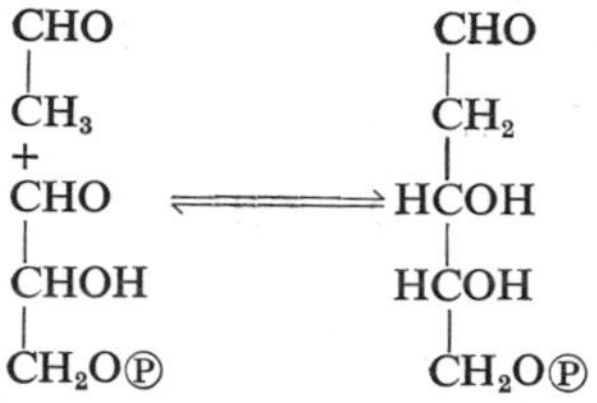

$$
\begin{array}{ccc}
\text{CHO} & & \text{CHO} \\
| & & | \\
\text{CH}_3 & & \text{CH}_2 \\
+ & & | \\
\text{CHO} & \rightleftharpoons & \text{HCOH} \\
| & & | \\
\text{CHOH} & & \text{HCOH} \\
| & & | \\
\text{CH}_2\text{O}\textcircled{P} & & \text{CH}_2\text{O}\textcircled{P}
\end{array}
$$

2-*deoxyribose*

Phosphoribomutase can probably catalyse the interconversion of the 5- and the 1-phosphates of this sugar.

Formation of nucleosides

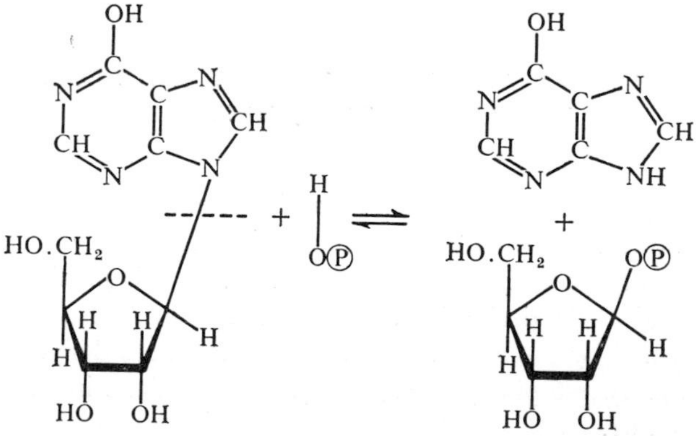

It has been known for some years that certain nucleosides can be reversibly broken down by hydrolytic enzymes, the so-called *nucleoside hydrolases*, but more recently it has been found that there also exist transferring enzymes that catalyse a process that yields β-ribose-1-phosphate together with the free purine, pyrimidine or other nitrogenous base. The reaction in this case is therefore one of *transribosylation* catalysed by *ribosyltransferases*. The first example of this kind to be discovered was that of inosine (hypoxanthine-9-N-β-D-riboside).

A number of similar reactions have since been described and it now seems probable that this reaction is typical for the synthesis and degradation of ribo- and deoxyribo-nucleosides alike. The following are some further examples catalysed by similar ribosyltransferases:

hypoxanthine-N-riboside + HO.℗ ⇌ hypoxanthine + ribose-1-phosphate,
nicotinamide-N-riboside + HO.℗ ⇌ nicotinamide + ribose-1-phosphate,
hypoxanthine-N-deoxyriboside + HO.℗ ⇌ hypoxanthine + deoxyribose-1-phosphate,
guanine-N-deoxyriboside + HO.℗ ⇌ guanine + deoxyribose-1-phosphate,
thymine-N-deoxyriboside + HO.℗ ⇌ thymine + deoxyribose-1-phosphate.

Little is known about the specificities of the ribosyltransferases concerned in these reactions, but always the nitrogenous and phosphatic aglycones seem to be freely interchangeable. Adenosine and cytosine are said not to react in this way however.

Phosphorylation: formation of nucleotides

Several examples of the formation of nucleotides from nucleosides have been demonstrated. In each case the source of the phosphate group lies in ATP and the phosphorylation is catalysed by an appropriate and specific kinase. For example:

adenine-N-riboside + ATP → adenosine-5′-phosphate + ADP,
riboflavin + ATP → riboflavin-5′-phosphate + ADP,
nicotinamide-N-riboside + ATP → nicotinamide-N-riboside-5′-phosphate + ADP.

In this manner adenylic acid (adenine mononucleotide), flavin mononucleotide and nicotinamide mononucleotide can be formed.

It would appear, then, that we are well on the way to an understanding of the problem of nucleotide formation and further advances are to be awaited, not only because of the outstanding interest and importance of nucleic acids and nucleoproteins at the present time, but also because of the enormous importance of certain mono- and dinucleotides in intermediary metabolism.

FUNCTIONS OF NUCLEOTIDES

Adenosine (adenine-9-β-D-ribofuranoside) is perhaps the longest known member of the family of nucleosides. By phosphorylation in position 5′ of the ribose residue by ATP it gives rise to the nucleotide, adenylic acid.

Adenylic acid, also known as *adenosine monophosphate* (AMP) and *adenine mononucleotide*, can be formed from adenosine by a kinase-catalysed reaction with ATP and has the structural formula shown in Fig. 62.

Further phosphate groups can be attached to that of adenosine mono-phosphate to yield *adenosine diphosphate* (ADP) and *adenosine triphosphate* (ATP). These substances play a fundamental part in the energetics of living systems and have been discussed in the chapter on biological energetics (Chap. 3).

The special importance of ATP arises from the fact that, under the in-fluence of the appropriate enzymes, the kinases, its terminal high-energy phosphate group, *together with the free energy with which it is associated,* can be transferred more or less intact to other substances, so that energy is, as it were, forced into the phosphate acceptor. This appears to be a fundamental operation in the synthesis of complex biological compounds from simpler

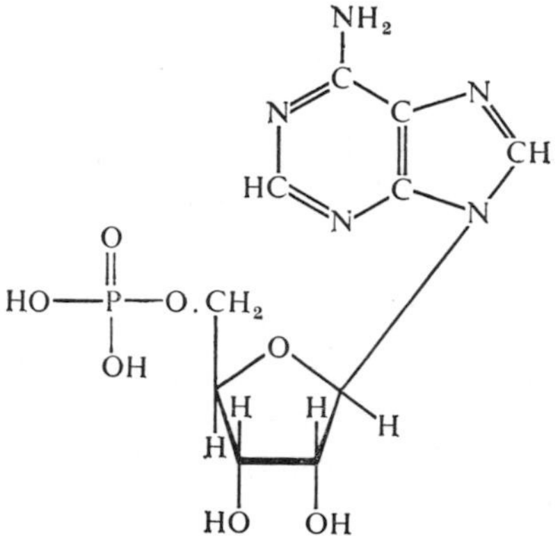

adenosine-5'-phosphate
(adenine mononucleotide)

Fig. 62. Structure of adenosine mononucleotide.

starting materials. With a few exceptions in which ATP is replaced by another nucleoside triphosphate such as guanosine triphosphate (GTP), inosine tri-phosphate (ITP) or uridine triphosphate (UTP), synthetic operations of this kind can only be accomplished at the expense of the terminal high-energy phosphate group of ATP. In cases where ATP is replaced by, say, GTP, the GDP left after its terminal $\sim\text{(P)}$ has been removed is rephosphorylated by ATP in a typical kinase reaction, so that the energy comes from ATP in the end:

$$\text{GDP} + \text{ATP} \rightleftharpoons \text{GTP} + \text{ADP}.$$

We must also mention the somewhat special case of adenylate kinase which catalyses the reaction

$$\text{ADP} + \text{ADP} \rightleftharpoons \text{ATP} + \text{AMP}.$$

Similar reactions, in the presence of other enzymes, can take place in which ATP, ADP and AMP can be replaced by the tri-, di- and monophosphates of uridine, cytosine, guanosine, and perhaps other nucleotides.

ATP itself can be re-formed at the expense of numerous katabolic processes leading to the generation of new high-energy phosphate groups, but the transmission of these to other substances can usually and perhaps always be accomplished only through the agency of the ATP $\rightleftharpoons$ ADP system.

ATP takes part in a great number and variety of biosynthetic operations. Most usually it transfers its terminal phosphate group to some one or other of many acceptors, numerous examples of which can be found in these pages, while a lengthy but by no means exhaustive list of kinases will be found on p. 89. One such reaction of special interest in the present context is the conversion of NAD to NADP by transference of the terminal phosphate group of ATP:

$$NAD + ATP \longrightarrow NADP + ADP.$$

But ATP can participate in a variety of other transfer reactions, the molecule being split in any of the several different ways shown in Fig. 63.

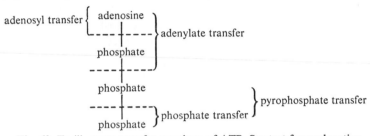

Fig. 63. To illustrate transfer reactions of ATP. See text for explanation.

Examples of *simple phosphate transfer* are common enough, but only a few well-authenticated cases of *pyrophosphate transfer* are known. One such is in the production of PRPP from ribose-5-phosphate catalysed by *ribose-5-phosphate pyrophosphotransferase*:

This is a fundamental starting reaction in the synthesis of nucleotides and, no doubt, nucleic acids.

Another such pyrophosphate transfer consists in the formation of thiamine di- or pyrophosphate from free thiamine by *thiamine pyrophosphotransferase* (p. 382).

Adenylate transfer is a very important process and we have already encountered several examples. For example, the first step in the activation of benzoic acid for the production of hippuric acid is:

benzoate + ATP $\longrightarrow$ benzoyladenylate + pyrophosphate.

Again, the initial stages in the activation of amino-acids for protein synthesis, and in that of fatty acids whether for breakdown or synthesis, also involves adenylate transfer:

amino-acids + ATP ⟶ amino-acyl adenylates + pyrophosphate;
fatty acids + ATP ⟶ fatty acyl adenylates + pyrophosphate.

In our more immediate context there are further cases of this kind of transfer reaction and to these we shall presently return, but for the sake of completing our list of examples we may mention one known case of *adenosyl transfer*. Methionine, important as a source of methyl groups, reacts not in its free form but as the *S*-adenosyl derivative (p. 111).

SYNTHESIS OF COENZYMES AND PROSTHETIC GROUPS

Adenine compounds play a fundamental part in many metabolic processes that lead to the generation of new ~ ℗, quite apart from their function as carriers of these groups. These reactions, for the most part, are oxidative in nature, and involve the participation of NAD, NADP and FAD. Adenylic acid enters into the composition of all three of these important compounds, in which it is present in combination with other nucleotides. These are, respectively, nicotinamide mononucleotide and the so-called flavin mono-nucleotide (Fig. 64). Adenylic acid also occurs in Co A (Fig. 65) and elsewhere.

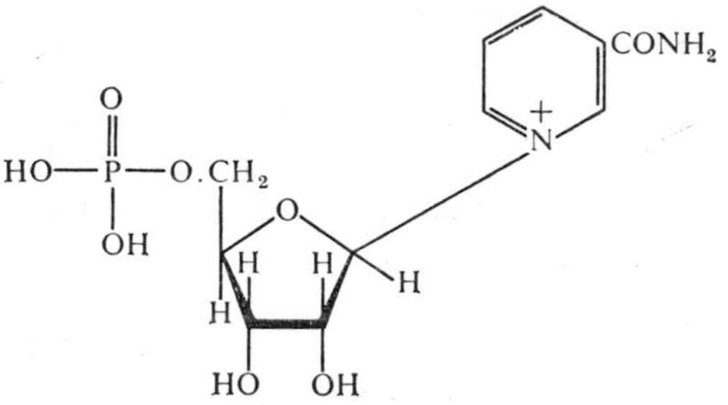

nicotinamide mononucleotide

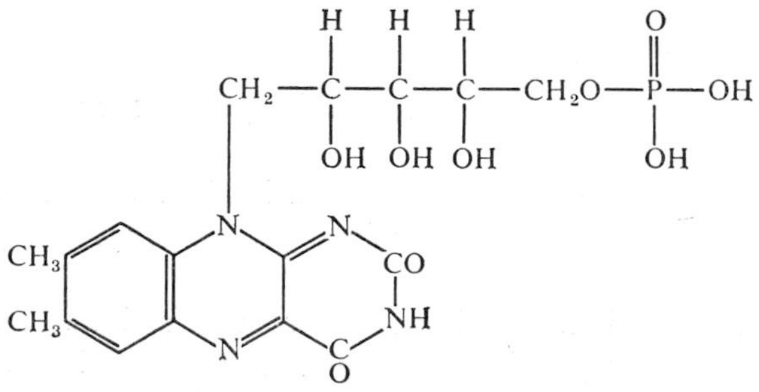

flavin mononucleotide

Fig. 64. Formulae of two important nucleotides.

Nicotinamide mononucleotide is a true nucleotide, the base present being nicotinamide and the pentose sugar D-ribofuranose. It can be formed from nicotinamide riboside by *simple phosphate transfer* from ATP.

Nicotinamide adenine dinucleotide (NAD) consists of a molecule of adenylic acid (adenine mononucleotide) and one of nicotinamide mononucleotide, the union between them being formed by condensation between their respective phosphate groups. It can be formed directly by an enzyme-catalysed *adenylate*

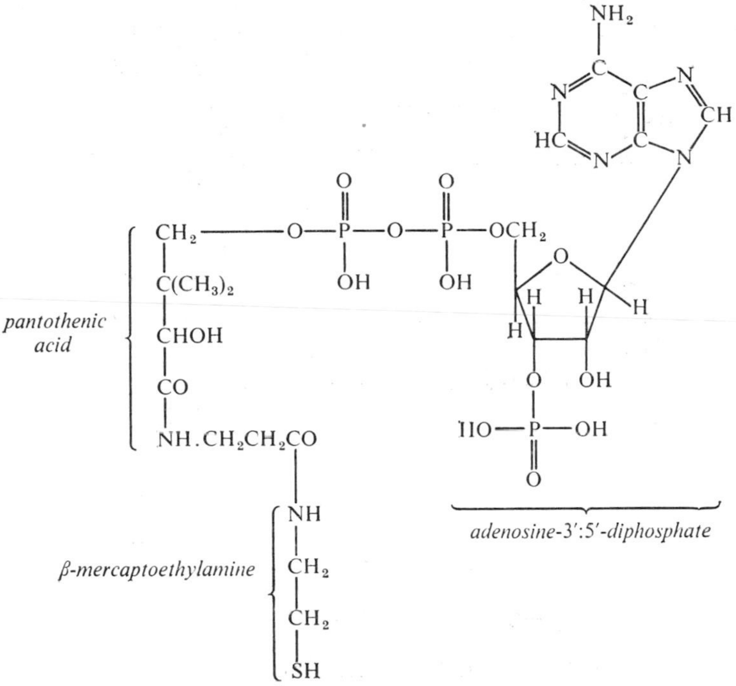

Fig. 65. Structure of coenzyme A.

transfer reaction between nicotinamide mononucleotide and ATP, which donates the adenine mononucleotide component:

nicotinamide mononucleotide + ATP $\longrightarrow$
nicotinamide-adenine dinucleotide + pyrophosphate.

By *simple phosphate transfer* from a further molecule of ATP, NAD can be converted into NADP:

$$\text{NAD} + \text{ATP} \longrightarrow \text{NADP} + \text{ADP}.$$

The third phosphate group present in NADP is probably attached at position 2' of the ribose ring of the adenosine moiety of NAD.

Flavinmononucleotide (FMN) is not a true nucleotide, although it is usually accorded that title. A nitrogenous base is present in the form of 6:7-dimethyl-

iso-alloxazine, but the place of the pentose sugar is taken by the corresponding sugar alcohol, D-ribitol. It can be formed from free riboflavin by *simple phosphate transfer* from ATP, catalysed by a flavin-specific kinase.

Flavin adenine dinucleotide (FAD) consists of a molecule of adenine mononucleotide and one of flavin mononucleotide, and again the dinucleotide can

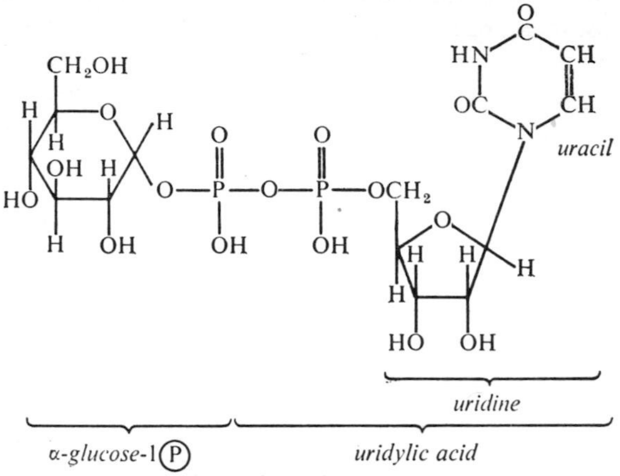

uridinediphosphate glucose (UDPG)

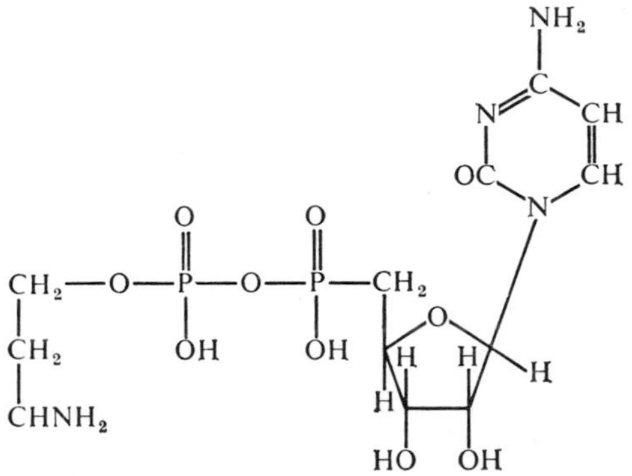

cytidine diphosphate-ethanolamine (CDP-*ethanolamine*)

Fig. 66. Formulae of UDPG and CDP-ethanolamine.

be formed from flavin mononucleotide by an enzyme-catalysed *adenylate transfer* from ATP:

flavin mononucleotide + ATP $\longrightarrow$ flavin adenine dinucleotide + pyrophosphate.

This dinucleotide is known to occur as the prosthetic group of certain oxidizing enzymes (oxidases and cytochrome reductases) which we have already considered (pp. 169–73).

Coenzyme A (Fig. 65) is the coenzyme concerned in transacetylation reactions and in oxidative decarboxylation (pp. 382–90). As we shall see later on, it is heavily involved in the metabolism of fatty acids too. It contains adenosine-3'-5'-diphosphate and may be considered as a dinucleotide-like substance in which the second 'nucleotide' is composed of β-mercaptoethylamine, pantothenic acid and phosphate. This substance reacts through its —SH groups and is often represented as CoA.SH or simply as A.SH for that reason.

Enzymes have been separated from liver which will catalyse (a) the phosphorylation of pantetheine (N-pantothenyl-β-mercaptoethylamine) by ATP (simple phosphate transfer), and (b) the reaction of the product, pantetheine-4-phosphate with a second molecule of ATP, this time by adenylate transfer. In this reaction the adenosine monophosphate group becomes attached through its 5'-phosphate group to that of the pantetheine phosphate. The synthesis is completed by a third enzyme which (c) catalyses the introduction of a phosphate group at 3' in the ribose ring at the expense of a third molecule of ATP (simple phosphate transfer).

Other nucleotides of importance include *guanosine, cytidine* and *uridine triphosphates*, already mentioned, and *uridinediphosphate glucose* (UDPG), one of the coenzymes of UDP-glucose epimerase (Fig. 66). This is another dinucleotide-like compound which plays a fundamental part in the synthesis of such polysaccharides as starch and glycogen and in that of many oligosaccharides, including sucrose (p. 329). The synthesis of UDPG is carried out by a reaction between UTP and α-glucose-1-phosphate:

$$\text{UTP} + \text{glucose-1-} \textcircled{P} \rightleftharpoons \text{UDPG} + \text{pyrophosphate}.$$

Finally, mention should be made of certain dinucleotide-like cytidine derivatives (see Fig. 66). These are formed by reactions between CTP and the phosphates of choline or ethanolamine, for example

$$\text{CTP} + \textcircled{P}\text{-choline} \rightleftharpoons \text{CDP-choline} + \text{pyrophosphate}.$$

These compounds play important parts in the synthesis of phospholipids by acting as donors of phosphoryl choline or phosphoryl ethanolamine (p. 439).

15

METABOLISM OF NUCLEIC ACIDS

NUCLEOPROTEINS

In this chapter we have to deal with a group of natural compounds called nucleoproteins and with certain products arising by their breakdown. The nucleoproteins are at the present time a subject of manifold interest. A full knowledge of their chemical constitution must necessarily await further developments in protein chemistry as far as their protein components are concerned, but, in the meantime, great strides are being made in the chemistry of these proteins and especially in that of their prosthetic components, the nucleic acids.

Nucleoproteins contribute only a small proportion to the total nitrogen of any average diet and account for only a small fraction of the total nitrogen of most cells and tissues but, as their name implies, they occur especially in cell nuclei. Any cell that contains a high proportion of nuclear material will form a good source of nucleoproteins, and glandular materials such as pancreas and thymus have been used extensively for their preparation. The richest source of all seems to be the heads of ripe spermatozoa: indeed, it has been estimated that the nucleoprotein content of fresh spermatozoa is from 50–80% of the total solid matter. Ripe fish milt (soft roes) therefore provides a valuable source of nucleoproteins and their derivatives.

It is only during the last 50 years that the nucleoproteins have attracted much interest. The early preparations were rather crude, insoluble substances unattractive to the pure chemist, and the present phase of interest only began when it was discovered that certain plant viruses can be isolated in crystalline form and are, in fact, nucleoproteins. Unlike the products formerly obtained, these viruses give beautiful crystals. A number of plant viruses have now been shown to consist of crystallizable nucleoproteins, and among animal viruses there are many, including vaccinia, the virus of cowpox, which contain, even if they do not consist entirely of, nucleoprotein material.

It had hitherto been supposed that viruses are living organisms, small enough to pass through the pores of bacterial filters and to be invisible under the microscope. Yet plant virus diseases can be transmitted to healthy plants either by inoculation with sap from an infected plant or with the crystalline virus. In many cases of virus infection however, *only the nucleic acid moiety enters the host-cells*, the protein component being left outside and contributing little, if anything at all, to the intracellular events evoked by the nucleic acids themselves.

If healthy plants are infected by means of pure virus nucleoprotein and allowed to develop the disease, the virus nucleoprotein can later be isolated in quantities very much in excess of those used for the original inoculation. These discoveries produced some shocks among biologists who, as a whole, had always supposed that there exists some sharp line of demarcation between things that are living and things that are not. Yet, in these viruses, we have something which can be crystallized and can at the same time transmit disease from a sick to a healthy organism and, moreover, reduplicates or 'reproduces' as the disease runs its course.

It has been shown too that mammalian and certain other chromosomes also consist largely of nucleoprotein material, so that we must look on the nucleoproteins, and particularly upon the nucleic acids, not only as the causative agents of a large number of infectious diseases, but also as the vehicles whereby hereditary characteristics are transmitted from parents to offspring. Viruses and chromosomes have several important features in common. Both, given the right kind of intracellular environment, can reduplicate. Furthermore, both show the phenomenon of mutation, either in the course of nature or under the artificial influence of X-rays or γ-radiation or any of a variety of chemical mutagenic agents. Just as new strains of organisms can suddenly appear as a result of genetic mutations, so, too, new virus strains can suddenly appear as a result of mutations in the old. This process of mutation among viruses is quite probably the reason for the hitherto inexplicable variability in the virulence of many virus-borne diseases, e.g. influenza.

Finally, there is every reason to believe that nucleic acids play a part in the all-important matter of protein synthesis.

DIGESTION OF NUCLEOPROTEINS

The salt-like union between the nucleic acid and the usually basic protein component of a typical nucleoprotein is disrupted by the acidic contents of the stomach, the protein fragment being digested along with the other food proteins. Nucleic acids are further split by hydrolytic enzymes contributed by the pancreatic and intestinal juices. Our knowledge of these enzymes is still far from complete, but they comprise (i) *nucleases*, which carry out a partial degradation of the nucleic acids, the products being further attacked by ill-defined mechanisms giving rise eventually to the component nucleotides; (ii) *nucleotidases*, phosphatases which catalyse the dephosphorylation of the nucleotides, yielding nucleosides, which are *N*-glycosides of certain nitrogenous bases (p. 291); and (iii) *nucleosidases*, which hydrolyse the nucleosides to liberate the basic and glycosidic components.

The nucleotides themselves contain nitrogenous bases of the purine and pyrimidine groups. A good deal is known about the fates of the pyrimidine bases but we shall not consider them here. That of the purines has already

pyrimidine 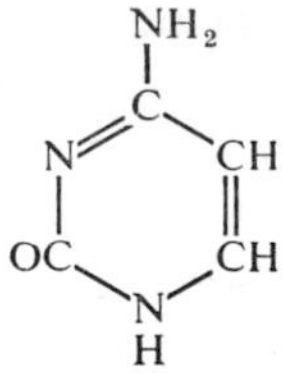   *cytosine* 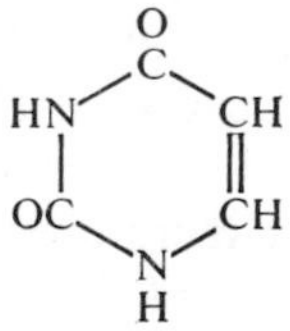   *uracil* 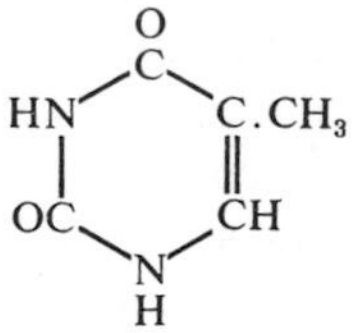   *thymine*

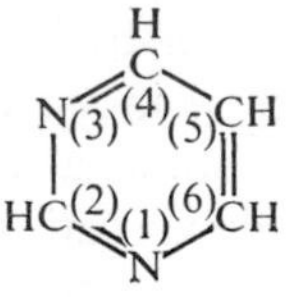

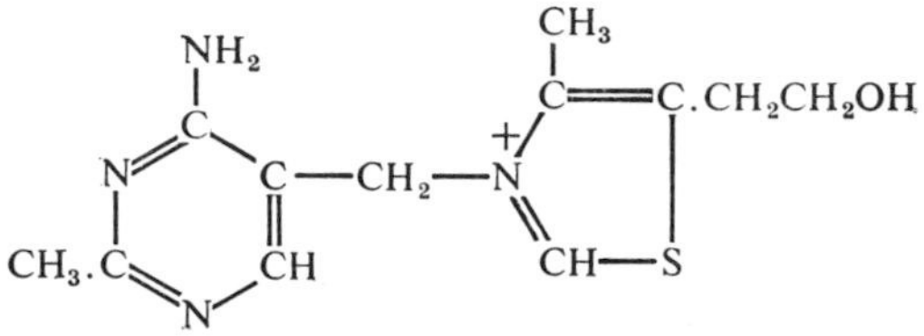

vitamin B$_1$ (=*aneurin*=*thiamine*)

been discussed (pp. 286–9). Their structures, together with that of pyrimidine itself, are appended as a matter of interest, and in passing it should be noticed that a large and important group of drugs, the barbiturates, are structurally related to the pyrimidine group, and that vitamin B$_1$ (thiamine, aneurin) also is a pyrimidine derivative. Similarly, it may be pointed out, caffein and a number of other important drugs are methylated purines. The parent of this group of compounds is purine, with the following structure:

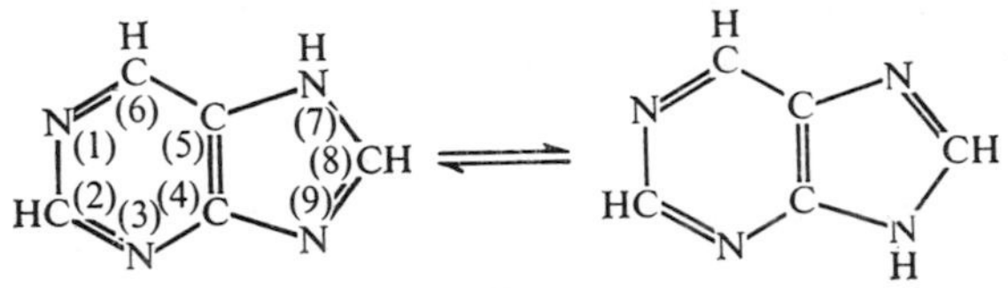

A   *purine*   B

The iminazole portion of the ring system is tautomeric, and it should be noticed that while form A is the one usually figured in text-books, B is probably the more biologically important because, when the purines enter into glycoside formation with pentoses, the union takes place at position (9) according to all available evidence. Finally it is worth noticing that the carbon and nitrogen atoms making up the pyrimidine and purine rings are coplanar, so that the molecules are flat plates.

NUCLEIC ACIDS

If thymus gland material is macerated with large volumes of water the thymus nucleoprotein is extracted, and the protein component, in this case a histone, can be precipitated by saturation of the extract with sodium chloride. On the further addition of ethyl alcohol, nucleic acid is precipitated as a fibrous mass.

Two types of nucleic acids have so far been recognized, one of which comprises the *deoxyribonucleic* (DNA) and the other the *ribonucleic acids* (RNA), which can be prepared from thymus gland and yeast respectively. Nucleoproteins of the DNA group are conjugated proteins formed by the union of a nucleic acid with a protein, often a protamine or a histone. These proteins usually contain large proportions of the basic amino-acids, arginine, lysine and histidine, but the protein components of the RNA group appear to be very much more complicated.

These acids, the structure of which is extremely complex, are built up from smaller units known as *nucleotides*, which yield on total hydrolysis one molecule each of a nitrogenous base, a pentose sugar and inorganic phosphate; they are, in fact, $5'$-phosphate esters of the N-glycosides of certain nitrogenous bases. Four such bases are usually found. Recent determinations of the molecular weights of the nucleic acids themselves, carried out by the method of ultracentrifugation, give values ranging from about 25,000 to many millions. At present therefore the tendency is to think of nucleic acids as being built up by the union of large numbers of nucleotide units, much in the way that proteins are built up by the union of large numbers of amino-acid units. The possible number of different nucleic acids, like that of proteins, is exceedingly large; DNA and RNA are, of course, collective titles. Many different DNA's are already known and characterized and the same is true of the RNA group.

The most striking differences between nucleic acids of the yeast and thymus types lie in the nature of the sugars involved. Nucleic acid prepared from yeast contains β-D-ribofuranose, while that from the thymus gland of animals contains β-D-2-deoxyribofuranose:

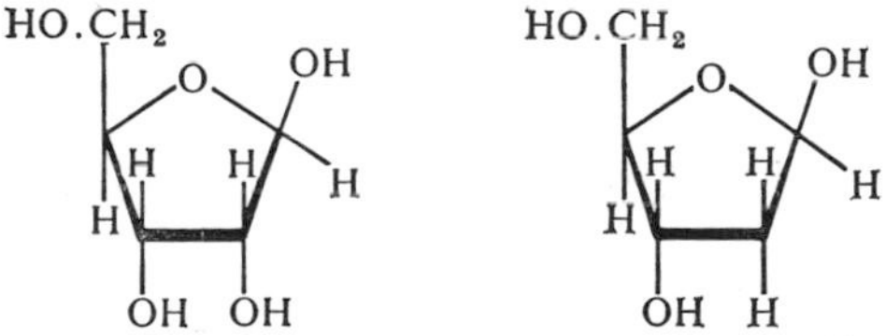

β-D-*ribofuranose* β-D-2-*deoxyribofuranose*

It was at one time believed that plant and animal cells always and only contain ribonucleic and deoxyribonucleic acids respectively. This view has turned out to be entirely erroneous. *All* cells contain nucleic acids of both types, the deoxyribose compounds preponderating in the nucleus and the ribose compounds in the cytoplasm generally, but particularly in the microsomes. Reproductive cells and micro-organisms are very rich in deoxyribo- and in ribonucleic acids respectively.

Both types of nucleic acids contain phosphate, and both contain bases belonging to the purine and pyrimidine groups. Hydrolysis of nucleic acids by means of dilute acids or appropriate enzymes yields the following recognized products:

	CYTOPLASM	NUCLEUS
	RIBONUCLEIC ACID	DEOXYRIBONUCLEIC ACID
Pentose:	D-ribose	D-2-deoxyribose
Pyrimidines:	cytosine; uracil	cytosine; thymine
Purines:	adenine; guanine	adenine; guanine

Small amounts of other bases are sometimes found, e.g. 5-methylcytosine in thymus nucleic acid.

The chief differences lie, therefore, in the nature of the pentose and in the presence of thymine and of uracil in deoxyribonucleic and ribonucleic acids respectively.

STRUCTURE OF NUCLEOTIDES

RNA can be split into its constituent nucleotides by the action of a hydrolase (ribonuclease) present in the venom of many snakes. Mild *acid* hydrolysis of nucleotides separates the base from the pentose phosphate, when it can readily be shown that the phosphate group occupies the 5'-position. Mild *alkaline* hydrolysis of nucleotides sets free only the phosphate group, leaving nucleosides. These have the properties of β-glycosides and there is good spectroscopic, synthetic and other evidence that the point of attachment is at 9 in the case of a purine or at 1 in that of a pyrimidine base. The structure of, say, the adenylic acid of RNA can therefore be written as shown in Fig. 67.

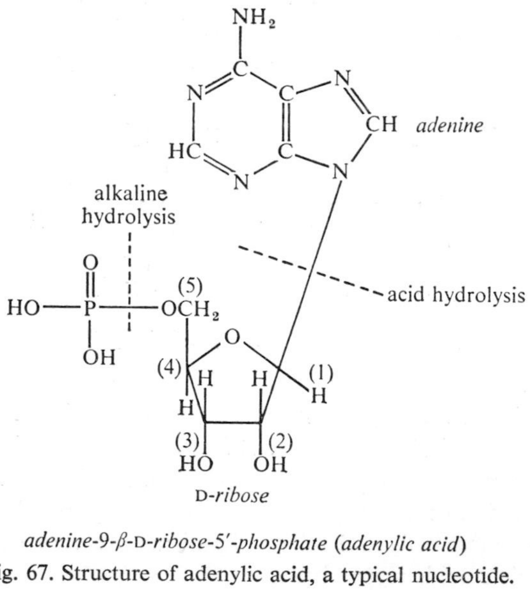

adenine-9-β-D-ribose-5'-phosphate (adenylic acid)

Fig. 67. Structure of adenylic acid, a typical nucleotide.

Incomplete hydrolysis of RNA with dilute alkali gives rise to cyclic 2′:3′-phosphodiesters of the following type:

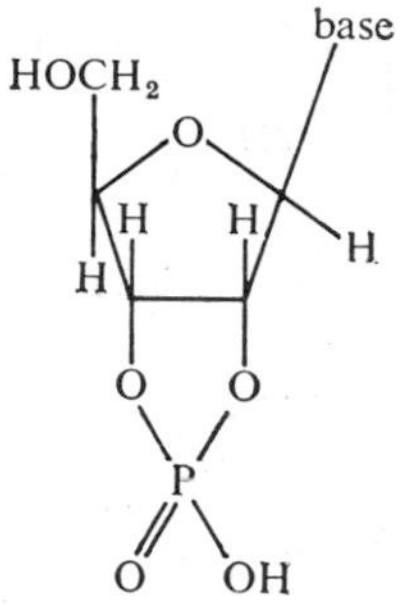

2′:3′-diester

If hydrolysis is pushed further a mixture of the 2′- and the 3′-esters is produced. It seems probable, therefore, that adjacent nucleotides must be joined by linking the pentose-5′-phosphate group of one *either* to the 2′- *or* to the 3′-position of the second pentose through a phosphodiester linkage. This problem does not arise in the case of DNA, the pentose of which carries no hydroxyl group at 2′ so that only 3′- and 5′- are available for esterification. In the case of RNA a decision can be reached by submitting polynucleotides to the action of the phosphodiesterase of snake venom, which is specific for 3′-linkages, and to that of another phosphodiesterase obtained from spleen, which acts only upon 5′-linkages. Neither attacks 2′-linkages but both enzymes split their substrates, the first giving the 5′-phosphates and the second the 3′-phosphates of the corresponding nucleotides. The existence of 3′:5′-phosphodiester linkages between adjacent nucleotide units is therefore now firmly established, so that polynucleotides of the RNA type can be formulated as shown in Fig. 68.

STRUCTURE OF NUCLEIC ACIDS

X-ray measurements on crystalline DNA have shown that the DNA molecule is a long, thin, rod-like structure of practically constant diameter. Attempts have been made to construct models which should satisfy these requirements and, at the same time, conform to established X-ray measurements. This led to the hypothesis originally proposed by Watson & Crick, and now generally accepted, that DNA consists of a twin spiral formed by the twisting one round the other of a pair of polynucleotide chains, like the strands of a piece of twin electric flex. The known dimensions can only be satisfied however if all the bases lie so as to form a core within the two pentose-phosphate spirals and, moreover, if pairs of bases lying opposite to each other are held together by hydrogen bonds. An example is shown in Fig. 69.

It will be recalled that the atoms in both types of rings are coplanar and they lie edge to edge, forming pairs, at right angles to the main axis of the twin spiral.

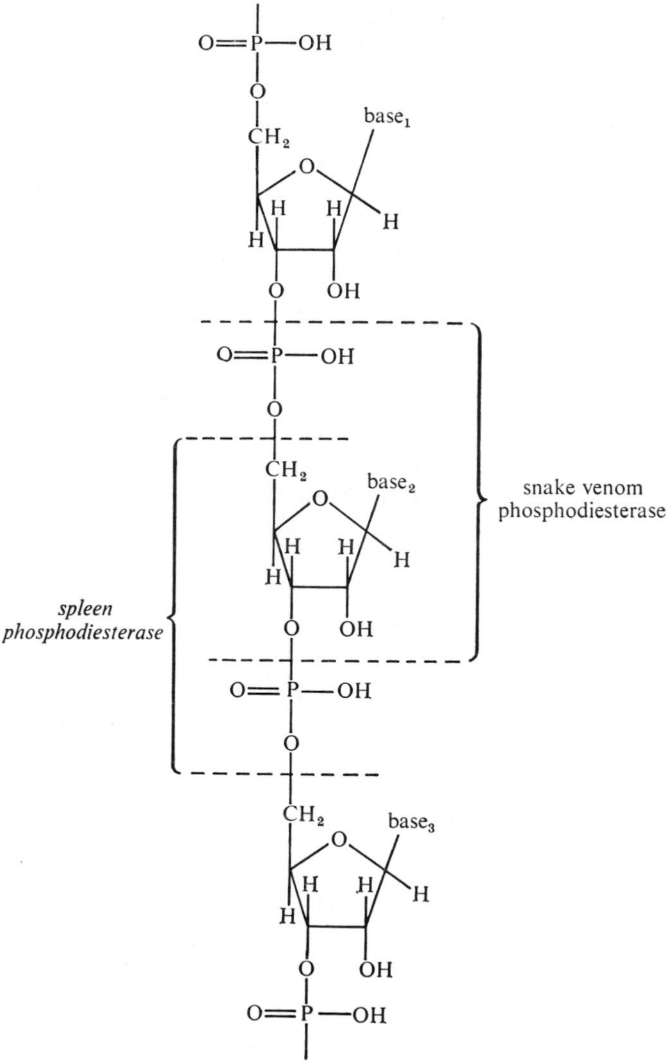

Fig. 68. Formula of part of an RNA molecule showing points of attack by diesterases. Polynucleotides of the DNA type are exactly similar except in the absence of the O atom at position 2′ in the pentose rings.

If all the bases pair off in this manner a purine must always lie opposite a pyrimidine and vice versa; otherwise the observed constancy of molecular diameter would not be possible because of the different sizes of the two ring systems. Finally, according to Watson & Crick, base pairing is not a random affair; on the contrary adenine (A) always lies opposite to thymine (T) and guanine (G) opposite to cytosine (C). This is confirmed by two experimental findings; first that the sum of the purines is about equal to the sum of the pyrimidines, and secondly that on analysis the amounts of adenine and thymine are approximately equivalent. The same is true of guanine and cytosine in hydrolysates of DNA.

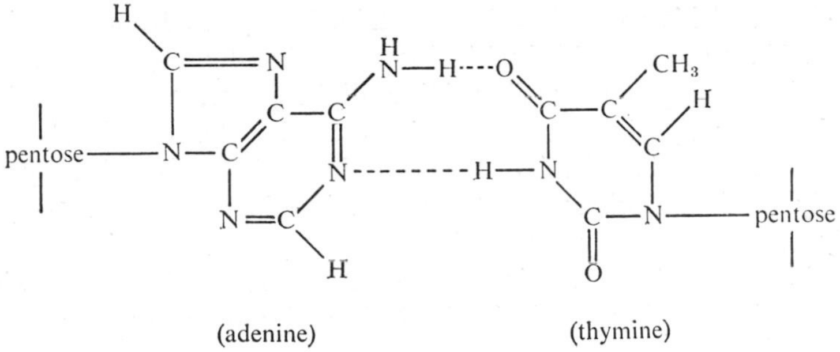

(adenine)   (thymine)

Fig. 69. Hydrogen bonding between adenine and thymine.

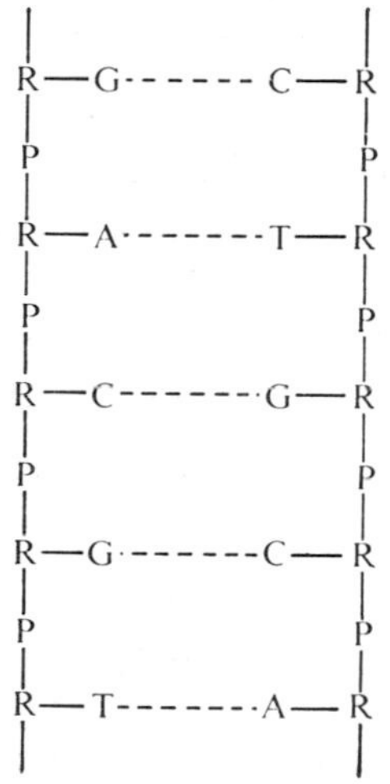

Fig. 70. Skeleton formula of part of a DNA molecule.

All these points can best be demonstrated by the construction of molecular models, but some appreciation of the structure of DNA can be gained if we represent it in two dimensions as the ladder-like structure shown in Fig. 70. If then the top of the ladder is twisted in a clockwise direction and the bottom anticlockwise, the ladder will be twisted to form a double spiral, similar to a spiral staircase, with the hydrogen-bonded base pairs forming the rungs or steps. In actuality there are 10 nucleotide units for each complete turn in the spiral.

So much for the molecular structure of DNA. As far as RNA is concerned less is known except that it has a highly complex structure in which the molecules fold back on themselves in places to form localized twin spiral-like regions, again by hydrogen bonding between pairs of opposite bases.

Synthesis of polynucleotides

Ochoa and his co-workers have obtained enzymes from several bacteria which, if they are provided with ribonucleoside diphosphates such as ribo-adenosinediphosphate or ribouridine diphosphate, will synthesize poly-adenine and polyuridine ribonucleotides respectively. Mixed polynucleotides too can be formed from mixtures of the diphosphates. Individually, these are usually random fibrillar structures but, if mixed together, 'poly-A' and 'poly-U' spontaneously form twin-stranded spirals which in many respects resemble those of DNA. This suggests that the formation of the twin spirals of DNA is probably a built-in, intrinsic and spontaneous property of the polynucleotide structure.

Still more interesting is Kornberg's discovery of what now appears to be a universally distributed enzyme capable of synthesizing DNA itself. In this case the starting materials are the deoxyribonucleoside triphosphates of adenine, guanine, cytosine and thymine, together with small amounts of pre-formed DNA to act as a 'template'. All four of the triphosphates are needed and all four of the corresponding nucleotides are incorporated into the newly formed DNA. Inorganic pyrophosphate is eliminated as the condensations take place.

A third synthetic enzyme has been discovered by Weiss. This is a so-called 'RNA polymerase'. It requires a mixture of the usual four ribonucleoside triphosphates but, like Kornberg's enzyme, needs priming with DNA.

Replication of nucleic acids

Lest the impression may have been given that RNA is always the same, or that one DNA is the same as any other, it may be emphasized here and now that this is not the case. Apart from much other evidence, many viruses, some containing DNA, some RNA and some containing both, are known to be highly specific towards their respective hosts, each producing its own charac-teristic symptoms and lesions, and reproducing (replicating) precisely and exactly in the process.

Subject only to the conditions that the numbers of the adenine and thymine nucleotides must be about equal and that the numbers of guanine and cytosine nucleotides must likewise be about equal, the individual nucleotides can be present in any proportions and arranged in any order, so that there is virtually unlimited scope for specific variations in the DNA molecule. Especially is this so when it is realized that not less than 20,000 and often many more nucleotide units may occur in a single molecule of DNA.

The fact that DNA can reduplicate is not in dispute, but the mechanism involved is less clear. The production of a new molecule which is an exact copy of the original DNA, with its complex geometry and a molecular weight running often into many millions must, we may well imagine, be a herculean task. Let us, however, think of the two strands of DNA as A and B respectively. A glance at Fig. 70 will make it clear that B is not an exact copy of A, nor is it even a mirror image of A, but it is nevertheless so specific in structure that, in the ordinary way, no entity other than B itself can form a double spiral with A. We may therefore say that B is the *complement* of A and that A is the complement of B. Notice the definite article.

Suppose now that A and B can separate by unwinding or in some other way. Then, in the presence of the requisite raw materials and the appropriate enzymes, we can suggest that complement for A will be built up and that complement for B will likewise be synthesized so that

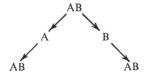

Since A and B are mutually complementary, the two daughter molecules will be exact replicas of the parent. The point that remains obscure is the mode of separation of A and B.

A beautiful demonstration of a process of this kind has been given by Meselson & Stahl, working on a bacterial organism, *Escherichia coli*. The cells were grown in a medium in which the sole source of nitrogen was the heavy isotope, ^{15}N, and cell multiplication was allowed to continue until the DNA contained virtually none of the ordinary isotope, ^{14}N. Next the cells were transferred to a ^{14}N medium and DNA was extracted from successive gnerations and examined in the ultracentrifuge. The parent cells gave a single zone corresponding to ^{15}N-DNA. Cells from the second generation again gave a single zone, this time of intermediate density between ^{15}N- and ^{14}N-DNA, while cells of the third generation gave intermediate and ^{14}N-DNA zones. In later generations the intensity of the ^{14}N-DNA zone increased as that of the intermediate zone diminished.

Evidently what is happening is as shown in Fig. 71, where heavy type corresponds to labelling with ^{15}N and ordinary type to ordinary ^{14}N. A remarkable feature of this process is the extreme stability of A and B; once formed they persist generation after generation, apparently unchanged.

It is difficult at present to visualize the process of complement formation which is, clearly, a very complex and highly specific process. However, since it appears that adenine must always lie opposite thymine, and guanine opposite cytosine, the pattern of the parent strand of DNA must to this extent dictate the disposition or siting of the pairing bases in its complement. Whether complete complement formation precedes spiral formation is doubtful, but we have evidence (p. 308) that poly-A and poly-U, while they each exist separately as single, random fibrils, spiralize spontaneously to give a double spiral when the two are mixed together. Probably, therefore, spiralization runs parallel to the synthesis of newly formed complement.

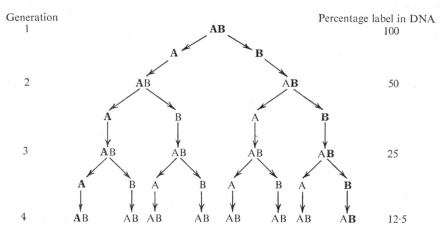

Fig. 71. Replication of isotopically labelled DNA.

DNA AND BIOCHEMICAL GENETICS

The science of genetics has been well established for many years. The transmission of hereditary characteristics is attributed to the handing-on of chromosomal material, each chromosome being a long, fibrous or filamentous structure made up of units called *genes*. Each gene or some small group of genes is responsible for the transmission of some particular characteristic; red hair, black skin, blue eyes and many more familiar and familial features are but a few examples of the end-products of inheritable characteristics transmitted by the genes.

It is characteristic of chromosomes as a whole that they reduplicate with complete accuracy, and from the time an ovum is fertilized and undergoes its first division, the chromosomes of every new cell produced are faithful and exact copies of those that went before.

Occasionally, however, accidents happen. Rarely in the course of nature but more frequently as the result of exposure to X-rays, ultra-violet or γ radiation or other mutagenic agents, one or more genes may be altered or deleted, and the change is referred to as a *mutation*. Most mutations are lethal but, in the rare cases where they are not, new features are transmitted by the altered chromosomes so that the daughter cells have properties that the parent did not possess.

It is usual in experimental genetics to work on very small organisms because they get through a very large number of generations in a comparatively short length of time. The fruit fly, *Drosophila*, has long been a favourite tool for geneticists and an unbelievable number of mutant forms have been discovered or invented. Mostly, however, the results of mutation in this animal have been described only in morphological terms with little obvious biochemical content. However, in more recent times a great quantity of biochemical work has been done on even smaller organisms, especially on the bread mould, *Neurospora crassa*, and on a variety of bacteria and protozoa. Results of the greatest biochemical interest and importance have emerged.

Very many mutant forms of micro-organisms, especially of *Neurospora*, have been produced and the results of an immense amount of work has led to the conclusion that, broadly speaking, *whenever a mutation takes place*, whether spontaneous or artificially induced, *the result is the loss, or at any rate an unfavourable modification, of one enzyme*.

Wild strains of *Neurospora* can grow on very simple media. Water, salts, a simple source of nitrogen such as potassium or sodium nitrate and a carbohydrate energy-source such as sucrose are required and, in addition to these, biotin is necessary as a growth factor. Mutant forms of this organism can commonly be detected because they will no longer grow on this basal medium but succeed only on more complex media. By sub-culturing on progressively simplified media it is usually possible, though often difficult, to find out what food or growth factor each mutant requires.

The results of mutation appear to be essentially the following. Suppose that in the wild strain some important metabolite such, for example, as an indispensable amino-acid or vitamin, is produced by a series of enzyme-catalysed reactions such as these:

$$A \longrightarrow B \longrightarrow C \longrightarrow D \longrightarrow .$$

In a particular mutant the enzyme for $C \rightarrow D$ may be lacking, in which case the mutant can only grow on the basal medium if D is added. A similar block between A and B would call for the addition of B, C or D to the basal medium, and if a large enough number of mutants can be discovered it is often possible to split up a long, stepwise metabolic process into separate stages with a block at each step in turn. One consequence has been the clarification of a good many previously unknown or uncertain pathways in intermediary metabolism, and by way of illustration the pathways leading to methionine and threonine in *N. crassa* are shown in Fig. 72. At every reaction shown in this diagram an enzyme is required and the numbers are used to identify the mutant strains lacking each particular enzyme. Here is a genetic

311

counterpart of the biochemical use of selective inhibitors, a field of genetics in which a biochemist can feel thoroughly at home.

Chromosomes, then, are the vehicles whereby hereditary characteristics are transmitted from parent to offspring, from one cell to its progeny. They are confined to cell nuclei, and so is DNA for the most part, though small amounts are sometimes present in mitochondria. Chromosomes and DNA alike exist as long, filamentous elements in the cell nucleus and both, given the right kind

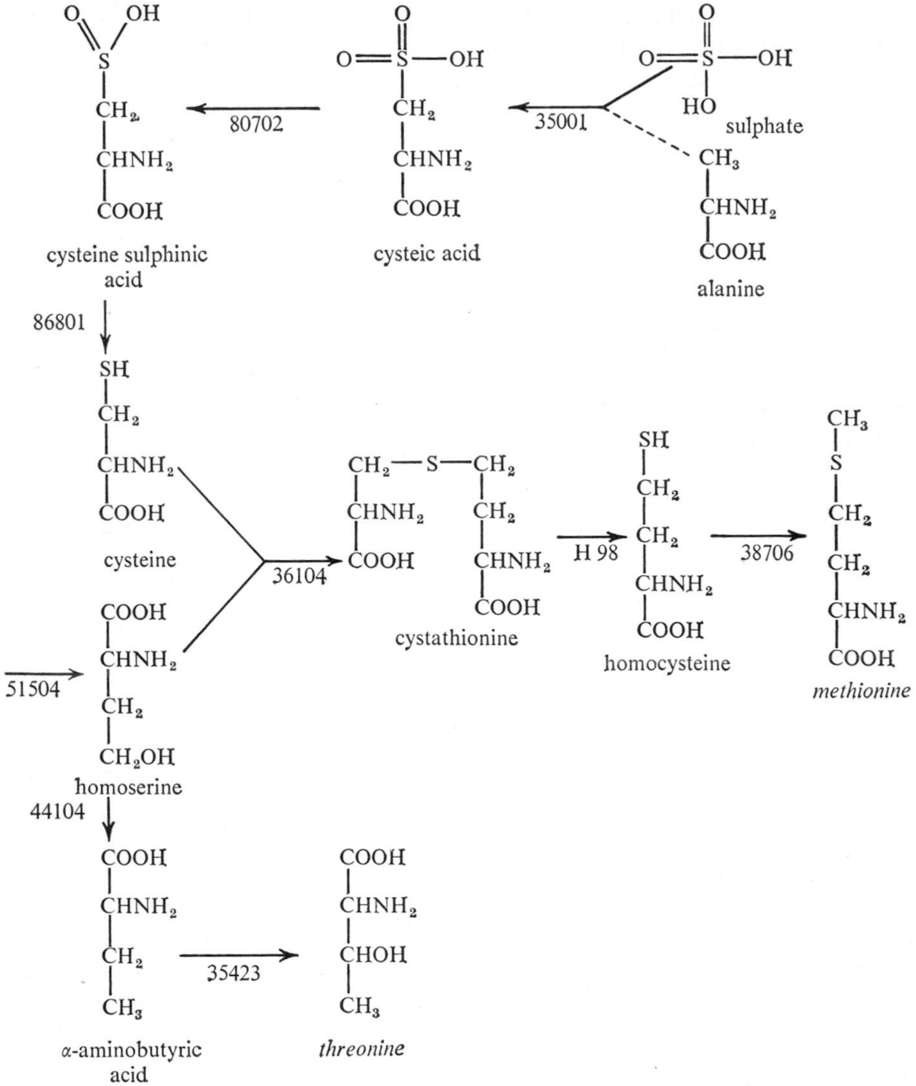

Fig. 72. Mutant blocks in synthesis of methionine and threonine by *N. crassa*.

of intra-nuclear environment, can reduplicate. Each new chromosome produced when a cell divides is, mutation apart, a faithful and exact replica of the last; DNA molecules can reduplicate in the same way. Finally, histochemical methods have established beyond all reasonable doubt that chromosomal material is nucleoprotein material and, more especially, that the nucleic acid concerned is DNA. Small wonder, then, that it has become axiomatic in genetics and biochemistry alike to regard DNA as the material basis of inheritance.

Mutations must presumably take place by errors or changes in the process of complement formation whereby new DNA molecules are produced and, while the majority of mutations have lethal results, some have advantageous consequences, and these are the basic ingredients of adaptation and evolutionary advancement. Evolution seems in fact to have proceeded by the accumulation of advantageous mutations and, if we wish to study evolution from the biochemical point of view, it will be well to study mutation from the same standpoint.

Inborn errors of metabolism

There are many inheritable human disorders and diseases which can be attributed to defects in enzymatic equipment. A familiar case is that of *albinism*; here the white hair, white skin and pink eyes testify to the absence of the melanin-forming enzyme, tyrosinase (p. 256). Another example, familiar enough to the physician, is a form of insanity called *imbecillitas* (*oligophrenia*) *phenylpyruvica*. While we cannot at present explain this mental lesion itself in biochemical terms, we do know that it is associated with the excretion of phenylpyruvate in the urine. This betokens a mutational block or lesion somewhere between phenylalanine and tyrosine. Further along the same pathway, homogentisic acid is formed and, in the ordinary way, undergoes oxidation, but in another hereditary disorder, *alkaptonuria*, the next enzyme in the sequence has been lost. Homogentisic acid therefore appears in the urine and the latter, on standing, goes inky black. Another case we have mentioned is the excretion of *argininosuccinate* by certain individuals (277), due apparently to some abnormality in ureogenesis.

Many other comparable mutational metabolic lesions are known. *Pentosuria*, in which L-xylulose is excreted, involves some sort of metabolic abnormality in carbohydrate metabolism. Again *goitrous cretinism* is associated with a block in the conversion of tyrosine into thyroid hormone. And there are many others, including a considerable number of chromosomal aberrations which lead to the production of *abnormal haemoglobins* in the human species.

NUCLEIC ACIDS AND PROTEIN SYNTHESIS

It seems certain from all this that DNA must in some way control the formation of enzymes and, by inference, the synthesis of proteins in general. Every species has its own collection of species–specific proteins and even individuals of the same species can differ in their enzymic equipment as a result of parental differences and mutations. Different species and different individuals, identical twins apart, each have their own molecular species of DNA, with or without superimposed mutational modifications.

Since it thus appears that *DNA in some way directs and controls enzyme and protein synthesis* in a very precise and exact fashion, we are driven to the assumption that DNA contains 'information' which, in some way or other, it communicates to the sites at which protein synthesis takes place. And it is improbable that the DNA, virtually confined as it is to the nucleus, can itself be directly and immediately responsible for the orderly and specific arrangement of the amino-acids involved in protein and therefore in enzyme synthesis in the cytoplasm.

But DNA can also serve as a mould or template for the formation of RNA, always with a degree of specificity so high that the daughter molecules reflect very accurately the nucleotide sequence of the parent DNA.

An illuminating example is found in the behaviour of a virus known as T-2 phage, which attacks *Esch. coli.* Although this phage is a DNA nucleoprotein, only the DNA component is required for infection, and many new complete phage molecules are produced by the time the host cell dies. Shortly after the viral DNA first gains access to the cell a new RNA, entirely foreign to *E. coli*, makes its appearance, and the intimate relationship between this new RNA and the infecting DNA is revealed by the fact that the two react together to form a complex. Neither, however, will do so with any other RNA or DNA that has ever been tested.

This case of viral activity deserves a little further attention. The replication of the viral DNA in the host cells, and likewise the production of the new RNA, calls for the provision of raw materials and enzymes. It appears that, by presenting the host with a new DNA template, the virus diverts the raw materials and enzymes from the formation of *E. coli* nucleoprotein to that of virus nucleoprotein instead. This process continues with the production of more and more virus DNA until the host is exhausted and finally bursts, liberating some 50–100 new virus particles to find new host cells in their turn. In the meantime the new RNA migrates to other parts of the host cell to direct and control numerous other synthetic processes, including the formation of the protein sheaths or envelopes that surround the newly formed DNA and complete the new virus molecules even before the host cells actually burst.

It is abundantly clear today that in the normal, healthy organism, the

314

extremely specific business of protein synthesis is under the control of 'information' originating in the nuclear DNA, yet the most active sites of protein synthesis lie, not in the nucleus, but in submicroscopic intracellular particles known as *ribosomes*, so called because they are particularly rich in RNA. The information contained in the nuclear DNA must evidently be communicated to these sites of protein synthesis in some way. Now some RNA is always found in association with the nuclear DNA. It may therefore be supposed that some fraction of the total RNA of the cell is synthesized in the nucleus, receives information from the nuclear DNA, and then migrates into the cytoplasm, where its information can be used to direct and control the sequential arrangement of amino acids in newly synthesized proteins. This hypothesis, that there is a *messenger RNA* of some kind which carries information from the nuclear DNA to the intracellular sites of protein formation, has gained widespread acceptance. There is much experimental evidence in its favour today.

There remains the problem of how an orderly and organized specific sequence of nucleotides in the messenger RNA (*m*-RNA) could lead to the formation of equally ordered and organized chains of amino-acids. Since 20 amino-acids have to be dealt with and only 4 individual nucleotides are usually available to marshal them, it is evident that several nucleotides must be required for the siting of each amino-acid species. This is a problem that has attracted much argument and speculation, too involved and too delicate to be considered here, but the bulk of informed opinion leads to the belief that, as a rule, 3 consecutive nucleotide units, known collectively as a *codon*, are required for the selection and positioning of each amino-acid unit.

Activation of amino-acids

It seems to be a general rule that the first step in the activation of organic acids consists in a reaction with ATP in which pyrophosphate is split off and the *acyl adenylate* is formed:

$$R.COOH + ATP \longrightarrow R.AMP + ⓅⓅ.$$

Then, in the activation of benzoic acid for the formation of hippuric acid for example (p. 98), the adenylate grouping is exchanged for coenzyme A. The same thing happens in the activation of fatty acids generally, for breakdown and synthesis alike. Amino-acids are similarly transformed into their adenylates as a first step in the direction of protein synthesis and the cell sap contains a group of enzymes each of which is highly specific towards some one particular individual member of the usual 20 amino-acids. These enzymes catalyse the formation of the amino-acyl adenylates at the expense of ATP.

Now in addition to the messenger and ribosomal fractions of RNA, cells in general contain a fraction of much lower molecular weight. This includes the so-called 'soluble' or *transfer RNA* (*t*-RNA), a number of polynucleotides which react specifically with specific amino-acyl adenylates to give amino-acyl-*t*-RNA compounds:

amino-acyl adenylate + *t*-RNA = amino-acyl-*t*-RNA + adenylate.

Again there is a specific *t*-RNA for each individual species of amino-acid, and it seems likely, in fact, that the amino-acid, the ATP and the *t*-RNA are accommodated at three adjacent sites on one and the same common enzyme protein.

Ribosomal RNA accounts for the greater part of all the RNA present in the cytoplasm and is of very high molecular weight. The ribosomes themselves can rather easily be broken up into two dissimilar parts with molecular weights of the order of 10^6 and 3×10^6 respectively, and hence the use of an inverted cottage loaf-like symbol (as in Fig. 94, p. 319), to represent them. Certainly the ribosomal RNA's play an important and indeed a vital part in protein synthesis but do not seem to have any specific information of their own. However, ribosomes are usually found in intimate association with longer or shorter fragments of messenger RNA and, as each amino-acid residue is added to the growing polypeptide chain, there is reason to think that the ribosomes move from one codon to the next along the *m*-RNA chain.

The t-RNA molecules are very different. Often, if not always, they contain nucleotides other than the four usually found in nucleic acids; each molecule contains only about 70–80 nucleotide units with a molecular weight of the order of about 25,000. It is known too that these molecules carry guanylic acid at one end of the chain and cytidyl-cytidyl-adenosine at the other. It is evidently to this latter sequence that the amino-acid units are attached (through their carboxyl groups), since if this —C—C—A— is in any way altered the *t*-RNA no longer reacts. The high specificity of these substances towards their respective amino-acyl adenylates must therefore be due to differences in the internal parts of their structures.

Synthesis of peptide bonds

Although there are still many gaps in our knowledge there are experimental and logical indications that each amino-acyl-*t*-RNA is directed to a specified and predetermined site by a codon of the *m*-RNA, which serves as an assembly template. We do not know in any certain detail how this specific siting is carried out except that a group of three adjacent nucleotides seem to be required to specify the siting, and that guanosine triphosphate (GTP) is implicated in some as yet unknown manner. The identity and sequence of these three nucleotides, collectively called a *codon*, presumably varies in a very specific manner, and each codon is specifically responsible for the choice and positioning of one particular species of amino-acyl-*t*-RNA.

Eventually the *t*-RNA molecules are released and the newly formed polypeptide or protein begins its existence as a free, structural or metabolic entity. These ideas can be represented in a diagram such as follows:

amino-acids ⟶ ATP ⟵ AMP ⟶ *t*-RNA-amino-acids ⟶ *m*-RNA ⟶ PROTEIN

ⓅⓅ ⟵ 'adenylates ⟵ *t*-RNA ⟵ *m*-RNA-peptides

activation cycle   transfer cycle   assembly cycle

Experimental evidence in support of these ideas is accumulating fast and one such piece of evidence may be mentioned here. Polyuridylic acid has been prepared by means of Ochoa's enzyme to act as a model of a messenger RNA. 'Transfer' RNA and ribosomes from *Escherichia coli* were added, together with a mixture of amino-acids. On incubation a polypeptide was formed and proved to consist entirely of phenylalanine units.

Triplet codes have now been established for a number of amino-acids, an operation calling for much experimental—and mental—ingenuity. Altogether 64 codons are possible if the 4 common nucleotides participate and 56 of these have positively been identified. This, of course, is substantially in excess of the 20 amino-acids to be manipulated and several alternative explanations have been advanced to account for this. Some amino-acids may be coded by more than one codon, and some codons may be used as signals for starting and stopping protein formation. Some may even be useless and totally redundant. Current conclusions about triplet requirements are summarized in Table 25.

Now, since different amino-acyl-*t*-RNA molecules are evidently recognized by different triplet codons on *m*-RNA it follows that they themselves must carry recognizable sites. It is probable that these 'recognition' sites are also triplets, usually called *anti-codons*. A given amino-acyl-*t*-RNA molecule will thus be 'recognized' by a given codon and so gain admission to its specific and predetermined site in the peptide chain. The identities of these anti-codons or recognition sites are imperfectly known at present.

At the start of protein synthesis—and we do not at present know how the start and finish of the process are controlled—a number of *t*-RNA-bound amino-acids will be present in the environment, each with its own specific anti-codon. One of these will be specifically recognized by the first codon in the *m*-RNA and so becomes attached to the ribosome upon which the synthesis presumably takes place. The ribosome now moves on to the next codon and a second amino-acyl-*t*-RNA molecule is recognized and admitted, but the new complex is unstable because the —NH₂ of the second amino-acid residue attacks and disrupts the bond joining the carboxyl group to the first *t*-RNA. An illustration of these reactions is shown in Fig. 73. The first amino-acid is separated from its *t*-RNA but remains in peptide linkage with the second. A third amino-acyl-*t*-RNA is now attached in the same specific manner, again the carboxyl-*t*-RNA bond is disrupted and the product is a

tripeptide in attachment to the third *t*-RNA molecule, which, in common
with its predecessors and those that come after, is released to seek out another
molecule of its specifically defined amino-acyl adenylate.

As more and more amino-acid residues are attached this cycle of events is
repeated over and over again, the newly synthesized peptide chain is released

Table 25. *The genetic code*

(Modified after Crick. Copyright © (Oct. 1966) by Scientific American, Inc.).

Second letter

		U	C	A	G	
First Letter	U	UUU UUC } Phe UUA UUG } Leu	UCU UCC UCA UCG } Ser	UAU UAC } Tyr UAA[1] ? UAG[1] ?	UGU UGC } Cys UGA ? UGG Tryp	U C A G
	C	CUU CUC CUA CUG } Leu	CCU CCC CCA CCG } Pro	CAU CAC } His CAA CAG } GluN[2]	CGU CGC CGA CGG } Arg	U C A G
	A	AUU AUC } Ileu[4] AUA AUG } Met	ACU ACC ACA ACG } Thr	AAU AAC } AspN[3] AAA AAG } Lys	AGU AGC } Ser AGA AGG } Arg	U C A G
	G	GUU GUC GUA GUG } Val	GCU GCC GCA GCG } Ala	GAU GAC } Asp GAA GAG } Glu	GGU GGC GGA GGG } Gly	U C A G

Third Letter

Amino acids are shown by their initial letters.
[1] May be 'stop' signals in protein synthesis.   [2] Glutamine.
[3] Asparagine.   [4] Isoleucine.

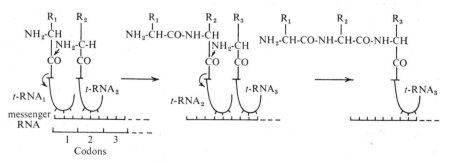

Fig. 73. Peptide-bond formation from amino-acyl-*t*-RNA.

from the ribosomal RNA step by step as it is formed until, eventually, the new protein molecule is completed, but remains still with its last, terminal residue attached to the last *t*-RNA molecule to be employed. How the final stage, the removal of this last *t*-RNA is accomplished, we do not know, but there is good experimental evidence that each of the attendant ribosomes moves, codon by codon, along the messenger RNA, and it may be that there are among the apparently redundant codons, specific nucleotide triplets for 'start' and 'stop'.

Not infrequently ribosomes as usually isolated occur not singly but in groups, all attached to a common molecule of *m*-RNA. These complexes are known as polyribosomes or *polysomes* and presumably each in turn works its way from the beginning to the end of the mutually shared *m*-RNA, each producing a new molecule of the same protein. (See Fig. 74.)

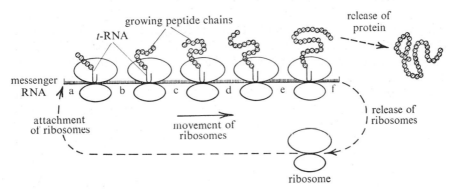

Fig. 74. Protein synthesis by polyribosomes.

Nuclear DNA, then, gets its information, presumably from parental DNA in the first place, and imprints it in code form on newly synthesized messenger RNA, which carries it then to the assembly shops in the ribosomes. Each codon in turn specifically admits an amino-acyl-*t*-RNA to the site of synthesis, then another and another until a long-chain polypeptide or protein is produced, and the recognition of each of the new *t*-RNA compounds is accomplished by some kind of pairing between mutually specific codons and anti-codons—a very remarkable case of biological specificity.

Even when the details of this story have been filled in there will still be many problems. DNA controls protein synthesis; in particular it controls enzyme synthesis and, moreover, enzymes are required for its own replication. We thus seem to be approaching something remarkably like the classical problem of the hen and the egg; which came first, DNA or the enzymes that make it?

16

CARBOHYDRATE PRODUCTION IN THE GREEN PLANT

INTRODUCTION

The green plants possess specialized machinery which enables them to draw energy for their synthetic operations from solar radiation. The photosynthetic formation of carbohydrates from CO_2 can be classified as an anaerobic process; indeed, while it is going on oxygen is actually evolved. In the presence of light, carbon dioxide, water and a handful of simple inorganic salts the green plant can synthesize an astonishing variety of organic products. Everything an animal can produce a plant can produce, and animals depend upon the synthetic powers of the plants for their supplies of essential amino-acids, vitamins and other compounds which the plants, but not they themselves, can manufacture.

In this section we shall consider in particular the photosynthetic formation of carbohydrates, especially of starch and sucrose, probably the most abundant carbohydrates in the world, apart from cellulose. We have, in point of fact, already reviewed the enzymes and processes involved in the production of amylose and amylopectin from glucose and the glucose phosphates (pp. 93–7), and as we shall see shortly, any substance lying on the glycolytic reaction chain, or giving rise to any substance lying on that chain, is convertible into one or other of the glucose and fructose phosphates.

The overall equation for photosynthesis is the precise reverse of that for the oxidative metabolism of carbohydrate:

$$6CO_2 + 6H_2O \underset{\text{oxidation}}{\overset{\text{photosynthesis}}{\rightleftharpoons}} 6O_2 + C_6H_{12}O_6.$$

The most striking features of photosynthesis are first, that *carbon dioxide is reduced* to the carbohydrate level, secondly that *water is oxidized*, so that oxygen is evolved, and finally that *ATP is formed* from ADP and inorganic phosphate. It must be said at once that these processes, simple though they may appear on paper, are very complex in reality and still a long way from being fully known and understood. It follows that the account we can give here must necessarily be incomplete and in some ways superficial. A good deal is known about the fixation of CO_2 and about its conversion into carbohydrate, but the mechanisms lying behind the evolution of oxygen are still speculative for the most part and our knowledge of the detailed mechanisms that lie behind the synthesis of ATP is still far from complete.

Presumably the primary event in phosphosynthesis is the interaction of a photon with a chlorophyll molecule, thus raising an electron to a higher energy state. In solutions of chlorophyll this electron can return to the ground state and the excess energy then reappears in the form of fluorescent light; solutions of chlorophyll do, in fact, display a fine red fluorescence during, and for a brief period after, exposure to light.

In intact chloroplasts, however, the return to the ground state is not necessarily an all-in-one operation; more probably the high energy electrons are tapped off in some way and fed into a series of hydrogen or electron transporting reactions, comparable in many ways with the respiratory re-action chains of animal tissues. ATP is generated at several of the intermediate steps, just as it is in oxidative phosphorylation, and in the present context it is usual to speak of *photophosphorylation*. Eventually, the surplus energy having been either captured in the form of ATP or dissipated, the electron is free to go back to the chlorophyll molecule from which it came.

Now it is possible to consider the processes involved in photosynthesis in two parts, one concerned with the photolytic splitting of water, which requires light, and the fixation of carbon dioxide, which depends in its earliest stages upon the photolysis of water and is therefore light-dependent but which, once the earliest stages have been accomplished, can continue in the dark. Both are attended by the synthesis of ATP.

GENERATION OF ATP

The chloroplasts of green plants contain, apart of course from chlorophyll itself and a variety of carotenoid pigments, a considerable number of revers-ibly oxidizable and reducible compounds. These include (i) *ferredoxin*, a reddish brown, iron-containing protein, (ii) *plastoquinone* which is related to ubiquinone (coenzyme Q), (iii) *plastocyanin*, a copper-containing protein, together with (iv) flavoproteins, (v) NAD, NADP and (vi) two special cyto-chromes peculiar to green plants, *cytochromes f and b_6*. There is however *no cytochrome oxidase*.

It seems certain that ferredoxin is the first electron-acceptor in photosynthesis and that its burden of electrons flows along a chain of carriers of the following kind

$$\longrightarrow \text{ferredoxin} \longrightarrow \text{flavoprotein} \longrightarrow \text{NADP}$$

One or more of these transfer reactions is attended by a coupled synthesis of ATP by mechanisms resembling those of oxidative phosphorylation.

Photophosphorylation of this kind can account for much, but by no means for all, of the ATP generated in the course of photosynthesis. Incidentally ferredoxin, or at least substances closely resembling it, are known in bacterial and animal tissues; succinate dehydrogenase, for example, certainly contains 'non-haem' iron (p. 178), at least a part of which is thought to be a ferredoxin-like protein. But all this time the chlorophyll has been short of one or more electrons.

EVOLUTION OF OXYGEN

The origin of the oxygen evolved by an actively photosynthesizing system has been much discussed. For many years it was attributed to the CO_2 consumed in this process, but in more recent times CO_2 has been superceded by water as the most likely oxygen source. A new break-through came from some work by van Neil on certain photosynthetic bacteria. Some of these organisms, green in colour, contain a bacterial counterpart of plant chlorophyll. They can live under strictly anaerobic conditions, and in the presence of CO_2 they thrive on hydrogen sulphide, which they oxidize to free, elemental sulphur according to the following equation:

$$CO_2 + 2H_2S = (CH_2O) + H_2O + 2S.$$

Here (CH_2O) represents carbohydrate material, which becomes incorporated into the bacteria themselves. The free sulphur here produced must have come, and can only have come, from some kind of (presumably photolytic) splitting of H_2S.

There is a close and obvious parallel between this and the usual equation for photosynthesis in the green plant:

$$CO_2 + 2H_2O = (CH_2O) + H_2O + O_2 \quad \text{or} \quad CO_2 + H_2O = (CH_2O) + O_2.$$

The resemblance is clear enough and, if the facts and ideas of comparative biochemistry are any guide, it is but a short step to suggest that, just as sulphur arises by the splitting of hydrogen *sulphide* in the green bacteria, so likewise must oxygen arise by the splitting of hydrogen *oxide* in the green plants. In both cases the process requires light and hence we can speak of *photolysis*.

Some interesting experiments have been carried out in which green plants were allowed to photosynthesize in water enriched with heavy oxygen (^{18}O), when the oxygen evolved proved to contain the isotope. On the surface this seems to be proof enough, but because of the likelihood of exchange reactions involving the oxygen, these results do not provide any rigid proof.

However, as was first shown by R. Hill, oxygen is evolved by suspensions of chloroplasts isolated from green leaves. Preparations of this kind not only evolve oxygen; they will reduce any of a wide range of reagents at the same time. Hill used ferric oxalate and showed that there is a stoichiometric relation between the amounts of oxygen evolved and ferric oxalate reduced. But no CO_2 was fixed in these experiments.

322

REDUCTION OF CARBON DIOXIDE

The splitting of water involves light, chlorophyll and a number of other reactants. Certainly it is a complex process and, equally certainly, it is still far from being understood, despite its importance and fascination. Suffice it for our purposes to say that the photolytic process, in which water is oxidized to oxygen, must necessarily be associated with the reduction of some other substance. If we represent the substance undergoing reduction as X we can write:

$$12H_2O + 12X \longrightarrow 6O_2 + 12X.H_2.$$

The fixation of carbon dioxide is a reductive process for which we might write:

$$6CO_2 + 12XH_2 \longrightarrow C_6H_{12}O_6 + 12X + 6H_2O.$$

The resultant of the two equations is of course,

$$6CO_2 + 6H_2O \longrightarrow C_6H_{12}O_6 + 6O_2,$$

leaving X as an indeterminate compound or group of compounds.

Now it is known that the reductive stage in CO_2-fixation is catalysed by an NADP-specific triosephosphate dehydrogenase, working 'in reverse' and requiring a continuous supply of $NADP.H_2$ to continue in operation. Since this reaction ceases immediately illumination is cut off, it seems reasonable to think that the reduction of NADP must be coupled in some way to the photolytic splitting of water.

THE PHOTOLYTIC PROCESS

Activated electrons appear to be accepted first by ferredoxin and are then transferred along a chain of electron-transporting carriers, creating a 'reducing atmosphere' as they go. This, evidently, is because no cytochrome oxidase is present to create an oxidizing atmosphere of the kind with which we are so familiar in animal tissues. One consequence of this is that NADP is reduced to $NADP.H_2$, in which form it is required and utilized at an early stage in CO_2-fixation. We shall examine this process shortly. When activated electrons leave chlorophyll the latter must, at least for a time, be positively changed. Neutrality is restored by importing an electron derived from water through a chain of carriers, probably including plastocyanin, various quinones and the cytochromes.

One thing is certain; there are at least two, but probably not more than two, processes requiring light, one for the production of oxygen and a second for the reduction of ferredoxin, and these two processes are in some way linked by a chain of electron carriers. Clearly enough much has still to be done. But happily much more is known about the fixation, reduction and conversion into carbohydrate material of carbon dioxide itself. Over a great number of years attempts have been made to discover the first-formed product of CO_2-fixation and many theories have been propounded, but it is only in the last decade that any convincing evidence has been forthcoming. The newer work has only been made possible through the ready availability of isotopic carbon and the techniques of paper chromatography.

If suspensions of green algae such as *Chlorella* or *Scenedesmus* are illuminated in the presence of ^{14}C in the form of carbon dioxide or bicarbonate, the isotope is rapidly incorporated into a large number of different substances, including sugars, polysaccharides and amino-acids. But if the period of illumination is short enough, say 0·1 sec., one substance and one substance only contains the isotope, viz. 3-phosphoglycerate. *For each molecule of carbon dioxide fixed, 2 molecules of this product are formed.* With longer periods of illumination, e.g. 1–5 sec., it is possible to observe the passage of the isotope into other substances, notably triosephosphates, fructose mono- and diphosphates and certain other sugar derivatives, and it has in fact been possible to trace out the early stages of the photosynthetic pathway in this manner. Unfortunately, however, there are many interlinking metabolic pathways and the labelled carbon follows them all, so that picking out the paths that lead to carbohydrate formation has had to depend a good deal upon the separation of enzymes from plant materials and piecing together the reactions they catalyse.

Since one molecule of carbon dioxide gives rise to two molecules of phosphoglycerate, each with 3 carbon atoms, it follows that a 5-carbon acceptor of some kind must probably be involved. Such a precursor has indeed been detected and identified as the 1:5-diphosphate of the pentose sugar, ribulose. An enzyme, variously known as carboxydismutase and *ribulose diphosphate carboxylyase* has been purified from spinach leaves and from extracts of *Chlorella* and shown to catalyse the formation of 2 molecules of 3-phosphoglycerate from CO_2 and ribulose-1:5-diphosphate. The reaction is a curious one and the overall process is the following. A more detailed scheme is shown in Fig. 75.

$$
\text{A.} \quad
\begin{array}{c}
CH_2O\circled{P} \\
| \\
CO \\
| \\
CHOH \\
| \\
CHOH \\
| \\
CH_2O\circled{P} \\
\text{(keto-)}
\end{array}
\rightleftharpoons
\begin{array}{c}
CH_2O\circled{P} \\
| \\
C.OH \\
\| \\
C.OH \\
| \\
CHOH \\
| \\
CH_2O\circled{P} \\
\text{(enol-)}
\end{array}
+ H_2CO_3 \longrightarrow
\begin{array}{c}
CH_2O\circled{P} \\
| \\
CHOH \\
| \\
COOH \\
+ \\
COOH \\
| \\
CHOH \\
| \\
CH_2O\circled{P} \\
\textit{3-phosphoglycerate}
\end{array}
$$

ribulose-1:5-*diphosphate*

This is the first of a network of reactions that lead back eventually to the regeneration of ribulose diphosphate.

From phosphoglycerate the isotope passes on to triosephosphate. Green plants contain several *triosephosphate dehydrogenases* at least one of which resembles that found in animals (p. 164). In particular, however, there is an *NADP-specific triosephosphate dehydrogenase* that is confined to the green parts of plants and catalyses the reduction of phosphoglycerate to glyceraldehyde phosphate at the expense of reduced NADP. This reaction stops immediately illumination is cut off, indicating that the production of the

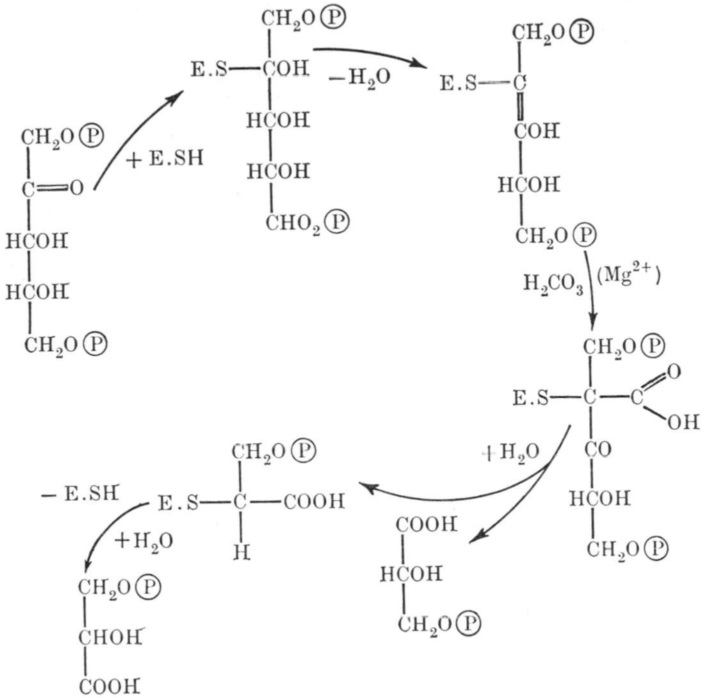

Fig. 75. Action of ribulose diphosphate carboxylyase. (E.SH represents the enzyme, which is —SH dependent.)

reduced NADP is light-dependent and is, in fact, coupled in some way to the photolytic splitting of water:

$$
\text{B.} \quad
\begin{array}{c}
CH_2O\textcircled{P} \\
| \\
CHOH \\
| \\
COOH
\end{array}
+ NADP.H_2 \longrightarrow
\begin{array}{c}
CH_2O\textcircled{P} \\
| \\
CHOH \\
| \\
CHO
\end{array}
+ NADP
$$

Then, under the influence of *triosephosphate isomerase*, equilibrium is set up between the two forms of triose phosphate:

$$
\text{C.} \quad
\begin{array}{c}
CH_2O\textcircled{P} \\
| \\
CHOH \\
| \\
CHO
\end{array}
\rightleftharpoons
\begin{array}{c}
CH_2O\textcircled{P} \\
| \\
CO \\
| \\
CH_2OH
\end{array}
$$

325

The triosephosphates lie on the direct lines of fermentation, glycolysis and carbohydrate synthesis and their fates in these reaction sequences will be discussed in later chapters. Their immediate fate in photosynthesis is at first the same as in the synthesis of carbohydrate in animal and other tissues, for fructose-1:6-diphosphate and fructose-6-phosphate are formed, but from this point on the processes become more involved. They can be followed by means of the chart shown in Fig. 76. All the enzymes concerned in catalysing these reactions, and all the intermediate substances, are detectable in plants and all the intermediates can be isotopically labelled. Most of the enzymes have been isolated in crystalline form and highly purified.

One molecule of each of the two forms of triosephosphate react together under the influence of a typical *aldolase* (reaction 1) to form fructose-1:6-diphosphate. This is a well-known reaction in all kinds of cells. The next step consists in the removal of one phosphate group (2) under the influence of a specific *phosphatase*, giving fructose-6-phosphate. This reaction too is well known and constitutes another step in the direction of carbohydrate synthesis, but now the pathways diverge.

Table 26. *Enzymes concerned in photosynthetic carbon-fixation*

Reaction	Enzyme
A	Ribulose diphosphate carboxylyase
B	Triosephosphate dehydrogenase (NADP-specific)
C	Triosephosphate isomerase
1	Aldolase
2	Specific phosphatase
3	Transketolase
4	Ribulose phosphate epimerase
5	Phosphoribulokinase
6	Aldolase
7	Specific phosphatase
8	Transketolase
9	Ribulose phosphate epimerase
10	Ribulose phosphate isomerase
11	Phosphoribulokinase

The next reaction (3) is catalysed by *transketolase*. The two terminal carbon atoms of fructose-6-phosphate are transferred to a molecule of glyceraldehyde phosphate to give xylulose-5-phosphate, which is transformed by *ribulose phosphate epimerase* into ribulose-5-phosphate (reaction 4), and the latter, under the influence of a specific *phosphoribulokinase*, reacts with ATP (5) to regenerate ribulose-1:5-diphosphate.

But this is not all. The second product of reaction (3) is the 4-phosphate of the tetrose sugar erythrose, and further supplies of ribulose diphosphate arise from this source in the following manner. The tetrose phosphate reacts with a molecule of dihydroxyacetone phosphate in a typical *aldolase* reaction (6), and

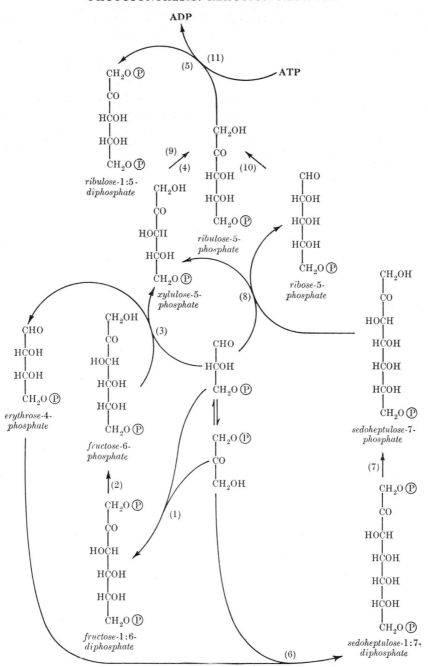

Fig. 76. Reaction network in CO_2 fixation. The formation of triosephosphate is described in the text. See Table 26 for summary of enzymes involved.

gives rise to sedoheptulose-1:7-diphosphate which is attacked by a specific *phosphatase* and yields sedoheptulose-7-phosphate (reaction 7). The product now reacts with another molecule of glyceraldehyde phosphate, with *transketolase* acting as catalyst (reaction 8). The products are: xylulose-5-phosphate, which epimerises (reaction 9) to give ribulose-5-phosphate: and ribose-5-phosphate, which is converted into ribulose-5-phosphate by *ribulose phosphate isomerase* (reaction 10), and the two molecules of ribulose-5-phosphate thus produced are further phosphorylated by *phosphoribulokinase* and ATP to form two molecules of ribulose-1:5-diphosphate (reaction 11).

In all, therefore, three molecules of ribulose-1:5-diphosphate are formed by the complete reaction network.

PRODUCTION OF STARCH

For the production of these 3 molecules of ribulose diphosphate 5 molecules of triosephosphate are consumed, 3 of the aldose and 2 of the ketose, the production of which would require the fixation of $2\frac{1}{2}$ molecules of CO_2 by $2\frac{1}{2}$ molecules of ribulose diphosphate in reaction A. Putting this into terms of whole molecules:

$$\text{5 pentose diphosphate} + 5CO_2 \longrightarrow \text{6 pentose diphosphate.}$$
$$\text{(25 carbon)} \qquad\qquad\qquad\qquad \text{(30 carbon)}$$

Presumably, however, the 5 additional carbon atoms, although they *could* pass on and give ribulose diphosphate, do not do so, since after a long enough period of illumination the plant would eventually consist almost entirely of this substance. One must therefore suppose that enough ribulose diphosphate is retained to allow of the optimal rate of carbon dioxide fixation under the conditions prevailing, and that the remainder of the carbon so fixed passes out of the reaction network through triosephosphate, fructose diphosphate or fructose monophosphate, all of which lie on the direct route for the synthesis of carbohydrate, whether as simple monosaccharides, di- or higher or even polysaccharides, particularly starch.

In the case of synthesis we know that the following chain of reactions can take place under the influence of enzymes known to occur in plants: all these enzymes are described in Part I of this book.

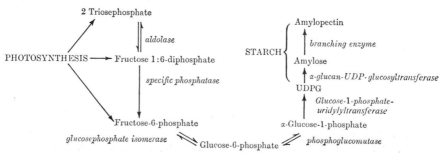

PRODUCTION OF SUCROSE

If photosynthesis is allowed to proceed in the presence of $^{14}CO_2$, sucrose appears fairly early and contains isotopic carbon. It has been established that sucrose can be formed in plant material (wheat germ) through a reaction in which uridine diphosphate glucose is involved. Fructose-6-phosphate, an intermediate in the photosynthetic network, also enters into the synthesis:

(a) Fructose-6-Ⓟ + UDPG → sucrose-Ⓟ + UDP,
(b) Sucrose-Ⓟ + H_2O → sucrose + HO.Ⓟ.

The enzymes concerned are *sucrose phosphate-UDP-glucosyltransferase* and a *sucrose phosphate-phosphohydrolase* respectively. Sucrose glucosyltransferase (p. 91) appears not to be present in plant materials.

It will be noticed that if this synthesis is to be a continuous process, UDP must be reconverted to UDPG. This takes place by two reactions, in the first of which UDP reacts with ATP:

$$UDP + ATP \rightarrow UTP + ADP.$$

This is followed by a reaction between UTP and α-glucose-1-phosphate, which can be formed from fructose-6-phosphate by way of glucose-6-phosphate:

UTP + glucose-1-phosphate ⇌ UDPG + pyrophosphate,

catalysed by *glucose*-1-*phosphate-uridylyltransferase*.

PRODUCTION OF OTHER CARBOHYDRATES

Plants are known to contain a variety of polysaccharides besides amylose and amylopectin. Members of the Compositae store inulin, a polyfructofuranoside, rather than starch. Other polysaccharides contain other sugars and sugar derivatives, e.g. galactose, N-acetylglucosamine, and, although we know that UDPG plays a part in the production of a number of the monosaccharide units from glucose, little is yet known about the details of the mechanisms concerned.

In the case of galactose, however, more is known. It is formed by an epimerase that occurs in animals and galactose-trained yeasts as well as in plants, and is discussed on p. 121.

UDP plays a part in a great many processes in which carbohydrates are involved and is known to form compounds not only with glucose and galactose but with N-acetylglucosamine and with glucuronic acid, with N-acetylgalactosamine, galactonic acid and even with glycerol, all of which enter into one or another of the numerous known plant oligo- and polysaccharides. Some workers believe that UDP plays a part in transglycosylation reactions generally, while UDP and UMP certainly play an important part in the synthesis of glycogen and starch, and that of many other polysaccharides too in all probability (p. 97). The many new discoveries concerning UDP and its functions suggest that a long new chapter in carbohydrate biochemistry is still being written.

FORMATION OF AMINO-ACIDS

In addition to undergoing synthetic reactions to produce polysaccharides, sucrose and other substances, triosephosphate can follow the katabolic pathways of glycolysis to produce pyruvate. Under oxidative conditions pyruvate enters into the citric acid cycle, which we shall deal with later, in the course of which α-ketoglutarate and oxaloacetate are formed. Like pyruvate these are α-keto-acids which can be aminated or transaminated to yield the corresponding α-amino-acids, alanine, glutamate and aspartate, respectively.

Aspartate and glutamate can act as transaminating agents for other α-keto-acids and the green plants can, apparently, produce α-keto-acids corresponding to all the naturally-occurring amino-acids and can transform these into the amino-acids themselves by transamination. Moreover they can store large amounts of nitrogen in the form of either asparagine, or glutamine, or both.

17

ANAEROBIC METABOLISM OF CARBOHYDRATES: ALCOHOLIC FERMENTATION

INTRODUCTION

A great deal was known about the anaerobic metabolism of carbohydrates in yeast and in muscle many years before much was learned about their aerobic breakdown. It may strike the reader as curious that these two kinds of cells, so different in their organization and function, should have been selected for examination rather than any others. Yeast, however, has long been an organism of great commercial importance for the production of alcoholic beverages and for the manufacture of industrial alcohol. Furthermore, various important by-products of fermentation, such as the components of fusel oil, find many important applications in chemical technology. No wonder, then, that alcoholic fermentation has been extensively studied. In the case of muscle, interest was originally aroused by more academic considerations. Muscle does mechanical work. Many muscles can be isolated and made to contract outside the body, and in these, beyond all other tissues, we have an opportunity of measuring the amount of work done by a biological system and attempting to correlate with it the amount of chemical change simultaneously taking place. It was rather late in the history of the subject before it was realized that, in spite of their many apparent differences, yeast and muscle both derive the energy they expend through very similar chemical manipulations of their carbohydrate starting-materials. It has, moreover, been discovered that the metabolism of carbohydrate follows substantially the same pathways in all kinds of animal and plant tissues, as well as in many micro-organisms, and it follows that knowledge gained by the study of fermentation has very wide applicability.

The enzymes concerned in the anaerobic degradation of carbohydrates can readily be obtained in particle-free cell extracts, whereas those concerned with their oxidative metabolism are intimately bound up with the particulate elements of the cell, especially the mitochondria. Anaerobic metabolism can therefore proceed and can be studied independently by the use of simple aqueous extracts of cells and tissues after the solid matter has been removed by filtration or centrifugation. We know now that the aerobic metabolism of starch, glycogen and glucose is, so to speak, a continuation of their anaerobic metabolism and, moreover, that it is much more complicated. Here, therefore, we shall deal first with anaerobic and with aerobic metabolism later. Starch, glycogen and glucose provide major sources of energy for plants and animals, to make no mention of the innumerable micro-organisms which likewise derive energy and employment from the breakdown of these substances. Their breakdown is attended by the liberation of some at least of the intrinsic energy of the carbohydrate molecule, but how much of the free, or available energy becomes *biologically* accessible to any given organism depends upon the nature of the chemical changes the organism is able to accomplish.

Now glucose can be broken down in other ways than by complete oxidation. Yeast can carry out an anaerobic *fermentation* of glucose, yielding ethyl alcohol and carbon dioxide:

$$C_6H_{12}O_6 = 2C_2H_5OH + 2CO_2.$$

Muscle cells, working under anaerobic conditions, can convert glucose into lactate, a process known as *glycolysis*:

$$C_6H_{12}O_6 = 2CH_3CH(OH)COOH.$$

In neither of these transformations is the loss of free energy as large as it is in complete combustion, for a large proportion of the total intrinsic energy of the starting material remains bound up in the products, and access to this can only be gained by further degradation of these substances. Aerobic metabolism is far more efficient than anaerobic.

To gain access to the same amount of energy, in fact, a muscle will need to glycolyse somewhere about 10–20 times as much glucose as it will if it oxidizes glucose to carbon dioxide and water completely. But in neither case does it necessarily follow that the cell or tissue can actually harness and utilize *all* the energy to which it gains access by oxidizing, glycolysing or fermenting its food materials. How this energy is trapped, and how much of it can be trapped, we shall see in ensuing chapters, but at the present time we have only a decidedly crude knowledge of biochemical energetics. One of the tasks confronting biochemistry is the invention and development of a new thermodynamics.

It is convenient to classify organisms as aerobic or anaerobic as the case may be. Relatively few living organisms are strictly anaerobic. Indeed, as far as we know, strict anaerobiosis is practically restricted to a few groups of bacteria, which are not merely unable to utilize oxygen but are actually poisoned by it. The vast majority of micro-organisms are facultative anaerobes, i.e. they can utilize oxygen when it is available and oxidize their foodstuffs completely, but can still survive under anaerobic conditions by catalysing a partial or 'fermentative' breakdown of the same food materials. Animals for the most part might almost be classified as 'strict aerobes', since few of them can live for long in complete absence of oxygen. Even so, animals of many kinds are known to live at the bottoms of the deep Canadian and other lakes, where oxygen is believed to be absent altogether, but certain 'fermentative' processes can still go on under anaerobic conditions even in the tissues of animals that normally live aerobically, provided that the 'oxygen debt' thus incurred can sufficiently soon be repaid.

332

ALCOHOLIC FERMENTATION OF GLUCOSE

Alcoholic fermentation has been familiar to the human species probably since prehistoric times, yet is was not until after 1857 that its cause was discovered. In that year Louis Pasteur was studying the lactic fermentation of milk and trying to discover its cause. The views held at that time look very strange by modern standards, for the great Liebig himself considered that the nitrogenous constituents of the fermenting mixture reacted with air, setting up 'unstabilizing vibrations' as they did so, and these vibrations were believed to rupture the fermenting molecules. The fact that a new fermentation could be initiated by inoculating the medium with a trace of an already fermenting fluid was attributed to the transference of vibrating material to the new medium.

Pasteur began his experiments with media containing very simple substances such as sugars, ammonium tartrate and mineral phosphates, none of which could reasonably be expected to develop 'unstabilizing vibrations'. His results were simple and clear-cut. Fermentation took place only in the presence of certain microscopic organisms, the lactate-producing bacteria of the present day. When precautions were taken to exclude these organisms, no fermentation occurred. Extending his studies to alcoholic fermentation in 1860, Pasteur showed that whenever it took place the appropriate micro-organism, in this case yeast, grew and multiplied. He therefore concluded that fermentation is a physiological process, intimately bound up with and wholly dependent upon the life of the yeast cell. In 1875, having shown that fermentation can take place in complete absence of oxygen, Pasteur defined fermentation as 'Life without oxygen'.

More than 20 years elapsed before the next major step forward, but in 1897 Hans and Eduard Buchner made the key discovery which opened the door not only to the investigation of the mechanisms of fermentation but to the whole of modern enzymology. Like many other great discoveries, that of the Buchners had in it an element of chance. They were primarily interested in making cell-free extracts of yeast for therapeutic purposes, and this they accomplished by grinding yeast with sand, mixing it with kieselguhr, and squeezing out the juice with a hydraulic press. There then arose the problem of preserving their product. Since it was to be used for experiments on animals the antiseptics of the day could not be used as preservatives, so they tried the method usual in kitchen-chemistry of adding large amounts of sucrose. This led to the momentous discovery that sucrose is rapidly fermented by yeast *juice*. Here, for the first time, fermentation was observed in the complete absence of living cells, and at last it was possible to study the process of alcoholic fermentation independently of all the other processes—growth, multiplication and excretion —which accompany fermentation in the living yeast cell.

The Buchners' work was soon followed by intensive studies of the properties of yeast juice. It was found capable of fermenting glucose, fructose, mannose, sucrose and maltose, all of which are fermented by living yeast. The disaccharides, sucrose and maltose, are broken down in some way to yield their constituent monosaccharides before being fermented. Glucose itself is almost quantitatively converted into ethyl alcohol and carbon dioxide according to the equation

$$C_6H_{12}O_6 = 2CO_2 + 2CH_3CH_2OH.$$

Fresh yeast juice is much less active than living yeast. The rate of fermentation can be followed by measurements of the rate of evolution of carbon dioxide, and experiments carried out in this way show that living yeast works 10–20 times as fast as an equivalent quantity of yeast juice. Moreover, the fermentative power of the yeast juice falls off rapidly with time. The juice is not inactivated by drying at 30–35° C. or by the addition of chloroform, but loses its activity if heated to 50° C., indicating that enzymes must be involved.

The first important step towards analysing the mode of action of yeast juice was made by Harden & Young in 1905. If fresh yeast juice is added to a solution of glucose at pH 5–6, fermentation begins almost at once. The rate of carbon dioxide production presently falls off, but the original rate can be restored by the addition of inorganic phosphate. The recovery is only temporary, however; the added phosphate disappears, and the rate of fermentation falls off as the concentration of free phosphate declines. The addition of more inorganic phosphate produces another burst of fermentation and so on.

The disappearance of added inorganic phosphate from fermenting mixtures suggested that organic phosphate esters must probably be formed and, as Harden & Young showed, this is indeed the case, for they were able to isolate such an ester in the form of fructofuranose-1:6-diphosphate ('hexose diphosphate'). This substance, like glucose, is fermented if added to an actively fermenting system, and must probably be an intermediate in the process of fermentation. Later Robison isolated another sugar phosphate, this time a monophosphate which, on detailed examination, proved to consist of an equilibrium mixture of glucopyranose-6-phosphate and fructofuranose-6-phosphate. Like hexose diphosphate these esters are fermentable. It seemed clear that these substances must arise by the coupling of inorganic phosphate with glucose, the respective esters probably arising in the order shown in Fig. 77: this could account for the effects of inorganic phosphate already mentioned. How these esters are formed, and in what way the fructose diphosphate is eventually converted into alcohol and carbon dioxide, are questions which were only answered over a period of decades and by the efforts of many workers in many different countries. Certain stages were first elucidated by studies of muscle extracts, for it became clear in time that the fermentation of glucose by yeast juice runs closely parallel to the glycolysis of glycogen by suitable muscle preparations.

The next fundamental step forward was also made by Harden & Young when they discovered that yeast juice loses its activity if dialysed. Activity could be restored to dialysed juice, either by adding the dialysate, or by means of small quantities of boiled juice. This showed that, in addition to enzymes, yeast juice contains dialysable, thermostable substances which function as coenzymes. Yeast juice thus came to be regarded as consisting of 'zymase', a non-dialysable, thermolabile enzyme, plus 'cozymase', a dialysable, thermostable fraction. We know now, of course, that zymase is in reality a complex system of enzymes and that cozymase consists not of one substance only but of several.

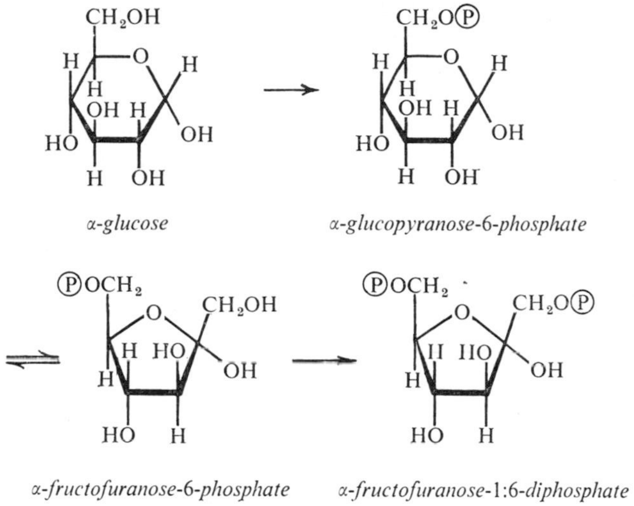

α-glucose   *α-glucopyranose-6-phosphate*

α-fructofuranose-6-phosphate *α-fructofuranose-1:6-diphosphate*

Fig. 77. Sugar esters in early stages of fermentation.

It is neither possible nor desirable here to give an historical account of subsequent work on the problem of fermentation. There were mistakes and gaps in the schemes that replaced one another in quick succession during the ensuing years, but one by one the mistakes were rectified and the gaps filled in until, at the present time, we have what we believe to be a clear picture of most of the details of the process. Before we can study this picture it is necessary to know something more about the composition of cozymase.

Cozymase comprises a number of factors. *Co-carboxylase*, now known to be identical with the pyrophosphate of vitamin B_1 (aneurin, thiamine), is the coenzyme of pyruvate decarboxylyase, an enzyme which catalyses the 'straight' decarboxylation of pyruvate to form acetaldehyde and carbon dioxide. (It is important that this enzyme should not be confused with the biotin-dependent pyruvate carboxylyase (p. 119).) *Nicotinamide adenine dinucleotide* (NAD), originally known as cozymase, we have discussed already. In fermentation,

335

as in respiration, this substance functions in collaboration with certain dehydrogenases as a hydrogen acceptor, donor and carrier. *Adenosine di-* and *triphosphates*, which we have discussed, act as phosphate carriers. *Magnesium ions*, too, are involved. They function as activators for many enzymes concerned with phosphate metabolism and, in particular, for enolase (phosphopyruvate hydratase), a lyase. In addition to these 'classical' coenzymes, dialysis removes other substances, including the ions of *inorganic phosphate*, *calcium* and *potassium*, and there is evidence that, for certain reactions at least, even these substances are of great importance and may strictly be classified as coenzymes of fermentation.

Since all of these co-substances are essential components of the fermenting system of yeast juice, it follows that the breakdown brought about by dialysis is due, not to the abolition of some one particular reaction, but of many. *Dialysed juice*, with and without the addition of one or more of the known cofactors, has therefore played a large part in unravelling the intricate reaction sequence that underlies fermentation. Much further information has been gained by taking advantage of the fact that certain substances have empirically been found to slow down or stop particular reactions. The addition of these *selective inhibitors* leads to the accumulation of intermediate products which can be isolated and identified. The reagents most widely used in the study of fermentation have been sodium bisulphite, sodium fluoride and sodium iodoacetate.

For purposes of discussion the reaction sequence of fermentation can be arbitrarily divided into several stages, each of which involves one or more individual chemical operations, but because we can dissect the whole process into stages and steps in this way it is not to be supposed that fermentation as such is a step-by-step process, catalysed by a mixture of enzymes. The living cell is something more than a mere bag full of enzymes; fermentation is a highly organized procession of chemical events, the overall result of which is the decomposition of glucose, with production of alcohol and carbon dioxide, all accompanied by the provision and capture of free energy which enables the cells to carry out the synthesis of the new tissue materials required for their maintenance, growth and reproduction.

(i) *Formation of phosphorylated sugars.* We have already seen that several sugar phosphates can be isolated from fermenting systems to which glucose and inorganic phosphate have been added. If glucose and inorganic phosphate are added to a dialysed juice, however, there is no fermentation and no sugar esters are formed, showing that one or another of the coenzymes must play a part in their synthesis. If ATP is added to the dialysed juice, phosphorylation of the sugar begins again and fructofuranose-1:6-diphosphate, fructofuranose-6-phosphate and glucopyranose-6-phosphate can be isolated from the system. Work with highly purified enzymes has resolved this stage into three separate reactions. First of all one phosphate group is transferred from ATP to glucose, yielding glucopyranose-6-phosphate and ADP, a process which is catalysed by *hexokinase* (reaction 1). Next the glucose ester is reversibly converted into fructofuranose-6-phosphate (reaction 2), the catalyst being *glucose-phosphate isomerase*. A phosphate group is then transferred from a second molecule of ATP to the fructose mono-ester, yielding fructofuranose-1:6-diphosphate. This reaction (reaction 3) is catalysed by *phosphofructokinase*. This group of reactions may be summarized as follows, *writing only the forward reactions for the sake of clarity*:

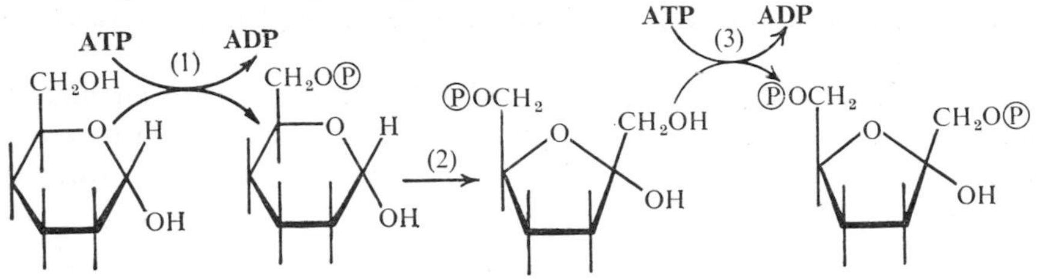

(ii) *Splitting of the hexose chain*. If glucose or one of the intermediate esters is added to yeast juice in the presence of iodoacetate, small amounts of 'triose phosphate' can be isolated, showing that hexose diphosphate is split into two 3-carbon fragments which, on isolation, prove to consist of an equilibrium mixture of D-glyceraldehyde-3-phosphate and dihydroxyacetone phosphate. The enzyme concerned, *aldolase*, has been isolated; and the reaction it catalyses (4) is known to be reversible:

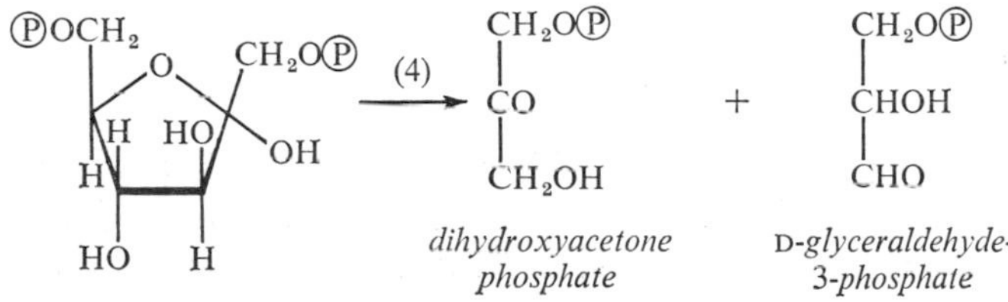

dihydroxyacetone phosphate *D-glyceraldehyde-3-phosphate*

Of the two components of triose phosphate, glyceraldehyde-3-phosphate is the more important from our present point of view, since its derivatives appear lower down in the reaction chain, whereas no direct derivatives of dihydroxyacetone phosphate are found under ordinary conditions. But dihydroxyacetone phosphate is not lost to the system, which contains a powerful *triosephosphate isomerase*. This enzyme catalyses the interconversion of the two triose phosphates (reaction 5): (Still writing only the forward reactions.)

CH_2OⓅ		CH_2OⓅ
|	(5)	|
CO	$\longrightarrow$	CHOH
|		|
CH_2OH		CHO
(96 %)		(4 %)

The original hexose has now been phosphorylated and quantitatively split into phosphorylated trioses.

(iii) *Oxidation of glyceraldehyde phosphate*. If hexose disphosphate or 'triose phosphate' is added to yeast juice in the presence of fluoride, two further phosphorylated derivatives of glycerol accumulate, viz. glycerol phosphate and glycerate-3-phosphate in equimolecular proportions. On isolation and examination the acid proves to be an equilibrium mixture of glycerate-2- and 3-phosphates. Of these the primary product must presumably be the 3-compound since it is formed from glyceraldehyde-3-phosphate. These products arise by the oxidation of a molecule of glyceraldehyde phosphate at the expense of the reduction of a molecule of dihydroxyacetone phosphate.

Yeast juice contains a powerful *triosephosphate dehydrogenase*, an enzyme which requires NAD, which is also present. But the amount of coenzyme is very small, and the whole would soon become reduced by acting as hydrogen acceptor for the oxidation of glyceraldehyde phosphate (reacting in its hydrated form):

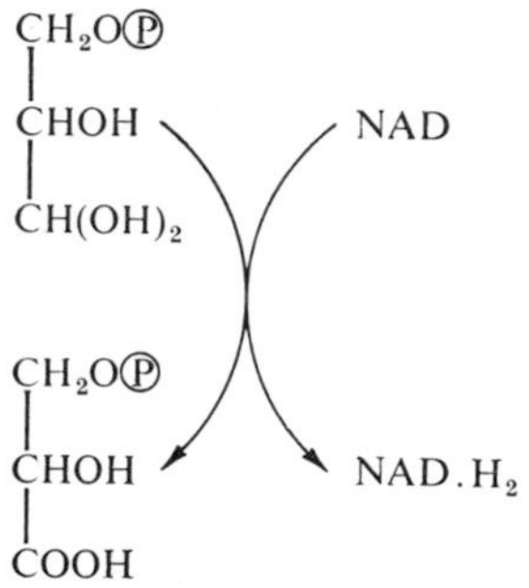

$$
\begin{array}{l}
CH_2O\textcircled{P} \\
|\\
CHOH \\
|\\
CH(OH)_2
\end{array}
\qquad\qquad NAD
$$

$$
\begin{array}{l}
CH_2O\textcircled{P} \\
|\\
CHOH \\
|\\
COOH
\end{array}
\qquad\qquad NAD.H_2
$$

Once all the available coenzyme had been reduced in this way, fermentation would come to an end. But yeast juice contains also a *soluble α-glycerol-phosphate dehydrogenase* (see p. 164). This dehydrogenase also co-operates with NAD and, like dehydrogenases generally, can act reversibly. The reduced coenzyme can therefore become reoxidized by passing on its 2H to a molecule of dihydroxyacetone phosphate, which is thereby reduced to L-glycerol-3-phosphate:

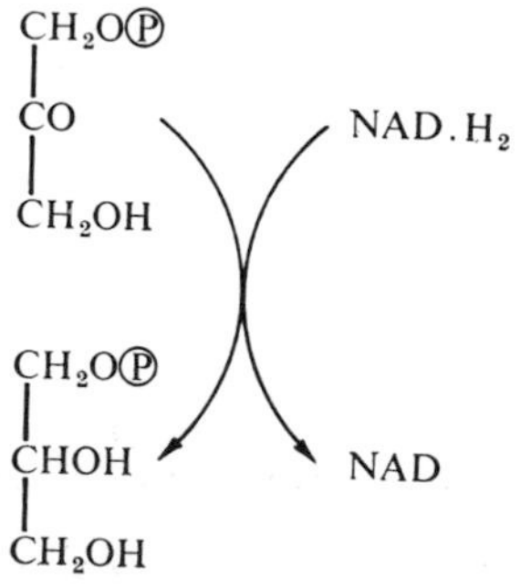

$$
\begin{array}{l}
CH_2O\textcircled{P} \\
|\\
CO \\
|\\
CH_2OH
\end{array}
\qquad\qquad NAD.H_2
$$

$$
\begin{array}{l}
CH_2O\textcircled{P} \\
|\\
CHOH \\
|\\
CH_2OH
\end{array}
\qquad\qquad NAD
$$

It should be noted that if D-glyceraldehyde-3-phosphate was reduced instead of the optically inactive dihydroxyacetone phosphate, D-glycerol phosphate would be formed.

As a result of this operation the reduced coenzyme becomes reoxidized and available, therefore, for the oxidation of another batch of molecules of glyceraldehyde phosphate.

It may be pointed out in passing that the reduction of the ketone to glycerol phosphate is not normally a large-scale process, but one that only takes place under unusual circumstances such, for instance, as when the system is poisoned with fluoride or blocked by sodium bisulphite, and the reasons for this we shall discover presently.

In more recent years it has been found that the oxidation of glyceraldehyde phosphate is a considerably more complex process than was formerly supposed. When glyceraldehyde phosphate is oxidized to glycerate phosphate in muscle extracts, one molecule of ATP is synthesized for every molecule of glycerate-3-phosphate formed. A similar phenomenon is also observed in yeast-juice fermentation, and attempts were made to discover its cause, using highly purified triosephosphate dehydrogenase. It was then found that no oxidation of glyceraldehyde phosphate takes place except in the presence of inorganic phosphate. When the latter is added, however, a brisk oxidation takes place, one molecule of inorganic phosphate disappearing for every molecule of glyceraldehyde phosphate oxidized. The product of oxidation now proved to be, not glycerate-3-phosphate, but a new compound, glycerate-1:3-diphosphate.

It now appears that the dehydrogenase unites with its substrate through an —SH group on the enzyme protein. This could account for the fact that the enzyme is sensitive to iodoacetate. Dehydrogenation follows with reduction of NAD and the oxidized product is split from the enzyme by inorganic phosphate, and this is why fermentation requires the presence of inorganic phosphate. The intermediate stages are shown on p. 164; for our present purposes they may be described as follows.

The substrate, an aldehyde, exists in solution in its hydrated form (reaction 6), and is then dehydrogenated under the influence of *triosephosphate dehydrogenase* and in the presence of NAD and inorganic phosphate, yielding glycerate-1:3-diphosphate. Reaction 7 is inhibited by iodoacetic acid, which blocks the —SH groups of the dehydrogenase, and by dialysis, which removes inorganic phosphate and the coenzyme. The next stage consists in the transference of the phosphate residue at position 1 to a molecule of ADP, so that glycerate-3-phosphate and ATP are formed (reaction 8), the process being catalysed by *phosphoglycerate kinase*. The 3-phosphate is then converted into the 2-ester (reaction 9) through the agency of *phosphoglyceromutase* (p. 123). The reactions, all of which are reversible, may be written in the following manner:

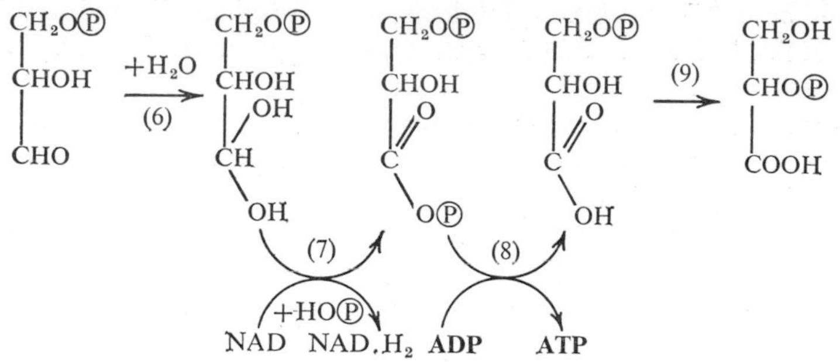

$$
\begin{array}{c}
CH_2O\textcircled{P} \\ | \\ CHOH \\ | \\ CHO
\end{array}
\xrightarrow[(6)]{+H_2O}
\begin{array}{c}
CH_2O\textcircled{P} \\ | \\ CHOH \\ | \\ CH \\ \diagdown OH
\end{array}
\begin{array}{c}
CH_2O\textcircled{P} \\ | \\ CHOH \\ | \\ C \\ \diagdown O\textcircled{P}
\end{array}
\begin{array}{c}
CH_2O\textcircled{P} \\ | \\ CHOH \\ | \\ C \\ \diagdown OH
\end{array}
\xrightarrow{(9)}
\begin{array}{c}
CH_2OH \\ | \\ CHO\textcircled{P} \\ | \\ COOH
\end{array}
$$

(7)   (8)

+HO\textcircled{P}

NAD   NAD.H$_2$ ADP   ATP

(iv) *Dephosphorylation of glycerate phosphate*. If one or both of the glycerate phosphates is added to whole yeast juice it undergoes fermentation. If, however, a dialysed juice is used there is no fermentation, but a new intermediate accumulates, viz. *enol*-pyruvate phosphate (phospho*enol*pyruvate). This arises by the dehydration of glycerate-2-phosphate (reaction 10) at the hands of a lysase, *enolase* (phosphopyruvate hydratase). It is at this point that fluoride inhibits fermentation; it does so because enolase requires magnesium ions for activity and is, apparently, a magnesium protein. Fluoride forms a complex magnesium fluorophosphate in the presence of inorganic phosphate. Dialysis, as ordinarily performed, does not stop enolase activity; it does, however, remove the coenzyme required for the next reaction, viz. the breakdown of phospho*enol*pyruvate.

If ADP is added to a dialysed juice containing phospho*enol*pyruvate, the latter begins to break down, and pyruvate appears. This reaction (11) is catalysed by *pyruvate kinase*, and the phosphate residue is transferred to ADP, yielding ATP once again (see next page). The *enol*-pyruvate liberated in reaction 11 presumably passes over into the more stable *keto*-form (reaction 12).

Pyruvate accumulates in a dialysed extract provided with ADP and phospho*enol*pyruvate. It does so because the next reaction, in which the pyruvate is decarboxylated, requires the presence of *co-carboxylase* (thiamine diphosphate), together with the enzyme *pyruvate decarboxylyase* (formerly known as *carboxylase*). The products of this reaction (13) are carbon dioxide and acetaldehyde, and the formation of the latter can be demonstrated by adding sodium bisulphite to a fermenting mixture, when the addition compound, acetaldehyde-sodium bisulphite, $CH_3CH(OH)SO_3Na$, is formed and can be isolated. The splitting of pyruvate is probably the only irreversible process in the whole fermentation sequence, apart from the initial phosphorylations (reactions 1, 3).

(v) *Production of alcohol*. The final stage of the process consists in the reduction of acetaldehyde to ethyl alcohol, and the mechanism of this reaction requires special consideration.

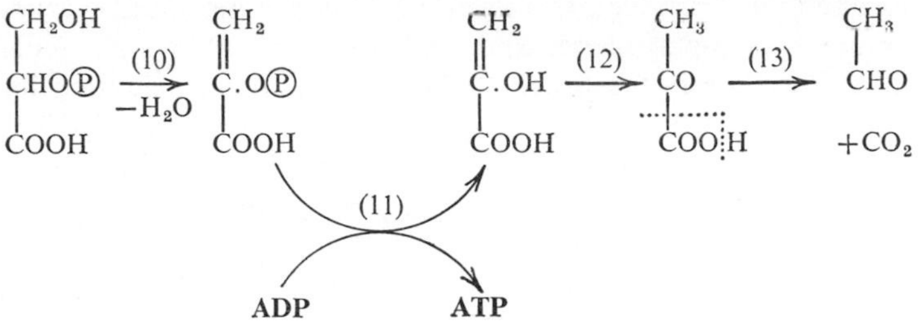

$$
\begin{array}{ccccccccc}
CH_2OH & & CH_2 & & CH_2 & & CH_3 & & CH_3 \\
| & (10) & \| & & \| & (12) & | & (13) & | \\
CHO\textcircled{P} & \xrightarrow[-H_2O]{} & C.O\textcircled{P} & & C.OH & \xrightarrow{} & CO & \xrightarrow{} & CHO \\
| & & | & & | & & | & & \\
COOH & & COOH & & COOH & & COO\vdots H & & +CO_2
\end{array}
$$

(11)

ADP   ATP

It will be remembered that, in the presence of fluoride, glycerate-3-phosphate and glycerol phosphate are produced. Glycerate phosphate is formed from glyceraldehyde phosphate by a reaction which involves the reduction of NAD. The reduced coenzyme passes on its 2H to a molecule of dihydroxyacetone phosphate and, without this, the whole of the available coenzyme would soon become and remain reduced. As no more glycerate phosphate could then be formed, fermentation would speedily come to an end.

In a normal as opposed to a fluoride fermentation, dihydroxyacetone phosphate is not required at this point, for there is available an alternative hydrogen acceptor in the form of acetaldehyde. Under the influence of *alcohol dehydrogenase*, working 'in reverse', the 2H of the reduced coenzyme are transferred instead to acetaldehyde, alcohol is formed, and the oxidized form of NAD is regenerated (reaction 14) and can be used over again (in reaction 7). This final operation can be written as follows:

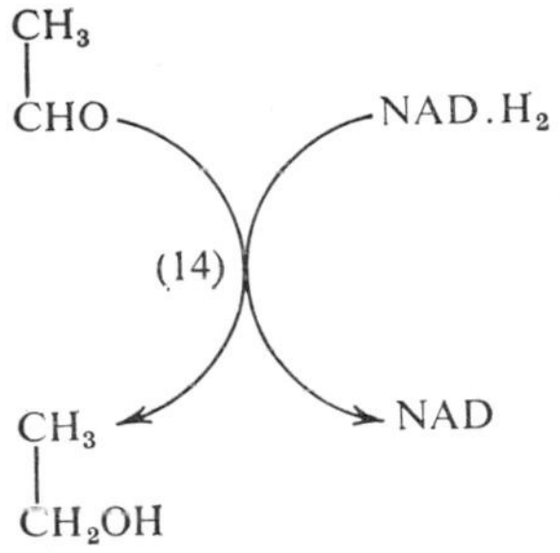

$$
\begin{array}{l}
CH_3 \\
| \\
CHO
\end{array}
\diagdown \qquad \diagup NAD.H_2
$$

(14)

$$
\begin{array}{l}
CH_3 \\
| \\
CH_2OH
\end{array}
\diagup \qquad \diagdown NAD
$$

Reaction 14, like 7, is inhibited by iodoacetate, which blocks the —SH groups of alcohol dehydrogenase.

The overall results of this reaction sequence, which is summarized in Fig. 78 are, first, that *for each molecule of glucose fermented, two molecules of alcohol and two of carbon dioxide are formed*. Secondly, for each molecule of glyceraldehyde phosphate oxidized, one molecule of NAD is reduced, and later reoxidized at the expense of the molecule of acetaldehyde formed from glyceraldehyde phosphate, so that *the coenzyme finishes in the oxidized condition in which it began*. Thirdly, two molecules of ATP are dephosphorylated in the phosphorylation of each molecule of glucose. Each molecule of

23-2

the phosphorylated product, fructose-1:6-diphosphate, yields two molecules of glyceraldehyde-3-phosphate, and each of these takes up a molecule of inorganic phosphate in the process of oxidation. After oxidation has taken

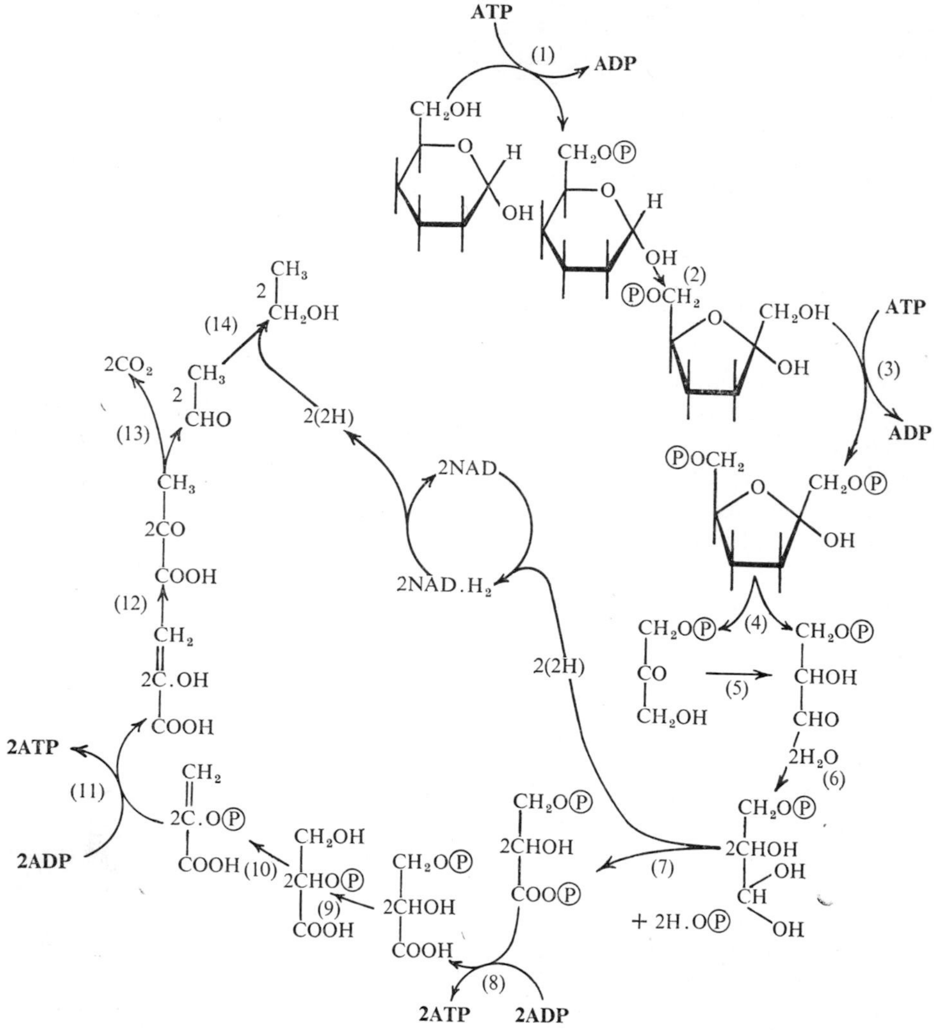

Fig. 78. Scheme to summarize reactions of alcoholic fermentation of glucose by yeast juice. Only the forward reactions are shown; they are numbered to correspond to the description given in the text. For names of enzymes, coenzymes, and inhibitors, see Table 27.

place, the phosphate groups, two for each molecule of glucose entering the system, are returned in the form of ATP in reaction 8, so that, at this stage in fermentation, the yeast has just recovered the amount of ATP used in the initial phosphorylations. Presently, however, two more molecules of ADP are taken in and, from these, two fresh molecules of ATP are formed in re-action 11. Thus, as far as the ADP/ATP system is concerned, *two new mole-cules of ATP are gained for each molecule of glucose fermented.*

Table 27. *Alcoholic fermentation: enzymes, coenzymes and inhibitors*
(See also Fig. 78)

Reaction	Enzyme	Coenzyme	Inhibited by
1	Hexokinase	ATP	Dialysis
2	Glucosephosphate isomerase	.	.
3	Phosphofructokinase	ATP	Dialysis
4	Aldolase	.	.
5	Triosephosphate isomerase	.	.
6	Spontaneous	.	.
7	Triosephosphate dehydro-genase	NAD; $-PO_3H_2$	Dialysis; $CH_2I.COOH$
8	Phosphoglycerate kinase	ADP	Dialysis
9	Phosphoglyceromutase	Glycerate-2:3-diphosphate	.
10	Phosphopyruvate hydratase	Mg ions	NaF
11	Pyruvate kinase	ADP	Dialysis
12	? Spontaneous	.	.
13	Pyruvate decarboxylyase	Cocarboxylase	Dialysis
14	Alcohol dehydrogenase	$NAD.H_2$	Dialysis; $NaHSO_3$; $CH_2I.COOH$

Note: Current synonyms are given in the text where this seems desirable.

ALCOHOLIC FERMENTATION OF OTHER SUGARS

Yeast juice is capable of fermenting glucose, fructose, mannose, sucrose and maltose, all of which are fermented by living yeast. The fermentation of the first three of these presents no special problems since *yeast hexokinase* cata-lyses the phosphorylation of all three to yield the corresponding 6-phosphate esters. Of these, glucose-6- and fructose-6-phosphates lie on the normal, direct line of fermentation, while mannose-6-phosphate is convertible into fructose-6-phosphate by a *mannosephosphate isomerase*.

Yeast cells, but not yeast juice, can be 'trained' to ferment galactose if it is grown on media in which this sugar is present. Lactose too is fermented by such a trained yeast. The training process induces the formation of enzyme systems which, in effect, convert galactose into glucose. Three reactions are involved. First, galactose is phosphorylated by a specific *galactokinase* to yield galactose-1-phosphate; secondly, galactose-1-phosphate is transformed into glucose-1-phosphate through the agency of the specific *UDPG-epimerase* together with its prosthetic group, uridine diphosphate glucose (p. 122) and NAD. The product can then either contribute to the polysaccharide reserves

343

of the cells through the intermediation of UDPG and an appropriate glucosyl transferase (cf. Fig. 22, p. 97) or α-glucose-1-phosphate can be set free and transformed into glucose-6-phosphate by *phosphoglucomutase* and so fermented.

The fermentation of the disaccharides, sucrose and maltose raises other problems and possibilities. Since yeast juice contains a *saccharase* and a *maltase* it seems likely that these sugars are first hydrolysed to yield the constituent monosaccharides (glucose and fructose; 2 glucose respectively) which can then be phosphorylated by hexokinase and fed into the fermentation reaction sequence. There are, however, other possibilities. There is present in another micro-organism, *Neisseria perflava*, a transglucosylase (amylosucrase) (p. 92) which acts upon sucrose as follows:

$$\text{sucrose} \rightarrow \text{fructose} + \frac{1}{n} \text{ ('glycogen'),}$$

while some strains of *Esch. coli* possess another transglucosylase (amylomaltase) (p. 93):

$$\text{maltose} \rightarrow \text{glucose} + \frac{1}{n} \text{ ('glycogen').}$$

In each case, one molecule of a fermentable sugar is set free and one monosaccharide unit is contributed to the polysaccharide reserves of the organism. It is not impossible that similar processes occur in yeast. Yeast cells can, in fact, remove fermentable sugars from their surroundings more rapidly than they ferment them, and lay down a glycogen-like substance as a reserve of carbohydrate upon which to fall back when no glucose is available in the surrounding medium.

ENERGETICS OF FERMENTATION

Now let us recall the overall equation of alcoholic fermentation:

$$C_6H_{12}O_6 = 2CO_2 + 2CH_3CH_2OH.$$

Two new $\sim ⓟ$ are formed at the cost of one molecule of glucose. That part of the energy which is not transferred to ATP is degraded, mainly in the form of heat, and this, probably, is why the temperature of a fermenting liquor is always rather higher than that of its surroundings. This, however, is not altogether disadvantageous, since, within limits, fermentation, growth and multiplication all proceed more rapidly at higher temperatures.

The question is often asked, why is it that yeast does not break up glucose into alcohol and carbon dioxide directly, instead of in this rather complicated manner? The *total* free-energy yield of the process would be the same, no matter how the sugar was broken down, but, by working in the way it does, the yeast is able, step by step, to transfer a large proportion of the total free energy of the process to the directly utilizable high-energy phosphate groups of ATP. If the glucose were *directly* split, even if enzymes existed that could catalyse this process, the chances are that the vast bulk of the free energy that fermentation renders available would be degraded as heat, and thus lost to the system.

Let us now see how this important transference of chemical energy from one substance to another is achieved. Living cells seem never to have discovered enzymes capable of catalysing a complete breakdown of the 6-carbon chain of unmodified glucose. The preliminary phosphorylation reactions seem, therefore, to be devices for getting glucose into a metabolizable form. To accomplish this, chemical work has to be done, and is in fact carried out at the expense of the terminal high-energy phosphate groups of two molecules of ATP (reactions 1, 3). Then, and only then apparently, the 6-carbon chain can be ruptured. In the subsequent metabolism of the products, the energy put in is recovered (reaction 8) and the energetic *status quo* is re-established. Later, still more energy becomes available (reaction 11), and may be used to start off the fermentation of fresh molecules of glucose, for example, or for the synthesis of new and complex tissue materials, so that the cells may grow and, in due time, divide.

The resynthesis of ATP from ADP requires the provision of free energy. This is provided by the generation of new $\sim \textcircled{P}$ in the partial breakdown products of glucose. Thus, in the phosphopyruvate hydratase (enolase) reaction (10) the removal of a molecule of water and the consequent rearrangement of molecular architecture is attended by a redistribution of the free energy of the molecule and the generation of a new high-energy phosphate group from one that formerly was of the low-energy type.

The process can be written as follows:

$$\begin{array}{ccc} CH_2OH & & CH_2 \\ | & & \| \\ CHO\textcircled{P} & \rightleftharpoons & C.O \sim \textcircled{P} + H_2O. \\ | & & | \\ COOH & & COOH \end{array}$$

Another high-energy phosphate is generated in reaction 7, in which triose-phosphate undergoes its phosphate-dependent oxidation to glycerate-1:3-diphosphate and here, as in the former case, the new $\sim \textcircled{P}$ is transferred to ADP to yield a new ATP molecule.

FERMENTATION BY LIVING YEAST

Yeast-juice fermentation differs from fermentation by live yeast cells in several noteworthy respects. In the first place, the juice is much less active than intact cells. This is probably because the enzymes and coenzymes are not arranged at random in the cell, as presumably they must be in the extract, but in some definite, orderly manner. It is probably safe to assume that, in the yeast cell, as in an efficient factory, the machinery is arranged in a manner calculated to yield the greatest possible degree of efficiency, and it may not be going too far to suggest that the organization is such that the substance produced by one enzyme in the series is passed immediately on to the next. We know relatively little about the submicroscopic internal organization of this or of any other kind of cell, but that an organization of a high order of complexity exists can hardly be doubted.

A second important difference between yeast and yeast juice lies in the effect of inorganic phosphate upon fermentation in the two cases. As Harden & Young first showed, yeast juice can only ferment sugar so long as there is free inorganic phosphate in the medium. The reason for this is clear from what we now know about the mechanisms involved, for inorganic phosphate is required for the removal of the oxidized substrate of triosephosphate dehydrogenase from the enzyme itself (p. 164). This is a component part of reaction 7, so that this important oxidative process ceases when inorganic phosphate is not available. Any free phosphate that is present is taken up and transferred by way of glycerate-1:3-diphosphate to ADP (reaction 8), and the ATP so formed may be used to esterify more glucose (reactions 1, 3). If inorganic phosphate is added to a juice fermentation, therefore, it disappears and is replaced by the organically bound phosphate of the sugar esters. But the addition of inorganic phosphate has no effect on the rate of fermentation of sugar by intact yeast cells. Once again the notion of intracellular organization has to be invoked: one must suppose that the interior of the cell is so arranged that inorganic phosphate is always available in the cell in the right place and at the right time.

In the intact cell, we must believe, ATP synthesized by the cell's fermentative activities is utilized for the performance of work of various kinds, especially for the synthesis of new tissue materials for growth and reproduction so that the terminal phosphate units of ATP are set free again in one way or another. This inorganic phosphate is caught up by the fermentation machine, recharged, so to speak, and again returned to ATP, and so on; a continual cycle of phosphate is built up and used to transfer energy obtained by fermentation to the places at which it is required and, presently, is actually put to employment.

Related to the effect of phosphate there is an interesting phenomenon known as the arsenate effect. If arsenate is added to a juice fermentation that has stopped through lack of inorganic phosphate, a long-continuing but very slow fermentation begins. This is because arsenate is able to replace phosphate in reaction 7, so that a glycerate arseno-phosphate is formed. The product fails, however, to react with ADP in reaction 8, but phosphoglycerate-1-arseno-3-phosphate is rather unstable and breaks down slowly, liberating arsenate, so that glycerate-3-phosphate is slowly produced. This re-enters the reaction sequence at reaction 9 so that a slow fermentation takes place. An important feature of this arsenate effect is that reaction 8 is by-passed, so that the high-energy phosphate normally generated at this stage is no longer available to the system.

The final products of fermentation, by yeast cells or by juice, always include small quantities of *glycerol* and other substances. The formation of glycerol can readily be accounted for. At the very beginning of fermentation, glucose is phosphorylated and split to yield glyceraldehyde phosphate and dihydroxyacetone phosphate. If fermentation is to proceed, the glyceraldehyde phosphate must be oxidized to form glycerate phosphate, a process in which NAD is reduced. As yet, no acetaldehyde has been formed by the reduction of which reduced NAD can be reoxidized and so put back into circulation. But, as we have learned from experiments on fluoride inhibition, dihydroxyacetone phosphate can be used instead of acetaldehyde, and this does normally take place until some acetaldehyde has been produced. Even when acetaldehyde is being formed, however, small amounts of dihydroxyacetone phosphate continue to be reduced, for the system contains a soluble α-glycerolphosphate dehydrogenase. A small proportion of the reduced coenzyme reacts with dihydroxyacetone phosphate so that a little glycerol phosphate is formed. This is the L-isomer, which proves that it is formed from the optically inactive dihydroxyacetone phosphate (see p. 161 and not, as was formerly supposed, from glyceraldehyde phosphate. The latter arises from hexose diphosphate in the D-form (p. 159) and would therefore yield D-glycerol phosphate on reduction.

Finally the glycerol phosphate is hydrolysed by a phosphatase that occurs in yeast, and glycerol itself is set free.

FERMENTATIVE MANUFACTURE OF GLYCEROL

Glycerol is a very important article of commerce, especially in time of war when large amounts are used in the manufacture of explosives. In ordinary times, the glycerol of commerce is a by-product from the manufacture of soaps by the saponification of fats, and fats are always in short supply in wartime. During the war of 1914–18 the British blockade led to a serious fat shortage in Germany, and the resulting shortage of glycerol meant a shortage also of explosives. The problem was met by making use of the ability of yeast to form glycerol.

High yields of glycerol can be obtained from sugar by modifying the course of normal fermentation in either of two ways. The two modified forms are known as Neuberg's 'second' and 'third' forms of fermentation respectively, the 'first' form being normal alcoholic fermentation. In Neuberg's second form, sodium bisulphite is introduced into the fermenting liquors. This gives an addition-compound with acetaldehyde, thus depriving the cells of their normal hydrogen acceptor for the reoxidation of reduced NAD. Its place is taken by dihydroxyacetone phosphate, and one molecule of glycerol phosphate is accordingly formed for each molecule of glycerate phosphate. The glycerol phosphate is hydrolysed by a yeast phosphatase, while the glycerate phosphate continues along its usual path until acetaldehyde is formed and reacts with bisulphite. Each molecule of glucose therefore yields one molecule

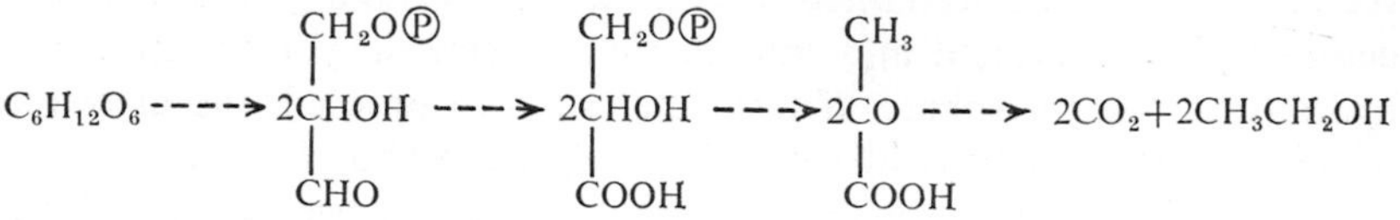

$$
C_6H_{12}O_6 \dashrightarrow 2
\begin{matrix} CH_2O\textcircled{P} \\ | \\ CHOH \\ | \\ CHO \end{matrix}
\dashrightarrow 2
\begin{matrix} CH_2O\textcircled{P} \\ | \\ CHOH \\ | \\ COOH \end{matrix}
\dashrightarrow 2
\begin{matrix} CH_3 \\ | \\ CO \\ | \\ COOH \end{matrix}
\dashrightarrow 2CO_2 + 2CH_3CH_2OH
$$

(a) *Neuberg's first form of fermentation*

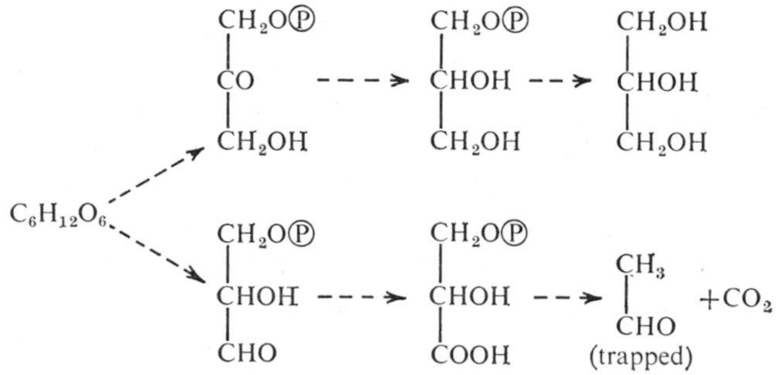

$$
C_6H_{12}O_6
\begin{cases}
\begin{matrix} CH_2O\textcircled{P} \\ | \\ CO \\ | \\ CH_2OH \end{matrix}
\dashrightarrow
\begin{matrix} CH_2O\textcircled{P} \\ | \\ CHOH \\ | \\ CH_2OH \end{matrix}
\dashrightarrow
\begin{matrix} CH_2OH \\ | \\ CHOH \\ | \\ CH_2OH \end{matrix}
\\[3em]
\begin{matrix} CH_2O\textcircled{P} \\ | \\ CHOH \\ | \\ CHO \end{matrix}
\dashrightarrow
\begin{matrix} CH_2O\textcircled{P} \\ | \\ CHOH \\ | \\ COOH \end{matrix}
\dashrightarrow
\begin{matrix} CH_3 \\ | \\ CHO \\ \text{(trapped)} \end{matrix} + CO_2
\end{cases}
$$

(b) *Neuberg's second form of fermentation*

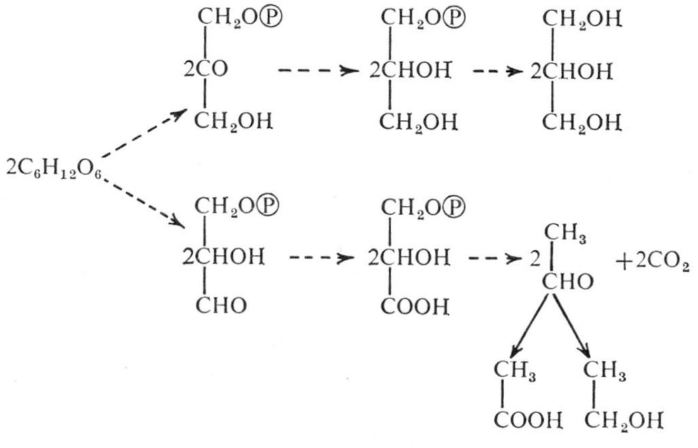

$$
2C_6H_{12}O_6
\begin{cases}
\begin{matrix} CH_2O\textcircled{P} \\ | \\ 2CO \\ | \\ CH_2OH \end{matrix}
\dashrightarrow 2
\begin{matrix} CH_2O\textcircled{P} \\ | \\ CHOH \\ | \\ CH_2OH \end{matrix}
\dashrightarrow 2
\begin{matrix} CH_2OH \\ | \\ CHOH \\ | \\ CH_2OH \end{matrix}
\\[3em]
\begin{matrix} CH_2O\textcircled{P} \\ | \\ 2CHOH \\ | \\ CHO \end{matrix}
\dashrightarrow 2
\begin{matrix} CH_2O\textcircled{P} \\ | \\ CHOH \\ | \\ COOH \end{matrix}
\dashrightarrow 2
\begin{matrix} CH_3 \\ | \\ CHO \end{matrix} + 2CO_2
\end{cases}
$$

$$
\begin{matrix} CH_3 \\ | \\ COOH \end{matrix}
\qquad
\begin{matrix} CH_3 \\ | \\ CH_2OH \end{matrix}
$$

(c) *Neuberg's third form of fermentation*

Fig. 79. Neuberg's three 'forms' of fermentation.

of glycerol and one each of carbon dioxide and the aldehyde-bisulphite addition compound. The process is sketched out in Fig. 79b, which may be compared with Fig. 79a, which represents normal fermentation in similar terms.

The third form of fermentation (Fig. 79 c) sets in if the fermenting liquors are made and kept alkaline. Under alkaline conditions, acetaldehyde is no longer reduced to alcohol in the normal manner, but instead undergoes a Cannizzaro reaction. One molecule is oxidized to acetate and a second simultaneously reduced to alcohol, and this takes place quite independently of the normal reactions. Again acetaldehyde is no longer available for the reoxidation of reduced NAD, and its place in that reaction is again taken by dihydroxy-acetone phosphate. In this case, therefore, each pair of glucose molecules gives rise to two of glycerol and two of acetaldehyde, one of which is further transformed into acetate and the other into ethyl alcohol.

In this third form of fermentation, which takes place only in alkaline solutions, the yeast cell changes its metabolism in such a manner as to produce acetate. Unless steps are taken to maintain the alkalinity of the medium, therefore, the pH falls until the medium becomes faintly acid, when the normal form of fermentation reasserts itself and no more glycerol is produced.

We are accustomed to the idea that changes in the environment of living organisms can bring about changes in those organisms, and in the present case the effect of alkalinity in the medium is to change the course of metabolism in the organism. But the organism reacts by producing acid, and we have therefore a case in which an organism produces changes in its environment. And this is not by any means the only example of its kind: many bacteria tend to produce acids when cultivated in alkaline media and strongly basic amines when the media are acid so that, in either case, the pH of the medium is changed in the direction of physiological neutrality.

PRODUCTION OF FUSEL OIL

Alcoholic fermentation carried out by live yeast is attended by the production of a number of alcoholic substances other than ethyl alcohol and glycerol, and to these the collective name of 'fusel oil' is applied. These substances usually account for less than 1 % of the total alcohols, but are of considerable industrial importance. They are interesting, too, because they are largely responsible for the characteristic flavours and bouquets of alcoholic beverages. Heavy wines, such as port, contain considerable amounts of higher alcohols and their esters, especially *iso*-amyl alcohol, and these are responsible not only for the taste and bouquet of the wine but also, in large measure, for the unpleasant effects of over-indulgence, since the higher alcohols are powerful narcotics. Another interesting product of the same kind is the bitter principle of beer: this again is an alcohol, in this case tyrosol.

These alcohols arise from amino-acids. The crude liquor contains amino-acids arising from the proteins present in grapes, hops and the like, and more are contributed by the autolysis of dead yeast cells. They are deaminated, apparently to furnish ammonia for the synthesis of the new yeast proteins

349

which are required as the cells grow and multiply, for if ammonium salts are added to the fermenting liquor there is a marked fall in the yield of fusel oil.

Yeast deaminates amino-acids in a peculiar manner that is perhaps unique, the process consisting in an apparently simultaneous decarboxylation and hydrolytic deamination:

$$R.CH(NH_2)COOH + H_2O = R.CH_2OH + NH_3 + CO_2.$$

In this way the leucines, for example, give rise to the corresponding amyl alcohols, while valine yields *iso*-butyl alcohol, for example:

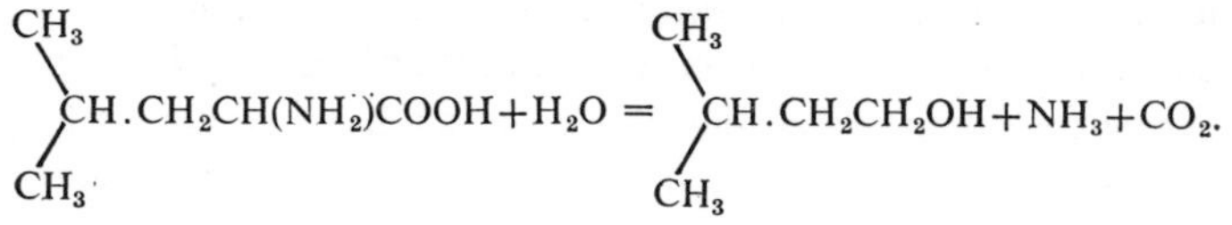

$$\begin{array}{c} CH_3 \\ \diagdown \\ CH.CH_2CH(NH_2)COOH + H_2O = \\ \diagup \\ CH_3 \end{array} \quad \begin{array}{c} CH_3 \\ \diagdown \\ CH.CH_2CH_2OH + NH_3 + CO_2. \\ \diagup \\ CH_3 \end{array}$$

leucine   iso-*amyl alcohol*

Tyrosol arises in the same way from tyrosine.

18

ANAEROBIC METABOLISM OF CARBOHYDRATES IN MUSCLE

INTRODUCTION

Modern muscle biochemistry was founded at the beginning of the present century. Before that time, muscular contraction or any other kind of cellular activity was thought to depend upon the sudden decomposition of large, unstable molecules of a hypothetical stuff called 'inogen'. In the case of muscle, this 'inogen' was supposed to give rise to carbon dioxide and L-(+)-lactate, and furnish the energy expended by the muscle. It was already well known that muscles produce carbon dioxide when they contract, and that lactate is produced in greater or smaller amounts at the same time.

'The justification for considering muscle tissue especially, out of all the active tissues of the organism, lies in the fact that only in muscle can we come near to comparing the chemical changes going on with the simultaneous work done or the energy set free as heat. It is difficult to assess the work performed by a secreting gland, and the metabolism of such an organ can only be studied in elaborate perfusion experiments; great advances have been made in the study of nerve tissue, but here the changes going on are so small as to make their detection only lately possible by modern methods. But certain muscles, and a variety of them, can be removed from the body with absolutely no injury, and can be kept functional for days' (D. M. Needham).

Many different methods have been used to elucidate the problems of muscular contraction. Histology, physiology, biochemistry, X-radiography and electron microscopy have all played a part. From the point of view of the histologist we can distinguish between three main types of muscle. These are: (1) the striped or striated voluntary muscles of the skeletal system, (2) the plain, unstriated or involuntary muscles of the visceral system, and (3) cardiac muscle. Most of the chemical work has been done on skeletal muscle, but there seem to be few differences between the different types from the point of view of the biochemistry of their contractile processes.

The structural unit of striated muscle is the long, thread-like *myofibril*. Some thousands of these go to make up a muscle fibre, and many fibres to make up a whole muscle. When a fresh muscle fibre is examined under the microscope, its image shows transverse alternating light and dark bands. This appearance is due to the existence, in each fibril, of alternating zones of higher and lower refractive index, these zones being in register across each fibre so as to form the transverse 'striations' (Fig. 80).

351

The higher refractive index of one set of bands, known as the A bands, indicates that the concentration of protein is higher there than in the others, known as I bands. The A bands are also doubly refracting, or optically anisotropic, which shows that the extra material there is in the form of sub-microscopic rodlets aligned parallel to the long axis of the fibre. When the fibre, or an isolated fibril, is watched during contraction, it is found that the I bands shorten while the width of the A bands stays unchanged.

Chemically speaking, the myofibrils consist chiefly of two proteins, actin and myosin. Experiments in which fibrils are treated with solvents which dissolve one but not the other of these proteins show that the thick filaments consist chiefly of myosin, and the thin filaments of actin.

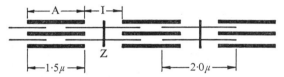

Fig. 80. Diagram to illustrate the structural arrangement of a muscle fibril.
(*After* A. F. Huxley.)

Actomyosin is a complex formed by the association of the two proteins, *actin* and *myosin*. A solution of actomyosin is singly refracting when at rest, when the molecules have a purely random distribution. But if the solution is made to flow along a glass tube, the thread-like molecules all become orientated in the same direction, pointing in the direction of flow, and the solution becomes doubly refracting. It is therefore probable that the anisotropic regions of the muscle fibre are double refracting because all the protein molecules point in the same direction.

This gives a rough picture of what is generally regarded as the contractile machinery. We now have two main problems to consider. First, what is the chemical source of the energy which is expended when the muscle machine does its work and, secondly, how is this energy transformed into the mechanical energy of contraction? At the present time we have a considerable amount of information on the first of these problems. We are less well informed about the second, but the structural basis for the contraction phenomena has been demonstrated by electron microscopy. Two sets of filaments are found, arranged in the way shown in Fig. 80. The filaments of the thicker type extend for the whole width of the A band, while the thinner filaments cross the I band and extend some way into each of the adjacent A bands. During contraction, there is no change in the length of either type of filament, and the overall shortening occurs by the filaments sliding between each other so that the amount of overlap between them increases. The mechanism whereby the sliding process is brought about is still quite unknown; what we do know in fact is that muscular contraction is very intimately associated with the breakdown of ATP.

The first really significant experiments on the chemistry of muscular contraction were carried out by Fletcher & Hopkins and published as long ago as 1907. Working on frog muscles, they showed that larger or smaller amounts of lactate are formed when muscle contracts. The general plan of the experiments was as follows. The sartorius or gastrocnemius muscles were removed from the hind legs of a frog and kept under identical and anaerobic conditions. One of the pair was made to do work by being stimulated, and this, the experimental muscle, and the unstimulated control were then dropped into ice-cold alcohol and finely ground with sand. These workers realized that a muscle is capable of doing a very large amount of work in a very short period of time, and that to injure a muscle amounts to stimulating it. By using small muscles, chilling and extracting them very quickly with ice-cold alcohol it was possible to inactivate the muscle enzymes very rapidly indeed, and so minimize the large-scale chemical changes which would otherwise result from injuries inflicted in the process of grinding. The chemical changes corresponding to the work done by the experimental muscle could then be found by analysing and comparing the extracts with those of control muscles. Other workers had done similar experiments already, but no precautions were then taken to cool the muscle before grinding, with the result that little difference could as a rule be detected between the experimental and control muscles, so grevious is the injury inflicted by grinding.

Fletcher & Hopkins, with their new technique, confirmed and extended the older observations that lactate is formed when muscle contracts, and their results demonstrated with beautiful clarity the following points. (i) Muscle can contract in a perfectly normal manner in complete absence of oxygen. (ii) Lactate is produced during anaerobic contraction, and piles up with continued stimulation until in the end, the muscle becomes fatigued. (iii) If the fatigued muscle is then put into oxygen it recovers its ability to contract, and lactate simultaneously disappears. (iv) Less lactate is formed in a muscle that is allowed access to oxygen than in one which works anaerobically.

Shortly afterwards it was shown by Meyerhof that the lactate is formed from glycogen and that, under anaerobic conditions, the amount of lactate formed is chemically equivalent to the quantity of glycogen broken down. A mass of later work made it clear that there is a strict proportionality between the amount of work done, the heat produced, the tension developed in a muscle, and the quantity of lactate formed; and by 1927 it had become certain that the energy expended in muscular contraction comes from the conversion of glycogen to lactate.

A good deal of interest centred round the phosphate compounds present in muscle, for it was already clear that phosphates play an important part in muscle glycolysis, just as they do in fermentation. The method in general use

353

for extracting phosphates from muscle tissue consisted in chilling the material thoroughly and extracting it by grinding with ice-cold trichloroacetic acid or some other protein precipitant. Ice-cold conditions were used here, as in the original work on lactate formation, in order to inactivate the muscle enzymes as rapidly as possible. Once the extract had been prepared it was allowed to warm up, and estimations of the phosphate content were subsequently made.

In 1927, however, it was discovered that *ice-cold filtrates* from trichloroacetic precipitation contain a hitherto undetected phosphate compound. This substance is exceedingly rapidly hydrolysed in acid solution and had not previously been noticed for that reason. In order to detect and estimate it, the trichloroacetic filtrate must be kept ice-cold until it has been neutralized to a pH of about 8, at which the new compound is fairly stable. In due course the new substance was isolated and shown to be *phosphocreatine*, to which the name of '*phosphagen*' was originally assigned.

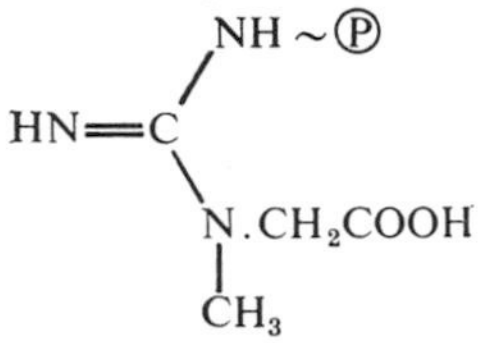

phosphocreatine 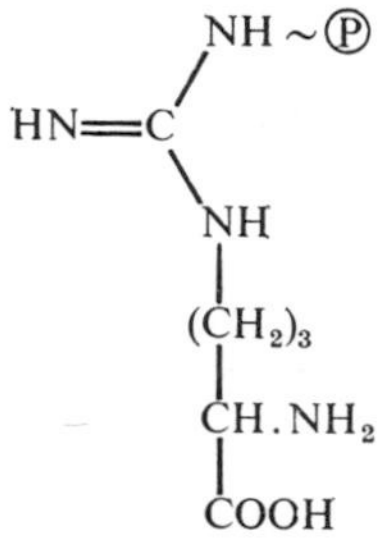　　　　　*phosphoarginine*

Later investigations showed that this compound is present in the striated, smooth and cardiac muscles of all classes of vertebrates, but absent from those of invertebrates with a few notable exceptions (see p. 369 *et seq.*). In its place, most invertebrate muscles contain an analogous derivative of arginine, *phosphoarginine*. In more recent years a number of yet other phosphoguanidine derivatives have been discovered, especially among the annelid worms.

In passing it should be noticed that all known phosphagens contain high-energy phosphate, a feature which turns out to be of great physiological significance. In what follows we shall discuss mainly the creatine compound, but it may reasonably be assumed that what goes for this substance is also true of the arginine and other analogues.

A wave of interest in phosphocreatine soon developed, and within a few years it became known that it plays an important part in the chemistry of muscular contraction. Phosphagen, it was shown, breaks down during activity and is resynthesized during rest, aerobically and anaerobically alike, and, moreover, it breaks down much more rapidly than does glycogen. It was suspected by some that, since the breakdown of phosphagen precedes that of glycogen, it must be a more immediate source of contraction energy, the more slowly acting process of glycolysis being used to resynthesize the phosphagen, rather as the lever of an air-gun is used to reset the spring after the trigger has been pulled. But this idea did not find much favour; muscle chemists were still too much wedded to the older lactic acid hypothesis.

In 1930, however, new evidence appeared. Lundsgaard, who was studying the pharmacological properties of iodoacetate, observed that in animals dying from iodoacetate poisoning the muscles went into rigor, and that, instead of becoming markedly acid as was to be expected, they actually became faintly alkaline. Closer examination showed that no lactate whatever had been produced. This discovery caused a good deal of surprise, since it demonstrated conclusively that muscle can contract *without producing any lactic acid at all*. Further work on iodoacetate-poisoned muscles showed that phosphagen breaks down when work is done, the amount split being strictly proportional to the amount of energy expended. Further, phosphagen was not resynthesized if an iodoacetate-poisoned muscle was allowed to rest, and, with repeated stimulation, the muscle went into rigor as soon as its stock of phosphagen was exhausted.

By this time it had begun to appear that muscle resembles yeast rather closely in its carbohydrate metabolism. Compounds such as the hexose phosphates could be detected as well in the one as in the other, while that all-important compound adenosine triphosphate was also detected in both. Most of the work so far described had been done on intact, isolated muscles, mostly of frogs, kept under anaerobic conditions, but in 1925 Meyerhof published a method for the preparation from muscle of extracts analogous to the yeast juice that had been so valuable in the study of fermentation. The method employed is roughly as follows.

The animal is anaesthetized and cooled to 0° C. The muscles are carefully cut away with the least possible injury, and care is taken to keep them cold. The tissue is put through an ice-cold mincer and allowed to stand for 30–60 min. with ice-water or isotonic KCl. After straining and centrifugation, a rather viscous liquid is obtained and stored in the refrigerator until required.

Extracts prepared in this way contain all the enzymes and coenzymes required for the production of lactate from added glycogen, and will also break down phosphocreatine and ATP if these are added. Most of our knowledge of muscle chemistry was first acquired with the aid of extracts of this kind. Dialysis removes the coenzymes, just as it does in the case of yeast juice, and many experiments have been carried out with extracts previously dialysed or treated with fluoride or iodoacetate. One most important point that should be noticed—and remembered—is that, although they can convert glycogen into lactate, these *extracts do not respire*, so that it is possible to work on extracts in the presence of air instead of having to take the elaborate precautions necessary to ensure anaerobiosis when isolated muscles are employed. Whereas the enzymes concerned in glycolysis are present in the cytoplasm and come out in the extract, cytochrome oxidase and most of the other enzymes essential for respiration are not soluble but remain attached to the insoluble cell debris, especially the mitochondria. Another curious but important feature of these extracts is that, unlike intact muscle, they have no action upon glucose though they attack glycogen vigorously enough.

FORMATION OF LACTATE

The reactions involved in glycolysis are very similar to those involved in alcoholic fermentation. The first step in muscle extract consists in the breakdown of glycogen to α-glucose-1-phosphate; the enzyme concerned is *α-glucan phosphorylase*. The product is converted into glucose-6-phosphate by *phosphoglucomutase*. Glucose-6-phosphate can also be formed by the action of muscle hexokinase upon free glucose. ATP is required in the usual way but fresh muscle extracts, while they contain hexokinase, contain no ATP because power-

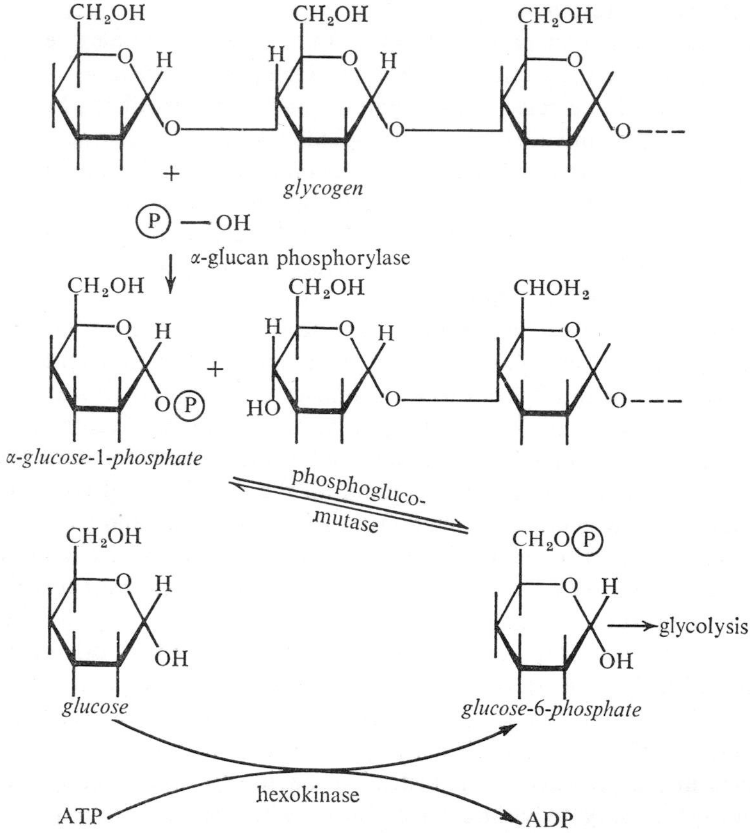

Fig. 81. Early reactions of glycolysis.

356

ful ATP'ases are also present. If the latter are inhibited, e.g. with fluoride or by prolonged dialysis at 0° C, the formation of glucose-6-phosphate from glucose can be demonstrated if ATP is added. It is worthwhile to notice in passing that whereas free glucose can be readily phosphorylated by ATP in the muscles, free glucose is not regenerated because the requisite enzyme, glucose-6-phosphate hydrolase, is absent. Muscle, then, takes up glucose from the blood but does not contribute anything to the glucose pool in return (Fig. 81).

From glucose-6-phosphate onwards, glycolysis and fermentation follow a common path until pyruvate is formed. Here the pathways diverge again, for muscle, unlike yeast, does not contain pyruvate decarboxylyase. Co-carboxylase is present, but, like NAD and NADP, it can collaborate with more enzymes than one, and its presence in muscle is in no way an indication that the decarboxylyase itself is present. In yeast, it will be remembered, pyruvate is split into carbon dioxide and acetaldehyde, the latter functioning as a hydrogen acceptor in the reoxidation of reduced NAD. In muscle extract, however, no pyruvate decarboxylyase being present, pyruvate itself discharges this function and is reduced asymmetrically to L-(+)-lactate, under the influence of *lactate dehydrogenase*;

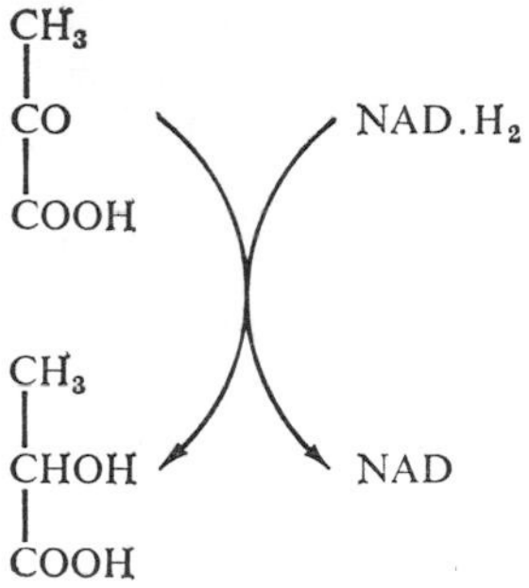

lactate dehydrogenase

The overall effect of this reaction sequence is that, on the carbohydrate side, *one 6-carbon unit of glycogen yields two molecules of lactate*. NAD is alternatively reduced and reoxidized just as it is in fermentation, but there are differences as far as the ATP/ADP system is concerned. In yeast juice and in muscle extract alike the sequence as a whole leads to the generation of four new high-energy phosphate groups for each 6-carbon unit metabolized. In fermentation, two of these new ∼ ℗ are used to compensate for those used in the preliminary phosphorylation of the glucose molecule, so that in this case there is a net gain of two molecules of ATP for each glucose molecule fermented.

Muscle extract, however, starts from glycogen, not from glucose, and the first stage in glycolysis consists in the splitting of glycogen by inorganic phosphate yielding glucose-6-phosphate by way of the 1-compound. *Only one*

357   24-2

molecule of ATP has to be used up in the production of each molecule of fructo-furanose diphosphate. Of the four molecules of ATP subsequently produced therefore, only one is required to restore the *status quo*, so that *in muscle glycolysis there is a net gain of three molecules of ATP for each 6-carbon unit of glycogen metabolized*, as compared with a gain of two molecules for each 6-carbon *glucose* unit metabolized in the case of fermentation.

PARTS PLAYED BY ATP AND PHOSPHAGEN

The reactions of glycolysis involve the breakdown and resynthesis of ATP, just as do those of fermentation, and the resemblances between the two processes are very striking indeed. There is, however, one very important difference between yeast and muscle, for whereas the latter contains phosphagen, yeast does not.

If an intact, isolated muscle is allowed to contract under anaerobic conditions there is a decrease in the amount of phosphagen present, together with corresponding increases in the amounts of free creatine and free inorganic phosphate. This suggests that muscle must contain an enzyme which catalyses the hydrolysis of phosphocreatine. However, no hydrolysis of phosphagen takes place if it is added to a dialysed extract, indicating that some dialysable factor is involved. This factor has been identified with ADP, and it was discovered that phosphagen is not hydrolysed, as had formerly been supposed, but that *it reacts with ADP*, a phosphate group being transferred to yield free creatine together with ATP. ATP is split by a powerful adenosine triphosphatase that is present in the extract, ADP is formed and rephosphorylated at the expense of the phosphagen and so on, until no more phosphagen remains:

$$ATP + H_2O \rightarrow ADP + HO\circledP$$
$$ADP + C\circledP \rightarrow ATP + C$$

Overall:
$$C\circledP + H_2O \rightarrow C + HO.\circledP$$

It was presently discovered that adenosine triphosphatase can be inactivated by prolonged dialysis at $0°C$. and, when the enzyme had thus been inactivated, it was possible to show that the reaction between phosphocreatine and ADP is freely reversible and catalysed by *creatine kinase*:

$$ADP + C\circledP \rightleftharpoons ATP + C.$$

Very little free energy change is involved and no inorganic phosphate is set free in this reaction. From the existence of this equilibrium system we can make certain deductions concerning the part played by phosphagen in the economy of the muscle. Anything that tends to decompose ATP will force the reaction over towards the right and phosphagen will be broken down. If, on the other hand, glycolysis is in progress so that new high-energy phosphate is being generated, fresh ATP will be formed, the reaction will swing towards the left and phosphagen will be resynthesized.

The most important outcome of this work was the realization that, until some ATP has been broken down to provide ADP, no decomposition of phosphagen takes place: in other words, *the breakdown of ATP must take place even earlier than that of phosphagen*. The breakdown of ATP, in fact, is the earliest reaction we have so far been able to detect and hence is the *most immediate known source of contraction energy*. This suggests that ATP must play some part in contraction, over and above that which it plays in glycolysis.

The probable nature of this additional function was first revealed by the work of Engelhart & Lubimova, who found that the adenosine triphosphatase of muscle is inseparable from and apparently identical with myosin. This was before the discovery of actin, and there is not much doubt that their 'myosin' consisted in reality of a complex of actin with myosin proper. Thus actomyosin, the actual contractile material of muscle, appears to be identical with the enzyme that catalyses the decomposition of ATP and leads to the liberation of the free energy of its terminal high-energy phosphate. Other experiments have given results which are completely in harmony with the supposition that, when actomyosin and ATP come into contact, a sudden shortening takes place, followed by decomposition of the ATP and a return of the protein molecules to their former length. These results were obtained by measurements of the viscosity and refractive indices of solutions of 'myosin'. Threads prepared from actomyosin similarly contract if treated with ATP. Again, it is possible, by treating muscle with the collagenase of *Clostridium welchii*, to obtain suspensions of isolated myofibrils, which similarly contract on addition of ATP.

The ability to contract in response to ATP is evidently a built-in property of the muscle proteins *in situ* for, if a few fibres are stripped from the *psoas* muscle of a rabbit, they contract when treated with ATP, and the same results are obtained even when the excised tissue has been preserved in 50 % glycerol for many months in the cold. The contracted fibres can be made to relax again if a preparation of creatine kinase is added. Exactly what happens in the process of shortening we still do not know, but the shortening of actomyosin and preparations containing it in an orientated form appears to be due to an ATP-induced dissociation or disaggregation of actomyosin into actin plus myosin. How such a process could lead to the 'sliding' of the interdigitated protein rodlets in the intact muscle we still do not know.

If we accept these results at their face value it follows that the first reaction known to take place when a muscle is stimulated consists in the decomposition of ATP, a process which is attended by a sudden shortening of the actin-myosin system and, in some manner which is still quite obscure at the present time, by the conversion of the free energy of its terminal $\sim \circled{P}$ into the mechanical energy of contraction. There is no doubt whatever that very intimate relationships exist between the source of contraction energy (ATP), the enzyme that catalyses the liberation of that energy (ATP'ase) and the contractile material itself (actomyosin).

It is difficult, in view of the evidence, to escape the conclusion that ATP plays a most intimate part in muscle contraction, but it has only quite recently been possible to demonstrate any direct decomposition of ATP in contraction. This vitally important demonstration was forthcoming from studies of muscles treated with 1-fluoro-2:4-dinitrobenzene. Like iodoacetate this compound puts a stop to glycolysis but *it stops the phosphorylation of ADP by phosphocreatine as well.*

On stimulation such muscles contract normally and their contraction is accompanied by the breakdown of ATP. Phosphocreatine remains unchanged. Here at long last is proof that the most immediate known source of contraction energy is ATP, and that the apparent constancy of ATP content observed during normal activity is attributable to a very rapid rephosphorylation of the residual ADP, catalysed by creatine kinase and at the expense of phosphagen.

If we accept this evidence the function of phosphagen becomes clear. The muscle contains relatively small amounts of ATP, and these would soon be exhausted but for the fact that the ADP formed by its decomposition can be rapidly rephosphorylated at the expense of phosphagen. This means that the muscle can go on acting several times longer than it could if no phosphagen was present, for the amount of phosphagen present in a typical striated muscle is several times greater than that of ATP. Phosphagen, then, may be regarded as a reserve of high-energy phosphate. It has in fact been shown that the *rectus abdominis* muscles of frogs can accomplish only two or three contractions after treatment with fluorodinitrobenzene as against about 30 in muscles treated with iodoacetate.

In all, three ways are known in which ATP can be resynthesized anaerobically from ADP in muscle. The first of these, the reaction with phosphagen, is probably used very early in contraction and enables the muscle to keep up a high level of immediately available energy in the form of ATP by drawing upon the stored $\sim \textcircled{P}$ of the phosphagen. The second source of supply consists in the new $\sim \textcircled{P}$ generated in the course of glycolysis and these are the main ultimate sources of energy when activity is prolonged. Glycolysis gets under way relatively slowly, however, and *phosphagen is used to tide the muscle over the interval between the onset of activity and the establishment of the glycolytic reaction sequence.*

There is yet a third possible source of ATP, though perhaps this is used only when the muscle is *in extremis.* In all the processes mentioned so far, only the terminal $\sim \textcircled{P}$ of ATP is involved. After this has been utilized there remains another such $\sim \textcircled{P}$ in each ADP molecule, but this is not directly accessible. Muscle, however, contains *adenylate kinase* (myokinase) and this enzyme, acting upon two molecules of ADP, can catalyse the transfer of phosphate from one molecule to the second, yielding a molecule of ATP, together with adenylic acid:

$$ADP + ADP \rightleftharpoons AMP + ATP.$$

As has been suggested, this way of producing ATP is probably only used as a last resort. The free adenylate produced is highly toxic and, once formed, rapidly undergoes deamination at the hands of the *adenylate deaminase* of muscle to yield inosinic acid, and the muscle goes into rigor. Inosinic acid cannot replace adenylic acid as a carrier of phosphate, nor does the action of the adenylate deaminase appear to be reversible.

Myosin ATP'ase splits not only ATP but the triphosphates of inosine, guanosine, uridine and cytidine as well, but there is no evidence that any of these can act as alternative sources of contraction energy.

CHEMICAL EVENTS IN ANAEROBIC CONTRACTION

We are now in a position to make some sort of picture of the course of events in anaerobic muscular contraction. It will be convenient first to consider what takes place during contraction in a muscle previously poisoned with iodo-acetate. This drug, it will be recalled, abolishes the activity of triosephosphate dehydrogenase and therefore puts a stop to glycolysis and to the oxidative breakdown of its products, so that only two anaerobic sources of ATP are now available to the cells.

On the arrival of a nerve impulse, ATP is broken down, giving rise, by way of unknown intermediary operations involving actomyosin or its components, to ADP and inorganic phosphate, and furnishing the contraction energy. The ADP is promptly converted again into ATP at the expense of phosphagen and no change in the ATP content of the muscle can be detected unless the creatine kinase has first been inhibited, e.g. by fluorodinitrobenzene; but in an iodoacetate-treated muscle some phosphagen disappears and is replaced by free creatine and free inorganic phosphate. If repeated stimuli are applied to the muscle, these processes continue until, eventually, no phosphagen remains. At a last resort the adenylate kinase of the muscle is called into play and the last traces of ATP are decomposed, giving, in the end, adenylic acid, which is deaminated. The muscle goes into rigor, and the ammonia produced by the deaminase can be detected in the cells.

In the case of a *normal, unpoisoned muscle* working anaerobically, glycolysis also comes into the picture and the following phenomena can be observed during *the period of anaerobic activity*:

ATP remains unchanged.
Phosphagen disappears.
Free creatine appears.
Free inorganic phosphate appears.
Some glycogen disappears.
Lactate is formed.

361

These changes continue as long as the muscle is active. There then follows a short *period of anaerobic recovery* and during this interval, which amounts to perhaps 30 sec., the following further changes take place:

ATP remains unchanged.
Phosphagen is resynthesized.
Free creatine disappears.
Free inorganic phosphate disappears.
Some glycogen disappears.
Lactate is formed.

Thus glycolysis continues for a short time even after the cessation of muscular activity, and this 'glycolysis of recovery' is attended by resynthesis of phosphagen and a return to the *status quo* of the resting muscle apart, of course, from the conversion of some glycogen into lactate.

All these phenomena can be accounted for in terms of the reactions we have considered. While activity lasts, ATP is broken down to provide the energy expended by the muscle. Phosphagen is used to maintain the level of ATP and corresponding amounts of creatine and inorganic phosphate are set free. The free phosphate is taken up, for the breakdown of glycogen in the first instance, and later for the conversion of glyceraldehyde-3-phosphate into glycerate-1:3-diphosphate. This is followed by the generation of new high-energy phosphate in the usual way, the new $\sim \circledP$ being transferred to ADP and fresh ATP synthesized. This relieves the drain on the phosphagen stores of the muscle and, indeed, once glycolysis is well established, new high-energy phosphate is usually generated somewhat more rapidly than it is expended in the breakdown of ATP. The surplus $\sim \circledP$ is transferred through ATP to free creatine accordingly, and phosphagen begins to be resynthesized even while activity is still in progress, and it has been shown that during a short period of moderate activity there is at first a steep fall in the phosphagen content of the muscle, followed by a rise to a new, steady level.

During the period of anaerobic recovery, glycolysis continues as long as free inorganic phosphate is available, glycogen accordingly disappearing and lactate being formed. ATP is synthesized as long as glycolysis is in progress and, since it is no longer being broken down for energy production, tends to accumulate so that the remaining free creatine is phosphorylated.

That two types of cells so different in morphology and function as yeast and muscle should make use of practically the same reactions for the breakdown of carbohydrate could hardly have been anticipated. The discovery that such close parallels exist between the two seems to be a hint that the reactions we have been considering may perhaps form a part of the fundamental metabolic equipment of living cells in general, and this does in fact seem highly probable. There is now evidence that tissues other than muscle, tissues such as liver, kidney, brain and so on, make use of reactions which are essentially the same as those employed in muscle. It is now certain that many bacteria and even plants also contain many or most of the enzymes involved in fermentation and in muscle glycolysis, and that ATP is an almost universal go-between which gathers up available energy from carbohydrate and other sources and stores it up in an immediately accessible form.

Table 28. *Concentration of phosphagen in various tissues*
(mg. P per 100 g.)

Organ and animal	Phospho-creatine	Phospho-arginine	Inorganic phosphate	Total phos-phagen + inorganic-P
Striated Muscle:				
Rabbit	62	.	26	88
Guinea-pig	22	.	58	80
Frog	50	.	30	80
Sea-urchin	11	18	9	38
Scallop	0	42	25	67
Electric tissue:				
Electric ray	37	.	25	62
Cardiac muscle:				
Rat	5	0	31	36
Unstriated muscle:				
Rat (stomach)	3	0	13	16
Sea-cucumber (body-wall)	.	28	5	33
Nerve:				
Dog (brain)	12	.	.	>12
Rabbit (sciatic)	6	.	9	15
Frog (sciatic)	7	.	10	17
Testis:				
Rabbit	1·4	.	11·6	13
Jensen sarcoma	1·2	.	22	23·2

The principal differences between different animal tissues from the point of view of their glycolytic mechanisms lie in the amounts of phosphagen they contain (Table 28). The larger and more powerful skeletal muscles usually seem to contain larger amounts of phosphagen, while the slow-acting, smooth muscle of the gastro-intestinal canal, for example, contains only a fraction, amounting to perhaps one-fifth, of the amount present in the average striated muscle. Cardiac muscle also contains relatively little. The only tissues known to contain phosphagen in concentrations comparable with those of striated muscle are the electric organs of certain fishes, e.g. *Torpedo*. Like striated muscle, these organs are capable of going into activity almost instantaneously and of dissipating very large amounts of energy in very short periods of time. There can be little doubt that the phosphagen mechanism is an energy-storing device which makes it possible for an organ to go rapidly into action and to do a large amount of work in a short time. Our knowledge of the part it plays in muscle certainly confirms this idea.

Phosphagen is also present in spermatozoa in appreciable amounts, and there is evidence that these cells draw the energy for their locomotion from typical glycolytic processes involving ATP and phosphagen. Traces of phosphagen are found in nerve, brain and various glandular structures, and it has been suggested that it plays a part in their activity also. Tissues of this kind are usually richly vascularized, however, and it may be that the phosphagen present is chiefly associated with the muscle cells that enter into the make-up of the walls of the blood vessels.

The phosphagen system apart, glycolysis seems to be an almost universal feature of the make-up of living cells, at any rate under anaerobic conditions, and is seems tolerably certain that ATP likewise occurs almost universally. In view of the evidently great importance of these processes, therefore, the reaction-sequence of glycolysis is summarized for reference in Fig. 82, though the part played by phosphagen has been omitted from the scheme for the sake of simplicity. The enzymes and their coenzymes and inhibitors are tabulated in Table 29 for reference. The accuracy of the glycolytic system as we know it is attested by the fact that it is possible to prepare all the enzymes and coenzymes in pure form and, when all these are mixed together, the preparation will carry out all the reactions of glycolysis from beginning to end

Emphasis must now be placed upon one feature of the glycolytic reaction sequence that hitherto has been but little mentioned, namely that, as far as the carbohydrate side is concerned, *all the reactions involved can be reversed*, though some of the 'forward' reactions are irreversible so that different reactions are required for the 'backward' process. It was formerly believed that the phosphate-transfer reaction between phospho*enol*pyruvate and ADP is irreversible, but it has been shown that ATP can phosphorylate pyruvate, given certain conditions, viz. (*a*) a high concentration of ATP and (*b*) the presence of potassium ions. Another more roundabout route for the re-synthesis of phospho*enol*pyruvate is known; this involves oxaloacetate, GTP and a biotin-dependent phospho*enol*pyruvate carboxylyase and is mentioned again later in another context (see p. 117).

Two other stages in the reverse sequence require special comment, first the dephosphorylation of fructofuranose-1:6-diphosphate to yield the 6-mono-phosphate. This does not proceed by reaction with ADP, since the sugar ester contains no high-energy phosphate such as is required for the phosphorylation of ADP. This dephosphorylation is carried out hydrolytically with the aid of a phosphatase specific for fructose diphosphate. Secondly, it is also true that the final stage of glycogen formation from α-glucose-1-phosphate requires UTP and the appropriate transferring enzymes, but taken as a whole it is true to say that *glycolysis is a reversible operation*.

According to the work of Meyerhof on frog muscle, some 80 % of the lactate formed during anaerobic activity is reconverted into glycogen in the muscle when the latter is allowed to rest in oxygen. The remaining 20 % is oxidized

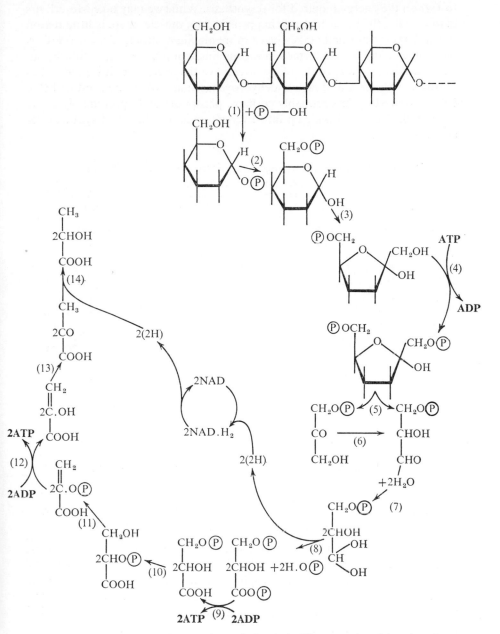

Fig. 82. Scheme to summarize reactions of glycolysis. The part played by phosphagen (where present) is omitted. For names of enzymes, coenzymes and inhibitors, see Table 29. (See also Fig. 78, p. 342.)

to furnish the energy required for resynthesis. While we may take Meyerhof's authority for the belief that this happens in frog muscle, there is little reason to think that it normally takes place in mammalian muscle. As Cori & Cori have shown, any lactate formed in the mammalian muscle *in situ* diffuses out and is carried, by way of the blood stream, to the liver. Here it is oxidized to pyruvate with the aid of lactate dehydrogenase and NAD, the product being either oxidized or, less probably perhaps, phosphorylated, presumably at the expense of ATP, and built up into liver glycogen by the usual synthetic reaction sequence.

Table 29. *Glycolysis: enzymes, coenzymes and inhibitors*
(See also Table 27, p. 343)

Reaction	Enzyme	Coenzyme	Inhibited by
1	α-Glucan phosphorylase	—PO_3H_2	Dialysis
2	Phosphoglucomutase	Glucose-1:6-diphosphate	.
3	Glucosephosphate isomerase	.	.
4	Phosphofructokinase	ATP	Dialysis
5	Aldolase	.	.
6	Triosephosphate isomerase	.	.
7	Spontaneous	.	.
8	Triosephosphate dehydrogenase	NAD; —PO_3H_2	Dialysis; $CH_2I.COOH$
9	Phosphoglycerate kinase	ADP	Dialysis
10	Phosphoglyceromutase	Glyceric acid-2:3-diphosphate	.
11	Phosphopyruvate hydratase	Mg ions	NaF
12	Pyruvate kinase	ADP	Dialysis
13	? Spontaneous	.	.
14	Lactate dehydrogenase	NAD.H_2	Dialysis

Note: Current synonyms are given in the text where this seems desirable.

DISTRIBUTION AND METABOLISM OF CREATINE AND OTHER GUANIDINE DERIVATIVES

The predominant members of this group are arginine and creatine. These bases are mainly confined to muscular tissues, in which they are present in the form of their very labile phospho-derivatives, the *phosphagens*. With certain notable exceptions to which we shall refer presently it may be said that phosphocreatine is found only among vertebrates, and phosphoarginine only among invertebrates.

CREATINE

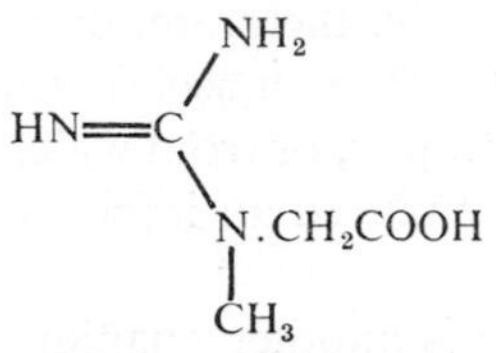

$$HN{=}C\begin{array}{c} \diagup NH_2 \\ \diagdown N.CH_2COOH \\ | \\ CH_3 \end{array}$$

Creatine itself is an amino-acid, though not of the usual α-amino-type, nor does it occur in proteins. The fact that it replaces arginine in the muscles of vertebrates led early to the belief that it must be formed metabolically from arginine, a view which received some *a priori* support from the common presence of an amidine group in both substances.

For many years it was believed that creatine must arise from arginine, and we have in fact already described the synthetic route via glycocyamine (p. 201).

Creatine itself does not usually appear in mammalian urine, except during childhood and sporadically in adult females, but phosphocreatine undergoes a slow and apparently spontaneous conversion under physiological conditions of temperature and pH into *creatinine*, a normal excretory product among mammals,

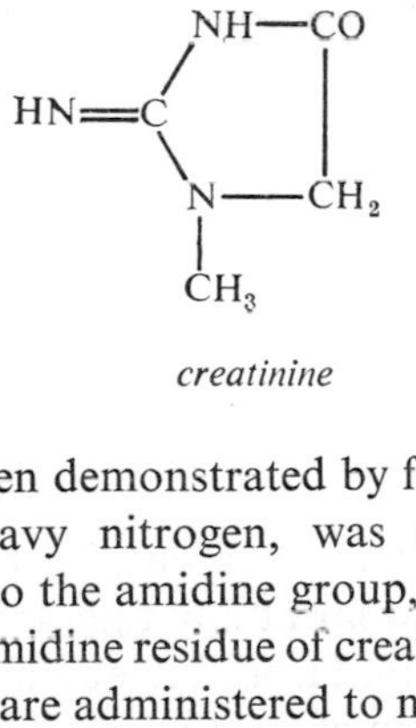

$$HN{=}C\begin{array}{c} \diagup NH{-}CO \\ \diagdown \quad | \\ N{-}CH_2 \\ | \\ CH_3 \end{array}$$

creatinine

Its origin in creatine has been demonstrated by feeding experiments in which creatine, labelled with heavy nitrogen, was administered. When heavy nitrogen was introduced into the amidine group, the isotope was recoverable in the same position in the amidine residue of creatinine isolated from the urine.

If small doses of creatine are administered to normal animals the substance is temporarily accommodated in the muscles, presumably as the phospho-compound, as may be demonstrated by comparing the muscles of animals thus fed with those of controls. Later, increased outputs of creatinine are observed. Larger doses of creatine give rise to a definite creatinuria, however, together with an increased excretion of creatinine.

In certain pathological conditions of the muscles, the muscular dystrophies, which are characterized by wasting of the muscular tissues, creatine and creatine kinase appear in the blood. Creatine is still produced in the organism

367

at the usual rate, but, as the muscles are no longer able to accommodate it, a large proportion is excreted in the urine. Creatine is also excreted after the amputation of a limb. Evidently, therefore, creatine is made elsewhere than in the muscles, and is excreted if the muscles cannot make use of it. Animals suffering from muscular dystrophy, or from which major muscle masses have been removed, have therefore played an important part in studies of creatine formation.

Nearly related to creatine is another guanidine base, known as *creatone*. This substance has been isolated from many vertebrate materials but never from invertebrate sources. It is an artefact derived from creatine during the process of isolation. The methods formerly used in working up the nitrogenous bases of biological materials often involved the use of mercuric salts as precipitants, and it has been shown that, in the presence of mercuric compounds, creatine readily undergoes atmospheric oxidation to creatone:

$$\text{HN}=\!\!\!\!\text{C}\!\!\begin{array}{l} \diagup\text{NH}_2 \\ \diagdown\text{N.CH}_2\text{COOH} \\ \quad\,|\\ \quad\text{CH}_3 \end{array} \xrightarrow{+\,O_2} \text{HN}=\!\!\!\!\text{C}\!\!\begin{array}{l} \diagup\text{NH}_2 \\ \diagdown\text{N.CO.COOH} \\ \quad\,|\\ \quad\text{CH}_3 \end{array} + \text{ H}_2\text{O}$$

creatine   *creatone*

Creatone itself is readily hydrolysed to give oxalic acid and another artefact, *methylguanidine*:

$$\text{HN}=\!\!\!\!\text{C}\!\!\begin{array}{l} \diagup\text{NH}_2 \\ \diagdown\text{N.CO.COOH} \\ \quad\,|\\ \quad\text{CH}_3 \end{array} + \text{ H}_2\text{O} = \text{HN}=\!\!\!\!\text{C}\!\!\begin{array}{l} \diagup\text{NH}_2 \\ \diagdown\text{NH} \\ \quad|\\ \quad\text{CH}_3 \end{array} + \begin{array}{l}\text{COOH}\\ |\\ \text{COOH}\end{array}$$

creatone   *methylguanidine*

Methylguanidine, like creatine and creatone, has been isolated from numerous vertebrate muscles but not, so far at any rate, from any invertebrate source. *Guanidine* itself has been obtained from a few animal sources, and may perhaps arise by the further degradation of creatine, arginine, or some other guanidine derivative. Other nearly related compounds to which reference has already been made are *taurocyamine* (p. 102 and *asterubin* (p. 103).

ARGININE

Quite apart from the special part it plays in muscle metabolism in invertebrates, arginine enters into the composition of the tissue proteins of animals of every kind. It will be recalled that the rate at which arginine can be synthesized *de novo* in young rats appears to be limited and can act as a limiting factor upon their growth-rate. Its synthesis from ornithine forms a major part of the 'ornithine cycle' (pp. 274–6). Whether invertebrates are less or more adept at the synthetic production of arginine we do not know: for present purposes we may reasonably assume that it is normally forthcoming in sufficient quantities in their diet.

Distribution of the phosphagens. Although arginine has been detected in, and in many cases actually isolated from, the tissues of representative members of almost every phylum and class among the invertebrates (excepting the annelid worms, from which it is virtually absent), there is no reason to believe that it occurs in vertebrates except, of course, as a constituent of proteins and also, in small concentrations, as a transitory metabolic intermediate.

Creatine has been isolated from representative members of every vertebrate group but never, except occasionally and in traces, from invertebrates. To these generalities, however, there are a few noteworthy exceptions. Of all the invertebrate types studied, the Echinodermata are among the few invertebrate groups in the muscles of which the presence of creatine has been satisfactorily demonstrated. In several echinoids (sea-urchins) arginine and creatine occur side by side, while in the ophiuroids (brittle stars) only creatine has been detected.

No comparable cases have been found in the Vertebrata proper but among the Protochordata, a group of creatures which resemble the true vertebrates in their possession of a primitive notochord, one similar case has been recorded. The Protochordata comprise three groups, the Tunicata (sea-squirts), the Enteropneusta (a small group of worm-like animals) and the Cephalochorda (lancelets). These Protochordata are of special interest since they are regarded on morphological grounds as lying on the border-line between the true Vertebrata on the one hand and the invertebrate phyla on the other.

Of the Cephalochorda it may be said that they resemble the vertebrates rather than the invertebrates, for not only do they possess a well-developed notochord but, chemically speaking, the relationship is clear from the fact that they contain creatine but not arginine. Superficially, the adult Tunicata resemble the invertebrates rather than the vertebrates in their general appearance and mode of life, but their tadpole-like larvae possess a well-developed notochord which is lost however at metamorphosis. It has been shown that the adults, nevertheless contain phosphocreatine and that, contrary to previous opinion, phosphoarginine is absent. In the Enteropneusta, however, there is one species in which there is evidence for the presence of arginine and creatine side by side in the musculature.†

The Enteropneusta thus appear to be related to the Vertebrata in that they

† In *Balanoglossus salmoneus*. The closely related *B. clavigerus* and two species of *Saccoglossus* contain creatine but no arginine.

contain creatine and, at the same time, to the invertebrate phyla in that they contain arginine. In particular, they show an affinity with the Echinodermata, for only in these two groups have arginine and creatine been found to co-exist in significant quantities. Similar relationships were established on purely morphological grounds by the classical investigations of Müller, Metschnikow

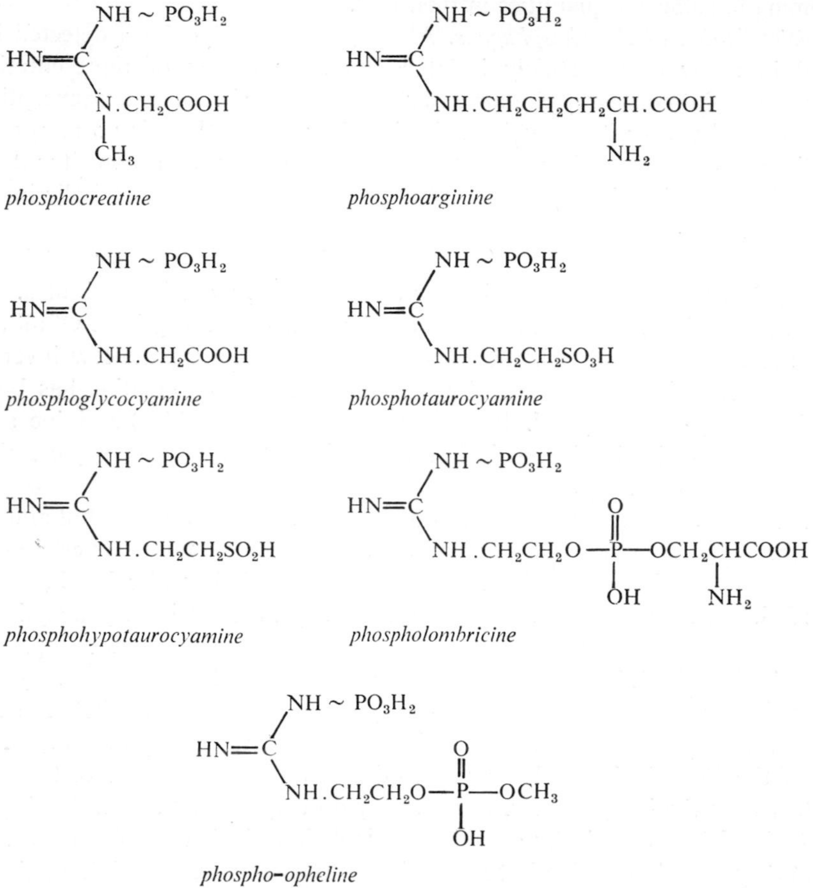

phosphocreatine

phosphoarginine

phosphoglycocyamine

phosphotaurocyamine

phosphohypotaurocyamine

phospholombricine

phospho-opheline

Fig. 83. Phosphagens found in annelid worms.

and Bateson. As Bateson showed, in addition to their gill-slits the Entero-pneusta possess a short but well-defined notochord, features which establish their relationship to the Vertebrata. But in their larval forms the Entero-pneusta so closely resemble the echinoderms that their larvae were classified with those of the Echinodermata before the adult forms were discovered. It was Metschnikow who, some time later, showed that the *Tornaria* larva, far from being an echinoderm larva as was previously believed, gives rise in the adult form to *Balanoglossus,* an enteropneust. From the phylogenetic stand-

370

point the natural distribution of these two bases, arginine and creatine, is clearly of considerable interest.

Special reference must here be made to the polychaete worms. These are most versatile animals. They possess sometimes one, sometimes two and sometimes more of the phosphagens listed in Fig. 83 but there seems to be no relationship between these phospho-compounds and the habits, habitats or morphology of the worms themselves.

There are a number of other guanidine bases which occur in nature and have clear structural relationships to arginine, from which they are in all probability derived. These include *agmatine*, which has been isolated from a number of invertebrates, and arises by the (probably bacterial) decarboxylation of arginine. It has no known function.

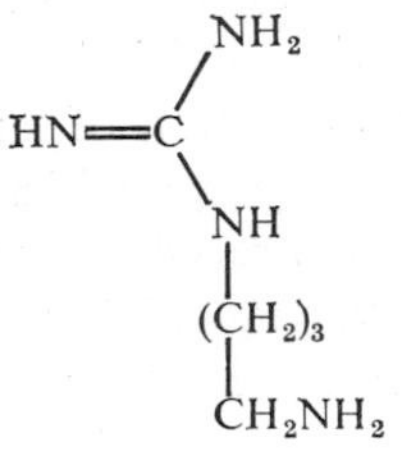

$$
\begin{array}{ccc}
\text{HN==C} \diagup^{\displaystyle NH_2}_{\diagdown NH} & \text{HN==C} \diagup^{\displaystyle NH_2}_{\diagdown NH} & \text{HN==C} \diagup^{\displaystyle NH_2}_{\diagdown NH} \\[2mm]
(CH_2)_3 & (CH_2)_4 & (CH_2)_3 \quad CH_3 \\[1mm]
CH_2NH_2 & NH & CH.NH\text{---}CH \\[1mm]
 & HN==C \diagdown NH_2 & COOH \quad COOH \\[2mm]
\textit{agmatine} & \textit{arcain} & \textit{octopine}
\end{array}
$$

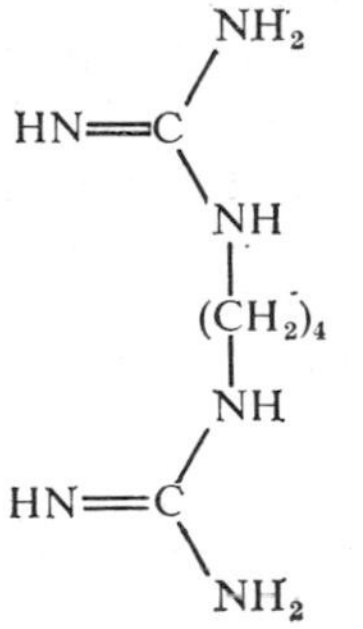

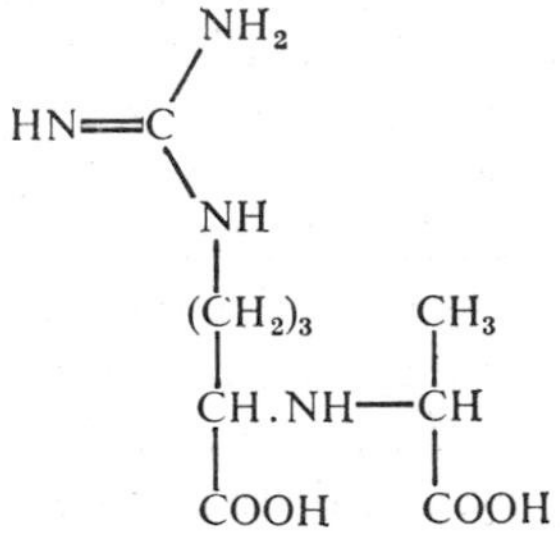

Arcain again has occasionally been isolated from invertebrate materials but neither its functions, if any, nor the mode of its synthesis is known. Conceivably it might arise by a transamidination of agmatine.

Octopine is more entertaining. This interesting substance was first isolated from the muscles of *Octopus*, but has since been obtained from several other cephalopod species and a few lamellibranchs. It is not present in perfectly fresh muscle but arises as a post-mortem product, arginine disappearing as octopine is formed. Its constitution has been established by synthesis, which can be accomplished by the reductive condensation of arginine with pyruvate. It is probable that octopine arises in this manner in *Octopus* muscle, for it has been known for many years that this, unlike most muscular tissues, contains very little lactate in fatigue or at death. Lactate, of course, is formed as a general rule by the reduction of pyruvate, but in the case of *Octopus* it seems reasonably certain that pyruvate condenses with arginine to yield an intermediate complex, which then undergoes reduction in place of pyruvate itself. Alternatively, octopine might arise directly by the condensation of arginine with lactate, but present evidence is insufficient to allow a definite decision on this point.

19

AEROBIC METABOLISM OF CARBOHYDRATES

INTRODUCTION

The bulk of our knowledge of the mechanisms of muscle glycolysis under anaerobic conditions was gained by studies of muscle extracts, which function anaerobically, or else of intact, isolated muscles kept under anaerobic conditions. Now a typical muscle *in situ* enjoys an excellent blood supply, and there is physiological evidence that this blood supply is actually augmented when the muscle goes into action. Analysis of the blood entering and leaving a perfused muscle shows that activity is attended by the utilization of large amounts of oxygen and the formation of correspondingly large quantities of carbon dioxide. Moreover, many muscles contain a special intracellular store of oxygen in the form of muscle oxyhaemoglobin (oxymyoglobin), upon which they can draw for additional oxygen during the interval between the onset of activity and the physiological augmentation of the normal blood supply. For these reasons we must enquire whether we have perhaps been led astray by studying muscle metabolism only under anaerobic conditions so far, and whether anaerobic contraction has any real biological significance at all.

It is characteristic of the anaerobic metabolism of muscle that glycogen is broken down and lactate formed. Lactate arises by the reduction of pyruvate, and its formation provides a mechanism for the reoxidation of reduced NAD. Unless the reduced coenzyme were in some way reoxidized, no further production of glycerate phosphate could take place and the generation of new high-energy phosphate for the resynthesis of ATP would come to an end. Under aerobic conditions, however, the reduced coenzyme is rapidly reoxidized through the flavoprotein/cytochrome/cytochrome oxidase reaction chain, and no lactate need be produced at all; instead, it might be anticipated, pyruvate would accumulate in the muscle. But there is no evidence that pyruvate does so accumulate in normal muscle, nor is there evidence that pyruvate escapes from the muscle into the blood. We are left, therefore, with the conclusion that if pyruvate is formed under aerobic conditions, it must be oxidized, and that it is the source of much or all of the carbon dioxide produced by an active muscle. We know that the amount of energy that becomes accessible to an organ or cell that oxidizes its metabolites completely is greatly in excess of that obtained by partial, anaerobic breakdown and, since the muscle is evidently capable of using oxygen and is provided with elaborate devices for supplying that oxygen, it becomes more than ever necessary to

372

find out whether anaerobic contraction has any biological function, and whether lactate is ever formed *in vivo*.

Careful estimations show that small quantities of lactate are normally present in the blood even at rest, and that the amount is somewhat increased as a result of moderate exercise. After very violent exertion, however, there is a sharp rise in the blood lactate, but the level soon begins to fall again as the lactate is taken up by the liver and either oxidized or converted into glycogen. A typical curve is shown in Fig. 84, for moderate (*a*) and for strenuous work (*b*).

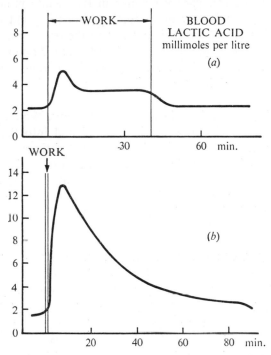

Fig. 84. Changes of blood lactate during and after exercise. (Modified after Lipmann, from data by Banga.) (*a*) Moderate work, (*b*) strenuous work.

If now we consider the conditions prevailing in an intact muscle *in vivo*, these phenomena can be accounted for. In mild or moderate exercise, oxygen is brought into the cells fast enough to reoxidize the $NAD.H_2$ as rapidly as it is formed, so that little or no lactate is produced, and pyruvate, instead of being reduced, is completely oxidized. If now the degree of exertion is increased, glycogen will be more rapidly broken down and NAD proportionately more rapidly reduced. Eventually, with increasing severity of exercise, a point will be reached at which the oxygen supplied by the circulatory apparatus can only just keep pace with the reduction of the coenzyme. But the muscle can work still harder by (*a*) utilizing oxygen as fast as it is made

25-2

available by the circulatory system, *and* (*b*) reoxidizing any coenzyme that still remains in the reduced condition by using the anaerobic device of lactate formation. In the case of an antelope running for its life from a pursuing lion, the ability to use its muscles anaerobically, above and beyond the limits set by the efficiency of its circulatory apparatus, may allow of the extra turn of speed that saves the antelope's life: but the lion, unfortunately, can make use of the same trick!

As has been pointed out already, when lactate is formed a good deal of potentially available free energy remains in the lactate molecules, so that anaerobic metabolism is very unproductive from the point of view of the muscle. But this residual energy is not lost to the organism, for any lactate formed rapidly escapes from the muscles into the blood and is transported to that maid-of-all-work, the liver, which proceeds to transform it again into glycogen or else to oxidize it, according to the conditions prevailing in the liver at the time.

These considerations raise a host of new problems. We must enquire into the mechanisms whereby pyruvate is oxidized and into the origin of the respiratory carbon dioxide which is so characteristic an end-product of aerobic, as opposed to anaerobic, metabolism. However, since it is from the liver, the central storehouse for carbohydrates, that the muscles derive their own local stocks of glycogen, we shall do well to look into the mechanisms of glycogen synthesis in the liver itself.

THE LIVER: GLYCOGENESIS, GLYCOGENOLYSIS AND GLYCONEOGENESIS

The carbohydrate metabolism of animals centres round the processes of glycogenesis and glycogenolysis, i.e. the production and breakdown of glycogen. These processes take place mainly in the liver. Glycogen can be formed from carbohydrate materials, in which case we speak of *glycogenesis* but, as we have already seen, it can also be formed by the liver from non-carbohydrate sources such, for example, as lactate, certain amino-acids, glycerol, pyruvate and propionate and many other simple substances. In this case, therefore, we speak of *glyconeogenesis*. The term *glycogenolysis* is used to refer to the breakdown of glycogen to glucose, as opposed to the more extensive process of anaerobic disintegration which we call *glycolysis*.

As far as is known, the carbohydrates of the food do not contribute any materials that are 'essential' in the sense that certain of the fatty- and amino-acids are essential. Their function is pre-eminently that of furnishing a readily metabolized source of energy. The principal monosaccharide formed by the digestion of an average meal is glucose. Other sugars play a relatively small part as a rule, but in the infant mammal, whose sole food carbohydrate is lactose, one-half of the lactose molecule gives rise to galactose on hydrolysis.

374

Other monosaccharides that arise from food include fructose, formed from sucrose and to some extent from fructofuranosans such as inulin.

Pentoses, if injected, are not utilized, but appear largely unchanged in the urine, while disaccharides similarly undergo excretion if injected into the blood stream. Glucose, together with fructose, galactose and the rarer sugar mannose, all lead to the deposition of glycogen in the liver. *Glucose* itself is phosphorylated by hexokinase at the expense of ATP and the product, glucose-6-phosphate, is transformed by phosphoglucomutase into glucose-1-phosphate, the raw material for the synthesis of glycogen via UDPG, the appropriate transglucosylases and a branching factor. *Fructose* also can be phosphorylated at the expense of ATP under the influence of hexokinase to yield fructofuranose-6-monophosphate which, in turn, is convertible into glucose-6-phosphate by glucosephosphate isomerase and hence, by way of glucose-1-phosphate and UDPG, gives rise to glycogen. *Mannose* too can be phosphorylated by hexokinase, yielding mannose-6-phosphate which, in the presence of mannosephosphate isomerase and glucosephosphate isomerase, gives an equilibrium mixture of the 6-phosphates of glucose and fructose, from which α-glucose-1-phosphate and then UDPG can be formed in the usual way. *Galactose* can be phosphorylated at the expense of ATP by a specific galacto-kinase to yield galactose-1-phosphate and the latter, under the influence of a second enzyme, UDP-glucose epimerase, which requires NAD as cofactor, gives rise to UDPG directly. No matter which of the utilizable mono-saccharides is administered, the polysaccharide formed in the liver is always glycogen and the routes of conversion are summarized in Fig. 85 for convenience of reference.

From the standpoint of general metabolism it is important to realize that, even under strictly normal conditions, the liver's capacity to form and store glycogen is by no means unlimited. Rabbits, for example, can be literally crammed with foods rich in carbohydrate, but a glycogen content of more than 18–20 % in the liver is seldom or never realized. Carbohydrate, if administered in excess of the storage capacity of the liver, gives rise to fat, which is deposited in the fat depots of the body to await metabolism when harder times come.

Glycogen can also be formed in the liver by *glyconeogenesis*, and many different substances are known to be able to contribute to this process. In a starving, hepatectomized 10 kilo. dog with no accessible stores of carbohydrate as such, the blood sugar level falls, but can be held at the normal level by injecting *ca.* 250 mg. glucose/kilo./hr. or *ca.* 60 g. glucose a day. It follows that a normal dog of similar weight must be capable of producing *at least* 60 g. glucose in 24 hr. since it can maintain a normal level of blood glucose for days on end, even if starved. If the animal is starved, this glucose must arise through glyconeogenesis which, therefore, is evidently a not inconsiderable source of carbohydrate formation.

An early technique for detecting the formation of carbohydrate from non-carbohydrate sources consisted in administering the suspected substances to a diabetic or to a phlorrhizinized animal. Alternatively the substances can be administered to starving animals to see whether or not they give rise to the deposition of glycogen in the liver. Isotopic labelling of suspected precursors has largely replaced the use of diabetic and phlorrhizinized animals and has given confirmatory results in many cases. For example ^{14}C-labelled lactate gives rise to correspondingly labelled glycogen.

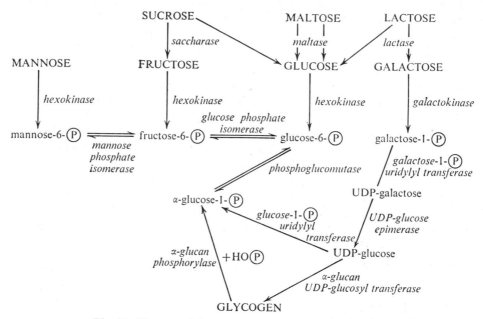

Fig. 85. Glycogenesis from common mono- and disaccharides.

We have already seen that the deaminated residues of certain amino-acids can give rise to pyruvate (Chapter 12), and, this the liver can convert into glycogen. *Any substance that lies on the glycolytic route from glycogen to pyruvate or which gives rise to any substance lying on that route, can be converted into glycogen*, and we can account for most of the reactions involved in the conversions of such substances. Thus glycerol, a well-known glucose-former, can be phosphorylated through the agency of ATP and a specific glycerol kinase to yield glycerol-α-phosphate, which can then be oxidized to triosephosphate by the α-glycerolphosphate dehydrogenases which are present in animal tissues generally. Triosephosphate lies directly on the route leading back to glycogen and we can therefore give a reasonable account of the processes which underlie the formation of glycogen from glycerol. Propionate again is capable of giving rise to glycogen (p. 119), and is of considerable importance, for it is one of the predominant short-chain fatty acids produced from cellulose when the latter is attacked by the symbiotic micro-organisms of the rumen-contents of the sheep. Similar processes go on in many herbivores, and it is likely that much of the carbohydrate produced and laid down in such animals arises, not directly from the carbohydrates of the food, but by glyconeogenesis from propionate.

Glucose does not lie directly on the pathway of glycolysis and its free interconvertibility with glycogen has therefore to be specially considered. There is now abundant evidence that *glucose must be phosphorylated before it can be stored as glycogen or can enter into the glycolytic reaction sequence.* This indispensable preliminary phosphorylation is catalysed by hexokinase. Most animal tissues, including muscle, contain enzymes analogous to the hexokinase of yeast, which catalyses the phosphorylation of free glucose at the expense of ATP, yielding glucose-6-monophosphate and hence the 1-phosphate, from which glycogen is synthesized by the tissue transglucosylases with UDPG as an intermediate.

Apart from the liver, which is the central storehouse for glycogen, the muscles contain considerable quantities of this polysaccharide. Other tissues contain only small quantities. These peripheral stores of glycogen are built up mainly and perhaps entirely at the expense of the liver glycogen, by way of the circulating glucose of the blood. The interconnexions are illustrated in Fig. 86. Liver glycogen undergoes breakdown to yield α-glucose-1-phosphate and hence the 6-phosphate, which is readily dephosphorylated by a specific liver phosphatase to give free glucose. In these breakdown processes UDPG is not involved. It is generally believed that the blood glucose must arise mainly in this way, since glucose-1-phosphate, which is a glycoside and not an ester, is attacked relatively slowly or not at all by the liver phosphatase. The free glucose passes into the blood stream. Other tissues such, for instance, as the muscles, appear not to possess any enzyme corresponding to the liver glucose-6-phosphate phosphatase and so cannot contribute free glucose to the blood. They can, however, take up free glucose from the blood and phosphorylate it at the expense of ATP by means of their hexokinase. The product, glucose-6-phosphate, is transformed into glucose-1-phosphate by phospho-glucomutase in the usual way and hence, with the aid of UDPG and the usual enzymes and branching factors, into glycogen. The latter is then held in readiness for use when the need arises. Essentially similar reactions are involved in the synthesis and breakdown of starch in plant tissues.

The concentration of free glucose in mammalian blood is very finely adjusted at a level of about 100 mg. per 100 ml., the precise concentration varying somewhat from species to species. Any rise in the blood-sugar level is compensated by the deposition of glycogen in the tissues, mainly in the liver, and any fall by the mobilization of more liver glycogen. Precisely how the blood concentration is controlled is still highly uncertain however, but it is perhaps significant that the liver α-glucan phosphorylase, which is intimately involved in the process, is powerfully inhibited by free glucose. Moreover, its activity rises in rats rendered diabetic by administration of alloxan and is depressed by administration of insulin.

The maintenance of the normal level of the blood sugar and the normal storage of glycogen in the liver are profoundly influenced by a number of

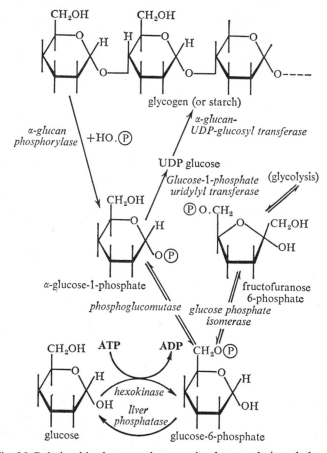

Fig. 86. Relationships between glycogenesis, glycogenolysis and glycolysis.

hormones. *Insulin*, the internal secretion of the islets of Langerhans in the pancreas, encourages the deposition of glycogen in the liver at the expense of the blood sugar. *Glucagon*, also produced in special cells in the pancreas, has the reverse effect. The '*anti-insulin hormone*' (growth hormone) of the anterior pituitary body, on the other hand, encourages the mobilization of glycogen and tends therefore to raise the level of the blood sugar.

Clinical diabetes, a condition which is characterized by intense hyperglycaemia and low glycogen storage on the part of the liver, can be due either to insufficiency of insulin secretion or to excessive production of anti-insulin factors. It has been discovered that certain extracts of the anterior pituitary have a powerful inhibitory action upon hexokinase, an effect that is antagonized by insulin. This discovery goes a long way towards explaining the chemical features of some types of diabetes. If glucose is to be stored and metabolized in the normal manner it must first of all be phosphorylated. A relative preponderance of the hexokinase-inhibitory factors over insulin will lead to suppression of hexokinase activity and this, in turn, to subnormal storage and subnormal metabolism of glucose. The other characteristic features of diabetes (ketosis and ketonuria) are metabolic consequences of the suppression of carbohydrate metabolism Although the 'anti-insulin hormone' and the hexokinase-inhibitory factor are both present in some extracts of the anterior pituitary there is little evidence that the two substances are identical.

Apparently the pituitary produces more than one 'anti-insulin hormone'. The *growth hormone* has pronounced anti-insulin properties, while the *adrenocorticotrophic hormone* (ACTH) leads to a transient glucosuria in normal rats and intensifies the glucosuria of diabetic animals. Repeated injections of either of these hormones reduce the tolerance towards carbohydrate and susceptibility to insulin, leading to total diabetes if administration is sufficiently prolonged, apparently by exhaustion of the pancreatic islet tissue. ACTH is believed to act by stimulating the secretion of cortical hormones, and certain of these promote the deposition of glycogen in liver slices while inhibiting it in diaphragm muscle. Clearly much remains to be done before the phenomena of the diabetic state can be fully explained in purely biochemical terms, but at least a promising start in this direction has been made.

Other hormones also, notably *adrenaline*, influence the level of the blood sugar, but their effects are usually short-lived and not to be compared with the long-term control exercised by insulin and the 'anti-insulin' factors. It would profit us little at this stage to go further into the relationships that exist between liver glycogen, blood glucose, and the secretions of the various endocrine organs that have influence upon them, and for further discussion of these problems the reader is referred to the standard physiological textbooks and to the monographs that have been devoted to them.

ORIGINS OF RESPIRATORY CARBON DIOXIDE

We have already given a good deal of consideration to the mechanisms whereby oxygen is consumed in the tissues and, before we go further, it may be well to make a brief review of the reactions which generate carbon dioxide.

Carbon dioxide can be formed anaerobically as well as aerobically; indeed, it is one of the two chief products of alcoholic fermentation. It is formed in this case by the action of pyruvate decarboxylyase and diphosphothiamine upon pyruvate. Animal tissues, however, contain no pyruvate decarboxylyase, although co-carboxylase (diphosphothiamine) is present among them, and we must therefore look for other possible sources of carbon dioxide.

Certain amino-acids are decarboxylated in animal tissues with formation of the corresponding amines, together with carbon dioxide:

$$R.CH(NH_2) \overline{COO} H \longrightarrow R.CH_2NH_2 + CO_2.$$

Reactions of this kind take place commonly in bacteria, but they are rare and probably small-scale processes among animals. Histidine and tyrosine, for example, are decarboxylated by specific decarboxylyases in animals, but the products are immensely powerful pharmacologically and are formed only in traces, so that the yields of carbon dioxide from these sources are comparatively trivial. Not many enzymes of this kind have been detected: those at present known include the histidine and tyrosine decarboxylyases already mentioned, together with specific glutamic and cysteic acid decarboxylyases, but we cannot look to these for the origin of the large volumes of carbon dioxide produced by animal tissues.

One important potential source of carbon dioxide lies in the dicarboxylic acids, succinate, fumarate, malate and oxaloacetate. These compounds are mutually interconvertible through succinate and malate dehydrogenases together with fumarase, a group of enzymes which appear to be universally distributed:

$$
\begin{array}{cccccc}
\text{COOH} & \text{COOH} & \text{COOH} & \text{COOH} \\
| & | & | & | & & \text{CH}_3 \\
\text{CH}_2 \; \pm 2\text{H} & \text{CH} \; \pm \text{H}_2\text{O} & \text{CH}_2 \; \pm 2\text{H} & \text{CH}_2 & & | \\
| \rightleftharpoons & \| \rightleftharpoons & | \rightleftharpoons & | \rightleftharpoons \text{CO}_2 + & \text{CO} \\
\text{CH}_2 & \text{CH} & \text{CHOH} & \text{CO} & & | \\
| & | & | & | & & \text{COOH} \\
\text{COOH} & \text{COOH} & \text{COOH} & \text{COOH} \\
\textit{succinic} & \textit{fumaric} & \textit{malic} & \textit{oxaloacetic} & & \textit{pyruvic} \\
\textit{acid} & \textit{acid} & \textit{acid} & \textit{acid} & & \textit{acid}
\end{array}
$$

Malate also undergoes a curious reaction under the influence of an enzyme which is particularly abundant in pigeon liver and requires NADP and Mn^{2+} for activity. This *malate decarboxylyase* catalyses a simultaneous oxidation and decarboxylation of its substrate, and the reaction is reversible:

$$
\begin{array}{ccccccc}
\text{COOH} & & & \text{CH}_3 \\
| & & & | \\
\text{CH}_2 & + \text{ NADP} \rightleftharpoons & \text{CO} & + & \text{CO}_2 & + & \text{NADP.H}_2 \\
| & & & | \\
\text{CHOH} & & & \text{COOH} \\
| \\
\text{COOH}
\end{array}
$$

Oxaloacetate is an unstable substance which, even *in vitro*, undergoes slow spontaneous decarboxylation in the β-position to give pyruvate. The same reaction, which is reversible, is *catalysed in liver*, though not apparently in other tissues, by a specific enzyme known as *pyruvate carboxylyase*. In this reaction ADP and inorganic phosphate are required and the enzyme is biotin-dependent:

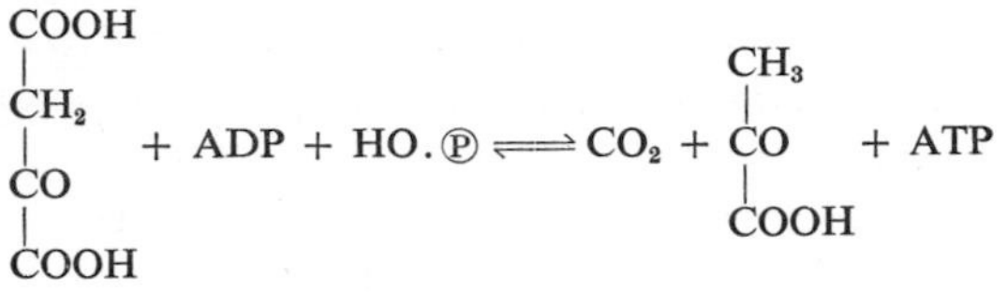

$$
\begin{array}{c}
\text{COOH} \\
| \\
\text{CH}_2 \\
| \\
\text{CO} \\
| \\
\text{COOH}
\end{array}
+ \text{ADP} + \text{HO.}\textcircled{P} \rightleftharpoons \text{CO}_2 +
\begin{array}{c}
\text{CH}_3 \\
| \\
\text{CO} \\
| \\
\text{COOH}
\end{array}
+ \text{ATP}
$$

Oxaloacetate can also undergo a reaction with guanosine triphosphate (GTP) to yield carbon dioxide together with phospho*enol*-pyruvate catalysed by *phospho*enol*pyruvate carboxylyase*;

$$
\begin{array}{c}
\text{COOH} \\
| \\
\text{CH}_2 \\
| \\
\text{CO} \\
| \\
\text{COOH}
\end{array}
+ \text{GTP} \rightleftharpoons
\begin{array}{c}
\text{CH}_2 \\
\| \\
\text{C.O} \sim \textcircled{P} \\
| \\
\text{COOH}
\end{array}
+ \text{CO}_2 + \text{GDP}
$$

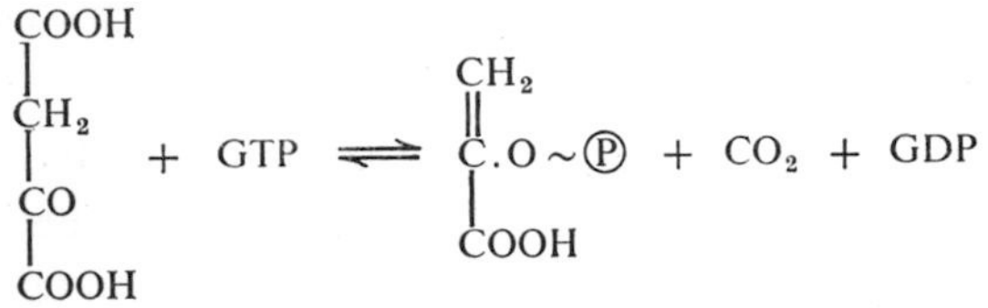

GTP can be replaced by inosine triphosphate (ITP) but not by ATP.

Another interesting decarboxylating enzyme is *oxalosuccinate decarboxylyase*, which is very widely distributed:

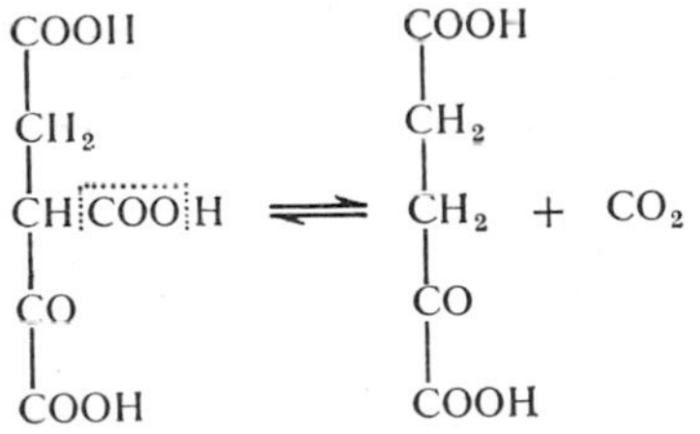

$$
\begin{array}{c}
\text{COOH} \\
| \\
\text{CH}_2 \\
| \\
\text{CH} \vdots \text{COO} \vdots \text{H} \\
| \\
\text{CO} \\
| \\
\text{COOH}
\end{array}
\rightleftharpoons
\begin{array}{c}
\text{COOH} \\
| \\
\text{CH}_2 \\
| \\
\text{CH}_2 \\
| \\
\text{CO} \\
| \\
\text{COOH}
\end{array}
+ \text{CO}_2
$$

The enzyme concerned is an *iso*-citrate dehydrogenase (see p. 166), which catalyses the dehydrogenation of *iso*-citrate *and* the decarboxylation of the product, oxalosuccinate.

These are interesting but atypical processes: the usual fate of α-keto-acids such as pyruvate is that they undergo *oxidative decarboxylation* at the α-carbon atom according to the following overall equation:

$$\text{R.CO} \vdots \text{COO} \vdots \text{H} + \tfrac{1}{2}\text{O}_2 = \text{R.COOH} + \text{CO}_2.$$

This reaction appears to be irreversible in animal tissues and is strongly inhibited by arsenite. Two points may be emphasized in connexion with this, which is an important source of respiratory carbon dioxide. First, it is essentially an oxidative process and therefore differs sharply from the 'straight' decarboxylations catalysed by pyruvate decarboxylyase and by the amino-acid decarboxylyases. The second important feature is that this reaction, though general for α-keto-acids, does not extend to β-keto-acids.

OXIDATIVE DECARBOXYLATION OF PYRUVATE

Now that we have reviewed the mechanisms that lead to the deposition of glycogen in the liver, the mechanisms for its mobilization as glucose and of the fermentative or glycolytic breakdown of glucose to pyruvate it is time to consider the fate of pyruvate itself. The first step in this direction consists of oxidative decarboxylation.

Although oxidative decarboxylation differs in most respects from the 'straight' decarboxylation catalysed by pyruvate decarboxylyase, it does resemble it closely in one important particular, namely that it requires the participation of co-carboxylase. This substance is identical with the pyrophosphate of vitamin B_1 and is usually referred to as *thiamine diphosphate* or aneurin diphosphate, the formula of which is shown on p. 387. It arises from free thiamine by a reaction with ATP, catalysed by *thiamine pyrophosphokinase*:

$$\text{thiamine} + \text{ATP} \longrightarrow \text{thiamine diphosphate} + \text{AMP}.$$

The relationship between the oxidation of pyruvate and thiamine diphosphate was brought out very clearly indeed by the work of Peters on the brain tissue of vitamin B_1-deficient pigeons. Normal brain tissue metabolizes little but glucose, which it oxidizes completely. Peters found, however, that the brain tissue of pigeons deprived of thiamine carries the breakdown of glucose as far as pyruvate, but can go no further. If a little thiamine is added it rapidly undergoes pyrophosphorylation in the tissue, yielding co-carboxylase, and pyruvate begins to disappear. Under aerobic conditions it is largely oxidized but gives rise in part to acetate. Under anaerobic conditions one molecule of pyruvate can be oxidatively decarboxylated, yielding acetate and CO_2, while a second molecule is reduced to lactate:

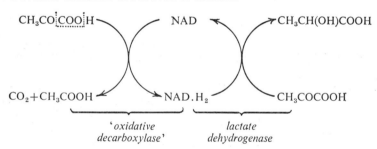

Several important deductions can be made from these observations. They show in the first place that thiamine diphosphate is indispensable for oxidative decarboxylation, and it may be mentioned in passing that the presence of pyruvate in the blood and its excretion in the urine are regular features of advanced vitamin B_1 deficiency in animals. Peters's observation that the oxidative decarboxylation of pyruvate is associated with the formation of lactate probably indicates that NAD must be involved, since the lactate dehydrogenase that catalyses the reduction requires the collaboration of NAD, and it is known that such a system can only couple with one requiring the same coenzyme (p. 175). Finally, since the reaction still takes place in an oxidative manner, even under anaerobic conditions, Peters's work serves to emphasize the essentially *oxidative* nature of this particular type of decarboxylation. Oxidative decarboxylation is not only important in the metabolism of pyruvate and carbohydrate but also in protein metabolism, since the oxidative deamination or transdeamination of amino-acids leads invariably to the formation of the corresponding α-keto acids.

The first detailed study of the oxidative decarboxylation of pyruvate we owe to Lipmann, who examined the process with the aid of enzymes prepared from a micro-organism, *B. delbrückii*. Lipmann observed that, in crude extracts of this organism, the oxidative decarboxylation of pyruvate was attended by the disappearance of inorganic phosphate from the mixture and by a simultaneous synthesis of ATP. Suspecting therefore that a phosphorylated intermediate of some kind must be involved he went further and was able to analyse the process into two distinct stages which may be represented as follows:

(i) $CH_3CO \vdots C\ddot{O}\ddot{O} \vdots H + HO\text{\textcircled{P}} + \frac{1}{2}O_2 = CH_3COO \sim \text{\textcircled{P}} + CO_2 + H_2O,$

(ii) $CH_3COO \sim \text{\textcircled{P}} + ADP = CH_3COOH + ATP.$

The intermediate body, acetyl phosphate, could be isolated in the form of its silver salt. Further work with partially purified enzyme preparations showed that the process as a whole requires the presence of a considerable group of small-molecular materials including *thiamine diphosphate, NAD, ADP, inorganic phosphate* and *magnesium ions*. The last of these requirements might perhaps have been predicted since Mg^{2+} is necessary whenever thiamine diphosphate is required: indeed, gross magnesium deficiency leads to symptoms similar to those of vitamin B_1 deficiency in experimental animals.

Now, the production of acetyl phosphate in bacteria seems to have no parallel in animal tissues although, as we have seen, acetate can be formed by the oxidative decarboxylation of pyruvate by enzymes of animal origin. Acetate formed in this way is very reactive indeed in biological systems, so much so, indeed, that it is commonly referred to as 'active acetate'. The chemical nature of 'active acetate' was an outstanding biochemical problem for several years. It was thought at first that 'active acetate' might be identical with acetyl phosphate, especially since the latter contains high-energy phosphate and might therefore be expected to be highly reactive. It had moreover, been isolated from *B. delbrückii* as an intermediate in the oxidative decarboxylation of pyruvate. But many experiments with synthetic acetyl phosphate have gone to show that this compound is relatively inert in animal systems and has few, if any, of the characteristic properties of 'active acetate' itself.

Later work threw new light upon the identity of this elusive compound. 'Active acetate' generated from pyruvate can bring about the acetylation of a considerable number of substances in the presence of acetyltransferases extracted from liver: free acetate and acetylphosphate cannot. Now biological acetylation is by no means a rare phenomenon (pp. 103 ff.). Many substances such as sulphanilamide and aniline undergo acetylation when given by mouth and appear in acetylated form in the urine, for example:

$$CH_3CO.HN\text{—}\bigcirc\text{—}SO_2NH_2 \qquad\qquad CH_3CO.HN\text{—}\bigcirc$$

N-*acetylsulphanilamide*   *acetyl aniline (acetanilide)*

Acetylation is also involved in the elaboration of the mercapturic acids (p. 246) and in the formation of acetylcholine from choline, and we may also refer to the acetylation of acetate itself to yield acetoacetate. The acetylation of sulphanilamide and choline by acetate itself takes place readily in liver slices, but is dependent upon concomitant respiration, which provides the free energy necessary for these endergonic processes. Free energy can also be effectively provided by the addition of ATP, when acetylation will proceed anaerobically, but in none of these cases will acetyl phosphate replace free acetate plus ATP.

The further study by Lipmann of biological acetylation, this time in homogenates and cell-free tissue extracts, showed, that, even in the presence of ATP and free acetate, a further thermostable factor is required as coenzyme. Thus, the coenzyme of acetylation, or *coenzyme A*, seems always to be involved in biological acetylations, and is very generally present in cells and tissues. It is derived from the virtually omnipresent pantothenic acid, and contains adenylic acid and β-mercaptoethylamine (see p. 297 for formula). The —SH group of the latter plays a very important part in the activity of CoA, which is inactivated by exposure to mild oxidizing conditions, reactivated by cysteine or by the reduced form of glutathione, and irreversibly inactivated by iodoacetate.

Subsequent work by Lynen on yeast led to the isolation of an acetylated form of CoA. The acetyl group appears to be attached to the —SH group because, while acetyl-CoA is capable of acetylating sulphanilamide in liver extracts and is not affected by iodoacetate, the latter inactivates the free coenzyme completely. If we represent CoA itself by A.SH, this new acetyl compound may be formulated as $A.S\sim OC.CH_3$ since acetyl$\sim$CoA can give rise to ATP from ADP and inorganic phosphate (e.g. on p. 104).

Now the acetylation of sulphanilamide by free acetate requires the presence of CoA and ATP, but *acetyl-CoA can replace all three*, suggesting that *acetyl-CoA is identical with 'active acetate' itself*. In addition, acetyl-CoA will react with oxaloacetate in the presence of citrate synthetase to form citrate—another characteristic property of 'active acetate'. Here again, free acetate fails to react except in the presence of ATP and Co A. But in neither of these cases can acetyl phosphate replace acetate plus ATP plus Co A, nor can it replace acetyl-Co A. It seems certain, then, that acetyl-Co A can arise either from pyruvate through reactions which we still have to consider, or from free acetate through reactions involving both ATP and Co A but not acetyl phosphate. In all probability the following reactions are concerned when the starting material is acetate:

$$\text{ATP} + \text{acetate} \rightleftharpoons \text{adenylacetate} + \text{pyrophosphate},$$
$$\text{adenylacetate} + \text{Co A} \rightleftharpoons \text{acetyl-Co A} + \text{AMP}.$$

Adenylacetate has been isolated and shown to react with Co A, forming acetyl-Co A in the presence of the appropriate acetyltransferase. But the *formation of acetyl-Co A from pyruvate* is much more complicated.

Since it is known that the evidently energy-rich acetyl-Co A can acetylate sulphanilamide or react with oxaloacetate without the participation of ATP, it is clear that this compound qualifies for identification with 'active acetate'. One important point should be noted before we proceed further. The acetyl moiety of acetyl-Co A can react either through its carboxyl or though its methyl residue, as is clear from the following equations:

Carboxyl:

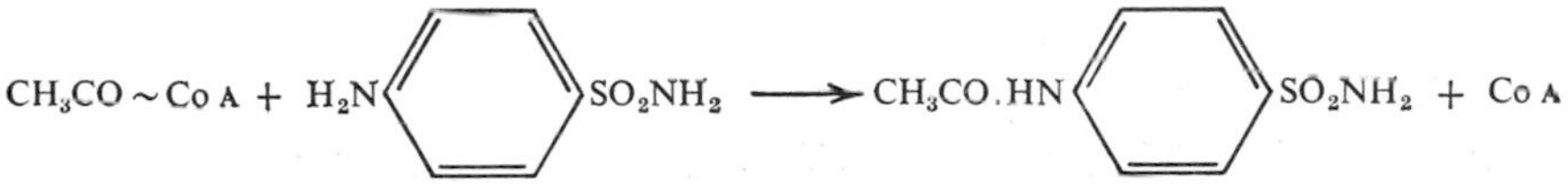

$$CH_3CO \sim Co A + H_2N\text{——}SO_2NH_2 \longrightarrow CH_3CO.HN\text{——}SO_2NH_2 + Co A$$

sulphanilamide

Methyl:

$$
\begin{array}{c}
COOH \\
| \\
CH \\
\| \\
C(OH)COOH \\
+ \\
CH_3 \\
| \\
CO \sim Co A
\end{array}
+ H_2O
\longrightarrow
\begin{array}{c}
COOH \\
| \\
CH_2 \\
| \\
C(OH)COOH \\
| \\
CH_2 \\
| \\
COOH
\end{array}
+ Co A
$$

citrate

This last reaction is catalysed by *citrate synthetase*, formerly and most unfortunately called 'condensing enzyme'.

Apparently, therefore, the oxidative decarboxylation of pyruvate in animal tissues gives rise to acetate by way of an 'active' acetate in the form of acetyl-Co A. But there is now reason to think that *acetyl-Co A is an intermediate in the oxidative decarboxylation of pyruvate* by bacteria as well as by animal tissues. This information came out of careful studies of the oxidative decarboxylation of pyruvate by cell-free extracts of *Escherichia coli*, another micro-organism which, like *B. delbrückii*, forms acetyl phosphate. These extracts will carry out the following coupled oxido-reduction:

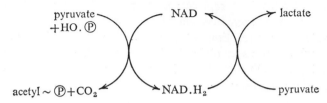

The reductive reaction only requires the presence of lactate dehydrogenase and NAD for its accomplishment, but the oxidative process is much more complicated. It requires at least three enzymes, one of which is an acetyl-transferase, together with CoA and magnesium ions, NAD and free phosphate: none of these can be omitted. The requirement for CoA suggests that acetyl-CoA must be involved in some way. Proof that 'active acetate' is formed as an intermediary comes from observations that (*a*) no acetyl phosphate is formed if inorganic phosphate is omitted from the reaction system, but that (*b*), if oxaloacetate and citrate synthetase are added instead, citrate is formed:

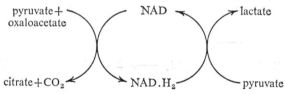

Citrate formation requires the provision of acetyl-CoA, and acetylphosphate cannot replace it.

Now let us sum up the information we have so far about oxidative decarboxylation:

$$CH_3CO.COOH + \tfrac{1}{2}O_2 \longrightarrow CH_3COOH + CO_2,$$

or, more realistically,

$$CH_3CO.COOH + H_2O \xrightarrow{\;-2H\;} CH_3COOH + CO_2.$$

We know, first of all, that this must be a complex process because it requires so many cofactors. These are: magnesium ions, inorganic phosphate, thiamine diphosphate, ADP, NAD and CoA among others. Our equation might therefore be expanded to read

$$CH_3CO.COOH + NAD + ADP + HO.\textcircled{P} \longrightarrow CH_3COOH + CO_2 + NAD.H_2 + ATP.$$

Our second piece of information is that acetyl-CoA is formed as an intermediate in the oxidative decarboxylation of pyruvate. Equations given earlier have accounted for some of the coenzyme requirements, but the need for thiamine diphosphate, the first of the cofactors to be discovered, must next be explained.

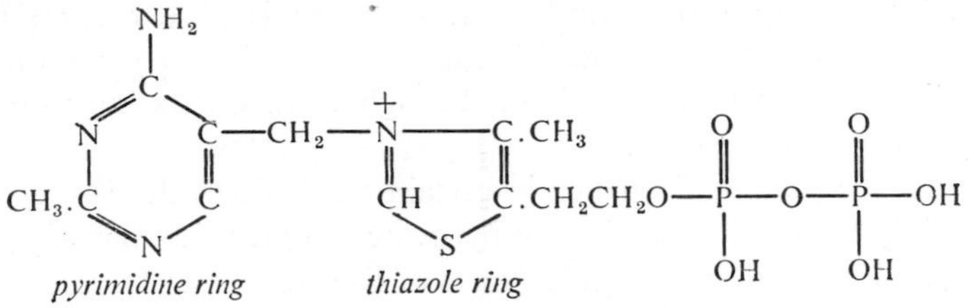

Formula of thiamine diphosphate.

The part played by thiamine diphosphate in 'straight' and in oxidative decarboxylation alike depends on the behaviour of the thiazole ring and, to concentrate attention on it we can write the formula in the following abbreviated manner (a):

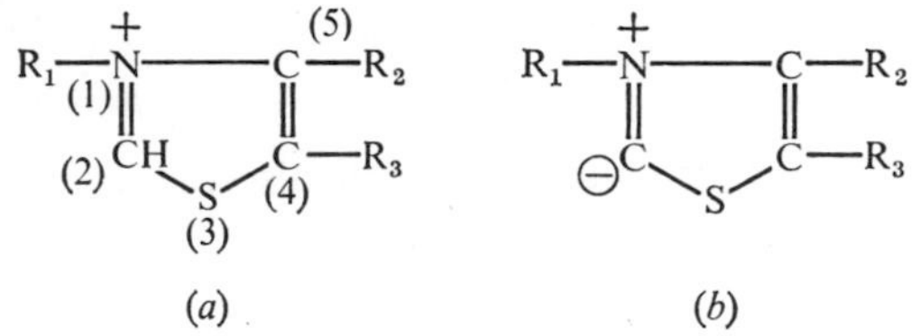

(a)   (b)

Because of its structure the carbon atom at 2 can lose a proton by ionization and is negatively charged as shown in (b).

This enables thiamine diphosphate to react with carbonyl compounds such as pyruvate to form an addition compound (reaction 1), which then (2) loses carbon dioxide, leaving a complex between acetaldehyde and thiamine diphosphate. The latter, in the presence of pyruvate decarboxylyase can be decomposed (3) to give free acetaldehyde and thiamine diphosphate (Fig. 87).

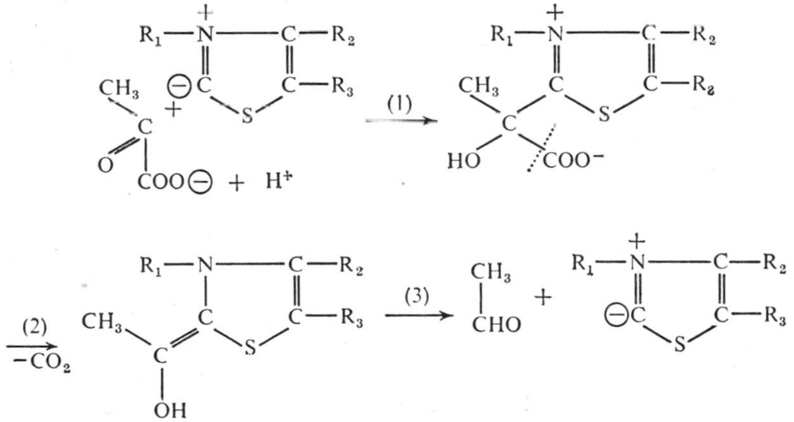

Fig. 87. Decarboxylation of pyruvate by pyruvate decarboxylyase.

In oxidative decarboxylation things are more complicated however. Reactions (1) and (2) effect the decarboxylation as before but the product, instead of decomposing (pyruvate decarboxylyase is absent from animal tissues), undergoes oxidation, and this calls for provision of yet another cofactor. This substance was discovered through work on the nutritional requirements of certain bacteria.

If *Streptococcus faecalis* is grown in certain synthetic media it will oxidize glucose rapidly but has little action on pyruvate. Rapid oxidation of pyruvate can be induced by adding suitable extracts of liver or yeast. A search for the responsible factor in these extracts led to the concentration, purification, isolation and eventual synthesis of a hitherto unsuspected cofactor now known as *α-lipoic acid.* Chemically this substance is the oxidized form of 6:8-dimercapto-octanoic acid ('thioctic acid'):

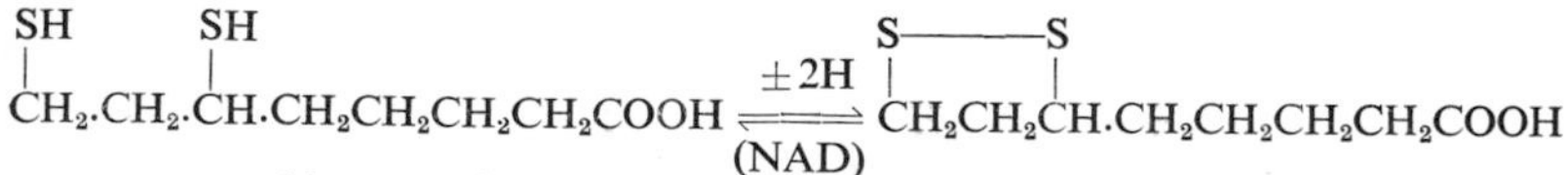
SH
SH
CH₂.CH₂.CH.CH₂CH₂CH₂CH₂COOH
±2H
(NAD)
S
S
CH₂CH₂CH.CH₂CH₂CH₂CH₂COOH

thioctic acid *α-lipoic acid*

Other micro-organisms that oxidize pyruvate also require this new co-factor. It is convenient to abbreviate the structure as

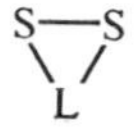
S
S
L

The isolation of α-lipoic acid led in turn to the demonstration that, in animal tissues as well as in bacteria, α-lipoate must be added to the list of cofactors required for oxidative decarboxylation.

The acetaldehyde-thiamine complex is formed as before and the following further reaction takes place:

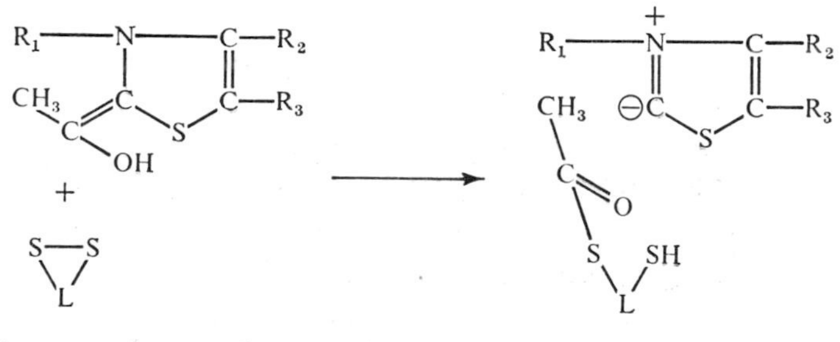
R₁
N
C
R₂
CH₃
C
C
R₃
S
OH
+
S
S
L
+
O
SH

In effect an acetyl group has now been transferred to lipoate. In the next reaction the acetyl group is transferred to co-enzyme A (*a*) leaving lipoate in the reduced form, and the latter is then reoxidized (*b*) at the expense of NAD:

(*a*)

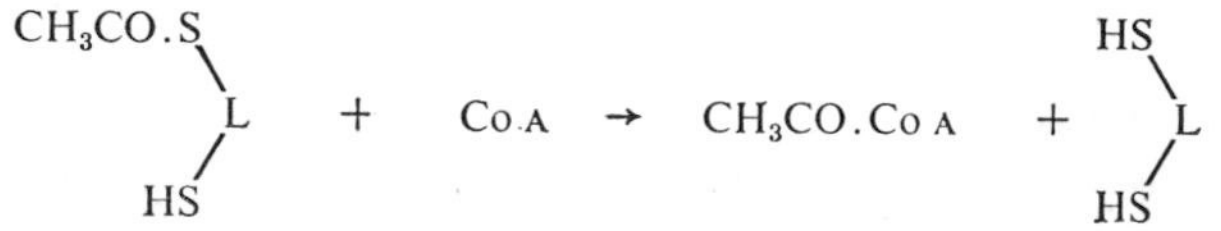
CH₃CO.S
L
HS
+
Co A
CH₃CO.Co A
+
HS
L
HS

(b)

$$\underset{\text{SH}}{\overset{\text{SH}}{\diagdown}}\text{L} \quad + \quad \text{NAD} \quad \rightarrow \quad \underset{\text{S}}{\overset{\text{S}}{\diagdown}}\text{L} \quad + \quad \text{NAD.H}_2$$

Starting, then, from pyruvate, diphosphothiamine acts as the decarboxylating agent and lipoate as a carrier of acetyl groups from diphosphothiamine to CoA, forming a new compound, i.e. acetyl-CoA. The oxidative component of this complex performance is achieved by NAD and the overall reaction is:

$$CH_3CO.COOH + NAD + CoA \longrightarrow CH_3CO.CoA + CO_2 + NAD.H_2.$$

The whole system is summarized in Fig. 88. The process as a whole is strongly *inhibited by arsenite*.

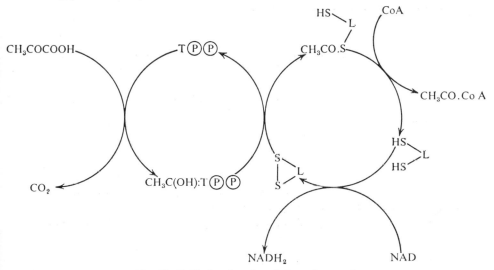

Fig. 88. Oxidative decarboxylation of pyruvate.

The acetyl-CoA formed in oxidative decarboxylation reacts with inorganic phosphate in many bacteria in the presence of a bacterial *phosphoacetyltransferase* to form acetylphosphate and regenerate CoA:

$$CH_3CO \sim CoA + H.O\circledP \rightleftharpoons CH_3CO \sim O\circledP + CoA.$$

In animal tissues acetyl-CoA does not yield acetylphosphate because no phosphoacetyltransferase is present but it can react with ADP and inorganic phosphate thus:

$$CH_3CO \sim CoA + ADP + HO.\circledP \rightleftharpoons CH_3COOH + ATP + CoA.$$

The intermediate steps in this reaction are still uncertain. In either case the CoA is regenerated and can be used over again. Acetyl-CoA most usually reacts with oxaloacetate to form citrate, enters into biological acetylation, or

26-2

forms acetoacetate. This last case touches the fringe of the field of fat meta-bolism, in which, as we shall see in a later chapter, acetyl-CoA plays a large and important part. Nevertheless it is worth remarking that the *production of a molecule of acetyl-CoA* or, more generally, any acyl-CoA compound, *is tantamount to the production of a molecule of ATP from one of ADP*, at any rate as far as biological energetics is concerned.

All α-keto-acids are liable to undergo oxidative decarboxylation and through similar mechanisms. In every case it is likely that an acyl $\sim$ CoA compound will be formed as an intermediate. The only case so far studied in detail, apart from that of pyruvate, is that of α-ketoglutarate. The enzyme concerned in this case is not identical with 'pyruvate oxidase' but the mechan-ism is similar and succinyl $\sim$ CoA is formed as an intermediate:

$$
\begin{array}{l}
\text{COOH} \\
\mid \\
\text{CH}_2 \\
\mid \\
\text{CH}_2 \quad + \text{CoA} + \text{NAD} \longrightarrow \\
\mid \\
\text{CO} \\
\mid \\
\text{COOH}
\end{array}
\qquad
\begin{array}{l}
\text{COOH} \\
\mid \\
\text{CH}_2 \\
\mid \\
\text{CH}_2 \\
\mid \\
\text{CO} \sim \text{CoA}
\end{array}
\qquad + \text{CO}_2 + \text{NAD} . \text{H}_2
$$

This is a reaction of prime importance for reasons that will appear later.

Again lipoate is concerned in the process. In this particular case it is known that the succinyl $\sim$ CoA can give rise to ATP through the following series of reactions:

(i) succinyl $\sim$ CoA + GDP + HO . $\circledP$ $\rightleftharpoons$ succinate + CoA + GTP,

(ii) GTP + ADP $\rightleftharpoons$ GDP + ATP,

Overall: succinyl $\sim$ CoA + ADP + HO . $\circledP$ $\longrightarrow$ succinate + CoA + ATP,

where GDP and GTP are guanosine-5'-di- and guanosine-5'-tri-phosphates respectively. Inosine di- and triphosphates can replace GDP and GTP but are less effective.

GENERAL METABOLISM OF PYRUVATE

It will be clear from what has gone before that pyruvate occupies a central position in the metabolism of carbohydrates. It can be formed either from glucose (after phosphorylation) or from glycogen by the normal reactions of glycolysis. By reversal of these processes it can contribute to the synthetic formation of carbohydrate. It constitutes, moreover, a link between the metabolism of carbohydrates and that of proteins, for it arises more or less directly from the products of deamination of a number of amino-acids (Chap. 12). These are only a few of the processes into which pyruvate enters, and a summary of some of the more important of its known origins and fates is presented in Fig. 89.

Under anaerobic conditions pyruvate can be broken down by 'straight' decarboxylation, as it is in yeast, or it may undergo reduction to lactate, as it does in muscle. If aerobic conditions prevail it can be completely oxidized, as in muscle, kidney, brain and other tissues, by oxidative decarboxylation in the first instance to yield acetyl-CoA or, in some bacteria, acetyl phosphate. Since acetyl-CoA is a potential source of fat, pyruvate may be regarded as a link between the metabolism of carbohydrates and that of fat, as well as that of proteins.

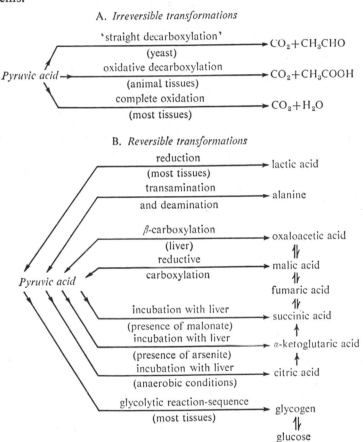

Fig. 89. Formation and fate of pyruvate. Note that the synthesis and breakdown, e.g. of succinate, *do not necessarily follow the same route.*

We can already give a tolerably satisfactory account of most of the processes listed in Fig. 89. Pyruvate can give rise, when incubated with liver tissue under suitable conditions, to citrate, α-ketoglutarate, succinate, fumarate, malate and oxaloacetate among other compounds, and there is evidence that carbon dioxide is 'fixed' when these synthetic reactions take place. Radioactive samples

391

of citrate and α-ketoglutarate, for example, have been obtained by the incubation of liver tissue with pyruvate in the presence of radioactive carbon dioxide. This is mainly due in the first instance to the β-carboxylation of pyruvate, a reaction which is catalysed by the biotin-dependent pyruvate carboxylyase of liver:

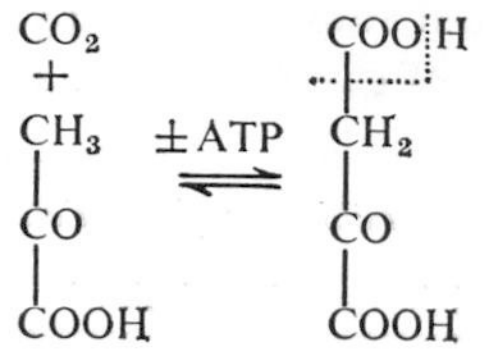

$$
\begin{array}{ccc}
CO_2 & & COOH \\
+ & & \\
CH_3 & \pm ATP & CH_2 \\
| & \rightleftharpoons & | \\
CO & & CO \\
| & & | \\
COOH & & COOH
\end{array}
$$

The responsible enzyme was first detected in bacteria, from which it has since been isolated, and there is now evidence for its occurrence in mammalian liver. It appears, however, to be absent from animal tissues other than the liver.

Oxaloacetate itself is a very reactive compound under biological conditions, and attempts to demonstrate the production of its isotopic form in liver provided with pyruvate and isotopic carbon dioxide have not so far proved fruitful. Other substances to which oxaloacetate itself gives rise (α-ketoglutarate, citrate) have, however, been isolated and shown to contain the isotopic carbon.

AEROBIC METABOLISM OF CARBOHYDRATES

We know of no oxidizing enzyme or enzymes capable of catalysing directly the complete oxidation of glucose or glycogen on the large scale. The liver contains a glucose dehydrogenase (p. 163) which catalyses the oxidation of glucose to the corresponding gluconate. Other oxidized sugar derivatives, e.g. the uronic acids, also occur biologically. In particular, glucuronate is used by mammals in the detoxication of phenolic and other substances, but there is no evidence that it arises from glucose: indeed, such evidence as we have suggests that it does not.

There occurs in the red blood cells and elsewhere a glucose-6-phosphate dehydrogenase (p. 163) which catalyses the oxidation of glucose-6-monophosphate to gluconate-6-phosphate. We know too of the existence of further enzymes that can remove one carbon atom from gluconate-6-phosphate to give ribulose-5-phosphate (p. 291).

Ribulose-5-phosphate can be converted into ribose-5-phosphate and into xylulose-5-phosphate by ribosephosphate isomerase and ribulosephosphate epimerase respectively, and the three pentose phosphates can enter into a complicated maze of reactions, catalysed largely by transketolase, transaldolase and aldolase. Although many such reactions are known to take place in animal tissues we know little about the way in which they are organizcd. Some at least lead to the formation of triosephosphate, and one possible route is the reverse of that followed in photosynthesis (see Fig. 76, p. 327). In this way 3 molecules of ribulose-5-phosphate could give rise to 5 molecules of triosephosphate. The latter could then run along the usual lines of glycolysis to form pyruvate and this, in turn, could be oxidized in the manner usual for pyruvate.

This last, however, is only one of many possibilities, for it is possible to account in terms of known reactions for the complete oxidation of carbohydrate by way of the pentose route (Fig. 90).

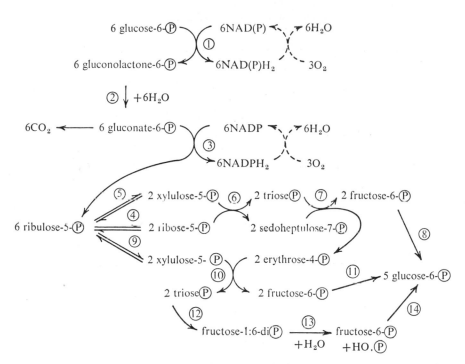

Fig. 90. Pentose pathway of carbohydrate oxidation.
Overall: 6 glucose-6-(P) + 6O_2 → 5 glucose-6-(P) + 6 CO_2 + 5H_2O + HO(P).

(12 molecules of water are formed of which 6 are consumed in reaction 2 and one more in reaction 13 leaving a balance of 5H_2O.)

All the necessary enzymes are present in animal tissues. They are tabulated in Table 30. Reaction (2) calls specifically for the provision of NADP, which is reduced to NADP.H_2. Generous supplies of NADP.H_2 are a *sine qua non* for the synthesis of fatty acids and it is probably not without significance, therefore, that the actively secreting mammary gland oxidizes most of its carbohydrate by way of this so-called 'shunt' or 'pentose pathway'.

393

Table 30. *Enzymes involved in the pentose pathway*

Reaction No.	Enzyme
1	Glucose-6-P dehydrogenase (can use NAD or NADP)
2	Gluconolactonase
3	Gluconate-6-P decarboxylating dehydrogenase (uses NADP)
4	Ribose-5-P-isomerase
5	Ribulose-5-P-3-epimerase
6	Transketolase
7	Transaldolase
8	Glucose-6-P-isomerase
9	Ribulose-5-P-3-epimerase
10	Transketolase
11	Glucose-6-P-isomerase
12	Aldolase
13	Fructose-1:6-di-P-1-P-hydrolase
14	Glucose-6-P-isomerase

To what extent this somewhat circuitous route is followed in the ordinary way it is difficult to be certain, but these reactions make it possible to account for the fact, which has been realized for a long time, that a part of the total carbohydrate oxidation that takes place in animal tissues goes on by some route other than glycolysis followed by oxidation of pyruvate.

20

THE CITRIC ACID CYCLE

INTRODUCTION

Having given some account of the 'pentose pathway' we must now turn our attention to the quantitatively more important oxidation of the pyruvate produced by glycolysis. Pyruvate is known to be totally oxidized to CO_2 and water by most tissues, as long as conditions are fully aerobic, and we have already dealt with the glycolytic reactions which lead to its formation and with the first step in its aerobic degradation which is, of course, oxidative decarboxylation.

Now the muscles are the biggest consumers of carbohydrate and it is natural therefore that most of the early work on the aerobic metabolism of carbohydrates should have been done with muscle tissue. Muscle extracts cannot be used for this purpose because, as will be remembered, they contain only the soluble enzymes of the tissues and do not respire. This fortunate circumstance made it possible to obtain a clear picture of anaerobic glycolysis before the more complex operations of aerobic glycolysis plus oxidation were investigated.

The foundations of our present knowledge of aerobic metabolism were laid by Szent-Györgyi, using suspensions of minced pigeon-breast muscle as his material. This is a very active muscle and one which contains a good deal of myoglobin so that, in the minced form, it can keep itself well supplied with oxygen. Indeed, Szent-Györgyi found that minced pigeon-breast muscle respires very actively and produces little or no lactate. He studied the rate of respiration of his preparations under a variety of experimental conditions and made the following fundamental observations:

(a) The rate of respiration is very high at first but falls off slowly with time.

(b) The fall in the rate of respiration is paralleled by the rate at which succinate disappears from the mince.

(c) The respiratory rate of a failing preparation can be restored to its original high level by the addition of catalytic amounts of succinate or fumarate but is depressed again by the addition of malonate.

(d) For each volume of oxygen consumed by the tissue an equal volume of carbon dioxide is formed, indicating that the material undergoing oxidation is of carbohydrate origin:[†]

$$(CH_2O)_n + nO_2 = nCO_2 + nH_2O.$$

[†] The ratio carbon dioxide produced/oxygen (by volume) consumed is called the respiratory quotient and, in theory, is characteristic of the kind of fuel being oxidized. For carbohydrate oxidation R.Q. = 1, while for fat and protein the values are about 0·7 and 0·8 respectively.

From these observations Szent-Györgyi concluded that the oxidation of carbohydrate is in some way catalysed by succinate and fumarate and, since it was already well known that the two are interconvertible through the activity of succinate dehydrogenase, he concluded that this enzyme must be intimately concerned in the oxidation of carbohydrate materials. This led to another important discovery. It had already been established that malonate is a powerful competitive inhibitor of succinate dehydrogenase, and Szent-Györgyi was able to show that malonate inhibits the catalytic effect of succinate and fumarate in failing preparations and, moreover, that it has a powerfully depressant action upon the respiration of the fresh mince.

To account for these observations, Szent-Györgyi brought forward the suggestion that succinate and fumarate act together as a carrier system for hydrogen removed from carbohydrate materials of some kind. Although it has long been discarded, this scheme paved the way for new discoveries of absolutely fundamental importance as we shall see. But let us stay with Szent-Györgyi a little longer.

If we represent the unknown substrates by AH_2, his hypothesis can be schematically represented as follows:

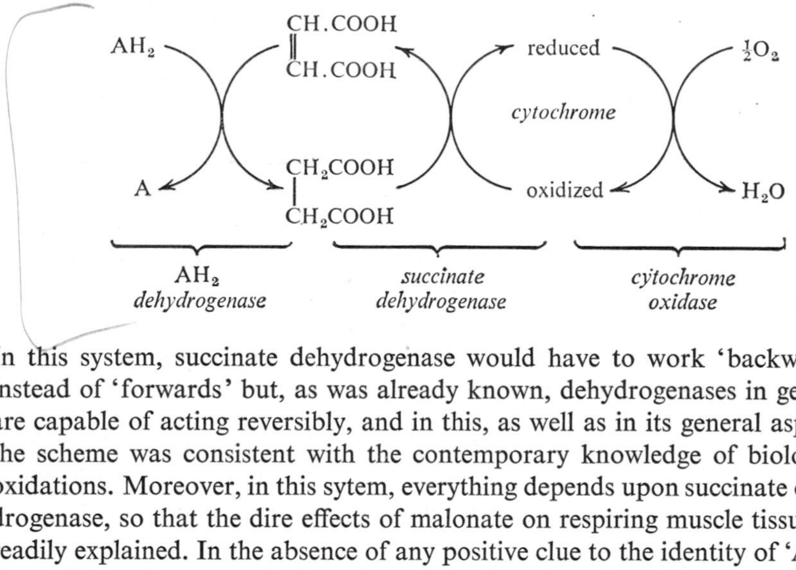

$$AH_2 \quad\quad \begin{array}{c} CH.COOH \\ \| \\ CH.COOH \end{array} \quad\quad \text{reduced} \quad\quad \tfrac{1}{2}O_2$$

$$\text{\textit{cytochrome}}$$

$$A \quad\quad \begin{array}{c} CH_2COOH \\ | \\ CH_2COOH \end{array} \quad\quad \text{oxidized} \quad\quad H_2O$$

AH_2 *dehydrogenase*	*succinate dehydrogenase*	*cytochrome oxidase*

In this system, succinate dehydrogenase would have to work 'backwards' instead of 'forwards' but, as was already known, dehydrogenases in general are capable of acting reversibly, and in this, as well as in its general aspects, the scheme was consistent with the contemporary knowledge of biological oxidations. Moreover, in this sytem, everything depends upon succinate dehydrogenase, so that the dire effects of malonate on respiring muscle tissue are readily explained. In the absence of any positive clue to the identity of 'AH_2', Szent-Györgyi suggested that this might be triosephosphate or, perhaps, pyruvate.

These results stimulated other investigators to study other materials and it soon became clear that Szent-Györgyi's malonate-sensitive system must be as widely distributed in living tissues as are cytochrome and cytochrome oxidase, for cells and tissues of many kinds were found to behave in the same manner towards succinate, fumarate and the inhibitory substance malonate. Presently it was discovered by Szent-Györgyi himself that two other C_4-dicarboxylic acids act in the same way as succinate and fumarate, viz. malate and oxalo-acetate both of which can give rise to succinate and fumarate by long familiar reactions. Shortly afterwards Krebs announced the discovery that, in addition to succinate, fumarate, malate and oxaloacetate, α-ketoglutarate and citrate also act catalytically on the respiration of minced muscle, and that the effects of all these substances are inhibited by malonate.

The behaviour of α-ketoglutarate could be explained readily enough, for it was already known that, being an α-keto-acid, α-ketoglutarate can undergo oxidative decarboxylation to yield succinate, and thus leads directly to Szent-Györgyi's catalytic cycle:

$$
\begin{array}{c}
\text{COOH} \\
| \\
\text{CH}_2 \\
| \\
\text{CH}_2 \\
| \\
\text{CO} \\
| \\
\text{COOH}
\end{array}
\quad
\begin{array}{c}
+H_2O \\
\longrightarrow \\
-2H
\end{array}
\quad
\begin{array}{c}
\text{COOH} \\
| \\
\text{CH}_2 \\
| \\
\text{CH}_2 \\
| \\
\text{COOH}
\end{array}
\quad + CO_2
$$

α-*ketoglutaric acid*

The behaviour of citrate too was explicable in terms of the widely distributed citrate dehydrogenase, for this enzyme was believed to convert citrate into α-ketoglutarate, from which succinate can then be formed. If, however, the formulae of citrate and α-ketoglutarate are compared, it will be noticed that whereas the ketonic oxygen of the latter is in the α-position, citrate contains a β-oxygen atom, so that the direct conversion of citrate into α-ketoglutarate seemed very improbable. This phenomenon was explained, however, by the discovery by Martius & Knoop of a new enzyme, aconitase, which catalyses the interconversion of citrate into *iso*-citrate, probably by way of a common intermediary in the form of aconitate:

$$
\begin{array}{c}
\text{COOH} \\
| \\
\text{CH}_2 \\
| \\
\text{C(OH)COOH} \\
| \\
\text{CH}_2 \\
| \\
\text{COOH}
\end{array}
\underset{}{\overset{\pm H_2O}{\rightleftharpoons}}
\begin{array}{c}
\text{COOH} \\
| \\
\text{CH} \\
|| \\
\text{C.COOH} \\
| \\
\text{CH}_2 \\
| \\
\text{COOH}
\end{array}
\underset{}{\overset{\pm H_2O}{\rightleftharpoons}}
\begin{array}{c}
\text{COOH} \\
| \\
\text{CHOH} \\
| \\
\text{CH.COOH} \\
| \\
\text{CH}_2 \\
| \\
\text{COOH}
\end{array}
$$

citric acid   *aconitic acid*   iso-*citric acid*

Reinvestigation of the old citrate dehydrogenase now showed that it is specific for *iso*-citrate, and has no action upon citrate itself except in the presence of aconitase, with which the early preparations were invariably contaminated.

iso-Citrate dehydrogenase, which collaborates with NADP, catalyses the dehydrogenation of *iso*-citrate to yield oxalosuccinate (α-keto-β-carboxy-glutarate). This compound then, under the influence of an oxalosuccinate decarboxylyase, loses carbon dioxide and gives rise to α-ketoglutarate. *iso*-Citrate dehydrogenase and oxalosuccinate decarboxylyase appear to be a single enzyme-protein, but the two reactions can be studied separately because the decarboxylyase but not the dehydrogenase requires the presence of manganese ions for activity:

$$
\begin{array}{ccc}
\text{COOH} & \text{COOH} & \text{COOH} \\
| & | & | \\
\text{CHOH} & \text{CO} & \text{CO} \\
| \quad \pm 2\text{H} & | \quad \pm\text{CO}_2 & | \\
\text{CH.COOH} \rightleftharpoons & \text{CH:COO:H} \rightleftharpoons & \text{CH}_2 \\
| \quad \overline{\text{NADP}} & | \quad \overline{(\text{Mn}^{2+})} & | \\
\text{CH}_2 & \text{CH}_2 & \text{CH}_2 \\
| & | & | \\
\text{COOH} & \text{COOH} & \text{COOH} \\
\textit{iso-citric} & \textit{oxalosuccinic} & \textit{α-ketoglutaric} \\
\textit{acid} & \textit{acid} & \textit{acid}
\end{array}
$$

Another *iso*-citrate dehydrogenase is also known and this is NAD-specific. It catalyses the same overall reaction as the NADP-specific enzyme but does not decarboxylate added oxalosuccinate. It is therefore open to question whether or not oxalosuccinate qualifies as a genuine intermediate, but here we shall assume that it does. The product, α-ketoglutarate, can then be oxidatively decarboxylated to yield succinate, and thus leads again to succinate, one of the primary catalysts of the Szent-Györgyi system.

It became possible, therefore, to trace out metabolic connexions between all the substances known to act catalytically on the oxidation of carbohydrates by minced pigeon-breast muscle and by many other kinds of cells and tissues. Their action could be explained by their convertibility into succinate and fumarate, the two primary catalysts which, according to Szent-Györgyi, function as a hydrogen-carrying system. The effect of malonate was explained because malonate specifically inhibits succinate dehydrogenase, upon which the carrier activity of the succinate-fumarate system depends. Thus all the phenomena observed in Szent-Györgyi's original experiments, together with a number of later observations, could be accounted for, apart only from the slow decline that takes place in the respiration of minced pigeon-breast muscle. Even this last phenomenon can be explained, however, by the slow and, in muscle, probably spontaneous decarboxylation of oxaloacetate which takes place under physiological conditions, yielding pyruvate, which does not act catalytically. Oxaloacetate and the other catalytic substances therefore drain slowly away and, as their concentrations decline, the rate of respiration of the preparation falls off.

In passing it should be noticed that the reaction chain leading from α-ketoglutarate, through succinate, fumarate, malate and oxaloacetate to pyruvate is an important link between carbohydrate metabolism and that of proteins, for α-ketoglutarate, oxaloacetate and pyruvate respectively are formed by the deamination of three of the commonest non-essential amino-acids, viz. glutamate, aspartate and alanine.

Before going on to consider more recent developments, the reader will do well to study Fig. 91, in which the reactions just discussed are collected together in schematic form and the compounds are shown in relation to some other metabolic products.

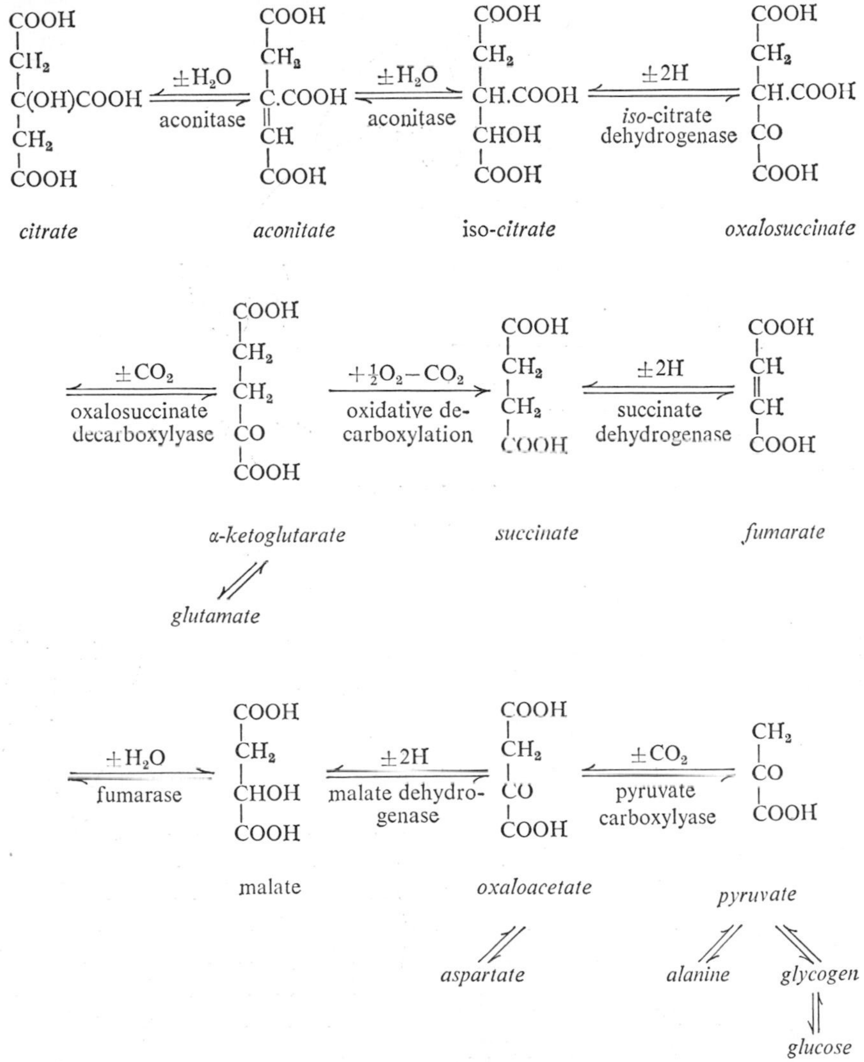

$$
\begin{array}{c}
\text{COOH} \\
| \\
\text{CH}_2 \\
| \\
\text{C(OH)COOH} \\
| \\
\text{CH}_2 \\
| \\
\text{COOH}
\end{array}
\underset{\text{aconitase}}{\overset{\pm\text{H}_2\text{O}}{\rightleftarrows}}
\begin{array}{c}
\text{COOH} \\
| \\
\text{CH}_2 \\
| \\
\text{C.COOH} \\
\| \\
\text{CH} \\
| \\
\text{COOH}
\end{array}
\underset{\text{aconitase}}{\overset{\pm\text{H}_2\text{O}}{\rightleftarrows}}
\begin{array}{c}
\text{COOH} \\
| \\
\text{CH}_2 \\
| \\
\text{CH.COOH} \\
| \\
\text{CHOH} \\
| \\
\text{COOH}
\end{array}
\underset{\substack{\textit{iso}\text{-citrate} \\ \text{dehydrogenase}}}{\overset{\pm 2\text{H}}{\rightleftarrows}}
\begin{array}{c}
\text{COOH} \\
| \\
\text{CH}_2 \\
| \\
\text{CH.COOH} \\
| \\
\text{CO} \\
| \\
\text{COOH}
\end{array}
$$

citrate aconitate iso-*citrate* oxalosuccinate

$$
\underset{\substack{\text{oxalosuccinate} \\ \text{decarboxylyase}}}{\overset{\pm\text{CO}_2}{\rightleftarrows}}
\begin{array}{c}
\text{COOH} \\
| \\
\text{CH}_2 \\
| \\
\text{CH}_2 \\
| \\
\text{CO} \\
| \\
\text{COOH}
\end{array}
\underset{\substack{\text{oxidative de-} \\ \text{carboxylation}}}{\overset{+\frac{1}{2}\text{O}_2-\text{CO}_2}{\longrightarrow}}
\begin{array}{c}
\text{COOH} \\
| \\
\text{CH}_2 \\
| \\
\text{CH}_2 \\
| \\
\text{COOH}
\end{array}
\underset{\substack{\text{succinate} \\ \text{dehydrogenase}}}{\overset{\pm 2\text{H}}{\rightleftarrows}}
\begin{array}{c}
\text{COOH} \\
| \\
\text{CH} \\
\| \\
\text{CH} \\
| \\
\text{COOH}
\end{array}
$$

α-ketoglutarate   succinate   fumarate

glutamate

$$
\underset{\text{fumarase}}{\overset{+\text{H}_2\text{O}}{\rightleftarrows}}
\begin{array}{c}
\text{COOH} \\
| \\
\text{CH}_2 \\
| \\
\text{CHOH} \\
| \\
\text{COOH}
\end{array}
\underset{\substack{\text{malate dehydro-} \\ \text{genase}}}{\overset{\pm 2\text{H}}{\rightleftarrows}}
\begin{array}{c}
\text{COOH} \\
| \\
\text{CH}_2 \\
| \\
\text{CO} \\
| \\
\text{COOH}
\end{array}
\underset{\substack{\text{pyruvate} \\ \text{carboxylyase}}}{\overset{\pm\text{CO}_2}{\rightleftarrows}}
\begin{array}{c}
\text{CH}_3 \\
| \\
\text{CO} \\
| \\
\text{COOH}
\end{array}
$$

malate   oxaloacetate   pyruvate

aspartate   alanine   glycogen

glucose

Fig. 91. Summary of known reactions leading from citrate to pyruvate, showing some metabolic interrelationships.

THE CITRIC ACID CYCLE

The whole picture took on an entirely new aspect with the suggestion by Krebs that pyruvate and oxaloacetate might react together to form a 7-carbon compound, from which citrate (6C), α-ketoglutarate (5C), and the 4C dicarboxylic acids were then re-formed. What had formerly been considered simply as a chain or series of reactions was now visualized as a cycle. In its simplest form this hypothetical scheme can be written as in Fig. 92.

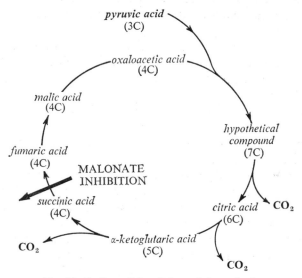

Fig. 92. Outline of the citric cycle hypothesis.

Pyruvate, with its three carbon atoms, enters this cycle by reacting with oxaloacetate. With each turn of the wheel one pyruvate molecule enters, three molecules of carbon dioxide are produced, and oxaloacetate is regenerated to take up a further molecule of pyruvate. Since a single molecule of oxaloacetate can be used over and over again in the oxidation of a theoretically unlimited number of molecules of pyruvate, it follows that oxaloacetate, or any substance lying on the cycle, can act catalytically in the oxidation of pyruvate to carbon dioxide. The mechanisms postulated involve succinate, fumarate, malate, oxaloacetate, α-ketoglutarate and citrate, all of which are known to act catalytically in the oxidative breakdown of carbohydrate. Succinate dehydrogenase is directly involved in the cycle in the oxidation of succinate to fumarate, so that the effect of malonate on the oxidation of carbohydrate is readily explained, while the slow failure of respiration in minced muscle preparations can again be explained in terms of a slow breakdown of oxaloacetate to pyruvate, which latter does not act catalytically.

400

Krebs's hypothesis therefore goes much further than that of Szent-Györgyi. It accounts at once for the catalytic activity of all the di- and tricarboxylic acids, it explains the malonate effect and, above all, it accounts for the complete oxidation of pyruvate, a known and important intermediate in the oxidation of carbohydrate and many amino-acids, to carbon dioxide. Glucose and glycogen, we know, can be metabolically converted into pyruvate, and Krebs's scheme, taken together with the known reactions of glycolysis, can account therefore for the complete oxidation of glucose and glycogen. Szent-Györgyi's scheme, by contrast, could only account for the dehydrogenation of the unidentified carbohydrate intermediate which we have described here as 'AH$_2$'.

In order clearly to understand what follows it is necessary to point out and, indeed, to emphasize the fact that several possible fates await pyruvate in the liver and are involved in citrate formation. First of all, pyruvate can be carboxylated to give oxaloacetate by the biotin-dependent *pyruvate carboxylyase* in the presence of ATP (p. 119). This enzyme is present in liver but apparently not in other organs. A second likely fate of pyruvate is its *oxidative decarboxylation* to give acetyl-CoA. There are other possibilities too but these can wait for the time being.

Krebs himself supplied the first evidence in favour of his 'citric acid cycle'. Pyruvate and oxaloacetate were incubated together with minced muscle under strictly anaerobic conditions. By working anaerobically it was expected that, as the cycle is essentially an aerobic system, some product might accumulate instead of being oxidized if, in fact, pyruvate and oxaloacetate do react together. After incubation, Krebs was able to demonstrate the formation of substantial amounts of citrate. Many workers have since repeated and confirmed Krebs's observations on a variety of tissues, and there is no shadow of doubt that some reaction involving pyruvate and oxaloacetate does indeed take place.

Other significant experiments have been carried out on liver tissue. If pyruvate is incubated with liver tissue it is possible to demonstrate the formation of considerable yields of succinate, together with smaller quantities of α-ketoglutarate. The fact that pyruvate can give rise to oxaloacetate under the influence of *pyruvate carboxylyase*, an enzyme that occurs in the liver but not in the extra-hepatic tissues, suggested a possible route for the synthesis of succinate from pyruvate. If pyruvate yields oxaloacetate, succinate can then be formed by the reversed actions of malate and succinate dehydrogenases, together with that of fumarase (Fig. 91, p. 399). In this case, however, the formation of α-ketoglutarate is left unexplained. Krebs pointed out that if this latter route were followed it should, since it involves succinate dehydrogenase, be inhibited by malonate. He therefore incubated pyruvate with liver tissue in the presence of malonate and found that even larger yields of succinate were obtained. Hence the synthesis must proceed by some route that does not involve succinate dehydrogenase. This, of course, is entirely in keeping with the cyclical hypothesis, and the evidence was further strengthened by the later demonstration that, in addition to succinate, substantial amounts of α-ketoglutarate and traces of citrate are formed at the same time.

Much illuminating evidence regarding the Krebs cycle was later obtained by employing radioactive carbon as a 'tracer'. If pyruvate is incubated with liver in the presence of radioactive carbon dioxide, radioactive intermediates including citrate and α-ketoglutarate can subsequently be isolated. This and other similar evidence again argues in favour of the cycle, and more precise information has been obtained by finding out precisely where, in the α-ketoglutarate molecule, the radioactive carbon was located.

There is now good evidence in favour of every step in the cycle, with an exception in the case of the supposed 7C intermediate. Critical experiments designed to detect the formation of such a compound have failed and, indeed, there has never at any time been positive evidence that such a substance is formed.

Probably, therefore, citrate must arise directly, presumably by a reaction between oxaloacetate and some 2-carbon substance, rather than the 3-carbon compound pyruvate. Citrate arises in fact, not directly from pyruvate plus oxaloacetate, but from oxaloacetate, which can itself be formed by the *carboxylation of pyruvate*, plus acetyl-CoA produced by the *oxidative decarboxylation of pyruvate*. It has been found that the aerobic oxidation of acetate itself is inhibited by malonate, a feature which indicates that acetate is probably oxidized by way of the catalytic cycle. Moreover, isotopically labelled acetate can be incorporated into the cycle and yields correspondingly labelled products but, as we now know, free acetate must be transformed into acetyl-CoA before it can react; free acetate plus ATP plus CoA can, however, lead to the synthesis of acetyl-CoA itself.

We have already considered some of the evidence for the occurrence of the reactions leading from citrate to oxaloacetate (summarized on p. 399) and the enzymes concerned have been dealt with in some detail in Part I of this book. Special consideration must now be given to the fundamental synthetic reaction which leads to the formation of citrate.

The formation of citrate from acetate and oxaloacetate has been demonstrated many times in tissue minces, homogenates and mitochondrial suspensions, and such a synthesis has also been demonstrated by means of a purified *citrate synthetase*. Until this was achieved it was open to question whether citrate itself is the first product to be formed. Citrate might arise by an addition reaction between acetate and oxaloacetate; condensation, on the other hand, would yield aconitate, from which citrate could arise indirectly as a side product through the action of aconitase:

Addition:

$$
\begin{array}{c}
\text{COOH} \\
| \\
\text{CH} \\
\parallel \\
\text{C(OH)COOH} \\
+ \\
\text{CH}_3 \\
| \\
\text{COCoA}
\end{array}
\qquad
\longrightarrow
\begin{array}{c}
\text{COOH} \\
| \\
\text{CH}_2 \\
| \\
\text{C(OH)COOH} + \text{CoA} \\
| \\
\text{CH}_2 \\
| \\
\text{COOH}
\end{array}
$$

Condensation:

$$
\begin{array}{c}
\text{COOH} \\
| \\
\text{CH} \\
\parallel \\
\text{C(OH)COOH} \\
+ \\
\text{CH}_3 \\
| \\
\text{COCoA}
\end{array}
\qquad
\longrightarrow
\begin{array}{c}
\text{COOH} \\
| \\
\text{CH} \\
\parallel \\
\text{C.COOH} + \text{H}_2\text{O} + \text{CoA} \\
| \\
\text{CH}_2 \\
| \\
\text{COOH}
\end{array}
$$

Ochoa was able to prepare from pigeon liver a preparation of his so-called 'condensing enzyme' (citrate synthetase), which was virtually free from aconitase and which, under suitable conditions, formed citrate in large yield from added acetate and oxaloacetate, thus demonstrating for the first time that *citrate arises directly* and not by way of aconitate. The precise mechanism of the reaction was still at the time unknown, but the achievement of the synthesis of citrate from acetate and oxaloacetate set the final seal of acceptance upon the citric acid cycle itself.

To achieve this synthesis from free acetate, ATP and CoA must be present, just as acetylation by acetate requires the presence of ATP and CoA (p. 384). That ATP is involved reawakened the long-standing suspicion that 'active acetate' is none other than acetyl phosphate. Acetyl phosphate itself cannot however replace free acetate plus ATP, but clearly enough some form of 'active acetate' was generated under the conditions of Ochoa's experiments. The subsequent discovery of acetyl-CoA by Lynen finally resolved the problem, for Ochoa found that this product can react with oxaloacetate in the presence of purified citrate synthetase to give almost quantitative yields of citrate, this time without the participation of ATP. Since CoA is now known to be an indispensable factor in most of the biological processes in which 'active acetate' is involved, it was established once and for all that 'active acetate' is none other than acetyl-CoA.

The mechanism of the synthesis of acetyl-CoA from free acetate has been investigated by a number of workers and the best available evidence points to adenylacetate as an intermediary:

$$\text{CH}_3\text{COOH} + \text{ATP} \rightleftharpoons \text{CH}_3\text{CO.AMP} + \text{pyrophosphate,}$$
$$\text{CH}_3\text{CO.AMP} + \text{CoA} \rightleftharpoons \text{CH}_3\text{CO.CoA} + \text{AMP,}$$

Overall: $\text{CH}_3\text{COOH} + \text{ATP} + \text{CoA} \rightleftharpoons \text{CH}_3\text{CO.CoA} + \text{AMP} + \text{pyrophosphate.}$

In addition, the reader will hardly need to be reminded that acetyl-CoA arises as an intermediate in the oxidative decarboxylation of pyruvate (p. 389) and this of course explains the synthesis of citrate from pyruvate and oxaloacetate in the early experiments.

In the meantime the next reaction in the cycle had been studied extensively. *Citrate is a symmetrical substance* and as such would be expected to undergo dehydration equally at either of two points, one in that portion of the molecule which arises from oxaloacetate and the second in that part which arises from acetyl-CoA.

Now, if radioactive carbon dioxide and pyruvate are incubated together with liver tissue, terminally labelled citrate is formed and has been isolated. If it behaves symmetrically, this citrate would be expected to go on round the cycle to give an α-ketoglutarate carrying radioactive carbon in both its carboxyl groups. This is shown by the equations below, in which 50 % of the citrate would be expected to follow route (A) and the remaining 50 % route (B).

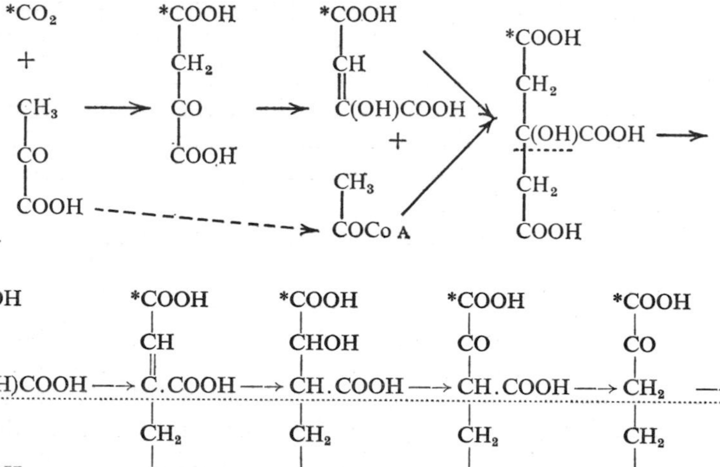

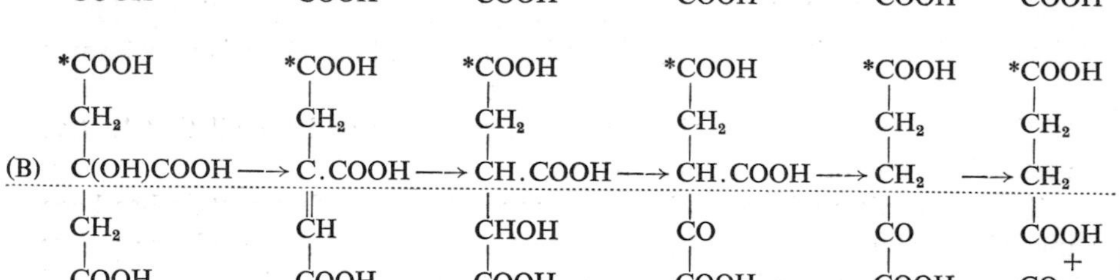

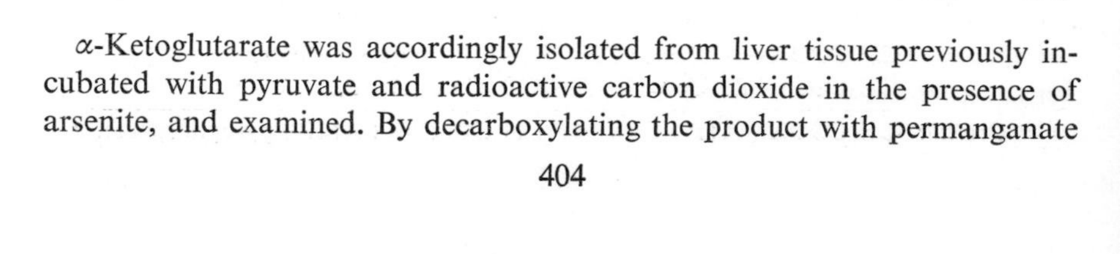

α-Ketoglutarate was accordingly isolated from liver tissue previously incubated with pyruvate and radioactive carbon dioxide in the presence of arsenite, and examined. By decarboxylating the product with permanganate

404

it was shown that all the radioactive carbon was present in the α-carboxyl group, for the carbon dioxide released was radioactive while the residual succinate was not. As an additional check, succinate also was isolated from the liver preparations and found not to contain radioactive carbon. These results show that *citrate does not react symmetrically* and that only route (A) is followed. However, as Ogston was the first to point out, it does not follow that because citrate is a symmetrical substance it will necessarily *behave* in a symmetrical manner when combined with its enzyme, if only because the enzyme-substrate complex is exceedingly unlikely to be symmetrical. Indeed, the foregoing evidence proves conclusively that citrate does not, in fact, behave as a symmetrical compound. This means that the dehydration of citrate by aconitase takes place in that part of the molecule that arises from oxaloacetate; or, to be more precise, between the α and β atoms of the original oxaloacetate (route A).

The same conclusion also follows from experiments in which citrate was formed from oxaloacetate and carboxyl-labelled acetate. In this case the succinate is radioactive and the carbon dioxide inert; again only route (A) is followed:

$$
\begin{array}{ccc}
\text{COOH} & & \text{COOH} \\
| & & | \\
\text{CH} & & \text{CH}_2 \\
\| & & | \\
\text{C(OH)COOH} & \longrightarrow & \text{C(OH)COOH} \\
+ & & | \\
\text{CH}_3 & & \text{CH}_2 \\
| & \nearrow & | \\
\text{*COCo A} & & \text{*COOH}
\end{array}
$$

$$
\text{(A)}\quad
\begin{array}{ccccccccccc}
\text{COOH} & & \text{COOH} & & \text{COOH} & & \text{COOH} & & \text{COOH} & & \text{CO}_2 \\
| & & | & & | & & | & & | & & + \\
\text{CH}_2 & & \text{CH} & & \text{CHOH} & & \text{CO} & & \text{CO} & & \text{COOH} \\
| & & \| & & | & & | & & | & & | \\
\text{C(OH)COOH} & \longrightarrow & \text{C.COOH} & \longrightarrow & \text{CH.COOH} & \longrightarrow & \text{CH.COOH} & \longrightarrow & \text{CH}_2 & \longrightarrow & \text{CH}_2 \\
| & & | & & | & & | & & | & & | \\
\text{CH}_2 & & \text{CH}_2 & & \text{CH}_2 & & \text{CH}_2 & & \text{CH}_2 & & \text{CH}_2 \\
| & & | & & | & & | & & | & & | \\
\text{*COOH} & & \text{*COOH} & & \text{*COOH} & & \text{*COOH} & & \text{*COOH} & & \text{*COOH}
\end{array}
$$

We have travelled some way since the first hypothetical scheme was suggested by Szent-Györgyi, and there is little doubt that many details remain even yet to be established. For the moment, however, it will be convenient to recapitulate and summarize the reactions of the 'tricarboxylic acid cycle' as they are now believed to take place. A schematic summary is presented in Fig. 93 and the individual stages will now be briefly reviewed.

(1) Pyruvate undergoes oxidative decarboxylation, yielding acetyl-CoA which (2) reacts with oxaloacetate, apparently in the *enol*-form. The product of this reaction is citrate which (3) loses water in the manner we have described

405

to yield aconitate, which (4) takes up water and yields *iso*-citrate under the influence of aconitase. Perhaps reactions (3) and (4) take place in concert. *iso*-Citrate is now dehydrogenated by its dehydrogenase (5), giving rise to oxalosuccinate which, in turn, (6) is decarboxylated and yields α-ketoglutarate. Oxidative decarboxylation of the latter (7) gives rise to succinyl-CoA and thence to succinate (8). This is dehydrogenated (9) to give fumarate which, under the influence of fumarase, is hydrated (10) to yield malate. A

$$C_3H_4O_3 + 5O = 3CO_2 + 2H_2O.$$

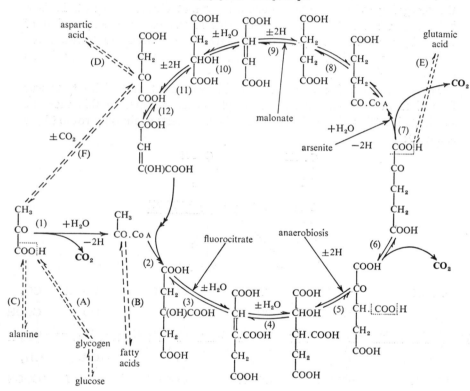

Fig. 93. The citric acid cycle, showing some important side-reactions and the effects of some inhibitors. For list of enzymes involved see Table 31.

further dehydrogenation (11) converts this into oxaloacetate which (12) enolizes and re-enters the cycle. One point must be made here, namely that the NADP-specific *iso*-citrate dehydrogenase catalyses the decarboxylation of oxalosuccinate as well as the dehydrogenation reaction (p. 166). Very probably both reactions take place in concert.

Pairs of hydrogen atoms are transferred to the appropriate hydrogen acceptors in reactions (1), (5), (7), (9) and (11), and pass on through the cytochrome system. In all, therefore, five pairs of H atoms are removed with each turn of the wheel, and three molecules of carbon dioxide are removed at the same time, one molecule of pyruvate being used up. These quantities fit the theoretical equation for the complete oxidation of pyruvate:

$$C_3H_4O_3 + 5O = 3CO_2 + 2H_2O.$$

Table 31. *Enzymes involved in the citric acid cycle* (*see Fig.* 93).

Reaction no.	Enzymes and coenzymes
1	Oxidative decarboxylation system*
2	Citrate synthetase
3	Aconitase
4	Aconitase
5	*iso*-Citrate dehydrogenase; NADP
6	Oxalosuccinate decarboxylyase; Mn^{2+}
7	Oxidative decarboxylation system*
8	Specific CoA transferase; GDP
9	Succinate dehydrogenase
10	Fumarase
11	Malate dehydrogenase; NAD
12	Spontaneous isomerization
A	Glycolytic systems† + phospho*enol*pyruvate carboxylyase
B	Fat-synthesizing enzyme systems‡
C	Transaminase§
D	Transaminase§
E	L-Glutamate dehydrogenase + NAD
F	Pyruvate carboxylyase

* See p. 389. † See p. 365. ‡ See pp. 432 *et seq.* § See pp. 227 *et seq.*

The five atoms of oxygen figured in this equation correspond to the five pairs of hydrogen atoms which, transferred through the cytochrome system, require five atoms of oxygen for their eventual conversion to water, three molecules of which are consumed in the cycle.

Although it is certain that most of the reactions are reversible it is probable that the cycle is unidirectional in actual operation, since the reactions involving oxidative decarboxylation (1, 7) are, in all probability, irreversible. This scheme, every step of which is now well established, provides us with an explanation for the complete oxidation of pyruvate itself, of acetate, and of any substance that gives rise to pyruvate or to acetate. As we have already pointed out, acetate only reacts in the form of acetyl-CoA, but this can be generated from free acetate if ATP and CoA are present, as they may be presumed to be in the tissues. It can arise also by the oxidative decarboxylation of pyruvate formed from carbohydrate sources and from many amino-acids, and, in addition, it is the principal product of the oxidation of fatty acids. The cycle accounts also for the complete oxidation of any substance lying on the cycle and for the complete oxidation of any substance that gives rise to a compound lying on the cycle.

If pyruvate is to be oxidized in liver tissue it not only provides the fuel, in the form of acetyl CoA; it also generates the primary catalyst for the oxidative process:

$$CO_2 + \begin{array}{c} CH_3 \\ | \\ CO \\ | \\ COOH \end{array} \quad \xrightarrow{+\,ATP} \quad \begin{array}{c} COOH \\ | \\ CH_2 \\ | \\ CO \\ | \\ COOH \end{array}$$

This also serves as a pathway whereby pyruvate can be converted again into oxaloacetate and so returned to the cycle. A biotin-dependent enzyme, *pyruvate carboxylase*, present in the liver, catalyses this, again a reversible reaction. Equilibrium conditions strongly favour the production of oxaloacetate.

We must not, however, lose sight of the cycle as a potential source of raw materials for the formation of glycogen, fatty acids and amino-acids. This is particularly important in the liver because it alone among the tissues contains *phospho*enol*pyruvate carboxylase*. If, for example, a large amount of succinate is added, only a catalytic amount remains in the cycle itself; the remainder passes round the usual reactions until it is converted into oxaloacetate, some of which may perhaps break down to yield carbon dioxide and pyruvate. But there is another pathway leading directly from oxaloacetate into the glycolytic reaction sequence. This is catalysed by *phospho*enol*pyruvate carboxylase*:

$$\begin{array}{c} COOH \\ | \\ CH_2 \\ | \\ CO \\ | \\ COOH \end{array} \quad \xrightarrow[+\,GTP]{} \quad CO_2 + \begin{array}{c} CH_2 \\ \| \\ CO\sim\textcircled{P} \\ | \\ COOH \end{array}$$

(GTP can be replaced by ITP but not by ATP.) This is a reversible reaction but the equilibrium conditions are strongly in favour of phospho*enol*pyruvate, so that this circumvents the otherwise apparently difficult step (p. 364) of phosphorylating pyruvate by ATP with the aid of pyruvate kinase.

It follows from all this that we have, in this catalytic system, a self-starting machine that can accomplish or complete the metabolic oxidation of a great diversity of important primary foodstuffs and the intermediary products of their metabolism; it can play an important role in glyconeogenesis and fatty acid formation at the same time.

Citrate itself tends to accumulate if the later reactions of the sequence are inhibited, e.g. by anaerobiosis, by arsenite or by malonate. It also accumulates in large quantities in the presence of *fluoroacetate*. The latter has no action upon citrate synthetase but is itself taken up by reacting with oxaloacetate to form a *fluorocitrate*, which has a powerful inhibitory action upon aconitase and so checks the further metabolism of citrate, but not its synthesis. Practically quantitative yields of citrate have also been obtained by incubating muscle tissue with pyruvate, or acetyl-CoA, and oxaloacetate under strictly anaerobic conditions, suggesting that *anaerobiosis* retards or suppresses the dehydrogenation of *iso*-citrate (reaction (5)). The next oxidative reaction in the sequence (7) is an oxidative decarboxylation and, as such, is inhibited by *arsenite*, so that α-ketoglutarate accumulates, but small amounts of citrate are found at the same time. Succinyl-CoA is formed as an intermediate product in (7). *Malonate* inhibits reaction (9), in which succinate dehydrogenase is the responsible catalyst, and leads to the accumulation of succinate, together with small amounts of α-ketoglutarate and traces of citrate.

It has been shown that the complex system of enzymes and co-factors involved in the cycle are exclusively associated with the mitochondria though some of the components are also present in the cytoplasm. Suspensions of intact, washed mitochondria can catalyse all the events which compose the cycle, but they can be disintegrated to some extent, e.g. by treatment with fat solvents, liberating some of the enzymes. All of the enzymic components have now been crystallized and highly purified, including citrate synthetase itself.

SOME SUPPLEMENTARY REACTIONS

In addition to the reactions composing the cycle proper, more careful attention must now be paid to a number of important side reactions. Very important among these is the glycolytic reaction sequence (A) which leads from glucose and glycogen to pyruvate and hence to acetyl-CoA by oxidative decarboxylation. The cycle therefore allows us now to account for the complete oxidation of glucose, glycogen, and all the known intermediates involved in glycolysis. Furthermore, it has been shown that the breakdown of fatty acids (B) also culminates in the production of acetyl-CoA, so that we can account for the complete oxidation of fat and its intermediary metabolites also.

This remarkable mechanism also provides a link between the metabolism of carbohydrates and that of proteins. Among the amino-acids that enter into the composition of proteins there are some that are non-essential and can be synthesized in the animal body in seemingly unlimited amounts. Three such are alanine (C), aspartate (D) and glutamate (E). The corresponding α-keto-acids, pyruvate, oxaloacetate and α-ketoglutarate, arise, according to this scheme, as intermediary products in the course of carbohydrate metabolism, and would only require to be aminated or transaminated to yield the amino-acids in question. Alternatively, of course, these amino-acids could, by trans-deamination, give rise to either pyruvate, oxaloacetate or α-ketoglutarate, any or all of which could give rise to glycogen if the cycle is overloaded with catalytic components.

Particularly important among the side reactions is the carboxylation of pyruvate to yield oxaloacetate (F). In muscle, which contains no pyruvate carboxylyase, oxaloacetate drifts slowly out of the system, probably by spontaneous β-decarboxylation, so that the rate of respiration of minced muscle declines only slowly. Presumably the maintenance of a high rate of respiration in intact muscle also requires the provision of a constant supply of some one or other of the catalytic dicarboxylic acids.

In liver, by contrast with muscle, *phospho*enol*pyruvate carboxylyase* is present, and here any excess of oxaloacetate will leave the system rapidly, giving phospho*enol*pyruvate, and, as has already been suggested, this provides a route of great importance for glyconeogenesis (see p. 364).

If pyruvate is suddenly produced, e.g. by the administration of pyruvate to the intact animal or, again, by a sudden burst of glycolysis, oxaloacetate will be formed by the action of pyruvate carboxylyase, aided and abetted by ATP, and the oxidation of pyruvate by way of acetyl-CoA can be initiated. Oxidation will continue so long as the concentration of pyruvate remains high, but when it falls again to a low level, oxaloacetate will once more leave the system and, as the catalytic carriers fall in concentration, respiration will again slow up. But oxidation will not be the only process tending to reduce the concentration of pyruvate: some will be converted into *glycogen*, by way of phospho*enol*pyruvate, some will be oxidatively decarboxylated and give rise to acetyl-CoA and hence to *ketone bodies or fat* instead of being immediately oxidized, while some may even be transaminated and converted into *alanine*.

The foregoing accessory reactions can be summarized in relation to the rest of the cycle as shown in Fig. 94.

While it is probable in the extreme that a great deal still remains to be discovered about this catalytic cycle and its various side reactions, there can be little doubt of the great and fundamental importance of the system as a whole, both as a clearing-house for the oxidation of the many products formed on the metabolic lines which converge upon it and as a meeting-place for the main metabolic pathways of carbohydrate, fat and protein. As has already been pointed out, there is evidence that this cycle, in the form envisaged by Szent-Györgyi, is as widely distributed as are cytochrome and cytochrome oxidase. The cells of animals, plants and micro-organisms appear, with very few exceptions, to contain succinate and malate dehydrogenases, together with fumarase, while *iso*-citrate dehydrogenase and aconitase, which together correspond to the citrate dehydrogenase originally described by Thunberg, seem likewise to have a very wide occurrence. Again, there is reason to believe that most cells have the ability to accomplish the oxidative decarboxylation of α-keto-acids. It seems not by any means impossible that the carbohydrate metabolism of many living cells is organized on the same fundamental lines, glycolysis leading to the formation of pyruvate from glycogen, the pyruvate being metabolized in its turn through acetyl-CoA and the tricarboxylic acid cycle under aerobic conditions, and linking up, as it does in mammalian liver, with the metabolism of protein and of fat. At the same time it is important to realize that if the cycle is overloaded, so that pyruvate or phospho*enol*-pyruvate is formed from excess oxaloacetate, glycogen can be formed.

That we shall find variations on the general, fundamental theme we may be sure, and, indeed, we have already discovered in the phosphagen of muscular tissues a specialized chemical adaptation which admirably subserves the highly

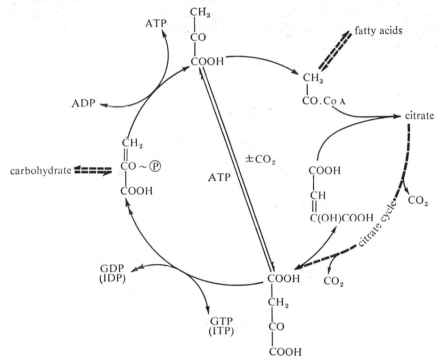

Fig. 94. Supplementary reactions of the citrate cycle.

specific functions which muscle is called upon to discharge. Similarly, we may consider that the pyruvate decarboxylyase of yeast is a specialization which permits that organism to live on carbohydrate even in the total absence of oxygen. It might, of course, be argued that fermentation is a simpler operation than respiration, and that the simpler must logically be the more primitive but, at the same time, there is evidence that evolutionary advancement and specialization can be attended by the loss of old enzymes, as well as by the acquisition of new, as is the case of the enzymes concerned with purine metabolism (p. 288).

THE GLYOXYLATE AND DICARBOXYLIC ACID CYCLES

Although the citric cycle is usually important as a major energy-producing device there are cases in which its main importance lies in its ability to produce substances of great metabolic importance such, for example, as α-ketoglutarate and succinate. If these are to be tapped off to take part in other synthetic processes a proportionate supply of pyruvate will be necessary, not only to provide carbon atoms but also to maintain the supply of oxaloacetate to keep the cycle in operation. This raises special problems for certain micro-organisms that live on 2-carbon compounds, such as acetate or glyoxylate, as their sole source of carbon.

A new cyclical mechanism, known as the *glyoxylate cycle*, has been found to operate in several such cases and the reactions concerned are shown in Fig. 95.

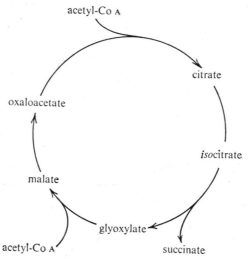

Fig. 95. The glyoxylate cycle (after Kornberg).

Here no carbon dioxide is produced so that this is not a degradative system, but some of the energy-yielding steps of the citrate cycle still operate. The special feature is the splitting of *iso*-citrate into succinate and glyoxylate:

$$
\begin{array}{c}
\text{COOH} \\
| \\
\text{CHOH} \\
--- \,|\, --- \qquad\qquad \text{COOH} \qquad \text{COOH} \\
\text{CH.COOH} \longrightarrow \quad | \qquad + \quad | \\
| \qquad\qquad\qquad \text{CHO} \qquad \text{CH}_2 \\
\text{CH}_2 \qquad\qquad\qquad\qquad\qquad | \\
| \qquad\qquad\qquad\qquad\qquad\qquad \text{CH}_2 \\
\text{COOH} \qquad\qquad\qquad\qquad\qquad | \\
\qquad\qquad\qquad\qquad\qquad\qquad \text{COOH}
\end{array}
$$

glyoxylic acid

412

Glyoxylate can react with acetyl CoA thus:

$$
\begin{array}{ccc}
\text{COOH} & & \text{COOH} \\
| & & | \\
\text{CHO} & & \text{CHOH} \\
+ & \longrightarrow & | \quad + \text{CoA} \\
\text{CH}_3 & & \text{CH}_2 \\
| & & | \\
\text{CO.CoA} & & \text{COOH} \\
& & \textit{malate}
\end{array}
$$

The other reactions are those of the usual citric acid cycle. At each turn of the cycle two 2-carbon acetate units are consumed and one 4-carbon molecule of succinate is synthesized. The glyoxylate cycle therefore provides for energy production and the synthesis of side-products as well.

Some micro-organisms can live with glycine as their sole carbon source and in some of these the glyoxylate cycle operates, since glycine can give rise to acetyl-CoA in many cells. But glycine can also give rise to glyoxylate by oxidative deamination:

$$
\begin{array}{ccc}
\text{CH}_2\text{NH}_2 & & \text{CHO} \\
| \quad + \tfrac{1}{2}\text{O}_2 & \longrightarrow & | \quad + \text{NH}_3 \\
\text{COOH} & & \text{COOH}
\end{array}
$$

Some bacterial cells can live and reproduce with glyoxylate as the sole carbon source. In such a case yet another new cyclical system is used, based this time on *di*carboxylic acids by contrast with the *tri*carboxylic acids that play so prominent a part in the citric cycle. The reactions of this, the *dicarboxylic acid cycle*, are shown in Fig. 96.

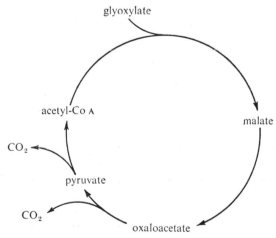

Fig. 96. The dicarboxylic acid cycle (modified after Kornberg).

This cycle allows of the degradation of glyoxylate to carbon dioxide but yields up energy in the process: several of the energy-producing reactions of the citrate cycle are incorporated in this cycle.

But beneath all these and the secondary and specific adaptations that we are likely to meet in the future, there is every reason to think that we shall discover evidence of the existence of a basic metabolic ground-plan to which living cells in general conform.

Evidence already available indicates that the citrate cycle is present in living organisms of many different kinds. Primarily perhaps, its function was that of providing supplies of citrate, α-ketoglutarate, succinate and the other components of the cycle. From these, other substances such as glutamate, glutamine and other side products, many of which are of enormous importance in intermediary metabolism, can then be produced. Again, as we have seen, *iso*citrate can be enzymatically split into succinate and glyoxylate, which latter can then give rise to glycine, serine and ethanolamine in that order. From glyoxylate, glycine and serine, 1-carbon units of formate can be formed, while ethanolamine can give rise to choline. And there are many more possibilities. But, in animal tissues at any rate, the main function of the cycle is apparently that of energy production, for here, quite apart from producing many substances of metabolic importance and dealing with the oxidative disposal of many more, the cycle operates fast enough to be capable of supplying the greater part of the free-energy requirements of animal cells and tissues. We shall accordingly proceed next to consider the energetic aspects of the cycle.

ENERGETICS OF CARBOHYDRATE OXIDATION

Under anaerobic conditions the glycolysis of 1 g.mol. of glucose gives a net yield of 2 $\sim$℗ (p. 344). Aerobically, however, no lactate need be formed; instead the reduced NAD can pass on its 2H to oxygen through the flavoprotein-cytochrome system, yielding 3 $\sim$℗ per molecule (see p. 181), i.e. 6 $\sim$℗ in all, since 2 molecules of triosephosphate arise from 1 molecule of glucose and reduce 2 molecules of the coenzyme. Thus the overall yield of $\sim$℗ in the aerobic formation of 2 molecules of pyruvate from one of glucose will amount to 8 $\sim$℗ in all:

$$C_6H_{12}O_6 + O_2 \rightarrow 2CH_3COCOOH + 2H_2O + 8 \sim ℗.$$

The subsequent energy-yielding stages may be summarized as follows (see p. 183):

Reaction	Primary H-acceptor	Estimated yield of $\sim$℗
Pyruvate $\rightarrow$ acetyl-CoA $\rightarrow$ citrate	NAD	3
iso-Citrate $\rightarrow$ oxalosuccinate	NADP	3
α-Ketoglutarate $\rightarrow$ succinate	NAD	4
Succinate $\rightarrow$ fumarate	Cytochrome	2
Malate $\rightarrow$ oxaloacetate	NAD	3
		Total 15

414

Since glucose yields 2 molecules of pyruvate the total yield of $\sim$Ⓟ from this part of the process will be 30, making a grand total of 38 $\sim$Ⓟ per g.mol. of glucose oxidized. These, of course, are imperfect calculations and the knowledge upon which they are based is still imperfect.

Many experiments have been carried out in order to determine directly the energy-yields of oxidative processes in various tissues. In one such series of experiments carried out on pigeon-breast muscle, the yields of ATP recorded ranged from 1 to 3 molecules of ATP for each atom of oxygen consumed. Since the oxidation of one molecule of glucose requires 12 atoms of oxygen, these results indicate an energy-yield of 12–36 $\sim$Ⓟ per molecule of glucose oxidized. Bearing in mind the probability that some ATP must have been broken down in the course of incubation we may assume that *at least* 36 $\sim$Ⓟ are formed for each molecule of glucose undergoing oxidation and that, given ideal experimental conditions, the yields might conceivably be greater even than this. We may therefore assume for the present that the true yield is probably *at least* 36 $\sim$Ⓟ for each g.mol. of glucose oxidized.

21

THE METABOLISM OF LIPIDS

TRANSPORT AND STORAGE OF LIPIDS

Triglycerides make up the bulk of the food lipids. Under the influence of pancreatic lipase these triglycerides are hydrolysed to form β-monoglycerides and free fatty acids. Triglycerides are resynthesized in the cells of the intestinal mucosa and enter the blood stream by way of the thoracic duct, in the form of microscopic oily droplets known as *chylomicrons*. This leads to a condition known as post-absorptive lipaemia. The chylomicrons are gradually taken up by the liver where they undergo hydrolysis, giving free glycerol and free fatty acids. The acids are then re-esterified and combined with proteins associated with the globulins of the blood to form *lipoproteins*, which are appreciably soluble in water, and it is in this form that triglyceride materials are transported from the liver to the fat depots, i.e. the adipose tissues, especially the mesenteries and the intramuscular and subcutaneous connective tissues.

Contrary to earlier belief, the adipose tissues, far from being inert, are in fact endowed with a high order of metabolic activity. Incoming lipoproteins are hydrolysed by a special *β-lipoprotein lipase*, which yields free fatty acids and free glycerol, then the triglycerides are again resynthesized and deposited as such. When the time comes for the depot fats to be mobilized for transport to other tissues, the triglycerides are hydrolysed yet again and the fatty acids now become associated with blood albumins, forming water soluble *albumin-bound fatty acids*, in which form they are distributed by the blood stream. A single albumin molecule can carry about 20 molecules of fatty acids in this way. The functions of the adipose tissue in lipid metabolism do not end here, for it is in this tissue especially that fatty acid synthesis from non-fatty precursors takes place, a matter to which we shall presently return. Another special feature is that adipose tissue shares with several others the ability to carry out the desaturation of some fatty acids, especially of stearic acid (C_{18}) to oleic (C_{18}), which has a double bond between C_9 and C_{10}.

Hydrolysis and resynthesis of triglycerides thus takes place in three different areas, the intestinal mucosa, the liver and the adipose tissue, and at each of these there is an opportunity for a 'reshuffling' of the individual fatty acid units so that fat, as deposited in the fat depots, is seldom if ever identical in composition with the original dietary fat. We shall consider the synthetic mechanisms presently.

Generally speaking, the kind of fat present in the fat depots of a given animal is fairly characteristic of the species. Beef fat is always much the same in composition, while mutton fat is always characteristically mutton fat. But it is not difficult to alter the composition of the depot fat considerably by feeding fats of a kind which the animal does not ordinarily consume. If for instance, a dog is given large amounts of linseed oil it will lay down a softer and much more unsaturated depot fat than is characteristic of dogs as a whole. But it is none the less true that, in the ordinary way, each species lays down its own kind of fat, with or without a previous 'reshuffling' of the fatty acids, just as each species lays down its own kind of tissue proteins. The reason for this constancy is only partly covered by the tendency of animals to select a diet which is fairly constant in composition; not all the fat found in the depots is merely food fat that has been transported thither from the alimentary canal, but fat which has been synthesized from non-fat sources, largely or mainly in adipose tissue and probably, to a lesser extent, in the liver.

As every stock-breeder knows, animals can be fattened cheaply by feeding them an abundance of carbohydrates, and fat can also be synthesized from protein to some extent. It seems likely that the nature of the newly formed fat from these non-fat sources will depend upon the metabolic make-up of the particular species concerned, different animals starting from much the same raw materials but each manufacturing its own kind of fat.

The synthesis of fats from non-fat sources is particularly important in cattle, sheep and other herbivorous animals, for here cellulose bulks large as a foodstuff. In these animals, cellulose is digested by symbiotic micro-organisms which produce from it high yields of short-chain, steam-volatile fatty acids, among which acetate and propionate predominate, together with some butyrate and small amounts of higher acids. Acetate and butyrate are fat-formers, whereas propionate can yield glycogen. In animals of this kind, therefore, it is probable that the main reserves of fat and carbohydrate are built up almost entirely from short-chain fatty acids.

The average animal is capable of laying down almost unlimited amounts of fat and, in point of fact, fat has certain definite advantages over protein and carbohydrate as a form of reserve fuel. Fat is far richer in carbon and hydrogen than the other primary foodstuffs, so that there is more combustible material in a gram of fat than in a gram of either protein or carbohydrate. From the point of view of energy, therefore, fat allows the greatest storage per gram of reserve material. If a gram of each of the three main types of food is burned in a bomb calorimeter the heat produced is approximately as follows:

1 g. protein	...	...	...	5600 cal.
1 g. carbohydrate		...	...	4200 cal.
1 g. fat...	...	...	...	9300 cal.

Closely bound up with this is the fact that fat, when burned, gives rise to about twice as much metabolic water as the other foodstuffs on account of its much higher content of hydrogen:

417

1 g. protein	...	...	...	0·41 g. water
1 g. carbohydrate		...	...	0·55 g. water
1 g. fat...	...	...	...	1·07 g. water

This is an important feature of fat metabolism, especially among terrestrial animals, many of which live under conditions of acute water shortage. In such cases there is commonly a heavy emphasis on the oxidation of fat; in this way the organism is better able to eke out its external supplies of water with metabolic water formed in its own tissues. As an example of this phenomenon we may refer to the developing chick embryo. At laying, the hen's egg is provided with a definite and limited amount of water, an amount which, by itself, would be insufficient to see the embryo through development. But during the 3 weeks of incubation, rather more than 90 % of all the material oxidized by the embryo consists of fat. Again, the mealworm, an insect larva that can live for long periods under the most arid conditions, metabolizes during starvation about 2½ parts of carbohydrate for every part of protein, and no less than 8 parts of fat. The almost legendary ability of the camel to travel for days in the desert without a drink is similarly attributable to heavy fat metabolism with a proportionately large-scale production of metabolic water.

Fat which is on its way from the gut to the liver is carried mainly in the form of chylomicrons, and a pronounced condition of lipaemia is regularly to be observed after the consumption of a fatty meal. Its further transport to the adipose tissues is, as we have seen, accomplished in the form of lipoproteins. When fat is being withdrawn from the depots to be metabolized elsewhere most of the fat transported travels in the form of albumin-bound fatty acids. These bound fatty acids, together with lipoproteins and a certain amount of cholesterol and cholesterol esters, are normally present in the blood, but free triglyceride fat is normally only present as such—and even then only in chylomicron form—while the condition of post-absorptive lipaemia persists.

The mobilization of depot lipids can conveniently be studied in starving animals. After a short period of starvation the glycogen reserves of the liver are used up, and no more glycogen is forthcoming except through glyconeogenesis. Presumably because of the shortage of carbohydrate, large amounts of fatty material are transported to the liver. This condition, which is known as 'fatty liver', can also be observed in a variety of conditions other than starvation and it seems certain that the accumulation of lipid material in the liver represents the first step towards its metabolic breakdown.

The lipids which appear in a fatty liver arise from the fat depots. The administration of deuterated fat to mice leads to the deposition of most of the heavy hydrogen in the fat depots to which it is transported after its preliminary adventures in the liver. But if the animals are allowed to starve for several days before being killed and the body fats are then worked up for heavy hydrogen, it is found that the fat content of the liver is now greater than at the beginning of starvation and, moreover, that the liver lipids contain twice or three times as much deuterium as those in the depots. After its absorption, therefore, most of the ingested fat goes to the fat depots, from which the fatty acids are withdrawn and transported to the liver as and when the need arises.

The condition of fatty liver can be established by any treatment that tends seriously to diminish the power of the liver to store, produce or metabolize carbohydrate. In diabetes, for instance, the storage powers of the liver are impaired, and fatty liver is one of the features of this disease. Furthermore, in severe cases, the blood may contain three or four times as much lipid material as normal. In the pseudo-diabetic condition induced by phlorrhizin the glycogen reserves of the liver are broken down and the glucose thus set free is excreted by way of the urine. Here again fatty liver is to be observed. Small doses of liver poisons such, for instance, as carbon tetrachloride, chloroform, phosphorus and diphtheria toxin, also lead to fatty liver, apparently because they interfere with the synthesis of lipoproteins. In all these cases the establishment of fatty liver is encouraged by feeding cholesterol and discouraged by the administration of choline or ethanolamine.

FUNCTIONS OF LIPIDS: CONSTANT AND VARIABLE ELEMENTS

It will be clear from what has been said that the lipid content of the depot tissues and of the liver—and the same is true of other organs—varies very widely with the nutritional condition of the organism. If an animal is allowed to starve for a long time, the amount of reserve fat in the body becomes very small, but, even at death from starvation, the tissues still contain large amounts of lipid material which, apparently, forms a part of the structural material of the tissues and is not available for use as fuel. This part of the total body lipids is sometimes referred to as the 'constant element'; constant because, being a part of the actual fabric of the organism, it is always present and always must be present. The remainder of the lipids comprise the 'variable element', so-called because they vary in amount with the nutritional state of the organism and the demands made upon them for energy production.

The contrast between the constant and variable elements can be appreciated by comparing the effects of different doses of liver poisons upon the lipid content of the liver. Small amounts of some liver poisons lead to the mobilization of fat from the depots and its deposition in the liver. Here we observe the movement of a part of the variable element from one place to another. If larger doses of the same poisons are administered, the liver cells suffer serious damage and we observe the condition known as fatty degeneration. Again the cells are rich in lipid materials, partly because the other cell constituents have been broken down and dispersed but mainly because the normal processes of lipid transport away from the liver have broken down.

Lipids, then, have two main functions. They act as fuel reserves, and they form an important part of the structure of living tissues, especially of cell membranes. In addition, there are indications that certain fatty acids have special and specific functions; for example it is known that rats kept on fat-free diets, or on diets from which particular fatty acids have been removed, become ill and suffer from caudal necrosis, but we shall return presently to consider this condition and its metabolic implications.

METABOLISM OF FATS: GENERAL

It has long been believed that the liver plays a predominant role in the metabolism of fat, for it is to this organ above all that fat is transported when carbohydrate metabolism is subnormal and an alternative energy source is required. It has usually been assumed, though never proved until fairly recently, that fats are hydrolytically split into glycerol and free fatty acids before any oxidation takes place and we have seen that this process takes place chiefly in adipose tissue, and that the resulting fatty acids reach the liver in an albumin-bound form.

Glycerol, if administered to a diabetic or phlorrhizinized animal, is converted almost quantitatively to glucose. It also leads to the deposition of glycogen in the liver if administered to a starving animal. In a normal animal, isotopically labelled glycerol gives rise to correspondingly labelled glycogen. The pathway for its conversion into carbohydrate is through glycerol phosphate, formed by a specific glycerol kinase at the expense of ATP, followed by the oxidation of glycerol phosphate to triosephosphate by a glycerol-phosphate dehydrogenase, and of course, triosephosphate lies directly on the normal lines of glycolysis and glycogenesis.

Fatty acids are normally completely oxidized to carbon dioxide and water. There is an abundance of evidence, which we shall presently review, to show that fatty chains are split into 2-carbon units consisting of some form of 'active acetate'. Acetate itself can be totally oxidized, e.g. by washed kidney-cortex homogenates in the presence of traces of succinate or oxaloacetate, and its oxidation is inhibited by malonate and other inhibitors of the citrate cycle, whence we may conclude that it enters and is oxidized by way of this cycle. Confirmation of this is found in the fact that isotopically labelled acetate can be incorporated into the cycle and gives rise to correspondingly labelled intermediates in the presence of the usual inhibitors (p. 405). Long-chain fatty acids also can be completely oxidized by a variety of tissue preparations, including slices, breis, homogenates and mitochondrial suspensions, and here again oxidation is dependent upon the presence of oxaloacetate or some other intermediate of the citrate cycle, and here too the oxidation is inhibited by malonate. These facts indicate that acetate and longer chains alike are probably metabolized through the same route and that acetate, and the 2-carbon units arising by the breakdown of longer chains, enter the cycle through a probably common intermediary.

Now the entry of acetate itself into the cycle depends upon the presence of ATP and Co A and we know (p. 104) that, in the presence of these substances, acetate is converted into acetyl-Co A. We also know that acetyl-Co A can react with oxaloacetate and so enter the citrate cycle. Since no form of 'active acetate' other than acetyl-Co A has so far been discovered in animal tissues we may reasonably assume that the 2-carbon units produced from fatty acids arise in this form. In confirmation of this we have the further fact that fatty acids can be synthesized from acetate, presumably by way of the same common 2-carbon intermediate. These points may be summarized:

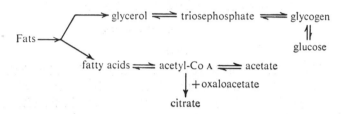

In the liver, under certain circumstances, large amounts of certain other compounds can accumulate. These are *acetoacetate*, together with *β-hydroxy-butyrate* and small amounts of *acetone*. It is generally agreed that the parent substance of this group is acetoacetate. Under the influence of the widely distributed β-hydroxybutyrate dehydrogenase and NAD, acetoacetate and β-hydroxybutyrate are freely interconvertible. Acetone arises from aceto-acetate by what appears to be a spontaneous decarboxylation, a process which takes place at an appreciable speed under physiological conditions of tempera-ture and pH, though an acetoacetate decarboxylase is known. The relation-ships between these three substances can briefly be summarized as follows:

$$CH_3COCH_2COOH$$
acetoacetic acid

$\pm 2H$

$$CH_3CH(OH)CH_2COOH \qquad\qquad CH_3COCH_3 + CO_2,$$
β-hydroxybutyric acid   *acetone*

Small quantities of ketone bodies are formed from ketogenic amino-acids also, but that they arise mainly from fat cannot be doubted.

Acetoacetate, the parent member of the group, can be formed from acetate, but the synthesis is an endergonic process, indicating that the starting material must be a high-energy substance of some kind. Moreover, the forma-tion of acetoacetate from acetate implies acetylation, and there is evidence both that acetyl-Co A is a high-energy compound and that it participates in the biological acetylation of a number of different substances (p. 104). Hence

28-2

we may suppose that acetoacetate arises from acetate and from acids with longer chains by way of acetyl-CoA. Now acetoacetate, like acetate itself and like the long-chain fatty acids, can be completely oxidized by various tissue preparations in the presence of oxaloacetate and other intermediates of the citrate cycle, and once again oxidation is inhibited by malonate. There is however an important exception in the case of the liver, for *liver tissue cannot oxidize acetoacetate* although it produces it readily enough.

Isotopically labelled acetoacetate can gain admission to the citrate cycle, yielding correspondingly labelled intermediates. Once again, therefore, we must suppose that acetyl-CoA is involved as an intermediate, and again we may summarize our conclusions:

$$\text{Acetoacetate} \rightleftharpoons \text{acetyl-CoA} \rightleftharpoons \text{acetate}$$

$$\downarrow + \text{oxaloacetate}$$

$$\text{citrate}$$

These considerations lead us to suspect that fatty acids and carbohydrates alike undergo final oxidation by way of a common 2-carbon intermediate in the form of acetyl-CoA, and by a common mechanism in the citrate cycle. We may therefore summarize our summaries in a united form as shown in Fig. 97.

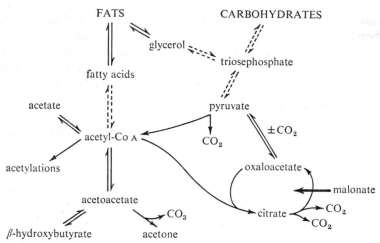

Fig. 97. Outlines of metabolism of fats and carbohydrates.

Traces of ketone bodies can be detected in the blood of normal animals, but the amounts present are greatly increased when there is a heavy emphasis upon fat metabolism. The energy requirements of a typical animal are normally met by metabolizing fatty acids and carbohydrate together, but if, for any reason, carbohydrate metabolism is subnormal, correspondingly more fat has to be metabolized and more 2-carbon units are produced and react together in pairs to give rise to acetoacetate. In fact ketone body production is always associated with unavailability or under-utilization of carbohydrate.

So intimate is the relationship between fat metabolism and that of the ketone bodies that any theory of fat metabolism which fails to explain the generation and metabolism of the ketone bodies is doomed in advance. Whether we regard the ketone bodies as direct or as secondary products of fatty acid metabolism, it is evident that the ability of the liver to oxidize fatty acids is limited and in some way linked with its carbohydrate metabolism, so that these ketone bodies fail to be completely oxidized when carbohydrate metabolism is subnormal.

Since it seems certain that acetyl-CoA is a common metabolite between fatty acids and ketone bodies alike and is able to enter and undergo metabolism by way of the citric cycle, it is clear that several possible fates are open to it. It may enter the cycle and be oxidized: it may yield acetoacetate, it may revert to acetate or fatty acids, or it may participate in acetylation reactions. It may be remarked, however, that these general arguments do not apply to tissues other than liver. The extrahepatic tissues can and do metabolize acetoacetate and β-hydroxybutyrate and, even when the liver is producing significant quantities of ketone bodies, these may pass into the blood and undergo oxidation in the extrahepatic tissues; no actual ketosis develops until the rate of formation of acetoacetate and the other ketone bodies in the liver outstrips that at which they can be oxidized by the other tissues.

Having now dealt in outline with the broad features of fat metabolism as we know them at present, it is time to give more detailed evidence in favour of and against the conclusions so far summarized.

DESATURATION

It has been known for many years that if fatty acids are incubated with liver tissue there is an increase in the iodine number of the sample, indicating that the liver is capable of introducing double bonds into the fatty chain. Leathes pointed out many years ago that the liver normally contains a rather high proportion of unsaturated fatty acids, and suggested that fatty acids might be broken down by the introduction of double bonds at more or less arbitrary points along the chain, the latter then being broken into fragments at the 'weak' linkages thus introduced.

That the liver can dehydrogenate fats is certain, but the notion that double bonds can be introduced *anywhere* in the chain is disproved by the following facts. If rats are given a diet from which all traces of fatty acids with a double bond at $C_{12:13}$ have been removed, they develop a condition known as fat-deficiency disease, characterized by a scaly condition of the tail (caudal necrosis), which may even drop off in severe cases. If the liver were able to introduce double bonds at *any* point in the fatty chain, this condition could hardly exist. The disorder is curable by the administration of linoleic acid, for example, and this could be made from oleic acid by desaturation in the 12:13 position. This the liver clearly cannot do, at any rate in rats.

Attempts have been made to discover precisely where double bonds *can* be inserted by incubating liver tissue with fatty acids already containing double bonds in known positions. When this is done, desaturation still takes place except in acids already containing a double bond at the $\alpha:\beta$ position. This indicates that desaturation can only be accomplished in the $\alpha:\beta$ position.

Evidence has been obtained in other ways to show that adipose tissue in particular contains an enzyme capable of introducing a double bond in the middle (9:10) of the stearic acid chain, giving rise to oleic acid.

β-OXIDATION

The introduction of an $\alpha:\beta$ double bond is the first step in the process known as β-oxidation, first postulated many years ago by Knoop. Knoop prepared a series of ω-phenyl fatty acids, i.e. fatty acids which had been 'labelled' by the introduction of a phenyl group at the carbon atom most remote from the carboxyl group. These acids were administered to dogs, the urine being collected and worked up for substances containing the phenyl group. In every case a phenylated compound was isolated, the products being benzoic and phenylacetic acids respectively. Both acids were actually eliminated in the form of their glycine conjugates, hippuric and phenaceturic acids, but for our present purposes the conjugation can be neglected. The important feature of Knoop's discoveries was the fact that administration of an acid with an even number of carbon atoms in the side chain led always and only to the formation of phenylacetic acid, odd-numbered chains giving rise always and only to benzoic acid. For example;

even: $C_6H_5CH_2CH_2 \vdots CH_2COOH \rightarrow C_6H_5CH_2COOH \rightarrow (C_6H_5CH_2CO.HN.CH_2COOH)$
    phenylbutyric acid   *phenylacetic acid*   *(phenaceturic acid)*
odd: $C_6H_5CH_2 \vdots CH_2COOH \rightarrow C_6H_5COOH \rightarrow (C_6H_5CO.HN.CH_2COOH)$
    phenylpropionic acid   *benzoic acid*   *(hippuric acid)*

This must mean that the organism is unable to remove carbon atoms one at a time from the chain: if it could do so, phenylacetic acid would be expected to be converted into benzoic acid, while the administration of either an odd- or an even-numbered chain would be expected to give either benzoic acid alone, or else a mixture of benzoic and phenylacetic acids. Presumably, then, *the carbon atoms must be split off in even numbers, most probably in pairs.* Knoop himself was of the opinion that these 2-carbon fragments consist either of acetate or, more probably, some highly reactive derivative of acetate; as we now know, these 2-carbon fragments are molecules of acetyl-CoA.

Now it happens that the naturally occurring fatty acids are, almost without exception, members of the even-numbered series. Any such acid, by losing two carbon atoms at a time, would pass eventually through the stage of butyrate which, undergoing β-oxidation in its turn, would give rise to aceto-acetate, an important point in favour of Knoop's scheme.

MECHANISMS OF FATTY ACID METABOLISM

At the time when Knoop's work was done nothing was known about the nature of the intermediate reactions and the nature of the 2-carbon 'active acetate' fragments was still unknown. The identity of the latter with acetyl-CoA, once established, soon led to the discovery that CoA is involved at every stage in the reaction sequence that makes up the process of β-oxidation.

There is abundant evidence that fatty acids undergo eventual oxidation by mechanisms involving the citrate cycle. Many experiments with a wide range of tissue preparations, including slices, breis, homogenates and suspensions of washed mitochondria, have shown that the oxidation of acetate and higher fatty acids alike depends upon the presence of one or other of the reactants of the citrate cycle, and that the oxidation is powerfully inhibited by malonate and other reagents known to inhibit the cyclic mechanism. We know that fatty chains undergo degradation to 2-carbon fragments in the form of acetyl-CoA, which can react directly with oxaloacetate to form citrate.

The oxidation of acetate and higher fatty acids in tissue preparations can be considerably accelerated by the addition of ATP and can also be facilitated by the addition of one or other of the reactants of the citrate cycle. The operation of the cycle leads, of course, to the production of ATP and might be thought to facilitate fatty acid oxidation for that reason. The effect of ATP led at first to the suggestion that the first step in the metabolic breakdown of fatty acids must probably consist in their phosphorylation by ATP. But in the case of acetate, such a reaction would lead to the formation of acetyl phosphate, which is probably never formed in animal tissues and, if it is, does not enter the citric cycle.

There is a more satisfactory explanation for the effect of ATP. In the case of acetate, both ATP and CoA are required before the acid can react with oxaloacetate and enter the citric cycle (p. 402), and it is now certain that ATP, CoA and free acetate first react together to yield acetyl-CoA, probably as follows:

(i) acetate + ATP $\rightleftharpoons$ adenyl acetate + pyrophosphate,

(ii) adenyl acetate + CoA $\rightleftharpoons$ acetyl-CoA + AMP.

This is a general reaction for the formation of the CoA derivatives of fatty acids, and this is in fact the first step in the metabolism of free fatty acids in general. Enzymes capable of producing acyl-CoA derivatives in the presence of ATP and CoA have been known for some years and are usually called *acyl-CoA synthetases*.

425

Another method appears to be possible in the case of some shorter chains: butyrate, for example, can react with succinyl-CoA arising in the citric acid cycle:

$$CH_3CH_2CH_2COOH + \begin{array}{c} COOH \\ | \\ CH_2 \\ | \\ CH_2 \\ | \\ CO\sim CoA \end{array} \rightleftharpoons CH_3CH_2CH_2CO\sim CoA + \begin{array}{c} COOH \\ | \\ CH_2 \\ | \\ CH_2 \\ | \\ COOH \end{array}$$

Acetoacetate undergoes a similar reaction.

It has now been established that fatty acids themselves, and likewise all the intermediates involved in their oxidation, react in the form of their CoA derivatives, rather than in the free condition, and enzymes are known which catalyse the whole series of reactions from fatty acyl-CoA to acetyl-CoA. The initial formation of the acyl-CoA from free fatty acids by the acyl-CoA synthetases we have already discussed These fatty acyl-CoA compounds are formed, apparently, in the cytoplasm and do not enter the mitochondria as such. Instead they enter into an exchange reaction with *carnitine* forming high-energy derivatives to which the mitochondria are permeable:

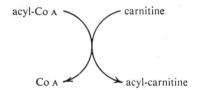

Once inside the mitochondria this reaction is reversed and the fatty acyl-CoA compounds are regenerated. We may pass on next to the reactions which ensue. All these and the enzymes which catalyse them are present in and apparently confined to the mitochondria. They are summarized in Fig. 98.

(1) ----------- $CH_2CH_2CH_2CO.CoA$

      $-2H \downarrow$ (ETF) *acyl-CoA dehydrogenases*

(2) ----------- $CH_2CH\!\!=\!\!CHCO.CoA$

      $+H_2O \downarrow$ *enoyl-CoA hydratases*

(3) ----------- $CH_2CH(OH)CH_2CO.CoA$

      $-2H \downarrow$ (NAD) *β-hydroxyacyl-CoA dehydrogenases*

(4) ----------- $CH_2CO.CH_2CO.CoA$

      $+CoA \downarrow$ *β-keto-acyl-CoA thiolases*

(5) ----------- $CH_2CO.CoA + CH_3CO.CoA$

Fig. 98. Enzymes involved in β-oxidation of fatty acids.

(1) *α:β Desaturation of acyl-CoA.* The introduction of an α:β double bond is catalysed by *acyl-CoA dehydrogenases*, obtainable from mitochondria. Several such enzymes are known (p. 172). Between them they act upon sub-

426

strates with fatty chains of 4–16 and possibly more carbon atoms. These enzymes are flavoproteins and require neither NAD nor NADP; instead the so-called 'electron-transporting' flavoprotein (ETF) is required. A similar enzyme acts upon substrates with 3–8 carbon atoms, with maximal activity towards butyrate. This enzyme is accordingly called *butyryl-CoA dehydrogenase.*

The reaction catalysed by this group of enzymes is, in general terms:

$$R.CH_2CH_2CO \sim CoA + ETF \rightleftharpoons R.CH:CH.CO \sim CoA + ETF.H_2.$$

Hydrogen is probably transferred from $ETF.H_2$ directly to cytochrome.

(2) *Hydration of unsaturated acyl-CoA.* This step is catalysed by *enoyl-CoA hydratases.* Enzymes of this kind have been isolated from heart, liver and elsewhere and act upon a wide range of substrates (p. 115). Either an α- or a β-hydroxyacid can be formed:

$$R.CH:CH.CO \sim CoA$$

$$\pm H_2O \qquad \qquad \pm H_2O$$

$$R.CH(OH)CH_2CO \sim CoA \qquad R.CH_2CH(OH)CO \sim CoA$$

Presumably the α-hydroxy-acids must pass back through the unsaturated to the β-hydroxy-compounds to be further metabolized.

(3) *Oxidation of β-hydroxyacyl-CoA* is catalysed by β-*hydroxyacyl-CoA dehydrogenase*, an NAD-specific enzyme (p. 165). Three such enzymes are known, acting upon substrates with 4–12 and probably more carbon atoms:

$$R.CH(OH)CH_2CO \sim CoA + NAD \rightleftharpoons R.CO.CH_2CO \sim CoA + NAD.H_2.$$

(4) *Breakdown of β-keto-derivatives.* We must now enquire how the β-keto-compounds are fragmented. We know that the products are a 2-carbon unit, now known to be acetyl-CoA, and a fatty acid derivative with 2C less than the starting material.

If the fatty acyl-CoA derivative undergoes oxidation in the β-position, the product could undergo fission, perhaps by hydrolysis, to yield acetyl-CoA:

$$HO \vdots H$$

$$---- CO \vdots .CH_2CO \sim CoA \longrightarrow --- COOH \mid CH_3CO \sim CoA$$

This product, as we know, is capable of direct reaction with oxaloacetate to yield citrate.

Now, if the β-oxidative split is hydrolytic in character, this reaction would lead to the dissipation of any energy associated with the $\alpha:\beta$ bond. The suggestion may be made that CoA is again involved, this time in the fission of β-keto-acyl-CoA compounds, perhaps as follows:

$$---- CH_2CO \vdots .CH_2CO \sim CoA + CoA \longrightarrow ---- CH_2CO \sim CoA + CH_3CO \sim CoA.$$

This so-called *thiolysis* of β-keto-acyl-CoA compounds can and does actually take place in the presence of enzymes known as β-*ketothiolases.*

This is an interesting operation since it means that the 'priming energy' originally provided from ATP through the mediation of coenzyme A is recovered in the high-energy $\sim$CoA of the acetyl-CoA unit split off, while the residual fatty chain is left already 'primed' for the next step in its degradation. Such a mechanism leads, moreover, to the eventually complete degradation of naturally occurring, even-numbered chains into 2-carbon units consisting of acetyl-CoA. Finally it has been shown that some of the acetyl-CoA thus produced leaves the mitochondria in the form not of acetyl-CoA as such but largely in the form of citrate. In the ordinary way, presumably, this citrate is formed and held in the mitochondria and enters the oxidative machinery of the citrate cycle.

If now we combine the ideas developed in the last few pages we can arrive at an overall picture of the process of β-oxidation as a whole. Such a picture

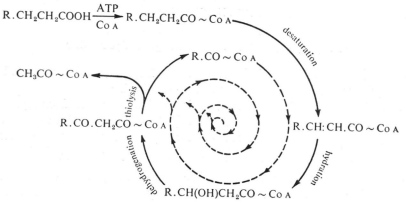

Fig. 99. β-Oxidative breakdown of fatty acids.

is presented in Fig. 99. It is not possible in this case to refer to a 'cycle' because the final product of each sequence of operations is not identical with the starting material but contains 2 carbons less. But the process can be likened to a clock spring. The free fatty acid is converted into its CoA derivative and enters the outermost coil of the spring and then undergoes desaturation, hydration, dehydrogenation and thiolysis under the influence of the enzymes we have already described. Acetyl-CoA is split off from the β-keto-compound and the residual acyl-CoA continues on along the next coil, undergoing the same sequence of reactions, loses acetyl-CoA again, and so on round the rest of the coils. Thus palmitic acid, with 16C, would require a spring of 7 coils and would yield 8 molecules of acetyl-CoA if completely broken down.

All the reactions involved in the β-oxidative process are reversible and it is, in fact, possible to start with acetyl-CoA and synthesize fatty acids by travelling backwards along the spring although, as we shall see, it is unlikely that fatty acids are normally synthesized in this way under biological conditions.

As so often happens, synthesis and breakdown follow different routes and call for the participation of different enzymes.

The acetyl-CoA units formed can, as we know, enter freely into and undergo rapid oxidation by the citric acid cycle and can enter freely into reactions of biological acetylation; they can also contribute to fatty acid synthesis as we shall see presently.

FORMATION OF ACETOACETATE

It has already been emphasized that no theory of fat metabolism stands much chance of survival unless it can account adequately for the formation of ketone bodies, and a great deal of attention has accordingly been paid to the problem of ketone-body formation. The tissue of choice is, of course, the liver since it produces acetoacetate readily but cannot oxidize it. Quantitative determinations of the yields of ketone bodies from a series of fatty acids, have shown that some fatty acids e.g. octanoic (8C) can give rise to more than one molecule of acetoacetate for each molecule of fatty acid metabolized.

According to the theory of β-oxidation, as formulated by Knoop, a C_8 acid would lose two pairs of carbon atoms in the form of presumptive acetate, leaving butyrate which, undergoing β-oxidation in its turn, could be converted into one molecule *and no more* of acetoacetate.

Another way of accounting for the observed high yields of ketone bodies is to invoke the alternative theory of 'multiple alternate oxidation'. According to this scheme the fatty chain was supposed to undergo oxidation to form keto-groups at the β-carbon atom and at every alternate carbon atom along the chain, thus:

$$\cdots\cdots CH_2CH_2CH_2CH_2CH_2\overset{\beta}{C}H_2\overset{\alpha}{C}H_2COOH$$
$$\downarrow$$
$$\cdots\cdots \vdots CH_2CO.CH_2CO \vdots .\overset{\beta}{C}H_2CO.\overset{\alpha}{C}H_2COOH$$

The chain is supposed then to split into 4-carbon fragments, each fragment coming away in the form of a molecule of acetoacetate. From every molecule of octanoate, therefore, we should expect to obtain two molecules of acetoacetate and this, of course, is what is actually observed. But the only evidence in favour of multiple alternate oxidation seems to consist of precisely those facts which it sets out to explain.

Clearly it is also important to consider the possibility that acetoacetate may arise primarily from acetyl-CoA. Many experiments have been carried out in order to elucidate the fate of acetate itself when added to various tissue preparations. Feeding, perfusion, tissue-slice, brei and isotope experiments have all been used and it has been established beyond all reasonable doubt that acetate is at any rate partially convertible into acetoacetate by liver tissue. That part which is not so converted disappears and is presumably oxidized.

Convincing evidence for the formation of acetoacetate from acetate has been obtained by some elegant experiments involving the use of isotopic carbon. If isotopically labelled acetate is incubated with liver tissue and the resulting acetoacetate is isolated, it is found that the isotopic carbon has been transferred to the acetoacetate. The behaviour of octanoate in particular was studied by this method. Octanoate was prepared with ^{14}C in the carboxyl group, the product was incubated with rat liver slices, and the acetoacetate formed was isolated and analysed for ^{14}C. Now, according to the classical theory of β-oxidation, acetoacetate would be formed only from the butyrate left after two 2-carbon units had been removed, and should therefore contain no ^{14}C (see Fig. 100). According to the theory of multiple alternate oxidation, the 8-carbon chain would give rise to two molecules of acetoacetate, of which only one would contain ^{14}C, located in the carboxyl group. If, however, pairs of carbon atoms were removed in the form of acetate or some highly reactive derivative thereof, followed by random condensation in pairs, ^{14}C should be present both in the carboxyl and the ketonic groups of the product, but not in the methyl or $-CH_2-$ residues as appears from Fig. 100.

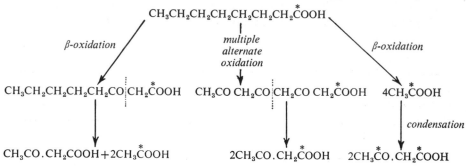

Fig. 100. Formation of acetoacetate from octanoate.

It was found that, in fact, isotopic carbon is present in the carboxylic and ketonic but not in the other two groups, indicating that the acetoacetate formed must have been produced from 2-carbon units split off from the molecules of octanoate by β-oxidation.

The formation of ketone bodies from free acetate has, of course, been repeatedly demonstrated in liver preparations of various kinds but is, nevertheless, an endergonic process. We know that acetyl-CoA is a high-energy compound and at the same time a natural acetylating agent, admirably qualified to serve as a source of acetoacetate:

$$CH_3CO\sim CoA + CH_3CO\sim CoA \longrightarrow CH_3CO.CH_2CO\sim CoA + CoA.$$

Several mechanisms exist for removal of the terminal $\sim CoA$ group from acetoacetyl-CoA; some molecules at least can transfer it to succinate, forming succinyl $\sim CoA$:

$$\text{acetoacetyl} \sim CoA + \text{succinate} \overset{GDP}{\rightleftharpoons} \text{acetoacetate} + \text{succinyl} \sim CoA.$$

430

This transferring enzyme is absent from liver however. According to more recent work, removal of the CoA can proceed through a more complex mechanism which can be outlined as follows and is catalysed by a specific hydrolase;

$$CH_3COCH_2CO.CoA + CH_3CO.CoA \xrightarrow{-CoA} \begin{array}{c} CH_2COOH \\ | \\ CH_3C(OH) \\ | \\ CH_2CO.CoA \end{array}$$

$$\longrightarrow CH_3CO.CH_2COOH + CH_3CO.CoA.$$

The intermediate product, β-hydroxy-β-methylglutaryl-CoA (HMG) is important in other connexions. In particular it lies on synthetic pathways leading to the synthesis of cholesterol (via mevalonic acid and squalene) and of terpenes, but these operations are too complex for discussion here.

OXIDATION OF ACETOACETATE AND β-HYDROXYBUTYRATE

Much attention has been given to the problem of the oxidative metabolism of acetoacetate and β-hydroxybutyrate, and it has long been known that they are interconvertible through the action of β-hydroxybutyrate dehydrogenase together with NAD, both of which are widely distributed in animal tissues.

Many tissues, but not liver, are able to oxidize added acetoacetate and β-hydroxybutyrate. Other compounds examined include the following:

Butyrate	$CH_3CH_2CH_2COOH$
Crotonate	$CH_3CH=CHCOOH$
β-Hydroxybutyrate	$CH_3CH(OH)CH_2COOH$
γ-Hydroxybutyrate	$CH_2(OH)CH_2CH_2COOH$
α-γ-Dihydroxybutyrate	$CH_2(OH)CH_2CH(OH)COOH$
Vinylacetate	$CH_2=CH.CH_2COOH$

All these substances are rapidly oxidized to carbon dioxide and in every case the oxidation is inhibited by malonate, which suggests that the oxidation of these substances, like that of the fatty acids themselves, must take place by way of the citrate cycle.

Now if isotopically carboxyl-labelled acetoacetate is incubated with oxaloacetate in muscle homogenates, correspondingly labelled citrate can be formed, and similarly labelled derivatives can be obtained by the use of the usual inhibitors of the cycle. In this respect acetoacetate behaves like other β-keto-acids, the two terminal carbon atoms being split off and incorporated into citrate.

Presumably, therefore, free acetoacetate first requires conversion into acetoacetyl-CoA, which, being a β-keto-compound could presumably undergo thiolysis in the usual way:

$$CH_3CO.CH_2CO\sim CoA + CoA \rightleftharpoons CH_3CO\sim CoA + CH_3CO\sim CoA.$$

This initial formation of acetoacetyl-CoA might take place through reactions with CoA and ATP similar to those involved in the formation of an acyl-CoA from a free fatty acid (p. 425), but if so the necessary enzymes(s) must presumably be lacking from the liver. To some extent, however, acetoacetate and some higher β-keto-acids can react with succinyl-CoA:

$$\text{acetoacetate} + \text{succinyl} \sim \text{CoA} \rightleftharpoons \text{acetoacetyl} \sim \text{CoA} + \text{succinate,}$$

but the necessary transferase is again absent from liver. Probably, however, the most important mechanism in the extrahepatic tissues depends upon acetoacetyl-CoA being split (by β-keto-thiolase) into two molecules of acetyl-CoA.

Since the oxidation of acetoacetate can be accounted for, it follows that the oxidation of substances which, like butyrate, crotonate and β-hydroxybutyrate, are known to give rise to acetoacetate, can also be accounted for.

If we accept the scheme just proposed to account for the formation of acetoacetate, together with that proposed for the intermediate stages of the β-oxidation of the CoA derivatives of fatty acids (Fig. 99, p. 428), we have a hypothesis that can account for the complete oxidative degradation of fatty acids by way of the citrate cycle, for the formation of ketone bodies by way of acetoacetate, and for the oxidative breakdown of any substance that can give rise either to fatty acids or to acetoacetate or to β-hydroxybutyrate. These are the salient features that have to be taken into account in any theory of fat metabolism and in all these operations acyl-CoA compounds are intimately concerned.

Evidence is accumulating meanwhile that the oxidative degradation of the fatty acids is coupled up to the generation of new high-energy phosphate and the synthesis thereby of ATP, probably at each of the dehydrogenations in β-oxidation, and certainly in the subsequent oxidative metabolism of the 2-carbon, acetyl-CoA units to which each cycle of events gives rise.

SYNTHESIS OF FATTY ACIDS

Long before the discovery of coenzyme A it was known that fat can be synthesized from fatty metabolites such as acetate and acetoacetate, and also from carbohydrate and protein sources. The naturally occurring fatty acids contain an even number of carbon atoms, practically without exception, and it follows that the starting materials for their synthesis must probably also contain even numbers of carbon atoms. It has from time to time been suggested that the raw material for the synthesis might be acetoacetate, with 4 carbon atoms, but in this case it would be anticipated that fatty acids with 4, 8, 12, 16 and 20 carbon atoms would predominate in nature. In fact, however, the 14-, 16- and 18-carbon acids are the most common, and all occur in similar proportions. It therefore seems probable that fatty acids must be synthesized from 2-carbon units.

Recent developments show conclusively that the starting material is acetyl-

CoA. The latter is known to be derivable from pyruvate by oxidative decarb-oxylation and, in addition, is the predominant product of the β-oxidative fission of fatty acid chains. It can also arise from free acetate by reacting with CoA and ATP (p. 425). Work carried out with the aid of isotopes bears out the truth of the supposition that fatty acids can be synthesized from 2-carbon units in the form of acetate, though without giving any indication of the intermediate processes. Acetate containing deuterium in the methyl group and ^{14}C in the carboxyl group, was fed to rats for 8 days. The liver fats were then collected and analysed, with the result that deuterium and ^{14}C were found in alternate —CH_2— residues throughout the chain, while ^{14}C also appeared in the terminal carboxyl group.

Fatty acid oxidation is localized in the mitochondria, but fatty acid syn-thesis has been demonstrated not only in mitochondrial preparations but also in particle-free extracts of that remarkable fat-producing organ, the active mammary gland. The routes and mechanisms of the synthesis in the two environments however are very different.

Synthesis in mitochondria

It has been shown that all the enzymes concerned with β-oxidation can act reversibly and that fatty acids can be synthesized from acetyl-CoA by revers-ing the processes of β-oxidation but this happens only in the mitochondria and is only a minor synthetic pathway. ATP is required as an energy source and synthesis is favoured by a high concentration of $NAD.H_2$ and also by removal of the fatty acids produced. This can be achieved by their conversion into neutral fats or into phospholipids (p. 439).

It does appear, however, that the $\alpha:\beta$-desaturating enzyme, which normally uses ETF as its hydrogen acceptor, acts as a bottle-neck on the route back to fatty acids. This synthetic pathway has been found to proceed much more rapidly if $NADP.H_2$ as well as $NAD.H_2$ is supplied to the system and there is, in fact, a mitochondrial enzyme which can be extracted, catalyses the saturation of $\alpha:\beta$-unsaturated acyl-CoA compounds and has an absolute re-quirement for $NADP.H_2$. This, however, is a minor pathway for fatty acid synthesis; synthesis on the large scale is localized in the particle-free cyto-plasm.

Synthesis in non-mitochondrial systems

If liver tissue is homogenized and centrifuged so as to remove the cell debris, nuclei, mitochondria and microsomes, the residual fraction contains soluble enzymes that will catalyse a rapid synthesis of fatty acids from acetyl-CoA. ATP, $NADP.H_2$ and Mg or Mn ions are required. *The system is absolutely dependent on CO_2 (or bicarbonate).*

The first-formed product was isolated by Wakil and identified as *malonyl-CoA*. This accounts for the CO_2-requirement:

$(a) \quad CH_3C.Co\,A + CO_2 \xrightarrow[Mn^{2+}]{ATP} \begin{array}{c} COOH \\ / \\ CH_2 \\ \backslash \\ CO.Co\,A \end{array}$

malonyl-Co A

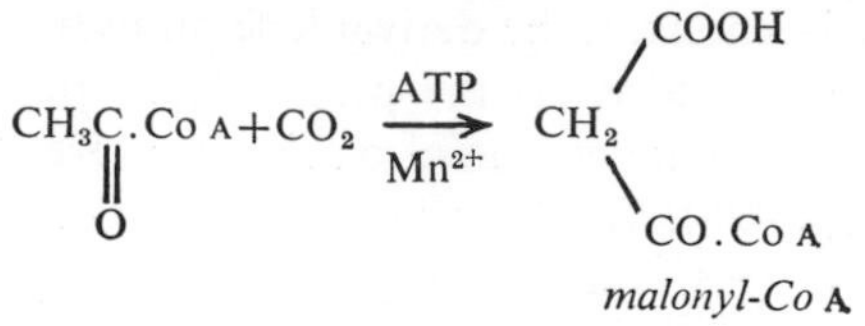

Later work has shown that *biotin* is intimately involved, together with ATP, in this synthesis, apparently as a tightly bound prosthetic group of the enzyme concerned. This prosthetic group binds and activates the CO_2 used in the reaction (see p. 118).

Once formed, malonyl-Co A can now react with a second molecule of acetyl-Co A and the product is decarboxylated:

$(b) \quad CH_3CO.Co\,A + \begin{array}{c} COOH \\ / \\ CH_2 \\ \backslash \\ CO.Co\,A \end{array} \longrightarrow \begin{array}{c} COOH \\ / \\ CH_3CO.CH \\ \backslash \\ CO.Co\,A \end{array} \longrightarrow CO_2 + CH_3COCH_2CO.Co\,A$

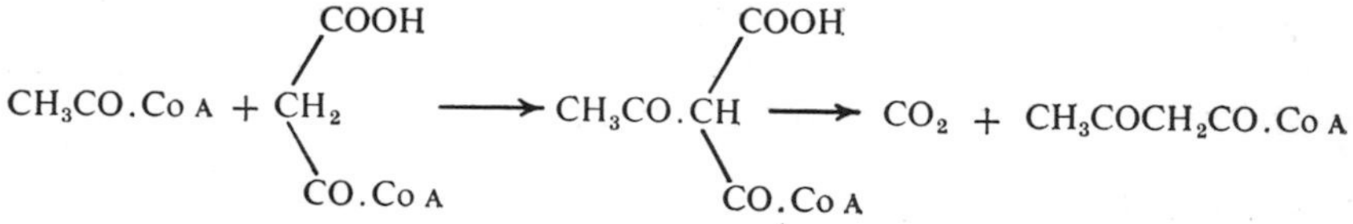

It seems likely that the two partial reactions in (b) take place simultaneously and are catalysed by a single enzyme protein. Enzymes have been obtained in a more or less pure state which will (a) catalyse the formation of malonyl-Co A from acetyl-Co A (malonyl-Co A carboxylyase) and (b) catalyse the reaction between acetyl-Co A and malonyl-Co A.

The part played by the CO_2 is essentially a catalytic one for if it is supplied in the form of $^{14}CO_2$, the isotope fails to appear in any of the products. It has the effect of activating the central —CH_2— group of malonyl-Co A, which can then react with a second molecule of acetyl-Co A and the CO_2, now no longer required, drops off, leaving acetoacetyl-Co A.

Now, if liver extracts or extracts of mammary gland are incubated with ATP, NADP.H_2, Mn^{2+}, together with acetyl-Co A in the presence of CO_2, large amounts of palmityl-Co A (16C) are formed. Until recently the formation of palmitate and other higher fatty acids found no explanation, but a turning point came with the discovery in bacterial systems of a new and highly specific protein which is involved in the synthesis of these higher fatty acids. This, the so-called *acyl-carrier protein* (ACP) can replace Co A by exchange reactions, e.g.;

acetyl-Co A + ACP $\rightleftharpoons$ acetyl-ACP + Co A
malonyl-Co A + ACP $\rightleftharpoons$ malonyl-ACP + Co A

Interestingly enough, analysis of ACP has shown that, like Co A itself, this protein contains β-alanine, β-mercaptoethanolamine and pantothenic acid, attached to a phosphoserine residue of the protein itself (Fig. 101). This should be compared with the structure of coenzyme A. *The functional regions of both are identical.* Like coenzyme A, ACP reacts with acyl residues by way of its SH group and it will be convenient if in order to emphasize the comparison in this particular context, we write coenzyme A as CoA.SH and the acyl carrier protein as ACP.SH.

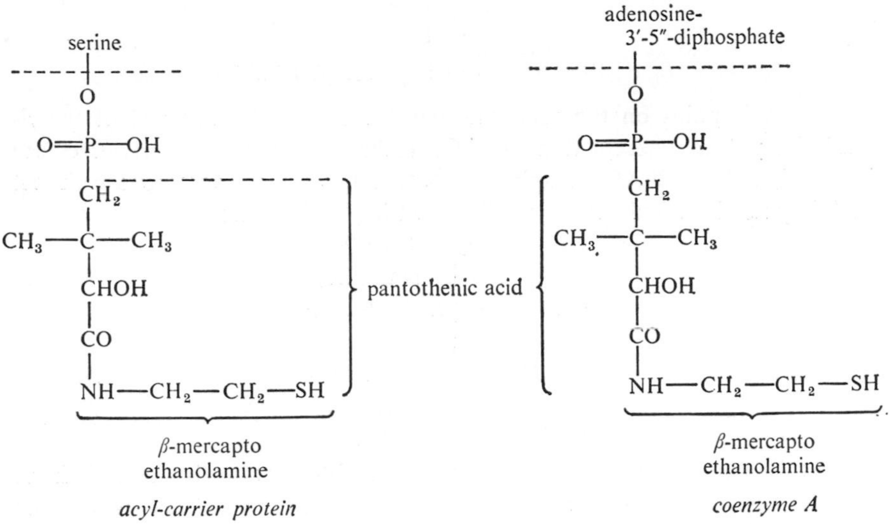

Fig. 101. Resemblances between ACP and coenzyme A.

Fatty acid synthesis in the presence of ACP starts with acetyl-S.CoA, which undergoes a transfer reaction with ACP giving acetyl-S.ACP (reaction (i)). Malonyl-CoA undergoes a similar reaction and gives malonyl-S.ACP (ii);

(i)

(ii)

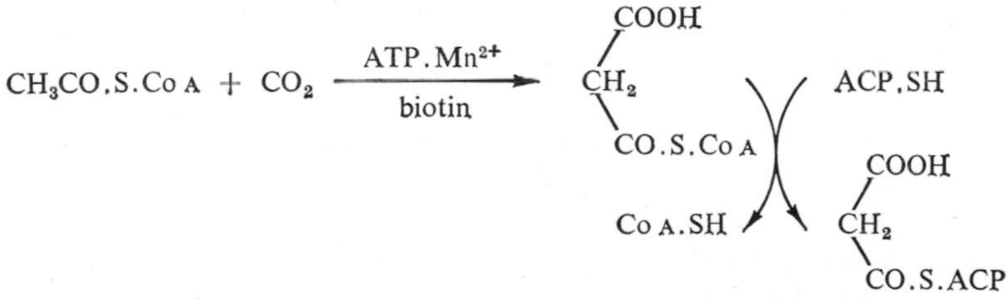

The product now reacts with a second molecule of acetyl-S.ACP, undergoes decarboxylation and yields acetoacetyl-S.ACP (iii):

$$(iii) \quad CH_3CO.S.ACP + \underset{\underset{CO.S.ACP}{\big\backslash}}{\overset{\overset{COOH}{\big/}}{CH_2}} \xrightarrow{-HS.ACP} \underset{\underset{CO.S.ACP}{\big|}}{\overset{\overset{COOH}{\big|}}{CH_3CO.CH}}$$

$$\xrightarrow{-CO_2} CH_3CO.CH_2CO.S.ACP$$

From this point on the reactions parallel those of a reversal of the β-oxidative pathway except that the ACP rather than the CoA derivatives are involved, and that $NADP.H_2$ replaces $NAD.H_2$, while the enzymes are almost certainly not identical with those involved in β-oxidation:

$$CH_3CO.CH_2CO.S.ACP$$
$$+2H \quad \big| \quad NADP.H_2$$
$$CH_3CH(OH)CH_2CO.S.ACP$$
$$\big| \quad -H_2O$$
$$CH_3CH{=}CH_2CO.S.ACP$$
$$+2H \quad \big| \quad NADP.H_2$$
$$CH_3CH_2CH_2CO.S.ACP$$

In this way 1 molecule equivalent of ACP-butyrate is formed from 2 molecule equivalents of acetate. But the matter does not end there. The butyryl-S.ACP complex can go on to react with another molecule of malonyl-S.ACP and the product is decarboxylated in the course of the reaction;

$$CH_3CH_2CH_2CO.S.ACP + \underset{\underset{CO.S.ACP}{\big|}}{\overset{\overset{COOH}{\big|}}{CH_2}} \longrightarrow \underset{\underset{CH_3CH_2CH_2CO.CH_2CO.S.ACP}{\big\downarrow \; -CO_2}}{\overset{\overset{COOH}{\big|}}{CH_3CH_2CH_2CO.CHCO.S.ACP}}$$

The resulting β-keto-derivative is then hydrogenated, dehydrated, and hydrogenated a second time, just as in the case of acetoacetyl-S.ACP, to give the —S.ACP derivative of hexanoic acid.

This sequence of reactions can be repeated a number of times with the addition of 2 carbons from malonyl-S.ACP at each operation until long-chain acids are formed. There is here a beautiful example of the way in which the biosynthesis and breakdown of substances of high biological significance take place by entirely different routes, catalysed by entirely different enzymes. The principal product seems to be the palmityl derivative (C_{16}) but myristic (C_{14}) and stearic (C_{18}) are probably formed as well. It is worth while to remember that acetyl-S.CoA contributes only the terminal CH_3CH_2— *directly*: the pairs of —CH_2— units added in the subsequent reactions also come from acetyl-S.CoA *but always and only by way of malonyl*-S.CoA, followed by malonyl-S.ACP, in that order.

The nature of the reactions leading from acetoacetyl-S.CoA to the long-chain fatty acids remained obscure for a long time. No intermediates were detectable at first, presumably because they are formed and transformed rapidly, but only in small amounts and firmly bound to this enzyme-like protein, ACP. Consequently it was not possible to demonstrate the nature of the intermediates until ACP itself was discovered, concentrated and considerably purified.

The part played by biotin in this cytoplasmic synthesis of fatty acids has been mentioned; every molecule of acetyl-CoA converted to the malonyl derivative calls for a biotin-activated molecule of CO_2. Biotin itself is activated by ATP in the first place. It is interesting to recall that avidin, a constituent of raw egg white, will react specifically with biotin to form a stable but inert complex; avidin in fact, completely blocks fatty acid synthesis by the complex of cytoplasmic enzymes we have just been considering.

It is worth while to notice in passing the heavy insistence on $NADP.H_2$ in fatty acid synthesis, whether in mitochondria or in the cytoplasm. The amounts of $NADP.H_2$ generated in other metabolic operations are small in comparison with those of $NAD.H_2$ in the ordinary way. However, the lactating mammary gland, a site of intense lipogenic activity, secures the $NADP.H_2$ it requires by diverting virtually all of its normal, resting carbohydrate metabolism from the usual glycolysis + oxidation pathway, which generates much $NAD.H_2$ and little $NADP.H_2$, to the so-called 'shunt' or pentose pathway (p. 393), which generates much $NADP.H_2$ and little $NAD.H_2$. Here is a fascinating metabolic adaptation towards a specific physiological function.

A further source of $NADPH_2$ which may well be of considerable importance for the synthesis of fatty acids depends upon a cyclical system of the kind shown in Fig. 102. Reactions (1) and (2) are catalysed by pyruvate

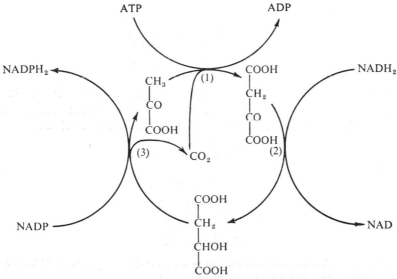

Fig. 102. Generation of $NADP.H_2$ from $NAD.H_2$.

carboxylyase and malate dehydrogenase respectively. The final step, reaction (3) is catalysed by the 'malate enzyme' described on p. 117 and to which no particular function has hitherto been ascribed. The overall equation can be re-expressed in the following manner:

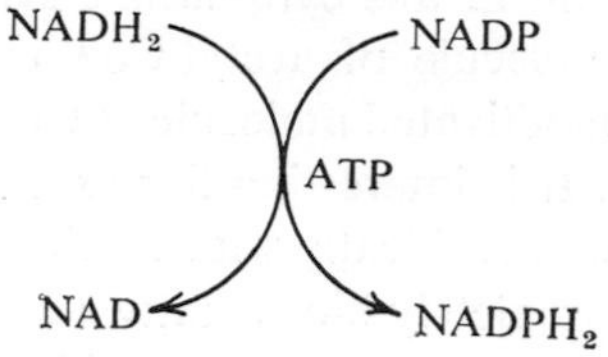

$$
\begin{array}{ccc}
\mathrm{NADH_2} \diagdown & \diagup & \mathrm{NADP} \\
 & \mathrm{ATP} & \\
\mathrm{NAD} \diagup & \diagdown & \mathrm{NADPH_2}
\end{array}
$$

$\mathrm{NAD.H_2}$, formed by the participation of NAD in many and diverse dehydrogenase reactions, becomes the source of $\mathrm{NADP.H_2}$, and may therefore well be important in fatty acid synthesis. ATP and the 'malic enzyme' play key roles in this interesting system.

SYNTHESIS OF GLYCERIDES

Triglycerides are formed by reactions between acyl-CoA compounds and α-glycerolphosphate, which is itself formed by a specific glycerolkinase:

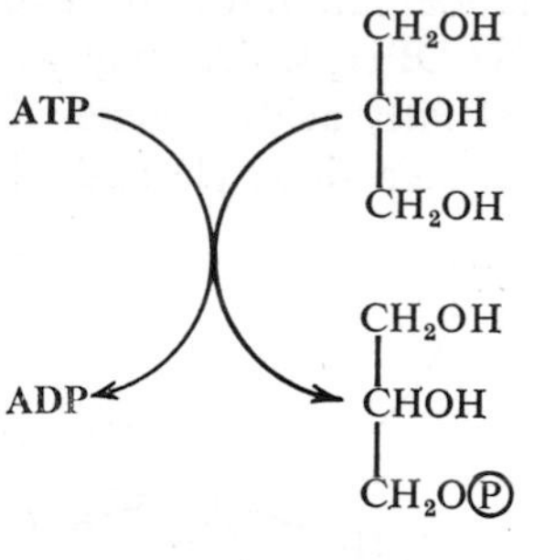

$$
\begin{array}{ccc}
 & & \mathrm{CH_2OH} \\
 & & | \\
\mathrm{ATP} \diagdown & \diagup & \mathrm{CHOH} \\
 & & | \\
 & & \mathrm{CH_2OH} \\
 & & \\
 & & \mathrm{CH_2OH} \\
 & & | \\
\mathrm{ADP} \diagup & \diagdown & \mathrm{CHOH} \\
 & & | \\
 & & \mathrm{CH_2O}\textcircled{P}
\end{array}
$$

Further enzymes are known (*glycerolphosphate acyl-transferases, monoglyceride acyl-transferases*) which will catalyse the formation of mono- and di-phosphatidic acids at the expense of fatty acyl CoA derivatives:

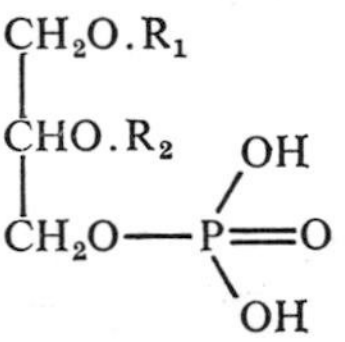

$$
\begin{array}{l}
\mathrm{CH_2O.R_1} \\
| \\
\mathrm{CHO.R_2} \quad \mathrm{OH} \\
| \qquad\quad \diagup \\
\mathrm{CH_2O\!-\!\!P\!=\!\!O} \\
\qquad\qquad \diagdown \\
\qquad\qquad \mathrm{OH}
\end{array}
$$

phosphatidic acids

e.g.   (i) stearyl $\sim$ CoA $+\alpha$-glycerolphosphate $\rightarrow$ monostearyl-α-phosphatidic acid $+$ CoA,

   (ii) oleyl $\sim$ CoA $+$ monostearylphosphatidic acid $\rightarrow$

α-stearyl-β-oleyl-phosphatidic acid $+$ CoA.

These enzymes seem to be fairly specific for chains of 16, 17 and 18 carbon atoms, which may be saturated or unsaturated.

The next stage involves removal of the phosphate group by a specific phosphatase and its replacement by a third fatty acyl residue, for example:

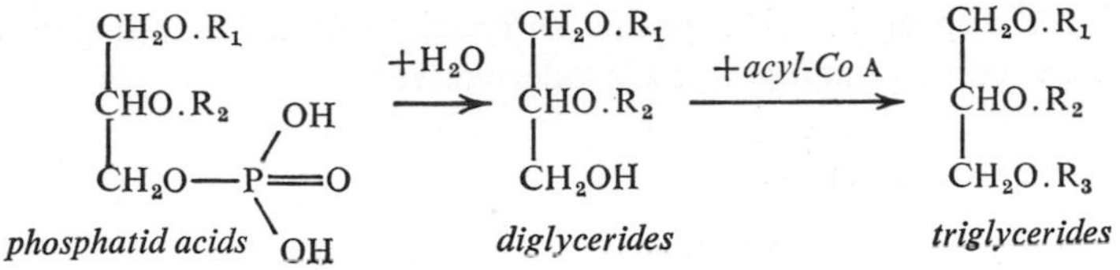

$$
\begin{array}{ccccc}
\begin{array}{l}
CH_2O.R_1 \\
| \\
CHO.R_2 \quad OH \\
| \quad\quad / \\
CH_2O\!-\!P\!=\!\!O \\
\end{array}
&
\xrightarrow{+H_2O}
&
\begin{array}{l}
CH_2O.R_1 \\
| \\
CHO.R_2 \\
| \\
CH_2OH \\
\end{array}
&
\xrightarrow{+acyl\text{-}Co\,A}
&
\begin{array}{l}
CH_2O.R_1 \\
| \\
CHO.R_2 \\
| \\
CH_2O.R_3 \\
\end{array}
\\
\textit{phosphatid acids} \;\; \text{OH} & & \textit{diglycerides} & & \textit{triglycerides}
\end{array}
$$

The enzyme concerned is a *diglyceride acyltransferase*.

FORMATION OF PHOSPHOLIPIDS

Alternatively the diacylphosphatidic acid, having lost its phosphate residue as before, can react with *cytidine diphosphate choline*, which transfers choline phosphate to the diglyceride to yield a phospholipid:

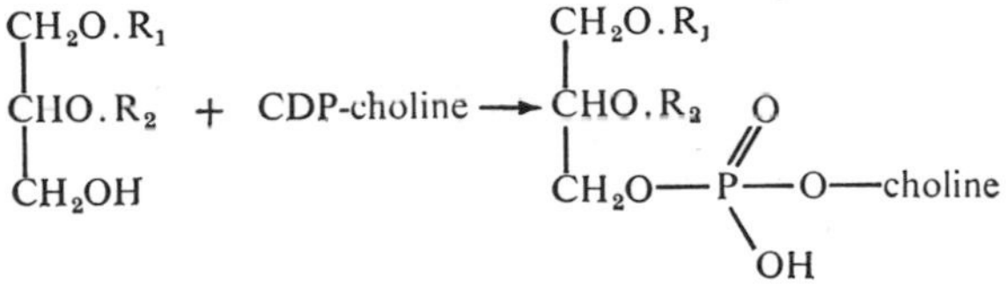

$$
\begin{array}{l}
CH_2O.R_1 \\
| \\
CHO.R_2 \quad + \quad CDP\text{-choline} \\
| \\
CH_2OH \\
\end{array}
\longrightarrow
\begin{array}{l}
CH_2O.R_1 \\
| \\
CHO.R_2 \quad O \\
| \quad\quad\quad \| \\
CH_2O\!-\!P\!-\!O\!-\!\text{choline} \\
\quad\quad\quad\;\; \backslash \\
\quad\quad\quad\;\;\; OH
\end{array}
$$

A similar reaction can take place in which *CDP-ethanolamine* replaces the choline derivative. Both CDP compounds arise by similar reactions:

CTP + choline phosphate $\rightleftharpoons$ CDP-choline + pyrophosphate,
CTP + ethanolamine phosphate $\rightleftharpoons$ CDP-ethanolamine + pyrophosphate.

The identities of R_1 and R_2 in phospholipids formed in the way we have just discussed is very much a matter of random chance but additional enzymes are known which can remodel the products and give rise to more specific groups of tissue phospholipids. These have the following general structure:

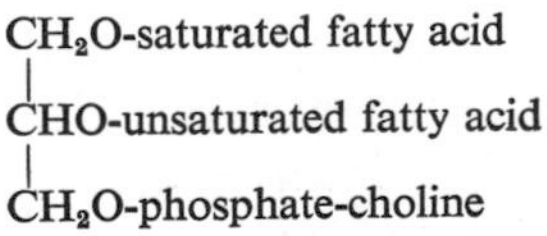

$$
\begin{array}{l}
CH_2O\text{-saturated fatty acid} \\
| \\
CHO\text{-unsaturated fatty acid} \\
| \\
CH_2O\text{-phosphate-choline}
\end{array}
$$

The commonest phospholipids ('lecithins' and 'kephalins') are those containing choline and ethanolamine residues respectively, but there are others in which these bases are replaced by residues of serine or, sometimes, inositol.

439

Phospholipases are known which can remove the non-specifically sited fatty acid chains from phospholipids formed by the 'random' synthetic pathway we have just discussed and produce products known as *lysolecithins*, i.e. lecithins lacking either the appropriate saturated or unsaturated fatty acid.

A further group of highly specific *acyl-CoA transferases* then catalyses the introduction of appropriate, i.e. specific, fatty acids into the vacant sites at the expense of the corresponding acyl-CoA derivatives;

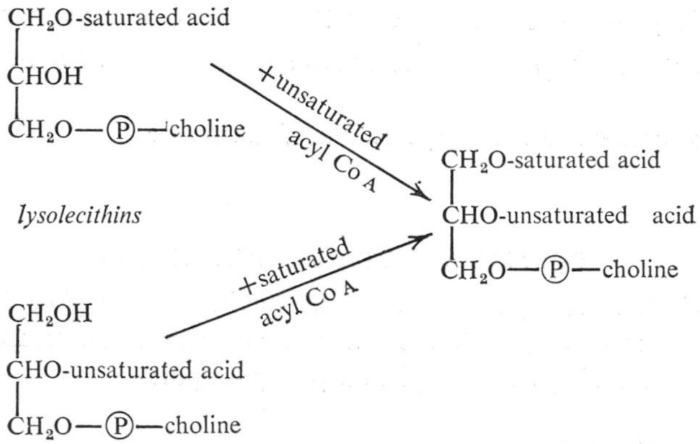

CH_2O-saturated acid
$|$
$CHOH$
$|$
CH_2O—$\textcircled{P}$—choline

+unsaturated acyl CoA

CH_2O-saturated acid
$|$
CHO-unsaturated   acid
$|$
CH_2O—$\textcircled{P}$—choline

lysolecithins

+saturated acyl CoA

CH_2OH
$|$
CHO-unsaturated acid
$|$
CH_2O—$\textcircled{P}$—choline

Probably similar 'remodelling enzymes' are concerned in the formation of other groups of glycerophospholipids.

INTERCONVERSION OF FAT AND CARBOHYDRATE

While there can be no doubt whatever that fats can be synthesized from carbohydrate sources, the reverse process takes place to a very limited extent only and, in fact, it has been denied that the conversion of fat into carbohydrate ever takes place at all. Inasmuch as they yield *glycerol* on hydrolysis, fats certainly are potential sources of glycogen, for glycerol is quantitatively converted into glucose when administered to diabetic or phlorrhizinized animals. Isotopically labelled glycerol likewise gives rise to correspondingly labelled glycogen in starving animals. Apart from glycerol, fats are not known to give rise to anything but acetyl-CoA and the free acetate and ketone bodies which can be derived from it.

One of the fatty acids produced by the bacterial degradation of cellulose in herbivorous animals does, however, give rise to carbohydrate material, namely *propionate*. The reactions involved are somewhat elaborate but begin with the formation of propionyl-CoA:

$$CH_3CH_2COOH \xrightarrow[\text{CoA}]{\text{ATP}} CH_3CH_2CO.CoA$$

This is followed by an addition reaction with CO_2, in which ATP and biotin are involved, giving methyl malonyl-CoA as an intermediate, and then by an intramolecular rearrangement in which vitamin B_{12} (cobalamine) is in some way concerned;

$$CH_3CH_2CO.CoA + CO_2 \xrightarrow[\text{biotin}]{\text{ATP}} CH_3CH.CO.CoA \xrightarrow{B_{12}} CH_2.CH_2CO.CoA$$
$$\qquad\qquad\qquad\qquad\qquad | \qquad\qquad\qquad\qquad |$$
$$\qquad\qquad\qquad\qquad\qquad COOH \qquad\qquad\qquad COOH$$

The product, succinyl-CoA, is of course an intermediate on the citrate cycle and can follow the usual route, succinate $\rightarrow$ fumarate $\rightarrow$ malate $\rightarrow$ oxalo-acetate $\rightarrow$ pyruvate $\cdots\rightarrow$ glycogen.

Experiments have been performed in which the incubation of sliced rat liver with butyrate gave small but apparently significant increases in the glycogen content of the tissue. But, as we now know, glycogen and fat alike are oxidized through a common metabolite in the form of acetyl-CoA and it may well be that the utilization of butyrate merely spares the oxidation of glycogen by giving rise to this common metabolite.

Since acetyl-CoA can be formed by the oxidative decarboxylation of pyruvate and can itself give rise to the synthesis of fat, there is not much doubt that this is an important route for the production of fat from carbohydrate sources. There is, however, no evidence that the stage of oxidative decarboxylation can be reversed, at any rate in animal tissues, so that, while pyruvate freely gives rise to acetyl-CoA, the latter is not convertible back into pyruvate. It has been shown that if lactate or pyruvate containing isotopic carbon is administered to animals, the isotope can be largely recovered in the liver glycogen. If acetate similarly 'labelled' is used instead, however, only a trivial transference of isotope to the liver glycogen can be detected. It may therefore reasonably be assumed that the conversion of acetate to pyruvate takes place to a very limited extent at most. In so far as labelled acetate carbon finds its way into glycogen it probably does so by incorporation into the constituents of the citrate cycle, going as far as oxaloacetate which, it will be recalled, is in near equilibrium with CO_2 and pyruvate. The latter, containing some labelled carbon, can revert to give labelled glycogen, but this does not seem to take place on any great scale.

In short, there is no convincing evidence for any large-scale production of carbohydrate from fatty sources and, in so far as such a conversion takes place at all, it must probably be attributed in the main if not entirely to the formation of glycerol as a product of fat katabolism and to the bacterial formation of propionate in herbivorous animals.

INDEX

Abalone, taurine in, 103
Abomasum, rennin from, 68
Absorption of products of digestion, (carbohydrate) 213–14, (fat) 216–18, (protein) 211
Absorption spectra, of catalase, and catalase
compound with azide, 22
— of CO– compounds: of *Atmungsferment*,
142; of chlorocruorin, 143; of cytochrome
oxidase, 148–9
— of cytochromes, 144, 145
— of FAD, 140
— of NAD and NADP, 154, 158, 175
— of peroxidase compound with H_2O_2, 22
Acetaldehyde, acetate from, 124, 349
— Cannizaro reaction of, 349
— and deoxyribose, 292
— from ethanol, 124, 156
— from pyruvate in fermentation, 35, 116, 335,
340–1, 387, 391
— from threonine, 243
Acetate kinase, 104
Acetic acid, from acetaldehyde, 124, 349
— acetoacetate from, 384, 421, 422, 429–30
— acetyl-Co A from, 104, 385, 402, 403, 422,
425
— acetylphosphate from, 104
— active, 383; identity of, with acetyl-Co A,
384–5, 403, 420–1
— from cellulose, 74, 417
— fat from, 417, 421, 433
— in fermentation, 349
— oxidation of, 402
— from pyruvate, 116, 382–3, 391
Acetic acid bacteria, cytochrome oxidase from,
148
Acetoacetate decarboxylase, 421
Acetoacetic acid, 429–31
— from acetate, 384, 421, 422, 429–30
— acetone from, 192, 421
— from amino-acids, 192, 255
— from fatty acids, 192, 421, 424, 428, 429–30
— interconversion of β-hydroxybutyrate and,
162, 421, 431
— oxidation of, 422, 431
— *see also* Ketone bodies
Acetoacetyl-coenzyme A, 254, 384, 430, 431
Acetokinase, *see* Acetate kinase
Acetone, from acetoacetate, 192, 421
— precipitation of enzymes with, 198
— *see also* Ketone bodies
Acetone powders, 198
Acetyl-coenzyme A, from acetate, 104, 385, 402,
403, 422, 425
— acetylation by, 103–5, 384, 421
— carboxylation of, to malonyl-Co A, 119, 434
— in citric acid cycle, 386, 405–8
— in dicarboxylic acid cycle, 413
— in fat metabolism, 409, 421–32

— in fatty acid synthesis, 105, 432–8
— in glyoxylic acid cycle, 412
— identity of, with active acetate, 384–5, 403,
420–1
— from metabolism of leucine, 249; of lysine,
254
— from pyruvate, by oxidative decarboxylation, 384–90, 391, 402, 404, 405, 407, 409,
411, 441
Acetyl-Co A carboxylase, 119
Acetyl transferase, 385
Acetylation, 103–5, 234, 384–5, 421
Acetylcholine, from choline, 104, 384
— hydrolysis of, 80
N-**Acetylgalactosamine,** 329
N-**Acetylglucosamine,** 75, 329
N-**Acetylglutamate,** in ornithine cycle, 103, 252,
275, 276
Acetylphosphate, from acetate, 104
— from acetyl-Co A, 104
— high-energy phosphate in, 53, 54
— from pyruvate, 383, 386
N-**Acetylsulphanilamide,** 384, 385
Acid phosphatases, 80
Aconitate hydratase (aconitase), 113, 120
— in citric acid cycle, 397, 399, 405, 407, 410
— inhibited by fluorocitrate, 113, 406
— specificity of, 10, 113
cis-**Aconitic acid,** 113, 120
— in citric acid cycle, 397, 399, 405, 406
trans-**Aconitic acid,** inhibits aconitate hydratase, 113
ACP, *see* Acyl carrier protein
ACTH, *see* Adrenocorticotrophic hormone
Actin, in muscle, 80
Activation, of amino-acids, 296, 315–16
— of enzymes, 32–7
— of substrates, 30–2
Actomyosin, in muscle, 352, 357
Acyl carrier protein (ACP), 434, 435
Acyl-Co A compounds, in fat metabolism, 425,
426, 432, 438
— in oxidative decarboxylation, 390
Acyl-Co A dehydrogenases, 159, 161, 172, 426
Acyl-Co A synthetases, 425, 426
Acyl-Co A transferases, 440
Acyl esterases, 78
Adenase, *see* Adenine aminohydrolase
Adenine, 83, 285, 286
— in nucleic acids, 304, 307, 308
Adenine aminohydrolase (adenase), 83, 286, 287
Adenine-*N*-riboside, 293
Adenosine, formation of, 286
— phosphorylation of, 89, 293
Adenosine diphosphate, 51, 286, 294
— high-energy phosphate in, 52
— *see also in references to* Adenosine tri-
phosphate

442

Cadaverine, bacterial formation of, 115, 254
— oxidation of, 138
Caffein, 302
Calcium carbonate, dissociation of, 21
Calcium ions, in fermentation, 336
Calcium phosphate, in bone formation, 81
Calorific value, of foodstuffs, 47, 417
Camel, fat metabolism of, 418
Cannizaro reaction, in fermentation, 349
Carbamoyl aspartic acid, 103, 250
— in pyrimidine synthesis, 250, 253, 276, 290
Carbamoyl phosphate, 103, 250, 253, 275
Carbamoyl phosphate synthetase (ligase), 275
Carbohydrases, 17, 39, 70–8
Carbohydrates
— amino-acids from, 223
— calorific value of, 417
— in diabetes, 192
— digestion and absorption of, 211–14
— fats from, 375, 417, 422, 440–1
— metabolism of: aerobic, 372–94, 395–415; anaerobic, (fermentation) 331–50, (in muscle) 351–71
— metabolic water from, 418
— production of, in plants, 320–30
— R.Q. of oxidation of, 395n
^{13}Carbon (radioactive), 200
^{14}Carbon (radioactive), 200
— in study: of carbohydrate metabolism, 324, 376, 392, 441; of citric acid cycle, 402, 404–5; of fat metabolism, 430, 433, 434; of methylation, 257
Carbon to carbon linkages, 118
Carbon dioxide, 'active', 118
— from carbonic acid, 116
— from cellulose, 74
— in fatty acid synthesis, 433
— from fermentation, 332, 340, 341, 342, 395
— fixation of: in liver, 117, 390, 391, 441; in photosynthesis, 321, 323–8
— from muscle, 351, 372
— in ornithine cycle, 103, 273
— in purine synthesis, 280, 283, 284
— respiratory origin of, 379–81, 400, 407, 413
Carbon monoxide, absorption spectra of compounds of: with Atmungsferment, 142; with chlorocruorin, 143; with cytochrome oxidase, 148–9
— inhibits cytochrome oxidase, 34, 142, 145, 146, 418
Carbon tetrachloride, as liver poison, 419
Carbonic acid, 116
Carbonic anhydrase, 115–16
Carbonyl phosphates, high-energy phosphate in, 54
Carboxydismutase, see Ribulose diphosphate carboxylase
Carboxylesterases, 78
Carboxylyases (carboxylases), 115–19
Carboxypeptidases, activation of, 63
— in digestion, 62, 66–8, 210
— esterase activity of, 68
— and kathepsin, 69
Carnitine, 112

— reaction of acyl-Co A compounds with, 426
Carnivores, 205–6
— in metabolic studies, 189
Carnosinase, 69, 241
Carnosine, 241, 257
— hydrolysis of, 69, 241
Carotenoid pigments, adsorption of, 21
— in chloroplasts, 321
Cartilage, ossification of, 81
Casein, 68, 242
Caseinogen, 68
Catalase, 139–40
— combination of, with azide, 22
— haem in, 30
— peroxidase activity of, 139–40
Catalysts, 3–7, 125–6, 140
— and energetics, 45
— turnover numbers of, 173
— see also Enzymes
Catechol, 131
Caudal necrosis, in rats, 420, 423
Cell-free extracts, preparation of, 198–9, 333, 355
Cellobiases, 76
Cellobiose, 76
Cellulase, 74–5
Cellulomonas, cellulase of, 74
Celluloses, digestion of, 74, 206, 211, 377, 417
Centrifugation, separation of cell components by, 176, 198
Cephalins, see Kephalins
Cephalochorda, phosphagens in, 369
Cerous hydroxide test, for H_2O_2, 138
Cetramide, and absorption of sugars, 214
Charcoal, as adsorbent, 21
— catalysis by, 141–2
Chelonia, arginase in, 272, 274
— nitrogen excretion of, 268–9, 271
Chemosynthetic bacteria, 203
Chick, glycine requirement of, 220
Chick embryo, fat metabolism of, 418
— nitrogen metabolism of, 270
Chimpanzee, phenylacetylglutamine in, 252
Chitin, 75
Chitinase, 75
Chlorella, photosynthesis in, 324
— respiration of, 152
Chloride ions, activation of amylase by, 34–5, 73
Chlorocruorin, 143, 147, 149
Chloroform, as liver poison, 419
Chlorophyll, bacterial counterparts of, 203
— in photosynthesis, 203, 231, 321
Chloroplasts, photolysis of water in, 322
— photophosphorylation in, 181
Cholesterol, in blood, 418
— in fat absorption, 215
— and fatty liver, 419
— synthesis of, 431
Cholic acid, in bile salts, 214
Choline, acetylation of, 104, 384
— and fatty liver, 419
— as methylating agent, 111, 247
— oxidation of, 112, 161
— as vitamin, 205

451

Hydrogen (*cont.*)
— oxidation of, by bacteria, 128
Hydrogen acceptors, 36–7, 38, 125, 140
Hydrogen bonds, in nucleic acids, 305
Hydrogen donors, 36, 125, 140
Hydrogen peroxide, combination of, with peroxidase, 22, 32
— fate of, 138–40
— reduction of oxygen to: by aerobic dehydrogenases, 130, 135; by reduced riboflavin, 169
Hydrogen sulphide, bacterial photolysis of, 322
— from cysteine, 243
— inhibits: catalase, 139–40; tissue respiration, 142, 145, 146
— reactivation of enzymes by, 69
Hydrolases, 8, 60–83
— group transfer by, 85–8
Hydrolyases, 113–15
Hydrolysis, 31, 32
Hydrolytic deamination, *see* Deamination
Hydroxy-acetylene diureine carboxylic acid, 134–5
β-Hydroxy-acyl-Co A dehydrogenase, 159, 165, 426, 427
3-Hydroxyanthranilic acid, 258, 259
β-Hydroxybutyric acid, from fats, 421
— dehydrogenation of, 162, 421
— interconversion of acetoacetate and, 162, 421, 431
— oxidation of, 431
— as substrate for studies of oxidative phosphorylation, 181, 182
— *see also* Ketone bodies
γ-Hydroxyglutamic acid, from hydroxyproline, 260–1
3-Hydroxykynurenine, 259
Hydroxymethyl groups, 239
β-Hydroxy-β-methylglutaryl-Co A, 249, 431
Hydroxymethyl-tetrahydrofolic acid, 239
p-Hydroxyphenyl-lactic acid, 191
p-Hydroxyphenylpyruvic acid, 191, 255
Hydroxyproline, glycogenic, 223
— non-essential, 219
— special metabolism of, 260–1, 262
Hydroxypyruvic acid, 106
Hypoxanthine, from adenine, 83, 286
— in nucleoside metabolism, 293
— oxidation of, 12, 136
— synthesis of, 282–6
Hypoxanthine-N-deoxyriboside, 293
— as intermediate in purine synthesis, 279
Hypoxanthine ribonucleotide, *see* Inosinic acid

IDP, *see* Inosine diphosphate
Imbecillitas (oligophrenia) phenylpyruvica, 253, 313
Iminazole ring, in purines, 302
Imino-acids, hydrolysis of, 135
— in oxidative deamination of amino-acids, 226–7
α-Iminopropionic acid, 114
Inborn errors of metabolism, 255, 313; *see also*

Albinism, Alcaptonuria, Cystinuria, Imbecillitas, Tyrosinosis
Indican, 246, 258
Indigo, 76, 260
Indole, 258
Indoxyl, detoxication of, 258
— β-glucoside of (indigo), 76, 260
Indoxyl sulphuric acid (indican), 246, 258
Inhibition, of absorption, 213–14, 217
— of enzymes: competitive, 28–32; by heavy metals, 18–20; by oxidation, 34; by products of reaction, 14, 28; selective (specific), 197, 199, 336
— of fermentation, 336, 343
— of glycolytic enzymes, 366
— of tissue respiration, 34, 142, 145, 146, 148
— *see also individual inhibitors*
Inosine, 280–1
Inosine diphosphate (IDP), 390
Inosine triphosphate (ITP), 294
— in carboxylation, 117, 384, 408
— phosphorylation by, 90, 390
Inosinic acid, from adenylic acid, 83
— in muscle, 361
— nucleotides from, 284–5
— synthesis of, 280–1, 282–3
Inositol, 205, 439
Insects, melanins in, 134
— nitrogen excretion of, 265, 289
Insulin, 220, 378
Intestinal juice, lipases in, 78, 215
— maltase in, 11, 75, 87, 212
— nucleases in, 301
— peptidases in, 62
Intestinal mucosa, 217, 416
Invertases, *see* Saccharases
Invertebrates, nitrogen excretion of, 263, 265–6
— phosphagens of, 275, 354, 369–71
Inulin, in Compositae, 93, 211, 329
— digestion of, 75, 212
Iodine coloration, with different polysaccharides, 70, 72, 95
Iodine number, of fatty acids, 423
Iodoacetate, blocking of SH- groups by, 245, 385
— inhibitions by: of absorption, 213–14, 217; of acetylation by Co A, 384; of alcohol dehydrogenase, 162, 341; of fermentation, 336, 337, 339; of succinate dehydrogenase, 160; of triosephosphate dehydrogenase, 165, 339, 361
— poisoning of muscle by, 355
Iodogorgoic acid, 257
Iridocytes, 286
Iron, in aerobic dehydrogenases, 130
— in *Atmungsferment*, 142
— in catalase, 139
— complexes of, with CO, 142
— in cytochrome oxidase, 135, 149
— in cytochromes, 148, 150
— in ETP, 177, 178
— in diaphorases, 171
— in flavoproteins, 171
— in haemoglobin, 30

Iron (*cont.*)
— in lactate dehydrogenase, 172
— in peroxidase, 22, 139
— in succinate dehydrogenase, 159, 172
— in xanthine (aldehyde) oxidase, 137, 141, 170
Isoelectric pH, 18, 21
Isomerases, 120–3
Isomerism, and enzyme specificity, 9–10
— optical, and nomenclature, 236–8
Isotope dilution, 201–2
Isotopes, use of, in metabolic studies, 188, 191, 199–202; *see also individual isotopes*
Isozymes, 162
ITP, *see* Inosine triphosphate

Jack-bean, urease in, 82
Janus green, as mitochondrial stain, 177

Katabolism, 45, 48, 209
— high-energy phosphate from, 57, 59
Kathepsins, 69, 207
Kephalins, 112, 439
Keratin, 244
α-**Keto-acids,** amination of, 192
— from D-amino-acids, 135, 225
— oxidative decarboxylation of, 36, 180, 361
β-**Keto-acids,** in oxidation of fatty acids, 426, 427
β-**Keto-acyl-Co A thiolases,** 426, 427
α-**Ketoadipic acid,** 254
α-**Ketobutyric acid,** from threonine, 243
α-**Keto-β-carboxyglutaric acid,** *see* Oxalosuccinic acid
α-**Ketoglutaric acid,** in citric acid cycle, 400, 402, 406
— from histidine, 257
— interconversions: of glutamate and, 163, 228, 250–1, 399, 409; of oxaloacetate and, 118, 166, 399
— oxidative decarboxylation of, 390, 397; ATP yield from, 181, 183
— from pyruvate, 391
— as respiratory catalyst, 397
— succinate from, 399
— in transamination, 100–1, 228, 229
α-**Keto-γ-hydroxyglutaric acid,** from hydroxyproline, 261
Ketone bodies, amino-acids giving rise to, 192, 222–3, 232
— in blood, 422
— in diabetes, 192, 379
— from fatty acids, 421, 429–31
— source of, 429–30
β-**Ketothiolase,** 426, 427, 432
Kidney, D-amino-acid oxidase in, 135
— L-amino-acid oxidase in, 226
— ammonia production in, 230
— arginase in, 272
— deamination in, 225
— of Elasmobranchs, 278
— glutamate dehydrogenase in, 163
— glycocyamine synthesis in, 200
— histidine decarboxylase in, 258

— β-hydroxybutyrate dehydrogenase in, 162
— kathepsin in, 69
— of pigeon, uric acid synthesis in, 279
— reabsorption of glucose in, 214
Kinases, 88–90, 293, 295
Krebs's physiological saline, 222
Kynurenic acid, 258, 259
Kynurenine, 258, 259

Laccase, 131
Lachrymators, 245
Lactase (β-galactosidase), 212, 376
D-**Lactate dehydrogenase,** in micro-organisms, 9–10
L-**Lactate dehydrogenase,** 128
— of *Bacillus delbrückii*, 10
— coupled reactions of, 174, 175, 382–3, 386
— of liver, 366
— of muscle, 9, 36, 156–8, 159, 162, 357, 366
— and NAD, 36, 38
— of yeast, 147, 159, 160–1, 172
Lacteals, transport of fat in, 217
D-**Lactic acid,** in blood, 373, 374
— glycogenic, 192, 233
— interconversion of pyruvate and, 156, 160, 364–5, 372, 373, 382, 386, 391
— from methyl glyoxal, 114
— in muscle, 332, 351, 353, 357, 364
Lactoflavin, 169
Lactose, 212, 374, 376
Lactosuria, 212
Lamellibranch molluscs, digestion in, 208
— nitrogen excretion of, 289
Lecithinase A, of snake venom, 79
Lecithinases, of *Clostridia*, 80
Lecithins, 439
— hydrolysis of, 79
Legumes, root nodule bacteria in, 232
Leucine, essential, 219
— ketogenic, 223
— special metabolism of, 248–9, 262
iso-**Leucine,** essential, 291
— glycogenic and ketogenic, 223
— special metabolism of, 249, 262
Leuconostoc dextranicum, dextran sucrase in, 92
Levan sucrase, 92
Levans, 75, 92, 93, 211, 212
Ligases, see *specific synth(et)ases*
Light energy, 57
— in photodissociation, 34, 38, 142
— in photosynthesis, 321, 323
Limit dextrin, 72
Linoleic acid, 424
Lipaemia, 217, 416, 418
Lipases, in digestion, 78–9, 214, 215, 416
— reversibility of action of, 46, 79
— in *Ricinus* seeds, 78, 79, 207
— specificity of, 10
Lipids, metabolism of, 416–41
Lipoamide dehydrogenases (diaphorases), 171
α-**Lipoic acid,** 205
— in oxidative decarboxylation of pyruvate, 388–9

457

Nicotinamide-adenine dinucleotide (*cont.*)
— phosphorylation of, to NADP, 89, 90, 295; by cyclical system, 437
— structure of, 296
Nicotinamide-adenine dinucleotide kinase, 90, 295
Nicotinamide-adenine dinucleotide phosphate (NADP, NADP.H₂), 37
— absorption spectrum of, 154
— as coenzyme of dehydrogenases, 131, 153–9, 162, 163, 165–7, 226, 291, 398
— in coupled reactions, 175
— in fatty acid synthesis, 433, 436, 437
— from NAD, 89, 90, 295; by cyclical system, 437
— oxidation of reduced form of, 171
— in pentose pathway, 393, 394, 437
— in photosynthesis, 321, 323, 325
Nicotinamide mononucleotide, 154, 293
Nicotinamide-*N*-riboside, 293
Nicotinic acid, detoxication of, 110, 240, 247
— from tryptophan, 258, 259
Nicotinuric acid, 110, 240
Nitrate, reduction of, in plants, 137
Nitrogen, excretion of, 187, 224, 263–71; 'premortal rise' in, 221
— metabolism of, 187; *see also* Amino-acids, Proteins
¹⁵Nitrogen (heavy nitrogen), 188, 200, 229, 309–10
Nitrogen glycosidases, 77, 293
Notatin (glucose oxidase), 137
Nucleases, 301
Nuclei of cells, isolation of, 176, 197–8
— nucleoproteins in, 300
Nucleic acids, hydrolysis of, 301
— metabolism of, 300–19
— molecular weight of, 303
— replication of, 308–10
Nucleoproteins, 300–1
Nucleosidases, 301
Nucleoside diphosphate kinases, 90
Nucleoside hydrolases, 292
Nucleosides, 97, 291
— formation of, 292–3
— hydrolysis of, 292, 301
— phosphorylation of, 90
— *see also individual nucleosides*
Nucleotidases, 301
Nucleotides, 97, 291
— functions of, 293–9
— hydrolysis of, 301, 304
— structure of, 304–5
— synthesis of, 283–5, 290–3
— *see also individual nucleotides*
Nutrition, 203–6
Nutritional status, of amino-acids, 219–20, 223

Octanoic acid, acetoacetate from, 429, 430
Octopine, 252, 371
Octopus, ink of, 134
— octopine in muscle of, 371
Oleic acid, absorption of, 216–17
— from stearic acid, 416, 424

Oligochaete worms, phospholombricine in, 370
Oligophrenia, *see* Imbecillitas
Oligosaccharides, 87–8
One-carbon transferases, 108–10
Ooflavin, 169
Ophiuroids, phosphagen in, 369
Ornithine, citrulline from, 103, 253, 275
— detoxication by, 98, 253
— glycogenic, 223
— interconversion of glutamate and proline and, 251, 253, 260
— in ornithine cycle, 9, 82, 252, 253, 273–7, 369
— special metabolism of, 229, 233, 253–4
— in transamidination, 102, 201
Ornithine carbamoyl transferase, 275, 277
Ornithine cycle, 252, 253, 273–7
Ornithuric acid, in urine of birds, 98, 253
Orotic acid, in pyrimidine synthesis, 290
Osmotic work, 58
Ossification, of cartilage, 81
Osteomalacia, 188
Ovoverdin, in lobster eggs, 38
Oxalic acid, from creatone, 368
Oxaloacetate decarboxylase, 117, 249
Oxaloacetic acid, in citric acid cycle, 183, 399, 400, 406
— citric acid from, 104, 385, 386, 402–3, 404, 405
— in dicarboxylic acid cycle, 413
— in glyoxylic acid cycle, 412
— inhibits succinate dehydrogenase, 160
— interconversions: of aspartate and, 249, 399, 409; of α-ketoglutarate and, 118, 166, 381, 399; of malate and, 156, 162, 399; of phospho*enol*pyruvate and, 117, 364, 381, 391, 408, 410; of pyruvate and, 119, 381, 391, 392, 399, 401, 402, 406, 408, 410, 437
— as respiratory catalyst, 397
— spontaneous decarboxylation of, 249
— transamination of, 100, 228, 229, 231–2
Oxalosuccinate decarboxylase, 118, 381, 399, 407
Oxalosuccinic acid, in citric acid cycle, 183, 406
— interconversions: of *iso*citrate and, 166, 398, 399; of α-ketoglutarate and, 118, 166, 398, 399
Oxidases, 129–41, 151
β-Oxidation of fats, 424–5
Oxidation-reduction reactions, 32, 125
— coupled, 174
Oxidative deamination, Oxidative decarboxylation, Oxidative phosphorylation, *see* Deamination, Decarboxylation, Phosphorylation
Oxidoreductases, 84, 124–84
Oxygen, cytochromes and, 150, 157
— evolution of, in photosynthesis, 322
— as hydrogen acceptor for oxidases, 129
— poisoning by, 160
— supply of: to muscle, 353, 372, 373; to tissue slices, 196
— as ultimate hydrogen acceptor for cell, 132
¹⁸Oxygen (heavy oxygen), 200, 322
Oxyhaemoglobin, 372

P-enzyme, 94
Palladium, as catalyst and hydrogen acceptor, 125–6
Palmityl-Co A, synthesis of, 434
Pancreas, as source of nucleoproteins, 300
Pancreatic juice, amylase in, 210, 212
lipase in, 78, 214, 215
— nucleic-acid-hydrolysing enzymes in, 301
— peptidases in, 63
Pantetheine, 299
Pantothenic acid, in ACP, 434–5
— bacterial synthesis of, 204
— in Co A, 205, 299, 384
— structure of, 241
Papain, from paw-paw, 17, 69
Parasitism, 206
Paw-paw, papain in, 69
Pectins, 75
Penicillium notatum, glucose oxidase in, 137
Pentose ('shunt') pathway, of carbohydrate metabolism, 392–4, 437
Pentoses, absorption of, 213
— fate of injected, 375
— origin of, 166, 291, 292, 392
— in photosynthesis, 328
Pentosuria, 313
Pepsin, in digestion, 62–6, 210
— and kathepsin, 69
— molecular weight of, 63
— optimum pH of, 17, 39, 210
— specificity of, 64, 66
Pepsinogen, activation of, 63, 210
— molecular weight of, 63
Peptidases, 61–70
— activation of, 33, 69
— esterase activity of, 10, 13, 68
— intracellular, 68–70, 99, 207
— specificity of, 64–8
— synthetic activity of, 69
Peptide groups, transfer of, 97–9
Peptides, digestion of, 63–70
— masking of pro-enzymes by, 33, 63
— syntheses of: biological, 69, 316–19; chemical, 61–2
— synthetic, action of peptidases on, 64–5
Peptone shock, 211
Peptones, 210–11
Perfusion of organs, in metabolic studies, 193–7
Peroxidase, 210–11
— absorption spectrum of, 22
— compound of, with peroxide, 22, 32, 139
— haematin in, 30, 38, 139
pH, and activity of enzymes, 17–18, 39, 73, 82, 83
— in digestion, 68, 210, 212
— and metabolism of micro-organisms, 349
Phage T-2, 314
Phagocytosis, 68, 207–8
Phenol, from tyrosine, 257
Phenol oxidases, 130–4
Phenol sulphuric acid, 246
Phenols, detoxication of, 246, 257
— oxidation of, 130–4
ω-Phenyl fatty acids, 190, 424

Phenylacetylglutamine, in man and chimpanzee, 252
Phenylalanine, essential, 219, 220–1
— gelatin lacking in, 222
— ketogenic, 223
— RNA coding for, 317
— special metabolism of, 255–7, 262
Phenylethylamine, detoxication of, 257
Phenylpyruvic acid, excretion of, 256, 313
Phlorrhizin, induction of pseudo-diabetes by, 192, 376, 419
— inhibition of absorption by, 213–14, 217
Phosphagens, concentration of, in various tissues, 363
— distribution of, 366–71
— in muscle, 58, 59, 358–62, 411
— *see also* Phosphoarginine, Phosphocreatine, etc.
Phosphatases, 78, 80–1
— in citrous fruits, 85
— for fructose-1:6-diphosphate, 364
— in liver, 96, 377, 378
— in photosynthesis, 326, 328
— transphosphorylation by, 85–6
— in yeast, 347
Phosphate, in absorption of fats, 216–17
— concentration of, in various tissues, 363
— in fermentation, 334, 336, 339, 345–6
— high-energy, 52–3, 294, 295, 345; in phosphagens, 354; production of, in tissue respiration, 180–4; *see also* ATP *and other high-energy compounds*
— in muscle, 353–4, 361–2, 363
— in oxidative decarboxylation, 383, 386
— in priming reactions, 50–2
— pyruvate carboxylase and, 381
— succinate dehydrogenase and, 160, 182
— transfer of, 295, 297, 299; *see also* phosphorylation
— triose dehydrogenase and, 164, 165
Phosphate acetyl transferase (phosphoacetyl transferase), 389
Phosphatidic acids, synthesis of, 438
Phosphoarginine, 354
— energy-rich phosphate in, 53, 54
— in muscle of invertebrates, 58, 252, 354, 363, 369–71
— synthesis of, 89
Phosphocholine, in lecithin, 79, 439
Phosphocreatine, 354
— distribution of, 354, 363–4, 366–71
— high-energy phosphate in, 53, 54
— in muscle, 358–62
— synthesis of, 54, 58, 88, 89, 358, 359
Phosphodiesters, 305
Phosphodiesterases, 80, 305, 306
Phosphofructokinase, 89, 90, 337, 343, 366
Phosphogalactoseisomerase, *see* UDP-glucose epimerase
Phosphoglucokinase, 89
Phosphoglucomutase, 122, 292
— in fermentation, 344
— glucose-1:6-diphosphate as coenzyme for, 122–3, 366

Protective synthesis, *see* Detoxication
Protein precipitants, and enzyme activity, 18–20
Proteinases, *see* Peptidases
Proteins, as ampholytes, 18, 21, 39
— biological value of, 222
— calorific value of, 417
— denaturation of, 16–17, 18, 207
— digestion and absorption of, 61–70, 207, 210–11
— enzymes as, 13–20, 39
— isoelectric pH of, 18, 21
— metabolic water from, 418
— metabolism of, 219–22
— minimum requirement of, 222
— nitrogenous end-products of, 187, 263–71
— in nucleoproteins, 303
— of plants and animals, 222
— replacement of, by amino-acids, 211
— R.Q. of oxidation of, 395*n*
— synthesis of, 314–19
Proteolytic enzymes, *see* Peptidases
Protocatechuic acid, 133
Protochordata, phosphagens in, 369
Protozoa, digestion in, 208
— symbiotic, 74
Pseudomonas saccharophila, synthesis of sucrose by, 91
Pteridine, in folic acid, 108
Pterocera, cellulase in, 74
Purines, metabolism of, 286–9
— in nucleic acids, 304, 306–7
— oxidation of, 135, 136
— structure of, 302
— synthesis of, 110, 239, 279–82, 290
Purpurogallin, 139
Putrescine, bacterial formation of, 115, 253–4
— oxidation of, 138
Pyocyanine, 127, 151, 174
Pyloric sphincter, 210, 212
Pyridine, detoxication of, 247
Pyridoxal, as vitamin, 115, 205
Pyridoxal phosphate, as coenzyme: of amino-acid decarboxylases, 115; of deaminases, 114, 115, 234, 242; of diamine oxidase, 138; of α-glucan phosphorylase, 94; of transaminases, 100–1, 228
— in transthiolation, 244
Pyridoxamine phosphate, 100–1
Pyrimidines, in nucleic acids, 304, 306–7
— structure of, 302
— synthesis of, 103, 110, 253, 290–1
Pyrogallol, 131, 139
Pyrophosphatase, 80
Pyrophosphates, formed in synthesis of hippuric acid, 98
— high-energy phosphate in, 53, 54
— transfer of, 295
Pyruvate carboxylase, of liver, conversion of pyruvate to oxaloacetate by, 119, 381, 391, 392, 399, 401, 408, 410, 437–8
Pyruvate decarboxylase, of yeast, 'straight' decarboxylation by, 36, 116, 335, 340, 343, 380, 387, 411
Pyruvate kinase, 89, 340, 343, 366, 408

'Pyruvate oxidase', 116
'Pyruvate oxidative decarboxylase', 382
Pyruvic acid, acetyl phosphate from, 104
— central position of, in metabolism, 390–2
— from cysteine, 243
— decarboxylation of, oxidative, to give acetyl-Co A, 382–90, 391, 402, 404, 405, 407, 409, 411, 441; 'straight', to give acetaldehyde, 36, 116, 335, 340, 387, 391
— glycogenic, 233, 376
— from hydroxyproline, 261
— interconversions: of alanine and, 101, 227 229, 241, 391, 399, 409; of lactate and, 156, 160, 364–5, 372, 373, 382, 386, 391; of malate and, 117, 166–7, 380, 391; of phospho*enol*pyruvate and, 89, 340, 342, 365, 379, 408, 410
— oxaloacetate from, 119, 381, 391, 392, 399, 401, 408, 410, 437–8
— reaction of, with cyanide, 156, 162
— from serine, 115, 242
— in thiamine deficiency, 382
— yield of ATP in production of, from glucose, 183

Q_{10}, *see* Temperature coefficient
Q-enzyme, 94, 95
Quinolinic acid, 258, 259
o-Quinones, 127
— melanins from, 133, 134
— from phenols, 130, 131

R-enzyme, of plants, 70, 73; *see also* Dextrin-1:6-glucosidase
Raffinase, 40
Raffinose, 40, 77
Raia, electric organ of, 57
Rats, fat-deficiency disease of, 420, 423
Rattlesnake, venom of, 79–80
Recapitulation, in ontogenesis of amphibia and reptiles, 268–9
Red blood corpuscles, *see* Erythrocytes
Rennet, 68
Rennin, 68
Reptiles, arginase in, 272
— nitrogen excretion of, 268–70, 271, 289
Respiration, free energy from, 180–4
— intracellular organization of enzymes of, 176–80
— of minced pigeon-breast muscle, 395–6
Respiratory chain, *see* Electron transfer
Respiratory quotient, 395*n*
Reticulo-endothelial system, 208
Retina, respiration of, 152
Reversibility of reactions, catalysts and, 4, 48–50
— of citric acid cycle, 407
— and coupling of dehydrogenase systems, 46, 168, 173–6, 396
— of glycolysis, 364
— of glycosidase action, 77–8
— of lipase action, 46, 79
— of β-oxidation, 428, 433
— of peptidase action, 69

Yeast (*cont.*)
— NAD from, 153
— nucleic acid of, 303
— nutritional requirements of, 204
— pyrophosphatase of, 80
— pyruvate decarboxylyase of, 36, 116, 335, 340, 343, 380, 387, 411
— riboflavin kinase of, 170
— saccharases of, 13, 23, 27, 32, 40, 344
— soluble α-glycerolphosphate dehydrogenase of, 161, 347

Zein, biological value of, 222
Zinc, as activator of enzymes, 34, 63
— in alcohol dehydrogenase, 162
— in carbonic anhydrase, 116
Zymase, 7, 335